Arguing About Bioethics

Arguing About Bioethics is a fresh and exciting collection of essential readings in bioethics, offering a comprehensive introduction to and overview of the field. Influential contributions from established philosophers and bioethicists, such as Peter Singer, Thomas Nagel, Judith Jarvis Thomson and Michael Sandel, are combined with the best recent work in the subject.

Organised into clear sections, readings have been chosen that engage with one another, and often take opposing views on the same question, helping students get to grips with the key areas of debate. All the core issues in bioethics are covered, alongside new controversies that are emerging in the field, including:

- embryo research
- selecting children and enhancing humans
- human cloning
- using animals for medical purposes
- organ donation
- consent and autonomy
- public health ethics
- resource allocation
- developing world bioethics
- assisted suicide.

Each extract selected is clear, stimulating and free from unnecessary jargon. The editor's accessible and engaging section introductions make *Arguing About Bioethics* ideal for those studying bioethics for the first time, while more advanced readers will be challenged by the rigorous and thought-provoking arguments presented in the readings.

Stephen Holland is a Senior Lecturer in the Departments of Philosophy and Health Sciences at the University of York, UK.

Arguing About Philosophy

This exciting and lively series introduces key subjects in philosophy with the help of a vibrant set of readings. In contrast to many standard anthologies which often reprint the same technical and remote extracts, each volume in the *Arguing About Philosophy* series is built around essential but fresher philosophical readings, designed to attract the curiosity of students coming to the subject for the first time. A key feature of the series is the inclusion of well-known yet often neglected readings from related fields, such as popular science, film and fiction. Each volume is edited by leading figures in their chosen field and each section carefully introduced and set in context, making the series an exciting starting point for those looking to get to grips with philosophy.

Arguing About Metaethics
Edited by Andrew Fisher and Simon Kirchin

Arguing About the Mind
Edited by Brie Gertler and Lawrence Shapiro

Arguing About Art 3rd Edition
Edited by Alex Neill and Aaron Ridley

Arguing About Knowledge
Edited by Duncan Pritchard and Ram Neta

Arguing About Law
Edited by John Oberdiek and Aileen Kavanagh

Arguing About Metaphysics
Edited by Michael Rea

Arguing About Religion
Edited by Kevin Timpe

Arguing About Political Philosophy
Edited by Matt Zwolinski

Arguing About Language
Edited by Darragh Byrne and Max Kolbel

Forthcoming titles:

Arguing About Science
Edited by Alexander Bird and James Ladyman

Arguing About Bioethics

Edited by

Stephen Holland

Routledge
Taylor & Francis Group

LONDON AND NEW YORK

First edition published 2012
by Routledge
2 Park Square, Milton Park, Abingdon, Oxon, OX14 4RN

Simultaneously published in the USA and Canada
by Routledge
711 Third Avenue, New York, NY 10017

Routledge is an imprint of the Taylor & Francis Group, an informa business

British Library Cataloguing in Publication Data
A catalogue record for this book is available from the British Library

Library of Congress Cataloging in Publication Data
    p. cm. – (Arguing about philosophy)
  Includes bibliographical references and index.
  1. Bioethics. 2. Medical ethics.
   I. Holland, Stephen (Stephen Michael), 1963–
  QH332.A74 2011
  174.2–dc23
                     2011042088

ISBN: 978–0–415–47632–4 (hbk)
ISBN: 978–0–415–47633–1 (pbk)

Typeset in Joanna and Bell Gothic
by RefineCatch Limited, Bungay, Suffolk

Contents

Acknowledgements

I would like to thank Richard Cookson and Martin Wilkinson for suggesting selections on resource allocation and organ donation, respectively; and the very helpful anonymous reviewers of early proposals influenced the structure and content of this volume. My biggest debt is to Louise Ballard for her patience and support during the compilation of this anthology.

The editor and publishers wish to thank the following for permission to use copyrighted material:

1 Oderberg, D. 2008. 'The metaphysical status of the embryo: some arguments revisited', *Journal of Applied Philosophy*, 25 (4), 263–76. Reproduced with permission from John Wiley and Sons.

2 Lizza, J. 2007. 'Potentiality and human embryos', *Bioethics*, 21 (7), 379–85. Reproduced with permission from John Wiley and Sons.

3 Reichlin, M. 1997. 'The argument from potential: a reappraisal', *Bioethics*, 11 (1), 1–23. Reproduced with permission from John Wiley and Sons.

4 Harris, J. 2003. 'Stem cells, sex, and procreation', *Cambridge Quarterly of Healthcare Ethics*, 12 (4), 353–71, reproduced with permission.

5 The Hastings Center Report by Hastings Center. Copyright 1999 Reproduced with permission of Hastings Center in the format Textbook via Copyright Clearance Center.

6 Savulescu, J. 2001. 'Procreative beneficence: why we should select the best children', *Bioethics*, 15 (5&6), 413–26. Reproduced with permission from John Wiley & Sons.

7 Robertson, J.A. 2001. 'Preconception gender selection', *American Journal of Bioethics*, 1 (1), 2–9. Taylor & Francis Ltd, www.taylorandfrancis.com, reprinted by permission of the publisher.

8 Sandel, M.J. 2004. 'The case against perfection', *The Atlantic Monthly*, 293 (3), 51–62.

9 Springer and *The Journal of Value Inquiry* 37 (4), 493–506, 2003, 'Human genetic enhancements: a transhumanist perspective', Bostrom, N. © 2004 Kluwer Academic Publications, with kind permission from Springer Science+Business Media B.V.

10 Chadwick, R. 1982. 'Cloning', *Philosophy*, 57 (220), 201–9, reproduced with permission.

11 Kass, L. 1997. 'The wisdom of repugnance', *New Republic*, 216 (22), 17–26. Reprinted by permission of the author.

12 Elliott, D. 1998. 'Uniqueness, individuality, and human cloning', *Journal of Applied Philosophy*, 15 (3), 217–30. Reproduced with permission from John Wiley & Sons.

13 Sparrow, R. 2006. 'Cloning, parenthood, and genetic relatedness', *Bioethics*, 20 (6), 308–18. Reproduced with permission from John Wiley & Sons.

14 Blackford, R. 2007. 'Slippery slopes to slippery slopes: therapeutic cloning and the criminal law', *The American Journal of Bioethics*, 7 (2), 63–4. Taylor & Francis Ltd, www.taylorandfrancis.com, reprinted by permission of the publisher.

15 Singer, P. 2003. 'Animal liberation at 30', *New York Review of Books*, 50 (8), May 15th, 23–6. Reprinted by kind permission of the author. Copyright © Peter Singer, 2011.

16 Harrison, P. 1991. 'Do animals feel pain?' *Philosophy*, 66 (255), 25–40, reproduced with permission.

17 Cohen, C. 1986. 'The case for the use of animals in biomedical research', *New England Journal of Medicine*, 315 (14), 865–70. © Massachusetts Medical Society.

18 This article was published in *Transplantation Proceedings*, 24, Caplan, 'Is xenografting morally wrong?', pp. 722–27. Copyright Elsevier (2002).

19 Nelson, J. 1993. 'Moral sensibilities and moral standing: Caplan on xenograft "donors"', *Bioethics*, 7 (4), 315–22. Reproduced with permission from John Wiley & Sons.

20 Kass, L.R. 1992. 'Organs for sale? Propriety, property, and the price of progress', *Public Interest*, 107, 65–86. Reproduced with permission from www.nationalaffairs.com.

21 Reprinted from *The Lancet*, 351 (9120), Radcliffe-Richards, J., Daar, A.S., Guttmann, R.D., Hoffenberg, R., Kennedy, I., Lock, M., Sells, R.A. and Tilney, N. for the International Forum for Transplant Ethics 1998. 'The case for allowing kidney sales.' 1950–2. © 1998 with permission from Elsevier.

22 This article was published in *Transplantation Proceedings*, 24, Cohen, 'The case for presumed consent to transplant human organs after death', pp. 2168–72. Copyright Elsevier (1992).

23 This article was published in *Transplantation Proceedings*, 27, Veatch and Pitt, 'The myth of presumed consent: ethical problems in new organ procurement strategies', pp. 1888–92. Copyright Elsevier (1995).

24 Spital, A. 1996. 'Mandated choice for organ donation: time to give it a try', *Annals of Internal Medicine*, 125 (1), 66–9. Used with permission from the American College of Physicians.

25 Reproduced from Beecham, L. 2000. 'Donors and relatives must place no conditions on organ use', *British Medical Journal*, 320 (7234), 534 with permission from BMJ Publishing Group Ltd.

26 Reproduced from Wilkinson, T.M. 2003. 'What's not wrong with conditional organ donation?' *Journal of Medical Ethics*, 29 (3), 163–4 with permission from BMJ Publishing Group Ltd.

27 Bernat, J.L. 2010. 'Point: Are donors after circulatory death really dead, and does it matter? Yes and yes', *Chest*, 138 (1), 13–16. Reproduced with permission from the American College of Chest Physicians.

28 Truog, R.D. and Miller, F.G. 2010. 'Counterpoint: Are donors after circulatory death really dead, and does it matter? No and not really', *Chest*, 138 (1), 16–18. Reproduced with permission from the American College of Chest Physicians.

29   Bernat, J.L. 2010. 'Rebuttal', *Chest*, 138 (1), 18–19. Reproduced with permission from the American College of Chest Physicians.

30   Lidz, C.W., Meise, A., Osterweis, M., Holden, J.L., Marx, J.H. and Munetz, M.R. 1983. 'Barriers to informed consent', *Annals of Internal Medicine*, 99 (4), 539–43. Used with permission from the American College of Physicians.

31   The Hastings Center Report by Hastings Center. Copyright 1990 Reproduced with permission of Hastings Center in the format Textbook via Copyright Clearance Center.

32   The Hastings Center Report by Hastings Center. Copyright 1995 Reproduced with permission of Hastings Center in the format Textbook via Copyright Clearance Center.

33   Reproduced from Savulescu, J. and Momeyer, R.W. 1997. 'Should informed consent be based on rational beliefs?' *Journal of Medical Ethics*, 23 (5), 282–8 with permission from BMJ Publishing Group Ltd.

34   Reproduced from O'Neill, O. 2003. 'Some limits of informed consent', *Journal of Medical Ethics*, 29 (1), 4–7 with permission from BMJ Publishing Group Ltd.

35   Whitney, S.N. *Medical Decision Making*, 23 (4) pp. 275–80, copyright © 2003 by SAGE. Reprinted by SAGE Publications.

36   Childress, J.F., Faden, R.R., Gaare, R.D., Gostin, L.O., Kahn, J., Bonnie, R.J., Kass, N.E., Mastroianni, A.C., Moreno, J.D. and Nieburg, P. 2002. 'Public health ethics: mapping the terrain', *Journal of Law, Medicine and Ethics*, 30 (2), 170–8. Reproduced with permission from John Wiley & Sons.

37   Copyrighted and published by Project HOPE/Health Affairs as Gostin, L.O. 2002. 'Public health law in an age of terrorism: rethinking individual rights and common goods', *Health Affairs*, 21 (66), 79–93. The published article is archived and available online at www.healthaffairs.org.

38   Thaler, R.H. and Sunstein, C.R. 2003. 'Libertarian paternalism', *The American Economic Review*, 93 (2), 175–9. Reproduced with permission from The American Economic Association.

39   Holm, S. 2007. 'Obesity interventions and ethics', *Obesity Reviews*, 8 (Supplement 1), 207–10. Reproduced with permission from John Wiley & Sons.

40   Isaacs, D., Kilham, H.A. and Marshall, H. 2004. 'Should routine childhood immunizations be compulsory?', *Journal of Paediatrics and Child Health*, 40 (7), 392–6. Reproduced with permission from John Wiley & Sons.

41   Reproduced from Chapman, S. 2000. 'Banning smoking outdoors is seldom ethically justifiable', *Tobacco Control*, 9 (1), 95–7 with permission from BMJ Publishing Group Ltd.

42   From Fairchild, A.L. and Bayer, R. 2004. 'Ethics and the conduct of public health surveillance', *Science*, 303 (5658), 631–2. Reprinted with permission from AAAS.

43   Williams, A. 1985. 'The value of QALYs', *Health and Social Service Journal* (Supplement), July 18th, 3–5. Reproduced with permission from EMAP.

44   Reproduced from Harris, J. 1987. 'QALYfying the value of life', *Journal of Medical Ethics*, 13 (3), 117–23 with permission from BMJ Publishing Group Ltd.

45   Reproduced from Williams, A. and Grimley Evans, J. 1997. 'The rationing debate. Rationing health care by age: the case for, and the case against', *British Medical Journal*, 314 (7083), 820–5 with permission from BMJ Publishing Group Ltd.

46   Reproduced from Hope, T. 2001. 'Rationing and life-saving treatments: should identifiable patients have higher priority?' *Journal of Medical Ethics*, 27 (3), 179–85 with permission from BMJ Publishing Group Ltd.

47   Daniels, N. 1981. 'Health-care needs and distributive justice', *Philosophy and Public Affairs*, 10 (2), 146–79. Copyright © John Wiley & Sons, Inc. Reproduced with permission of Blackwell Publishing Ltd.

48   Lurie, P. and Wolfe, S.M. 1997. 'Unethical trials of interventions to reduce perinatal transmission of the human immunodeficiency virus in developing countries', *The New England Journal of Medicine*, 337 (12), 853–6. © Massachusetts Medical Society.

49   Angell, M. 1997. 'The ethics of clinical research in the third world', *The New England Journal of Medicine*, 337 (12), 847–9. © Massachusetts Medical Society.

50   Varmus, H. and Satcher, D. 1997. 'Ethical complexities of conducting research in developing countries', *The New England Journal of Medicine*, 337 (14), 1003–5. © Massachusetts Medical Society.

51   IRB: Ethics and Human Research Copyright 1998 by Hastings Center. Reproduced with permission of Hastings Center in the format Textbook via Copyright Clearance Center.

52   Orentlicher, D. 2002. 'Universality and its limits: when research ethics can reflect local circumstances', *Journal of Law, Medicine and Ethics*, 30 (3), 403–10. Reproduced with permission from John Wiley & Sons.

53   The Hastings Center Report by Hastings Center. Copyright 2001 Reproduced with permission of Hastings Center in the format Textbook via Copyright Clearance Center.

54   The Hastings Center Report by Hastings Center. Copyright 2004 Reproduced with permission of Hastings Center in the format Textbook via Copyright Clearance Center.

55   Macklin, R. 2009. 'The Declaration of Helsinki: another revision', *Indian Journal of Medical Ethics*, 6 (1), 2–4. Reproduced with the permission of IJME.

56   An open letter to all Members of Parliament and of the House of Lords, from leaders of British faith communities of Buddhists, Christians, Hindus, Jews, Muslims and Sikhs, expressing grave concerns at continuing and renewed efforts to legalise euthanasia.

57   Dworkin, R., Nagel, T., Nozick, R., Rawls, J., Scanlon, T. and Thomson, J.J. 1997. 'Assisted suicide: the philosophers' brief', *New York Review of Books*, 44 (5), March 27th, 41–5. From The New York Review of Books. Copyright © 1997 NYREV, Inc.

58   The Hastings Center Report by Hastings Center. Copyright 1998 Reproduced with permission of Hastings Center in the format Textbook via Copyright Clearance Center.

59   Dieterle, J.M. 2007. 'Physician-assisted suicide: a new look at the arguments', *Bioethics*, 21 (3), 127–39. Reproduced with permission from John Wiley & Sons.

60   Thomson, J.J. 1999. 'Physician-assisted suicide: two moral arguments', *Ethics*, 109 (3), 497–518. Reprinted with permission from The University of Chicago Press and the author.

Every effort has been made to contact copyright holders for their permission to reprint material in this book. The publishers would be grateful to hear from any copyright holder who is not here acknowledged and will undertake to rectify any errors or omissions in future editions of this book.

GENERAL INTRODUCTION

THIS ANTHOLOGY HAS two main aims. The first is to introduce students of philosophy, medicine and related disciplines to the fundamental topics, arguments and debates in the field of bioethics. To this end, the book includes pieces that are clear and accessible, and each chapter starts with an introduction that explains the issue under discussion and contextualises the selections. The second main aim is to engage and challenge more advanced and experienced students. The pieces selected contain rigorous, influential and interesting arguments, arranged so as to allow the dialectic between bioethicists to develop. As a result, this anthology contains material that provides an excellent basis for both introductory, and more advanced, courses in bioethics.

The anthology covers all the main branches of the discipline. Central to bioethics are the ethical challenges created by biomedical sciences such as genetics, and by the practice of biomedical research. Bioethics also incorporates traditional medical ethics concerns about the way health care professionals and patients interact – for example, what it means in practice to respect autonomy and the right of patients and research subjects to consent – and public policy questions, such as how best to allocate scarce health resources. And bioethics covers 'beginning of life' issues – including the moral status of the human embryo, and selecting and enhancing our offspring – and 'end of life' issues, such as euthanasia and physician assisted suicide. All these main areas of bioethics are represented in this anthology.

Also included here are pieces that discuss new and exciting developments in bioethics. A case in point is public health ethics, an offshoot of bioethics that explores dilemmas arising from interventions aimed at protecting and promoting the health of populations. For example, to what extent is it legitimate for the state to coerce individuals for the sake of the public's health? Another example is developing world bioethics, which explores challenges created by practicing medicine and providing health care, and undertaking medical research, in non-Western contexts. A central issue here is whether the principles and paradigms devised in Western bioethics should be applied universally or adapted to fit the distinctive circumstances in the developing world. A chapter of this anthology is devoted to each of these new fields of bioethical enquiry.

The pieces selected here implicitly address the methodological question as to how bioethical problems are to be approached. One way of going about bioethics is to refine and apply a normative moral theory, such as utilitarianism. But this is bound to be unsatisfactory because it is unlikely that one theory will be appropriate to all bioethical questions, and very likely that bioethicists will disagree as to which is the best theory. By contrast, the methodology implicit in this anthology is philosophical, in two senses. First, the selections present clear and rigorous arguments and counter-arguments about bioethical problems. Second, a range of accessible but substantive philosophical theories informs discussions of, for example, the ontological status of the early human embryo, the content of animal minds, and the legitimacy of liberty-limiting interference by the state.

The anthology comprises ten parts, each of which addresses a pressing bioethical question. The selections are sufficiently diverse to represent the important and influential positions in these debates. Papers from better known bioethicists are combined with newer or less well known pieces that present arguments deserving of a wider audience.

Is it wrong to do research on human embryos?

INTRODUCTION TO PART ONE

H UMAN EMBRYOS ARE used and destroyed in a number of biomedical procedures. Notably, *in vitro* fertilisation (IVF) requires producing more embryos than
will be transferred to the uterus, creating so-called 'spare' embryos which are usually
either discarded or used and destroyed as research embryos. Another case in point is
stem cell research because stem cells extracted from human embryos remain the
most promising in developing stem cell therapies, despite ongoing attempts to find
alternatives, such as adult stem cells. Is such embryo destruction justifiable, or does
the moral status of the human embryo demand greater respect and protection?

The first way of addressing this issue represented in the selections in this chapter
centres on what the embryo is; i.e., on the identity of the embryo. If the embryo is an
individual human being then embryo destruction would seem to be very wrong. In
fact, it would seem to be tantamount to killing you or me for biomedical purposes.
Whether the embryo is an individual human being depends on when life – i.e., the life
of a human being – begins. A useful way of conceptualising this is to imagine how far
back through time and space one can trace oneself: from young adult to adolescent,
child, infant, foetus . . . what is the earliest stage at which what one still has in mind
is a younger, very different looking, version of oneself?

Oderberg suggests that there is a bioethical consensus on this matter. The embryo
is not an individual human being because our lives cannot be traced back to conception, only to some point afterwards. An influential view is that the life of a human
being begins about fourteen days after fertilisation, during the process known as
gastrulation. This is when a primitive streak forms, and the embryo develops a body
plan that is a clear precursor to the adult human being. It is on the basis of this influential view that research on embryos up to, but not beyond, fourteen days after fertilisation is legally permitted in some countries. Oderberg looks at three arguments for
this consensus and finds them wanting.

A second way of addressing the issue represented in this chapter is to argue not
from the basis of what the embryo is – not, that is, from its identity – but what it has
the potential to become. The pieces by Reichlin and Lizza capture the two main sets
of views about potentiality. Reichlin claims that the potentiality argument has been
misconstrued because the salient sort of potentiality is neither the possibility nor the
probability of X becoming Y but, rather, X's natural and inherent dynamic to become
Y. This is known as active, as opposed to passive, potential. To illustrate, a conker has
the active potential to become a horse chestnut tree because it has the natural and
inherent dynamic to do so; by contrast, the tree has only a passive potential to become
a table because such a transformation is the result of external manipulation by a

carpenter, not something for which the tree is naturally striving. Reichlin claims that a healthy human embryo has the active potential to become a person. This establishes the moral status of the embryo and, arguably, the impermissibility of embryo destruction for biomedical purposes.

Importantly, for Reichlin, the embryo's moral status is independent of contingent empirical facts about the likelihood of a creature's active potential being realised. Consider, for example, two healthy human embryos, one nestled in the wall of a uterus, the other in a laboratory Petri dish marked 'to be discarded'. According to Reichlin, these have the same, equally high, moral status despite the fact that only one of them has a realistic chance to develop, because they have the same active potential. Lizza disagrees. He reinstates possibility as central to our ordinary understanding of potentiality: 'if X is potentially Y, then it must be possible for X to be Y'. Furthermore, this is an empirical matter. Lizza distinguishes two sorts of contingent restrictions on X's development: physical restrictions due to, for example, defects of the entity in question or features of its environment; and choices or decisions, such as the intention to destroy the embryo in the laboratory Petri dish. Since the potential of two entities of the same kind can vary depending on circumstances, so can their moral status.

In the final selection, John Harris presents a different sort of argument. The premise combines some familiar 'facts of life' with attitudes and practices so common as to go unnoticed. A crucial fact of life is that innumerable embryos are naturally lost in the course of human sexual reproduction. Commonly, people feel no remorse about this, nor does it influence their attempts to have children. Harris suggests that embryo research is analogous in creating and destroying embryos for important purposes; therefore, embryo research is justifiable. Harris is alert to the rejoinder that this argument could be turned on its head; i.e., rather than common attitudes and practices justifying embryo research, the impermissibility of embryo research could be said to count against our common attitudes and practices. He also addresses more theoretical objections, for example, based on the doctrine of the double effect: embryo wastage is an unintended though foreseeable side-effect of human sexual reproduction, whereas it is an integral part of research involving embryos. Harris demurs, arguing that the doctrine is 'fallacious': consequences – 'what the agents knowingly and voluntarily bring about' – are what matter, not intentions.

Selections

Oderberg, D. 2008. 'The Metaphysical Status of the Embryo: Some Arguments Revisited', *Journal of Applied Philosophy*, 25 (4), 263–76.

Reichlin, M. 1997. 'The Argument from Potential: a reappraisal', *Bioethics*, 11 (1), 1–23.

Lizza, J. 2007. 'Potentiality and Human Embryos', *Bioethics*, 21 (7), 379–85.

Harris, J. 2003. 'Stem Cells, Sex, and Procreation', *Cambridge Quarterly of Healthcare Ethics*, 12 (4), 353–71.

Chapter 1

David S. Oderberg

THE METAPHYSICAL STATUS OF THE EMBRYO: SOME ARGUMENTS REVISITED

Abstract

This paper re-examines some well-known and commonly accepted arguments for the non-individuality of the embryo, due mainly to the work of John Harris. The first concerns the alleged non-differentiation of the embryoblast from the trophoblast. The second concerns monozygotic twinning and the relevance of the primitive streak. The third concerns the totipotency of the cells of the early embryo. I argue that on a proper analysis of both the empirical facts of embryological development, and the metaphysical importance or otherwise of those facts, all three arguments are found wanting. None of them establishes that the embryo is not an individual human being from the moment of conception.

1 Introduction

Despite the sounding of occasional dissident voices, there is now a settled view among the majority of bioethicists concerning the status of the embryo. Debates about abortion and embryo experimentation, cloning, 'designer' babies, embryonic stem cell research, and related matters all proceed on the basis that the embryo's metaphysical and moral standing is now largely agreed. At the very most, where differences over metaphysics exist among the majority, they do not significantly affect the moral conclusions reached.

Central to the development of the established consensus have been the contributions of bioethicists such as Peter Singer and John Harris. In particular, in books such as *The Value of Life*[1] and *Clones, Genes, and Immortality,*[2] as well as in many papers, Harris has consistently and vigorously defended the view that the human embryo is not an individual human being, and that even if it were it would not be a 'person', yet it is 'persons' who matter morally. His arguments, like those of Singer, have become standard fare among supporters of the non-individuality and/or non-personhood of the embryo.

In contemporary philosophy consensus is hard to come by. That is why bioethicists, with some justification, pride themselves on the fact that they can present a united front to the public and cause their leading ideas to be so quickly accepted by policy-makers as the deliverances of experts who have reflected on these difficult issues to the greatest extent possible. Nevertheless consensus, no matter how solid, does not make for truth. When it comes to both metaphysics and morality, the dominant view among bioethicists concerning the status of the embryo is wrong. My purpose in this paper is to revisit some of the metaphysical arguments given prominence by Harris (among others) in rebuttal of the proposition that the embryo is an individual human being. Although metaphysics rather than ethics will be my primary concern here, the former is an indispensable foundation for the latter, bioethics being no exception. Unless bioethicists get their ontology right, there can be no hope of proposing a plausible moral stance toward the embryo or anything else.

2 The embryoblast and the trophoblast

In several places,[3] Harris claims that human fertilization 'does not result in an individual' because the fertilized egg becomes a cell mass (the morula, three days post-fertilization[4]) that divides into the 'inner cell mass' or embryoblast, from which the embryo proper and foetus develop, and the trophoblast, from which extra-embryonic membranes, the placenta, umbilical cord, and other supporting structures develop.

Harris gives no details as to how the argument is supposed to work. The claim is that since the morula[5] is undifferentiated with respect to embryoblast and trophoblast, there can be no individual human being at this stage. This in turn appears to rely on the implicit claim that there can be no human individual prior to the stage at which what one might call the 'proto-body' of the embryo, the so-called 'embryo proper', comes into existence − this being the inner cell mass.[6]

But is Harris's implicit claim true? He omits to mention that the extra-embryonic structures (such as placenta and umbilical cord) developing from the trophoblast have exactly the same chromosomes as the embryoblast. They do not have the mother's DNA, moreover their development is directed by the embryo/foetus

and they support its functioning. For these reasons alone they are correctly regarded as parts of the embryo/foetus.[7] Furthermore, as a standard embryology textbook points out,[8] the idea of a sharp distinction between embryoblast and trophoblast is questionable. It seems that the hypoblast − the layer of cells adjacent to the epiblast (from which the embryo's body develops) on the side facing the blastocyst cavity − is displaced to extra-embryonic regions, and though it gives rise to extra-embryonic structures such as the yolk sac and allantois, part of the yolk sac is incorporated into the primordial gut of the embryo, and the allantois is incorporated into the embryo as the median umbilical ligament which connects the apex of the urinary bladder with the umbilicus.[9] Hence it seems that there is intermingling between hypoblast and epiblast. If so, why should we exclude the likelihood of intermingling between embryoblast and trophoblast?[10] (Or at least between embryoblast and other supporting structures, even though the trophectoderm, which forms directly from the trophoblast, is as far as we know fixed.) Whatever the details, the point is that one cannot simply assert that before differentiation into embryoblast and trophoblast there is no human individual if, as a matter of fact, after differentiation the embryo is still directing or controlling what cells become structures of its body proper and what do not.

Clearly the processes of development at this stage are far more complex than Harris ever indicates. Might he not object, though, that by speaking of the embryo/foetus in contradistinction to the trophoblast, this in fact lends support to his central point that no identification of an individual embryo/foetus can be made before the differentiation occurs? This objection carries no weight if we are clear about what we mean by expressions such as 'embryo/foetus' and 'embryo proper.' Thinking now of the foetus at a much later stage of development, when we use the term does it refer to the growing child with or without its placenta? Generally, 'foetus' refers to the child minus the parts that are discarded at birth and not required for normal functioning outside the womb.[11] But there is nothing improper, ontologically speaking, from using the term to refer to the child along with its placenta, umbilical cord, and so on: we might do so in order to distinguish it, with all its parts, from the mother. The former use is more common, though, since we are usually more interested in the persistence and development of the child inside and outside the womb; for which purpose we use terms such as 'foetus' and 'embryo' to refer to everything that is not discarded at birth, namely the body of the child minus the parts needed only for gestation. This we can call the 'embryo/foetus proper' in order to clarify what it is we are referring to.

To make such a reference, however, is not to concede Harris's point: for it does not follow from the fact that we can distinguish between the embryo proper and the trophoblast at a certain stage that we cannot identify an embryo before that stage. Before such differentiation,[12] we are constrained to use 'embryo' to refer to all the matter that will separate into embryoblast and trophoblast, i.e. to use the term in the same way as we might use 'foetus' at a later stage only to refer to everything that is not part of the mother, i.e. the body of the foetus plus its placenta, umbilical cord, and all other later-discarded matter. The fact that in the case of the morula the matter has not yet separated into embryoblast and trophoblast in no way implies that the morula is not an embryo, where 'embryo' is used of necessity in the restricted sense just mentioned. And where we have an embryo, we have an individual human being. The somewhat surprising fact that parts of the extra-embryonic membranes are later incorporated into the embryo proper only serves to reinforce the claim that there is a single human

individual prior to the early blastocyst stage, with its cells interacting in varied and complex ways, even though no 'embryo proper', i.e. embryonic body distinct from its later-discarded parts, can at that point be identified.

3 Monozygotic twinning and the 'primitive streak'

Early on in the debate about embryo experimentation, the onset of the 'primitive streak' at around 14 or 15 days became a kind of totem for supporters of such research. It formed the basis of the Warnock Committee's recommendation in 1984 that research on embryos be prohibited after the appearance of the primitive streak, which prohibition became part of the legislative regime established by the Human Fertilisation and Embryology Act 1990 (sec. 3). The idea that there is no human individual prior to the appearance of the primitive streak is now a commonplace among bioethicists.[13] Harris endorses it without argument: 'A further complication is that the fertilized egg cannot be considered a new individual because it may well become two individuals. This splitting to become "twins" can happen as late as two weeks after conception'.[14]

Perhaps it is too harsh to say that he gives no argument. If there is one, the implicit major premise must be that if the fertilized egg may well become two individuals, it cannot be considered a new individual before twinning is no longer possible. Before evaluating the claim, note that Harris is keen to observe in this context that fertilization does not always give rise to an embryo but may, when the process goes wrong, give rise to a tumour such as a hydatidiform mole.[15] The only interesting conclusion to draw from this observation is the biological one that not all fertilization leads to conception.

Nothing, however, follows about the normal course of embryogenesis from observations about what happens when things go wrong. And because of twinning, not all conception results (at least directly) from fertilization. But isn't conception supposed to be fertilization, especially in the minds of those who defend the human individuality of the embryo? This would be a mistake, for conception is a metaphysical phenomenon and fertilization only one kind of biological manifestation of it. Here is a definition of conception that captures what it is we should be looking at in considering the relation between the coming into existence of a human being and the biological event typically associated with it.[16]

Conception is that event, typically involving the union of sperm and egg, which consists in a change in the intrinsic nature of a cell or group of cells, where that change confers on the cell or group of cells, or on their descendants in the case of division, the intrinsic potential to develop, given the right extrinsic factors, into a mature human being.[17]

There is no room to address the issue of potentiality here, but for present purposes the important point to note is that the definition encapsulates the crucial distinction between the metaphysical phenomenon of coming into existence and whatever biological event is typically associated with and exemplifies it, thus leaving open the possibility that some other event may also fulfil that role. The slogan 'life begins at conception' can, then, be read in different ways. In non-philosophical use – what we might call 'ordinary parlance' – it means that the human being's coming into existence is typically manifested by the process of fertilization. A person uttering the slogan might not know about hydatidiform moles or even be aware of the way in which twinning occurs, but for them to avoid knowingly saying something false (or at least ambiguous) they must at least have an implicit grasp of how such phenomena are relevant to what they mean. In

philosophical use, on the other hand, 'life begins at conception' states a necessary truth: not a tautology or a proposition whose truth depends solely on the meanings of the words used, but one that defines coming into existence in terms of a metaphysical phenomenon, namely the emergence of a new nature, an organizational unity that is not a part of its host but an individual that uses its host for the purpose of its own self-directed development into a mature member of its kind.

Where, though, does this leave the embryo before twinning occurs? For Harris and most other bioethicists, it cannot be an individual precisely because it might become two. Yet it is hard to see how this is supposed to follow. Before going further down this track, however, we should recall some of the biological facts about twinning. First, it is simply not true that twinning can only occur before formation of the primitive streak. It also occasionally occurs after this time; though as far as is known all cases involve deformity, such as Siamese twins or, rarely, foetus-in-foetu, where one foetus grows for a time inside the other, eventually dying; the host foetus is sometimes born alive and goes on to lead a normal life. But if the defender of the argument from twinning wishes to deny the embryo individuality before the primitive streak appears, during which time twinning can occur normally, why would she not make the same denial in respect of a post-14-day embryo? Why should the fact that twinning after 14 days would be abnormal make any difference?

For Karen Dawson, who discusses this issue at length,[18] '[t]he possibility of conjoined twins and fetus-in-fetu occurring weakens the applicability of the concept of irreversible individuality, as defined, and similarly the validity of using the proposed argument for segmentation [twinning] in ascribing moral status from even 14 days after fertilization'.[19] For her, moral status must therefore be based on something other than 'irreversible individuality', and one assumes Harris and others would agree that when it comes to morality, something other than mere 'ontological individuality', to use Ford's expression, must be the determining factor. But for present purposes what matters is the metaphysical implication of post-primitive streak twinning. It appears we do not yet know just how far into development an embryo or foetus must be before twinning of any kind, normal or abnormal, is physically impossible. It must be the case, though, that twinning is possible when the embryo or foetus has reached a stage where to deny it individuality would be a case of biological and metaphysical blindness. Yet if individuality is ascribed in such a case, why not ascribe it before 14 days? It is irrelevant that twinning then would be normal; so what other criterion could one use – size or shape?

Secondly, the only sense in which it seems that twinning is abnormal is that it is rare;[20] it does not appear to warrant ranking alongside tumourous growth after fertilization, so we cannot easily reiterate, as in the latter case, that faulty development has no metaphysical implications for the normal course of events.[21] Thirdly, though, we are profoundly ignorant about the factors that give rise to the possibility or actuality of twinning. It might be, for all we currently know, that twinning is caused by a random genetic or environmental change occurring after fertilization and early embryonic development have begun. This would make it no different, ontologically, to something more dramatic such as a freakish discharge of lightning that struck the mother and caused her gestating embryo to divide. But should we say, in the fanciful case, that there was ipso facto no human individual prior to the lightning strike? It is hard to see what the argument for this could be.

According to C. Ward Kischer, however, there is 'strong' evidence that twinning is determined at fertilization itself, hence that the individuality of the embryo that does not twin is also determined at fertilization.[22] Although 65–70 per cent of cases of monozygotic twinning are due to fission of the inner cell mass, the other 30–35 per cent are explained by division at the two- to eight-cell stage of cleavage, occurring two to three days after fertilization.[23] Kischer takes this to mean that twinning may well be determined at fertilization or else in early cleavage,[24] but his interpretation of the evidence is confused. If twinning is determined to occur at all — and here I mean intrinsically determined, i.e. by the very nature of the embryo — it will have to be at fertilization, not at any stage later (e.g. first cleavage). That is, whenever the twinning actually occurs, that it is determined to occur at all must be settled at fertilization, when all the genetic information is in place for twinning to be determined by that information and (perhaps) environmental factors.[25] We do not even have to regard the embryo, for the purposes of the present point, as a human individual, merely as a new unit of genetic information such that twinning, if determined at all to occur at some time, is so determined by the coming into existence of that information with the possible co-operation of environmental factors.

For suppose twinning were determined to occur at, say, first cleavage: in other words, that it will occur at some future time is determined at first cleavage. Call the determining event E, where E is the acquisition of some property by the embryo at first cleavage such that it is then determined that it will split into twins at some future time. What caused (and hence determined) E?[26] Either something or nothing. If nothing, then E is uncaused, hence undetermined, hence random.[27] Since E is random, the twinning caused by E will also be random as far as its ultimate explanation is concerned, since it will have been caused by a random event. So it would not be accurate to state that in such a case the twinning, whenever it occurred, was truly determined at first cleavage, and if not determined then it could not have been determined at all. Suppose, on the other hand, that E was caused by some prior event E1; then for the twinning to be determined, E1 itself would have to be determined; and so on back in time until we reached a non-arbitrary point at which an event occurred that determined all the subsequent events leading to twinning. The only such principled point is fertilization, when all the intrinsic information is in place to determine whether the embryo will split into twins (with, perhaps, environmental events co-operating to produce the split).[28]

So twinning, if determined at all, will be determined at fertilization, though it may occur at various times, as the 30–35 per cent statistic quoted by Kischer and others demonstrates. That statistic is, however, not strong evidence that twinning is determined at all, since it is equally consistent with twinning's being a random event brought about by chance genetic or environmental changes at some time before 14 or 15 days, thus making the situation no different in principle to the freak lightning strike mentioned earlier. The mere fact that a certain proportion of twinning events occur very early in development does not favour determination over randomness; maybe it is a brute fact that 30–35 per cent of twinning randomly occurs in early cleavage.[29] Maybe there is another explanation altogether, namely that twinning is never intrinsically determined at fertilization, but is wholly extrinsically determined by some course of events beginning with the mother and affecting different embryos at different times after fertilization. Or perhaps some twinning is intrinsically determined at fertilization and some later by purely environmental events. All we know for certain, at present, is that during the period before the primitive streak appears, twinning is biologically possible. Whether it is random, or determined, and if the latter then by what, is a mystery.

The question is, what metaphysical inferences can be drawn from the various possibilities? It is worth making a jointly epistemic and ethical point first, namely that given our profound ignorance about twinning, we should give the benefit of the doubt to all embryos, even if there does exist an argument to the effect that embryos for which twinning is determined (a tiny minority, moreover) are not individual human beings. The ethical substance to the point is simple: if I am about to fire a gun at an unknown object, and am concerned only to avoid killing a human being, and there is at least a significant possibility that it is a human being, I should give the benefit of the doubt to the object. Similarly, if we are concerned to avoid killing human beings, and there is a significant possibility that some or all embryos are human beings, though some or all might not be if it is determined that they will twin, we should give the embryos the benefit of the doubt and not kill them. Of course Harris and others ultimately do not put much stock in an ethic that protects human beings as opposed to 'persons', but since he offers the twinning argument against individuality for those who think it matters, we may also show why, ethically, it does not matter given the current state of knowledge.

More importantly, though, there is no good argument for the metaphysical conclusion that an embryo is not an individual human either because it might or must twin.[30] The mere fact that it might twin makes the case no different from that of the lightning strike. An embryo cannot be considered to lack individuality because something might happen to it. Many plants are capable of being split into objects that are themselves plants and capable of continuing to grow as plants. Planarian flatworms can be divided and the divided halves continue to grow as individual worms. Cells that can divide are no less individual cells because of that possibility – they belong to exactly the same kind as their descendants. This includes cellular animals such as bacteria and amoebae.[31] Why, then, should human zygotes or early embryos be an exception?

Moreover, amoebae and most bacteria always reproduce by division, and yet they still all belong to the same kind – but we know that only a small minority of human embryos ever divide. So why should individuality be withheld from the latter whilst accorded to the former? We can be certain, moreover, that the reproduction by fission of amoebae and bacteria is determined rather than the result of a massive conjunction of chance events: it is part of their constitution. By parity of reasoning, therefore, even if humans always came into existence as twin descendants of embryos, and even if this were determined, the embryos would not thereby fail for human individuality. The parity could only be broken if it could be shown, for instance, that a given embryo (or class of embryos) did not share with its twin descendants the morphology or functionality of those descendants (mere difference of DNA would not be enough if the genotype was still human), such that the embryo was best classified as a kind of precursor to its twins, in much the same way as gametes are precursors to embryos but not themselves embryos. Yet this is precisely not true of human embryos, whatever the cause of their twinning.

Note that the twinning argument would receive no support from the idea that, at least for embryos determined to twin, and before such twinning occurs, there is not one human individual present at the same place and the same time, but two. Even if, *per impossibile*,[32] this were an available interpretation of the facts, the only ethical implication would be that destroying an embryo would, or at the very least might, involve destroying two human beings, not one. Again, the fact that we do not mourn the loss of a human being when twins are born gives us no

insight into ontology. What twinning shows is that there are more ways for a human being to go out of existence than by dying.[33] Harris makes much of the fact that many embryos are lost through unknown miscarriage during the course of natural reproduction,[34] but it is as misguided to infer that what is lost are not human beings with the same moral standing as the rest of us as it would be to draw the same inference from the fact that every day thousands of people die of hunger and poverty alone and unknown. I am not trying here to draw implausible ethical parallels between radically different states of affairs, only to point out that it is mistaken to use such facts as reasons for diminishing the metaphysical status of embryos any more than of other humans. It may be that natural miscarriages are inevitable, and there may be sound reasons in nature why they happen; but what is lost are still individual human beings, just as it may be that certain severely deformed babies have an inevitably short life expectancy, and that there may be sound reasons in nature why this is so. They are no less human beings for that.

4 Totipotency

In the debate over embryonic stem cell research, much is made by its supporters of the alleged fact that, at least up to the three- or four-day morula stage, human blastomeres are totipotent, that is, through culturing them as stem cells they can be caused to develop not only into any cell in the human body but into a new human being altogether, as happens in twinning. Hence at this stage, at least, there cannot be a human individual.[35] The argument goes along similar lines to the twinning argument, but has important differences that require separate consideration.

In particular, the fact that twinning is either possible or necessary for an embryo should not (as mentioned above) lead one to the absurd conclusion that there are two (or more) spatio-temporally co-located human individuals. Multiple occupancy in this sense is a non-starter. But if the totipotency argument works against the idea that there is a single embryo, it might be taken to support the positive view that there are several embryos, not precisely co-located but in close proximity to one another and enclosed within the zonapellucida. This might be one way of reading Harris when he states:[36] 'if the cells removed [from an early embryo] are totipotent (capable of becoming literally any part of the creature including the whole creature), then they are in effect separate zygotes, they are themselves "embryos", and so must be protected to whatever extent embryos are protected.' (Which will be hardly at all, according to his own theory.)

It is not clear why he places 'embryos' in scare quotes: either the totipotent cells are embryos or they are not. More importantly, his words are ambiguous inasmuch as he may be saying that the totipotent cells are separate zygotes only after they are removed, or he may be saying that even before they are removed they have this status. The latter interpretation would certainly lend support to the view held by him and others that the early embryo is not an individual. On this interpretation – and it may be the reason for the scare quotes around 'embryos' – what we take to be the embryo is really a kind of cluster of embryos, each member of which is capable of doing what the cluster is capable of doing, namely dividing and maturing. But then in what sense is the thing we take to be a single embryonic human individual really an individual at all? It may be a potential individual with human DNA, but that does not make it a human being any more than a fully differentiated somatic cell with human DNA is a human being. In short, the very totipotency of the embryonic cells belies the supposed unity and individuality of the embryo.

It is as well first to counter the misrepresentation of totipotency in the philosophical literature. Supporters of embryo experimentation represent totipotency as a state of affairs in which the zonapellucida is no more than a kind of fence within which are held a handful of cells, each undifferentiated, none of them 'knowing', as it were, what they want to be, yet all of them capable of developing into any kind of human cell or even a full human being. The characterization is superficial, and masks as well our continuing ignorance as to how and why cell differentiation occurs. The fact that the early embryonic cells do before long begin differentiating and specializing is consistent with the hypothesis that this is a random process (as I suggested earlier with regard to twinning): if there is randomness in nature, it may be random whether, for example, a given embryonic cell becomes or gives rises to a cell of the embryoblast or of the trophoblast. On the other hand, it is also consistent with the hypothesis that each cell is already programmed to develop one way or another, even though, if cultured outside the zonapellucida of that embryo, it might develop in some other direction. That is, given the actual setting within which the cell is located – within the particular zona, surrounded by other particular cells, all of them interacting with each other in ways we do not fully understand – that cell may well be determined to develop in one particular direction. In other words, 'totipotency' does not necessarily mean – and probably does not mean – that a cell is capable of becoming anything its genotype allows no matter what the conditions in which it is set. Within the embryo, each cell may have its own programmed role from the moment of fertilization, whatever it might be capable of in another setting. If true, this supports the unity of the embryo.

There is, moreover, an underlying metaphysical point here, namely that being totipotent does not mean being wholly undifferentiated or indeterminate. Even if it is random as to which developmental pathways are followed by which totipotent embryonic cells, no such cell actually can, short of twinning or artificial manipulation, develop into an embryo in its own right. As argued earlier, however, what might happen to an embryo does not tell us what it actually is, and the same applies to an embryonic cell. From the time of fertilization, and absent any intrinsic or extrinsic faults, the zygote begins an inexorable process of development, as a metaphysical unity, towards being a mature member of its kind. That there is cell differentiation at all after fertilization is hardly a random matter, even if it is perhaps random which cells follow which differential pathways. Far from being indeterminate, the zygote is wholly determinate and 'knows' exactly what it wants to be, whatever might happen to it in the future. Similarly, in its actual context, and whatever might happen to it in the future (whether by twinning, or by artificial manipulation, or by some freak event), each totipotent blastomere within the zonapellucida, as part of the embryo, 'knows' exactly what it wants to be – namely, a part that gives rise to some differentiated developmental pathway or other. It is no more an 'embryo within an embryo' than a mature, adult somatic cell – a cell in my calf muscle, for instance – is either an embryo or part of an embryo simply because, in other circumstances, it might be used in cloning by nuclear transfer. Again, the analogy with plants and other organisms is instructive. It is typical for plants to have totipotent parts in the sense that one can take a cutting from any part of a given plant and culture that cutting to develop into a full individual plant of the same kind as its parent. The same may be true of certain primitive organisms. But does that mean the plant is not an individual in its own right? What might the argument for this be? The organizational and developmental unity of the plant is fully given in the way its parts interact with each other, and is not undermined by the mere fact that, separated from the whole, a part could itself be cultured into a whole. Nor can the analogy be dismissed as highly imperfect or even irrelevant. Of course not all analogies between mammals and plants, or between plants and humans in particular, are going to have any relevance to a metaphysical issue. But in this case we are concerned with purely vegetative functions, i.e. the functions of growth and physical development, not with higher functions such as sentience, where the analogy with plants would break down, or rationality, where the analogy with other animals would break down. Therefore, if the proper interpretation of plant totipotency is correct, then the same interpretation should be given of human embryonic totipotency.

The same point can vividly be made by applying it to pluripotency. We now know that adult humans contain pluripotent stem cells throughout their bodies, and more sources of these are being found regularly. Yet it is incorrect to say, for example, that an adult olfactory stem cell that can be coaxed into becoming a nerve cell in a paralyzed patient[37] is now a nerve cell itself, or that a collection of such nasal cells found in close proximity is really part of the nervous system, or that such cells are anything other than olfactory cells contributing intrinsically to olfactory function as part of the nasal mucosa. All of which is compatible with their having the intrinsic potential, given the right circumstances, to develop into other kinds of cell. For the same reason, totipotent embryonic cells are not embryos within embryos, or nonspecific in their function: as parts of the embryo, they contribute essentially to its development, even if, at least at the morula stage, a given cell can be removed without destroying the embryo. (At the blastocyst stage removing such a cell does usually destroy the embryo.)

To make matters even worse for Harris and other deniers of embryo individuality, we have so far been speaking of totipotency in respect of embryonic cells as though it were a given, lumping together twinning and artificial manipulation, and engaging in a large amount of conjecture about just what embryonic cells can do outside their normal context. Such is often the way when philosophers – and for that matter theologians – discuss science. The truth of the matter, however, appears to be that embryonic stem cells are not totipotent in the sense in which Singer, Harris, Ford and others would have us believe. Embryonic stem cells do not 'have the capacity . . . to produce the whole (total) embryo and fetus with all its extraembryonic membranes and tissues'.[38] As molecular biologist and biochemist David M. Gilbert points out, embryonic stem cells are not totipotent because they have been removed from the supporting trophectoderm (the outer wall of the blastocyst from which the trophoblast is derived) and cannot recreate it themselves. Since the trophectoderm is essential for embryonic development, embryonic stem cells cannot themselves develop into a full embryo, even if they can develop into any tissue of the human body, thus making them merely pluripotent.[39] Hence when embryologists speak of 'totipotency', they should be understood as referring only to the capacity of early embryonic cells to form any kind of human tissue, not to their alleged ability to form on their own an entire human being. Absent this ability, whether we call such cells totipotent or merely pluripotent is a matter of indifference, as long as philosophers and scientists are both talking about the same thing and using the terms consistently. As far as the ability to form an entire individual is concerned – what we might call ultimate totipotency – that belongs only to the zygote and early embryo as a whole, by virtue of its intrinsic developmental capacity, which includes the capacity to twin.

Perhaps it will become technically possible to manipulate an extracted embryonic stem cell into fusing with a separately extracted or synthesized trophectoderm, complete with all necessary membranes and cellular material, and thereby cause it to develop into an individual human being. None of this, however, would undermine the individuality of the embryo from which the cell was taken, or the status of the cell when located within the embryo as a part of that embryo contributing to its overall development. Nor would such a process be anything like either natural or artificially stimulated twinning, which demonstrates only the potential of the whole embryo to split, cells, membranes and all, and does not militate against its individuality, as I have argued. So it seems that, even if the philosophical argument from totipotency against individuality had any merit as applied to other organisms – as I have claimed it does not – the scientific ground has in any case been taken away from it as far as human beings are concerned.

5 Conclusion

I have examined three common arguments, found in Harris and many other bioethicists, against the metaphysical status of the embryo as an individual human being. Neither the distinction between embryoblast and trophoblast, nor the potential for twinning, nor the argument from totipotency, undermines the proposition that the embryo is a human being, a full member of the same natural kind as both the author and reader of this paper. Moreover, there is good reason to think that philosophical discussions of totipotency are scientifically ill informed.

Yet these arguments reappear with disturbing frequency in the bioethical literature. Often they are barely presented as arguments, their conclusions being brought forward as established truths beyond dispute. There are other arguments as well, which I have no space to consider here. In particular, Harris and others regularly dismiss appeals to potentiality as providing no succour to defenders of the humanity of the embryo. In some respects they correctly identify mistaken appeals by such defenders, ones that are irrelevant or based on confusion about just what potentiality involves and how it affects both the metaphysics of the embryo and the proper ethical stance toward it. Full discussion of this issue, however, high-lighting as well the confusions and mistakes in the discussions of those who, on considerations of potentiality, argue against the humanity of the embryo and against according it the right to life, must await another occasion.[40]

So must a full discussion of the embryo's moral status.[41] John Harris and others raise interesting and provocative arguments against according the embryo (and foetus) the right to life – arguments that are largely independent of the question examined here, namely whether the embryo is an individual human being. The point of the present paper is to caution against the role that inadequate metaphysics can play in distorting ethical debate. Bioethicists who do not ultimately care, for ethical purposes, whether the embryo is an individual or not, are bound nevertheless to cease using specious metaphys-ical arguments against those who do. That such arguments do not work is, for those who fall into the latter camp (such as myself), a matter of no small importance.

Notes

1 (London: Routledge, 1985); hereafter VL.
2 (Oxford: Oxford University Press, 1998); origi-nally published as *Wonderwoman and Superman: The Ethics of Human Biotechnology* (Oxford: Oxford University Press, 1992); hereafter CGI.
3 CGI, p. 47; VL, p. 11; 'In vitro fertilization: The ethical issues', *Philosophical Quarterly* 33 (1983):

217–37, at 223. Note, however, that whereas in the second and third locations Harris speaks of the embryoblast/trophoblast distinction, by the time of CGI (and also the first edition, *Wonderwoman and Superman*), a number of years later, all reference to the embryoblast has disappeared in favour of the embryo/trophoblast distinction. It is not clear what new evidence prompted the change of terminology.
4 All references to numbered days of gestation should be taken as approximate.
5 Harris uses the term 'fertilized egg', but this is tendentious terminology and embryologically misleading when speaking of any stage later than that at which the sperm penetrates the egg. I will use the embryologically correct term for each stage, e.g. zygote, morula, blastocyst, etc. Harris's 'trophoblast argument' is supposed to apply to all stages prior to that at which the inner cell mass differentiates from the trophoblast, but for convenience I will speak of the morula only.
6 The claim is explicitly endorsed by Norman Ford, who asks: 'In short, how could the cluster of cells of the early embryo be an actual ontological human individual if it has not yet differentiated into the cells and tissues that will constitute the future embryo proper and those that will not be integral and constituent parts of the embryo proper?' Norman M. Ford, *When Did I Begin?* (Cambridge: Cambridge University Press, 1988), p. 156.
7 Ford makes the strange claim, responding to Bernard Towers, that the placenta and umbilical cord are not 'living parts of the eventual [sic] embryo and foetus that are discarded when no longer needed – somewhat like deciduous teeth' because they '[have] no nerves, [are] insentient and [have] always been regarded as extraembry-onic tissue': Ford op. cit., p. 156. But neither are there nerves or sentience in the foetus's hair or fingernails, yet these are parts of it. To add the qual-ification 'living' to 'part' just obscures the issue, since the hair and nails are growing, and there is no need for an object to consist entirely of living cells in order for it to be a genuine part of an organism. And 'extraembryonic' is ambiguous between

'located outside the embryo's body' and 'not part of embryo'. Ford evidently has the second meaning in mind, but this is just begging the question and anyway false about how embryologists regard the placenta and umbilical cord.

8   K. L. Moore, *The Developing Human* (Philadelphia: W.B. Saunders, 1982; 3rd edn.), ch. 7; the observation is repeated in the 5th edition (1993) by K. L. Moore & T. V. N. Persaud. The relevant passages from these editions are cited in C. W. Kischer & D. N. Irving, *The Human Development Hoax* (2nd edn., 1997; self-published, distributed by American Life League, Stafford, VA), pp. 41–3; hereafter HDH. Kischer is Associate Professor Emeritus, Cell Biology and Anatomy, University of Arizona College of Medicine. See also D. N. Irving, '"New Age" embryology text books: Implications for fetal research', *Linacre Quarterly* 61 (1994): 42–62. In the most recent edition of Moore and Persaud, it is again stated, though this time less explicitly, that there is not a sharp distinction between embryoblast and trophoblast, given intermingling between the cells of both: see Moore and Persaud, *The Developing Human*, 8th edn. (Philadelphia: Saunders Elsevier, 2008), p. 134. As Kischer and Irving point out, however, Moore/Moore and Persaud also, in their discussion of embryoblast and trophoblast early in the various editions of their book, somewhat contradictorily give the impression that such a clear distinction does exist; see the references in HDH and also *The Developing Human*, 8th edn., pp. 36ff.

9   *The Developing Human*, 8th edn., p. 134.

10 'Cell interactions occur between these two nascent populations of cells [embryoblast and trophoblast] that are essential for specifying their fate': G. C. Schoenwolf, S. B. Bleyl, P. R. Brauer, & P. H. Francis-West, *Larsen's Human Embryology* (Philadelphia: Churchill Livingstone Elsevier, 2008; 4th edn.), p. 43. This refers to what happens when differentiation occurs; it would be surprising if such interactions ceased after differentiation. On the embryoblast and trophoblast, see also P. Lee & R. P. George, *Body-Self Dualism in Contemporary Ethics and Politics* (New York: Cambridge University Press, 2008), pp. 126–7.

11 Let us leave aside for present purposes the question of ectogenesis, and confine our observations as to what is required for normal functioning to natural gestation.

12 Here I am assuming what is not presently known, namely that before separation into embryoblast and trophoblast no particular cells are already determined to become one or the other. There are good philosophical and empirical reasons for thinking the opposite to be the case, which would justify the claim that before observable separation there is still a real (if unidentifiable) distinction between the embryoblast or embryo proper (the mass of cells determined to come together and form the proto-body) and the trophoblast/trophoblastic cells. Needless to say, this would not aid Harris's case.

13 See, e.g., Ford, op. cit., pp. 170–7; P. Singer, *Practical Ethics* (Cambridge: Cambridge University Press, 1993; 2nd edn.), p. 137; M. Lockwood, 'Human Identity and the Primitive Streak', Hastings Center Report 25 (1995): 45; L. Silver, *Remaking Eden: Cloning and Beyond in a Brave New World* (New York: Avon Books, 1997), p. 43.

14 CGI, p. 47; also VL, p. 11; 'In vitro fertilization', p. 223.

15 Caused, inter alia, by two spermatozoa fertilizing the egg and then fusing their pronuclei with no contribution from the egg nucleus, leading to a watery cluster or mass (hydatidiform, Greek 'watery'; mole, Latin 'mola', circular cake). For references in Harris, see CGI, p. 47; VL, p. 10; 'In vitro fertilization', p. 223.

16 I am restricting the definition to the human case, but it can be suitably generalized, perhaps to cover all organisms.

17 For further discussion of this subject, see D. S. Oderberg, 'Modal properties, moral status, and identity', *Philosophy and Public Affairs* 26 (1997): 259–98, and *Applied Ethics* (Oxford: Blackwell, 2000), pp. 16ff. The quoted definition is from p. 21. See also Lee & George, op. cit., pp. 123–5.

18 K. Dawson, 'Segmentation and Moral Status: A Scientific Perspective', in P. Singer, H. Kuhse, S. Buckle, K. Dawson & P. Kasimba (eds.) *Embryo Experimentation* (Cambridge: Cambridge University Press, 1990), pp. 53–64, at 57–9.

19 Singer et al. op. cit., p. 59.

20 About 0.4 per cent for spontaneous twinning; the rate is at least twice that for cases of assisted reproduction: see M. Schachter, A. Raziel, et al., 'Monozygotic twinning after assisted reproductive techniques: A phenomenon independent of micromanipulation', *Human Reproduction* 16 (2001): 1264–9.

21 On the other hand, there is evidence of increased incidence of congenital abnormalities in monozygotic twins, so perhaps it is a mild case of faulty development: see 'Twin pregnancy' (author unknown), *Atlanta Maternal-Fetal Medicine* 2, 4 (1994) (at http://atlanta-mfm.com/body.cfm?id=48, accessed 13.6.08); G. Leblebicioglu, S. Balci & A. Üzümcügil, 'Variable expressivity of congenital longitudinal radial deficiency and spinal dysraphism in monozygotic twins', *Turkish Journal of Pediatrics* 47 (2005): 390–2.

22 HDH, p. 28.

23 J. S. and M. W. Thompson, *Genetics in Medicine* (Philadelphia: W.B. Saunders, 1986; 4th edn.), p. 274 and Moore, *The Developing Human* (4th edn., 1988): 122–6, both cited by Kischer in HDH, pp. 28, 34, put the figures at 30/70 per cent. These are confirmed in L. Scott, 'The origin of monozygotic twinning', *Reproductive BioMedicine Online* 5 (2002): 276–84. Moore and Persaud in *The Developing Human*, 8th edn., p. 138 put the figures at 35/65 per cent.

24 HDH, p. 28.

25 Perhaps it is already settled by some property of the sperm and/or egg that will eventually unite to form the embryo, but then this opens up the question of whether it is determined that a given sperm and egg will unite, and we can safely skirt this thorny broader question for present purposes. My point is simply that if twinning is determined to occur, it cannot be later than fertilization. If environmental factors are also necessary, and these are not present at fertilization but obtain later, the determination to twinning at fertilization will be conditional only. In other words, it will be the case that the embryo is determined intrinsically to twin on condition that certain environmental factors obtain at some later time.

26 For simplicity I am taking it that all biophysical causation is deterministic. The point I am making

can be enlarged to take account of probabilistic causation.

27 Again, omitting considerations of probability, I take it that a biophysical event is undetermined just in case it is random.

28 For simplicity's sake I allow for the possibility of some random event prior to fertilization that, say, causes the development of a gamete, which itself joins with another gamete at fertilization, and where twinning later follows, such that we could still legitimately say that the twinning was determined even though some random event existed prior to fertilization. As long as causation from fertilization onwards is transitive, and all the embryogenetic events are determined, this will be enough for us to hold that the twinning is determined at fertilization even though there might be some random event prior to fertilization: we would not be forced to say that that inexplicable event caused the twinning, hence that the twinning itself was undetermined. To go any more deeply into this question would take us too far into general issues to do with causation and explanation. The basic point is that it is non-arbitrary to choose fertilization as the point at which the future course of the embryo, including whether it will split, is determined, assuming its course to be determined at all.

29 Just as, were one to have a roulette wheel with 30 per cent of slots black and 70 per cent red, and the wheel physically operated like a normal one, it would be a brute fact that 30 per cent of all throws landed randomly on black. (As far as we can tell. Again, let us leave aside the question whether there is true randomness in nature at all.)

30 For detailed discussion, see my 'Modal properties, moral status, and identity'. There, however, I focus on cases where twinning is possible but not determined.

31 Ford (op. cit., pp. 121–2) sees the analogy between zygotic twinning and amoebic or bacterial fission, though he dismisses the analogy with plants. He does not, however, come up with an argument showing either that zygotes, or amoebae and bacteria, fail for individuality.

32 See D. S. Oderberg, 'Coincidence under a sortal', *Philosophical Review* 105 (1996): 145–71. For a

contrary view in respect of artefacts, see C. Hughes, 'Same-kind coincidence and the Ship of Theseus', *Mind* 106 (1997): 53–67. Even if one could, as one cannot, make the argument work for artefacts, this does not mean it works for natural objects such as embryos.

33 For the present discussion I am assuming the overwhelmingly likely position that a divided embryo ceases to exist altogether, since twinning before the primitive streak appears is generally symmetrical. There may be cases both before, and certainly after, where twinning is asymmetrical: perhaps some cases of foetus-in-foetu are like that. In these situations, no embryo or foetus ceases to exist, but there is a faulty or incomplete separation of an entity from it, where that entity may be more or less well formed as a human being.

34 Harris, 'The ethical use of human embryonic stem cells in research and therapy', in J. Burley & J. Harris (eds.) *A Companion to Genethics* (Oxford: Blackwell, 2002), pp. 158–74, at 164.

35 For a typical statement, see H. Kuhse & P. Singer, 'Individuals, humans and persons: The issue of moral status', in Singer et al., pp. 67–8. See also: Ford op. cit., ch. 5; B. Smith and B. Brogaard, 'Sixteen Days', *Journal of Medicine and Philosophy* 28 (2003): 45–78.

36 'Ethical use': 163.

37 As has been demonstrated by research in various parts of the world. See, for instance, the work of Dr Carlos Lima of Portugal: C. Lima, J. Pratas-Vital, P. Escada, A. Hasse-Ferreira, C. Capucho & J. D. Peduzzi, 'Olfactory mucosa autografts in human spinal cord injury: A pilot clinical study', *Journal of Spinal Cord Medicine* 29 (2006): 191–203.

38 Ford op. cit., p. 212; see also p. 119 where he claims of the zygote that its 'first two daughter cells are totipotent – each one can develop into a complete living human individual'.

39 D. M. Gilbert, 'The future of human embryonic stem cell research: Addressing ethical conflict with responsible scientific research', *Medical Science Monitor* 10 (2004): RA99–103. This important point is missed by R. P. George and C. Tollefsen in *Embryo: A Defense of Human Life* (New York: Doubleday, 2008), pp. 12–13, where they briefly reiterate the common misconception about totipotency.

40 For a broader discussion of potentiality, and of its uses and misuses as applied to the embryo, the foetus, and the person, see my *Moral Theory* (Oxford: Blackwell, 2000), pp. 177–83, and *Applied Ethics*, pp. 32–40.

41 See *Applied Ethics*, ch. 1, and 'Modal properties, moral status, and identity', for contributions to such a discussion. See also Lee & George op. cit., and George & Tollefsen op. cit.

Chapter 2

Massimo Reichlin

THE ARGUMENT FROM POTENTIAL:
A REAPPRAISAL

The argument from potential (AFP) does not have a good press in today's bioethical debate. A standard formulation of the argument can be reconstructed as follows:

(a) A human person is a human being who possesses the capacity to exercise the operations that are characteristic of the human species (i.e., consciousness, thought, language);

(b) The human embryo does not have this capacity, nor has it the physiological structures which support the capacity; however, it has the potential for them, that is, it will develop them in the future;

(c) The human embryo is a potential person and therefore has the rights of a person.

Most scholars are unsatisfied with this argument, some believing that it does not suffice in order to provide an effective foundation of the respect owed to the human embryo or fetus, many more believing that it proves too much, and that, once we should allow that it works in ruling out abortion and experimentation on pre-implantation embryos, we must be prepared to hold that it does rule out contraception too. While some may be willing to uphold this last consequence, most consider that, were this

consequence inevitable, the argument should be seriously revised or else abandoned.

Both groups tend to converge in affirming that the argument is in the end misleading and does not offer a theoretically helpful tool for the discussion of the bioethical issues concerning the beginnings of human life. My aim is to show that most presentations of the argument in the recent literature are themselves misguided and that, once the argument is adequately formulated, it should be recognized as the best way of arguing on a philosophical basis against those biomedical practices which threaten the vital integrity of human embryos.

Criticisms of the argument

The analysis of the criticisms put forward against the AFP in the debate on the human embryo or fetus shows that most criticisms are rooted in one of two main understandings of potentiality: the first equates potentiality with possibility, the other conflates potentiality and probability.

Potentiality as possibility

a) According to the first understanding of the concept, potentiality stands for the possibility for future change. On this view, an adequate example to explain the concept would be to say that the ingredients have the potentiality to be a cake, in that they can be assembled by an external cause in order to produce a cake.[1] For *a* to have the potential to become *b* is taken to mean that a particular entity *b* is a possible outcome of a process involving an entity *a*: this generic understanding of potentiality is evident in authors such as Mary Warnock, for whom 'if x is potentially y, this means that x may be y in certain circumstances'.[2] Still more explicit is Richard Hare's statement that 'the potential is the possibility (*potentia* is derived from *posse*) that its development [i.e., of the embryo] will, or would under favourable conditions, occur in the future'.[3] In this understanding, the relationship between the embryo and the person that will eventually emerge from the process rests altogether unexplained: in fact, many authors seem to suggest that there is no relationship at all. Consequently, the fact of not destroying the embryo is not believed to confer a benefit on the embryo itself, rather on the person into which it will subsequently develop. This idea denies any moral meaning to the embryo's *actually* possessing such a potential, and bases every moral evaluation on the consideration of the interests of the future grown person into which it may turn. On the basis of this barely empirical approach, there seems no reason to believe that stopping the process through which a person is formed in a certain moment could be morally more objectionable than in some other moment: if we have a moral duty to pursue the final result, anything preventing it from being accomplished should be equally banned. The difference between practices involving gametes, embryos, or fetuses may only be traced as a function of the empirical proximity to the outcome of the person, which is the only stage valuable *per se*. It is clear, however, that, on this broad understanding, the potentiality for a child begins with his/her parents' making love, so that, if we should accept the AFP, contraception would be as morally objectionable as abortion.[4] In fact,

Human semen is potential human material, if it fertilises an egg and is implanted. Human ova are likewise potential human beings if they are fertilised by semen, and implanted. Yet no-one supposes that sperm and egg are in themselves to be protected, although they are plainly of *value*, and are not to be thought of just like anything else.[5]

Besides, we could add that, if this understanding of potentiality is correct, a child begins with his/her parents *desire* to make love, since this desire does have the potential to produce a baby; if pressed to its consequences, the AFP should question the morality of not consenting to such a desire and should end in establishing a moral duty to have as many children as one can.[6] Most authors cut short with these embarrassing consequences by rejecting the AFP altogether.

The kind of potentiality these critics have in mind should indeed be termed 'possibility', since it is equal to the mere idea of non-contradiction of a certain development: in this view, something has the potential to become something else if this does not involve any contradiction. This corresponds to the general sense of potentiality as possibility which was also set out by Aristotle in *Metaphysics*:[7] in this generic sense of potentiality, 'if X has an active potentiality for giving rise to Y, and Y has an active potentiality for giving rise to Z, then it must follow that X itself has an active potentiality for giving rise to Z',[8] so that in the end everything is potentially something else. This understanding of potentiality is the premise that leads to speaking of potential persons as if 'there are ghostly persons somewhere just waiting to be given flesh and blood'.[9] It is thus important to preserve the distinction between potentiality and possibility; as we shall see in detail, this distinction can be expressed in Aristotelian terms as that between *active* potentiality — which means a being's inherent capacity to autonomously develop itself

— and *passive* potentiality — which only implies the capacity to undergo modifications from external agents. The case of human gametes shows an even more remote sense of potentiality, since what we have in the gametes is not an individual's capacity to undergo modifications, but rather the possibility that two entities unite in order to form a new individual which is distinct from the two originals. The difference between the gametes' and the embryo's potentials is thus as great as that between 'the potential to cause another entity to come into existence and the potential of an existing entity to realise its implicit nature'.[10] It should thus be regarded as seriously misleading to say that an unfertilized ovum is a potential person.

b) Moreover, the conflation of potentiality and possibility seems to involve a naive view of fertilization as the mere addition of material with no qualitative increase. As Singer and Wells put it,

> Everything that can be said about the potential of the embryo can also be said about the potential of the egg and sperm when separate but considered jointly. If we have the egg and sperm then what we have also has the potential to develop into a normal human being, with a degree of rationality, self-consciousness, autonomy, and so on. On the basis of our premise that the egg and sperm separately have no special moral status, it seems impossible to use the potential of the embryo as a ground for giving it special moral status. . . . That there is one more stage that the egg and sperm must go through, compared to the embryo, can scarcely make a decisive difference.[11]

The idea that the event of conception could be something like just 'considering jointly' the egg and sperm seems to simply overlook the fact of the 'biological roulette' through which a new member of the species (what the gametes were not) is formed, and one with an altogether new genetic identity. This makes a 'decisive difference' from an ontological viewpoint: in fact, 'taken separately, but considered jointly, they [i.e., the sperm and egg] are not the same ontological entity as the single-cell zygote that results from their fusion and which may be said to be a potential person'.[12] While it can be discussed whether this suffices in order to affirm the presence of a person, fertilization surely is *the* moment (or *the* process, if you consider that it takes a number of hours) of the beginning of a new biological individual: no other moment before this can legitimately make such a claim.

c) A similar understanding of potentiality seems to be at the basis of the recurrent analogy of the potentiality of the human embryo with that of a normal citizen to become the president. On this view, that makes what Feinberg called 'the logical point about potentiality',[13] personhood is clearly viewed as a kind of property, made up by a collection of characteristics, that will be acquired in the future and the AFP is reduced to some sort of a divination, or the projection of a future state. As affirmed by S. I. Benn,

> If *A* has rights only because he satisfies some condition P, it doesn't follow that B has the same rights now because he *could* have property P at some time in the future. It only follows that he *will* have rights *when* he has P. He is a potential bearer of rights, as he is a potential bearer of P. A potential president of the United States is not on that account Commander-in-chief.[14]

The idea is reinforced by Joel Feinberg's analogy:

> In 1930, when he was six years old, Jimmy Carter didn't know it, but he was a potential president of the United States. That gave him

no claim *then*, not even a very weak claim, to give commands to the U.S. Army and Navy. Franklin D. Roosevelt in 1930 was only two years away from the presidency, so he was a potential president in a much stronger way (the potentiality was much less remote) than was young Jimmy. Nevertheless he was not actually president, and he had no more of a claim to the prerogatives of the office than did Carter.[15]

The argument simply presupposes that talking of the embryo's potential for personal operations involves having already excluded its actual personhood, which is actually to demonstrate. Moreover, it shows an understanding of the concept that seems to imply that prior potentiality can only be subsequently induced from actuality. In fact, in 1930 Carter did not know he was a potential president: but did anyone else know? It seems that we should say that, only by knowing the fact that he subsequently became the president, we can also say that he was a potential one. And how can we say that in 1930 Roosevelt's potentiality for presidentship was much less remote than Carter's? His potentiality, as Carter's, was in fact dependent on a long series of external events, none of which could be considered as having an inherent teleology. It should be concluded that, on this account, being potential does not mean any thing more than being possible: in 1930 none of these presidents was in any sense more president than any other American child, who has a theoretical possibility to become the president.

The difference between this kind of potentiality and the one attributable to the human embryo does seem quite clear. The fact that I will become the president is in fact dependent on several external causes, such as social conventions and regulations, and in no way implies the kind of necessity shown by a natural development: moreover, the same external causes that

can promote my rising to the dignity of presidentship can also hasten my becoming a simple citizen again. On the contrary, the embryo's development does not depend on external causes, rather on an inherent teleology that only demands certain environmental factors to be displayed: the embryo has in itself the potential for full personhood, and does not receive it from outside. Even the king's son[16] will not be king on the basis of his mere 'nature': even a monarchy is in fact a social convention and may be overthrown by a revolution, or through a democratic process. Moreover, if, talking of personhood, we intend to refer to an essential condition of the beings referred to, we cannot intend personhood as a transient property, such as a social or political office: if this attribution is to have more than a functionalistic meaning, a person cannot lose her ontological dignity (at least, not in the general course of events),[17] she cannot abdicate such dignity and no political revolution can dispossess her of this 'office'.

This difference in the very nature of the dignity in question (presidential *vs.* personal) requires a similar difference in the rights with which this dignity is associated. In the person of the President it is the importance of the office and its symbolic significance for the life of us all which we honour: we thus attribute particular rights to a human individual not *per se*, but only in as much as he is the President and only as long as he holds the office. On the other hand, we do not value a human person on the basis of her office or significance: we value her *per se*, that is, we value the very individual she is. In this sense, the rights which are in question in the case of the human embryo are not conferred rights, but they are 'natural' ones: and the problem remains as to when the embryo can be said to possess its 'nature'.[18]

d) A similar confusion between potentiality and possibility seems to be at the basis of Michael Tooley's famous criticism of the AFP.[19]

In his unreal mental experiment, Tooley imagines that in the future it will be possible to inject kittens with a chemical that will enable them to develop intellectual capacities similar to those of human beings: on the basis of his construction of the proper subject of a right to life,[20] Tooley concludes that these 'intelligent kittens' would have a right to life. However, it would not be seriously wrong to refrain from injecting the chemical in a kitten and kill him: besides, on the basis of the principle of moral symmetry between actions and omissions,[21] it would not be seriously wrong even to interfere with the ongoing process. As a conclusion, if we do not want to say that the killing of newborn kittens is rendered morally wrong by the fact that they can become rational animals, we must conclude that it cannot be seriously wrong also to interfere with the process through which a member of the species *Homo Sapiens* develops the qualities which endow him with a right to life: in fact, the potential is the same in both cases.

I believe that this argument is not conclusive at all in favour of abortion, rather it can be regarded as telling in favour of contraception. In fact, to keep with this absurd mental experiment, injecting the 'rationalizing' substance into a kitten involves creating an altogether new being, that is, a kitten which has a quasi-natural capacity to develop those intellectual and symbolic properties which are, by hypothesis, the foundation of our respect for persons. This properly shows the inapplicability of the symmetry principle: in a world where kittens could be transformed into rational animals, to kill them as newborns (not yet transformed) would not be equal to aborting a fetus, but rather to preventing an embryo from forming, that is, it would be equal to performing a contraceptive action. On the other hand, if we are talking about already injected kittens, there is no ontological symmetry between 'intelligent kittens' and 'normal kittens' (i.e., kittens lacking the properties that are relevant for the right to life), since they are two different species of animals. In other words, the non-injected newborn kittens are inherently unable to acquire the additional property in the normal course of their development, while the injected kittens change their nature by way of an external intervention. The first kittens are to the second ones as the sperm (or the unfertilized ovum) is to the embryo: both in the injection of the kittens and in the formation of an embryo, what we have is a change in substance by way of an external intervention: and as the entities prior to and after such a change are radically different from the biological point of view, it is not possible to apply a principle of moral symmetry.

Tooley pretends not to notice this change in species, taking advantage from the emotive sensation created by his unreal example; however, if we take the example seriously, we must note that in the second case we are not talking about kittens at all, at least not in the sense usually intended with the word 'kitten', a sense which grounds our moral intuitions about it. What seems to give plausibility to Tooley's argument is that, being so difficult to take his example seriously, these 'normal' moral intuitions still linger in our reasoning when we consider the mental experiment, thus rendering ridiculous the idea of endowing a kitten with a right to life. What the AFP affirms can be affirmed on the basis of both theoretical reasoning and the moral experience of what is at the end of the process of embryo development, i.e. the human person: and obviously we have no experience of what an intelligent kitten could be like.

Potentiality as probability

Some authors are so distressed by the seemingly vague notion of potentiality that they try to reword the concept as to assure more ontological certainty: they are firmly convinced that something either is in a full sense or is not, but it cannot be only 'potential', for this seems to mean 'non-existing, though in some way existing'. This is especially true of the person: having defined the person as an entity possessing certain properties, it is clear that if these properties are present, so is the person; if the properties are absent, the person is absent too. These authors thus suggest that the concept of potentiality should be reformulated in terms of probability: this makes clear that a potential person is not to be considered a person at all, but at most an entity that has a certain probability to develop the properties that will make it a person.

H.T. Engelhardt Jr., for example, explicitly rejects the very idea of potentiality by saying that the concept 'is in itself misleading, for it is often taken to suggest that an X that is a potential Y in some mysterious fashion already possesses the being and significance of Y.'[22] This author offers the reinterpretation needed, by saying that:

> It is perhaps better to speak not of X's being a potential Y but rather of its having a certain probability of developing into Y. One can then assign a probability value to that outcome. Recent research concerning zygotes suggests that there is a great amount of zygote wastage. Since only 40–50 percent of zygotes survive to be persons (i.e., adult, competent human beings), it might be best to speak of human zygotes as 0.4 probable persons.[23]

A similar collapse of potentiality into probability can be seen in E.A. Langerak, who nevertheless claims to be an advocate of the potentiality principle. In fact, this author carefully distinguishes the concept of a *possible* person, as that of 'a being that could, under certain causally possible conditions, become an actual person (for example, a human sperm or egg)'[24] from the concept of a *potential* person, as that of a being 'that will

become an actual person in the normal course of its development (for example, a human fetus)';[25] however, Langerak does not ground the difference on the ontological distinction that exists between the gamete and the embryo, but rather on the difference in the probability of the outcome. In fact, reconsidering the analogy of the president, Langerak affirms that while every American child is a possible president, a potential president is a person who already won the elections but has not yet been inaugurated: what makes the difference is not something inherent to the being in question but rather the fact that, in the normal course of events, the outcome is much more likely to occur. In this sense, the predetermined tendency of the human embryo:

> Does distinguish the organism from possible persons by guaranteeing a dramatic shift in probabilities. . . . Even those of us who refuse to mythologize the predetermined tendency in potential persons must agree that this tendency makes it highly likely that, without outside interference, they will become persons.[26]

The most paradigmatic example of this reduction of a metaphysical concept to empirical considerations was stated in Noonan's explicitly biological way of arguing against abortion:

> Once spermatozoon and ovum meet and the conceptus is formed, such studies as have been made show that roughly in only 20 percent of the cases will spontaneous abortion occur. In other words the chances are about 4 out of 5 that this new being will develop. At this stage in the life of the being there is a sharp shift in probabilities, an immense jump in potentialities . . . If a spermatozoon is destroyed, one destroys a being which had a chance of far less than 1 in 200 million of developing into a reasoning

being, possessed of the genetic code, a heart and other organs, and capable of pain. If a fetus is destroyed, one destroys a being already possessed of the genetic code, organs, and sensitivity to pain, and one which had an 80 percent chance of developing further into a baby outside the womb who, in time, would reason.[27]

Lastly, this concept of potentiality as probability is often set out in relation to IVF embryos, whose potential personhood is considered by some authors directly proportional to the probability that they have to be reimplanted in the uterus. In this perspective, potentiality is viewed as resting altogether on external events, which are completely dependent on other people's decisions: once again, the potentiality of the embryo is thus equated to the mere possibility of the gametes. In Carol Tauer's words,

> If the zygote's normal conditions in the laboratory are essentially the same as those of the oocyte before fertilization, which appears to be the case, then the zygote will never develop into a person. Thus it might better be classified as a 'possible' person, one which could become a person only under certain causally possible (and deliberately chosen) conditions.[28]

This is quite far from what the concept of potentiality was originally meant to affirm. If we take seriously the idea that philosophical concepts arise from a whole practice of thought and must always be understood as a part of a larger tradition of inquiry,[29] we cannot overlook the fact that the concept of potentiality has its proper place in the domain of ontology. To speak of the potentiality of a being implies affirming something about the *nature* of that being, something about the kind of being it actually is: the biological facts about it cannot be but *a posteriori*

confirmations and empirical expressions of an underlying ontological structure. This structure is completely independent of other people's choices: these can prevent an embryo from realizing its inherent potential, but they cannot prevent it from having such potential, that is, they cannot prevent it from being what it is. This ontological framework is the proper context in which the AFP can show its plausibility; if this context is removed, the argument lends itself to several misinterpretations and may give rise to untenable consequences. In fact, to shift from potentiality to probability does not eliminate the central objection to the idea of potentiality as possibility. What this second reduction comes to is just to say that the moment of conception determines a high increase in probabilities: but this does not prevent the gametes themselves from having such a probability, although a very low one. No theoretical evidence is thus provided to deny the plausibility of shifting backwards the moment of the beginning of life, thus making the difference between abortion and contraception a mere matter of degree, and even reaching the *reductio ad absurdum* of questioning the morality of not consenting to the mere desire for sexual intercourse.

The argument reconsidered

All these formulations show basic misunderstandings of the original meaning of potentiality, since the concept does not refer to merely empirical facts in the future, but rather points at an ontological quality of the being considered. I will try to make clear in the following that a correct understanding of the embryo's potentiality shows that by progressively acquiring new capacities – including the capacity to perform the rational operations – the human individual develops and perfects the human nature it already possesses. This interpretation obviously presupposes an ontology, but not the 'naturalistic' ontology of traditional metaphysics: the notion of potentiality is in fact a dynamic one, basically designed to account for a being's continuity through change, so that that being's identity is preserved while it completes and perfects its nature, acquiring those capacities and qualities which did not show themselves in its early stages, but toward which its development actually pointed. For, as Aristotle put it, 'nature also is in the same genus as potency; for it is a principle of movement – not, however, in something else but in the thing itself qua itself[30]'. In trying to reconsider the argument, I will first recall the essential traits of the original Aristotelian meaning of potentiality and then try to show how this ontological insight can be useful for a contemporary understanding of the human person.

Potentiality in an Aristotelian sense

The notion of potentiality (but the most accurate word, by which the Greek 'dynamis' is correctly translated, is 'potency') was originally introduced by Aristotle in order to deal with the problem of becoming and to avoid the ontological immobilism of Parmenides and the Megarics: the proper place to look for the meaning of potency is thus Aristotle's Metaphysics.[31]

In book IX of the Metaphysics, Aristotle accurately distinguishes between two kinds of potencies, namely passive and active potency: the first is extrinsic, in that the principle of actualization comes from outside the body which is transformed. This kind of potency has to do with contingent qualities of the substance, which the substance can assume or not, without its perfection being affected: this meaning of potency is simply the disposition to receive modifications. Thus, a tree can be said to be a potential table, in so far as it can be the object of a craftsman's work through which a table can be produced; however, for a tree to be a potential

table does not mean in any sense that it is already a table.

This concept of passive potency is useful for us to understand the relationship between the gametes and the person. Far from being a potential person, in Aristotle's perspective a gamete is not even a potential embryo, at least not in the sense of an active potentiality; in fact, its potency for becoming an embryo is not inherent to itself and perfectly ordered as to produce an embryo, but rather depends essentially on external causes. The human sperm does not just need a proper place wherein to develop its inherent potentialities, but needs an external event which is going to change radically its identity and potentialities, that is, it needs a sexual act or an in vitro procedure that enables it to approach a human female gamete; and even when these acts are performed the process is far from being complete, since most male gametes die without having realized their fertilizing potential. As Aristotle notes in Metaphysics:

> The seed is not yet potentially a man; for it must be deposited in something other than itself and undergo a change. But when through its own motive principle it has already got such and such attributes, in this state it is already potentially a man; while in the former state it needs another motive principle, just as earth is not yet potentially a statue (for it must first change in order to become brass).[32]

By contrast, active potencies are those in which the principle of becoming is internal, in that it does not depend on external causes. A first kind of active potency is exemplified by those specific human qualities that are acquired through instruction and learning. The often-quoted example of the musician is instructive here: every person is a potential musician, in that every person has the specific capacity to develop her sensitivity for artistic values and become one. Whether she becomes one or not is mostly dependent on her personal preferences, inclinations and choices: in this sense, Francis C. Wade, following a suggestion from Max Scheler, considers these actions examples of an 'active *specifiable* potentiality', that is characterized by the fact that 'some action beyond its constitution is needed from the side of the agent for it to act'.[33] However, we should also note that whether the person becomes a musician is dependent upon contingent external factors as well, such as whether she is going to have money enough to pay for her studies or whether she is going to find a good teacher, and so on. Besides, what is most important, her ontology, that is her nature or essence, is not directly affected by the actualization or the non actualization of this specific potency. It is the very capacity of becoming a musician which is essential to the human being, i.e. the fact of being *open* to artistic values, but not the fact of actualizing this capacity: a person who will never be a musician is not in any sense less a person.

Active potencies in a literal sense are those inherent to the nature of the being, whose principle of actualization is the very nature of that being: 'a thing is potentially all those things which it will be of itself if nothing external hinders it'.[34] Here we do not simply have the capacity to be subject to transformation by an external agent in action, nor the capacity to act in order to specify a tendency, but rather the capacity to express and actualize inherent potentialities towards which the being in question has a natural tendency – i.e., a tendency which is dependent on its very nature. What we have here is the potentiality to complete oneself, to act in view of one's specific perfection. On this metaphysical perspective, Aristotle ties potency to act just as matter is tied to its form: it makes no sense to talk of potency without relationship to its act, because the act is defined in relation to potency. In this sense, the specific form of the act

is already active in order to realize the potency that will later enable specific new capacities: as noted by Wade, what distinguishes the notion of active potency is the tendency of the agent, that is, the dynamic factor that gets action going and keeps it going. Without this tendency, the agent would not be what it is. The nature of these active *natural* potentialities[35] is such that 'the agent's constitution is ready and prepared to act when conditions permit':[36] these potentialities are guarantees of the future as far as the agent is concerned. To say that the embryo has a potential for intellectual life is to say that, being what it is (i.e., being human), it has a natural tendency to develop those biological structures that will enable it to perform, at the right time, those operations that are distinctive of a thinking being. The concept of potentiality is thus meant to refer to this progressive actualization of the capacities which a being first possesses only in project; its ontological identity persists through difference, while the being passes from not performing certain specific operations to performing them. As affirmed by Aristotle in *The Generation of Animals*:

> When we are dealing with definite and ordered products of Nature, we must not say each is of a certain quality because it *becomes* so, but rather that they *become* so and so because they *are* so and so, for the process of Becoming or development attends upon Being and is for the sake of Being, not *vice versa*.[37]

When Aristotle says that the embryo only subsequently develops the intellectual faculty, he does not mean that this faculty is added from the external at a later moment, but that it is present in potency from the beginning as a capacity, or, as the old Scholastics said, *in actu primo*. As was noted by a distinguished historian of Greek philosophy, this is the point at which Aristotle's theory is divergent from Aquinas'.[38] In fact, unlike Aquinas, Aristotle did not have the opportunity to rely on a personal God as an external cause in order to explain the formation of the spiritual soul; so, while Aquinas introduces a direct intervention of God, who creates and infuses the spiritual soul in a well developed animal body (after 40 days for males and 90 days for females), Aristotle believes that the intellectual soul is already present from the beginning; this soul — intellectual, because human — is one and the same throughout the process, even though it firstly actualizes only its vegetative capacities, then the sensitive, and lastly the rational. The form is thus an inherent telos, that is actualized by its own virtue in absence of external obstacles. 'Whenever, then, we speak of a being actualizing itself, we recognize that there is more to it than its present factual condition, and that the being in some sense already is that which, once it is actualized, it will be fully'.[39] The sense in which the embryo is already what it will be is the project which it contains: it has all the information needed in order to accomplish the projected person it is. It is obviously true that this accomplishment is dependent on several external conditions as well. However, these conditions, including a uterus to develop and the oxygen and nutritional support from the mother, are not constitutive of the personal quality of the process: they are necessary conditions for the development of an already existing human vital process, not necessary conditions for the existence of a project of full personal life.

In other words, on this approach, the aim of a being 'is not simply at the end but is at the beginning of a being's development like a direction giver'.[40] The being's nature is not indifferent to the actualization of the potentialities for intellectual operations: rather, the whole body is going to die if these potentialities are not actualized, as is shown by the natural selection of fetuses and by the impossibility for anencephalic newborns to survive for more than a few days.[41] Thus, a man is still a man if he does not become a musician; but an embryo dies if it cannot actualize its potential for full personhood. Thus, a correct account of the Aristotelian concept of potency leads to the conclusion that 'it is not metaphorical to say that the fetus is in this sense personal: its whole natural thrust is to become a functioning person.'[42]

[. . .]

Conclusion

Our analysis shows that the concept of potentiality has undergone several misunderstandings in contemporary bioethical discussions. However, once the ontological meaning of potentiality is recovered, it clearly appears that a human embryo's potential does not refer to the mere possibility of acquiring new capacities by means of an external agent, nor to the probability rate of the realization of such empirical mutations; rather, it refers to an inherent dynamic principle which rules the process of development of that being towards the progressive and never ending actualization of its already present nature.

In the light of this analysis, the argument can be reformulated as follows:

(a) The first actualization of a person is the existence of a human body which has an inherent tendency to develop characteristically personal operations (i.e., consciousness, thought, language);

(b) The human embryo is a human being with such an inherent tendency or potential;

(c) The human embryo shares the nature of a person, even though in the very beginnings of its actualization, and thus must be treated as a person.

This human nature or essence is the basis of a very serious moral claim raised by the human embryo upon us in the context of contemporary biomedical practices. This conclusion does not by itself commit us to the view that no difference can be traced between the forms in which the respect due to the human being is made effective in the case of the human embryo and in that of a developed adult person: however, it emphasizes that strong and well grounded reasons are needed in order to put a human embryo's life at risk, and that practices like IVF and embryo-cryopreservation are in no way morally unproblematic. Another relevant conclusion that follows is that the phrase 'potential person' should be very carefully handled: this expression seems plausible, in that it may refer to the fact that the human embryo, while already sharing the human nature, is nonetheless at the very first stage of its actualization and possesses only in potency the structures and the operation that are distinctive of the person; on the other hand, the expression is subject to serious misunderstanding if it is taken to mean 'possible' or 'future', but in no way actual person. That being potential should mean 'not being at all', notwithstanding the common sense interpretation of the concept of potentiality, is in fact what the AFP correctly understood is aimed at denying.

Notes

1 In Jonathan Glover's words: 'if it is the cake you are interested in, it is equally a pity if the ingredients were thrown away before being mixed or afterwards' (*Causing Death and Saving Lives*, Harmondsworth: Penguin 1977, p. 122).

2 Mary Warnock, 'Do Human Cells Have Rights?', *Bioethics* 1:1, 1987, pp. 1–14, p. 12.

3 Richard M. Hare, 'When Does Potentiality Count? A Comment on Lockwood', *Bioethics* 2:3, 1988, pp. 214–226, p. 215.

4 And indeed, for those refusing any qualitative cut-off point before the grown person, infanticide will be on the same list with contraception, embryo experimentation, early and late abortion (see Hare, *op. cit.*).

5 Warnock *op. cit.*, p. 12.

6 See Rosalind Hursthouse, *Beginning Lives*, London: Blackwell, 1987. Somehow connected with this interpretation of potentiality is also the question of our duties to future possible people: in replying to Derek Parfit (*Reasons and Persons*, Oxford: Clarendon Press, 1984, pp. 381–390), Hare has tried to show that a 'total utilitarianism' approach to this question does not imply the acceptance of the counter-intuitive conclusion that we should increase population as far as possible (see 'Possible People', *Bioethics*, 2:4, 1988, pp. 279–293).

7 Aristotle, *Metaphysica*, trans. by A. Platt in W.D. Ross (ed.), *The Works of Aristotle Translated into English*, vol. 8, Oxford: Clarendon Press, 1966, 1047[a].

8 Michael Lockwood, 'Warnock versus Powell (and Harradine): when does potentiality count?', *Bioethics* 2:3, 1988, pp. 187–213, p. 197.

9 Roland Puccetti 'The Life of A Person', in W.B. Bondeson et al. (eds.), *Abortion and the Status of the Fetus*, Dordrecht, Holland: D. Reidel, 1983, pp. 169–182, p. 171.

10 Robert Larmer, 'Abortion, Personhood and the Potential for Consciousness', *Journal of Applied Philosophy* 12, 1995, pp. 241–251, p. 243.

11 Peter Singer and Deane Wells, *The Reproduction Revolution. New Ways of Making Babies*, Oxford: Oxford University Press 1984, p. 91. See also Glover's quoted analogy between fertilization and the making of a cake (*op. cit.*, p. 122).

12 Norman Ford, *When Did I Begin? Conception of the Human Individual in History, Philosophy and Science*, Cambridge: Cambridge University Press, 1991, p. 97.

13 Joel Feinberg, 'Potentiality, Development, and Rights', in J. Feinberg (ed.), *The Problem of Abortion*, Belmont, CA: Wadsworth, 1984, pp. 145–150, p. 145.

14 Stanley I. Benn, 'Abortion, Infanticide, and Respect for Persons', in J. Feinberg (ed.), *op. cit.*, pp. 135–144, p. 143.

15 Feinberg, 'Potentiality, Development, and Rights', pp. 147–8.

16 The analogy is set up in Peter Singer, *Practical Ethics*, Cambridge: Cambridge University Press, 1979, who notes that Prince Charles is a potential King of England, but he has not the rights of a King.

17 The clinical situation in which this could be said to happen is the case of persistent vegetative state: however, the loss of personal dignity caused by the disease that some authors claim to take place in these cases is admittedly different from the cessation of an office, which involves no ontological change.

18 The objection, put forward by an anonymous reviewer, that 'naturally' we are all potential corpses, but would not like to be treated now as having actualised this potential, does not appear to me as decisive against the AFP. On one side the analogy fails, since the potentialities are different in kind: an embryo's potentiality refers to the future development of an actual entity that will preserve its identity through this development; the person's potentiality to become a corpse is the possibility of an entity to cease to exist altogether. On the other hand, however, the implication can be regarded as true: we should be treated as beings who will be corpses, that is, we should be treated as mortals, since death is our extreme possibility, and indeed our destiny. Similarly, human embryos must be treated as persons, since personhood is their destiny.

19 Michael Tooley, 'Abortion and Infanticide', *Philosophy and Public Affairs*, 2, 1972, pp. 37–65.

20 A criticism of this construction is of course besides the scope of this article.

21 I cannot discuss such a principle in detail. However, I do believe that a morally relevant difference exists between acts and omissions, one grounded on the consideration of the symbolic meaning of human action, a meaning which places human actions on a different level from the causality of natural events. I tried to argue this contention in an article on Prof. Rachels' thesis of equivalence in the context of the moral dilemmas of euthanasia (see my 'L'eutanasia nella bioetica di impostazione utilitaristica' [Euthanasia in utilitarian bioethics], *Medicina e Morale* 43, 1993, pp. 331–361).

22 Hugo T. Engelhardt Jr., *The Foundations of Bioethics*, New York: Oxford University Press 1986, p. 111.

23 *Ibid.*

24 Edward A. Langerak, 'Abortion: Listening to the Middle', *Hastings Center Report* 9, no. 5, 1979, pp. 24–28, p. 25.

25 *Ibid.*

26 *Ibid.*

27 John T, Noonan, 'An Almost Absolute Value in History', in J. Feinberg (ed.), *op.cit.*, p. 12 and p. 13.

28 Carol A. Tauer, 'Personhood and Human Embryos and Fetuses', *Journal of Medicine and Philosophy* 10, 1985, pp. 253–266, p. 264; see also Peter Singer and Karen Dawson, 'IVF Technology and the Argument from Potential', *Philosophy and Public Affairs* 17, 1988, pp. 87–104.

29 See Alasdair MacIntyre, *Three Rival Versions of Moral Inquiry: Encyclopaedia, Genealogy and Tradition*, Notre Dame: University of Notre Dame Press, 1990.

30 Aristotle, *Metaphysica*, 1049[b]. It is noticeable that also the Catholic Magisterium, speaking of the uncertainty with reference to the moment of rational animation, seems to recognize the possibility of such a development by affirming that even 'supposing a later animation, there is still nothing less than a *human* life, preparing for and calling for a soul in which *the nature received from parents is completed*' (*Congregation for the Doctrine of the Faith, Declaration on Procured Abortion*, Vatican Polyglott Press, 1974, endnote 19, emphasis added).

31 It is important to make explicit reference to Aristotle's masterpiece, since also those who largely refer to Aristotle in today's bioethical discussion normally quote almost exclusively from the biological work *The Generation of Animals* (see also the important and well-documented book by Norman Ford, *op. cit.*, pp. 19–64). This is not to deny the obvious importance of *The Generation of Animals* in order to understand Aristotle's view on the matter, but to stress the philosophical relevance of the question of potentiality.

32 Aristotle *Metaphysica*, 1049[a].

33 Francis C. Wade, 'Potentiality in the Abortion Discussion', *Review of Metaphysics* 29, 1975, pp. 239–255, p. 243.

34 *Ibid.*

35 As distinguished from the active specifiable potentialities already referred to.

36 Wade, *op. cit.*, p. 244.

37 Aristotle, *De Generatione Animalium*, trans, by A. Platt in J.A. Smith and W.D. Ross (ed.), *The Works of Aristotle Translated into English*, vol. 5, Oxford: Clarendon Press, 1912, 778b.

38 Enrico Berti, 'Quando esiste l'uomo in potenza. La tesi di Aristotele' [On when a potential man exists: Aristotle's thesis], in S. Biolo (ed.), *Nascita e morte dell'uomo*, Genoa, Italy: Marietti, 1992, pp. 115–123.

39 John F. Crosby, 'The Personhood of the Human Embryo', *Journal of Medicine and Philosophy* 18, 1991, pp. 399–417, p. 406.

40 Sacred Heart Catholic University's Centre for Bioethics, 'Identity and Status of the Human Embryo', *Medicina e Morale* 39, no. 4 supplement, 1989, p. 20. In Aristotle's words: 'everything that comes to be moves towards a principle, i.e. an end (for that for the sake of which a thing is, is its principle, and the becoming is for the sake of the end), and the actuality is the end, and it is for the sake of this that the potency is acquired' (*Metaphysica*, 1050a).

41 The natural selection of embryos before implantation is probably better interpreted as the absence of the potentiality for full personhood, since this selection is in most part due to severe chromosomal defects. This is most clear in the case of embryos developing into hydatidiform moles: these embryos are predetermined to develop only into accessory tissues and lack the potential for rational acts: in this sense, as noted by Germain Grisez, they are not new human beings, for this entails having in oneself 'the epigenetic primordia of a human body normal enough to be the organic basis of at least some intellectual act' ('When Do People Begin?', *Proceeding of the American Catholic Philosophical Association* 63, 1990, pp. 27–47, p. 39).

42 Wade, *op. cit.*, p. 255.

Chapter 3

John P. Lizza

POTENTIALITY AND HUMAN EMBRYOS

In calling for a four-year moratorium on research on human embryos, the US President's Council on Bioethics called for greater public debate on the critical issue of the moral status of the embryo. The majority of the members of the Council held that the developing embryo was a being worthy of 'special respect' and claimed that those who deny the potentiality of the embryo to become a person lack an understanding of the meaning of potentiality. The majority stated that to treat the developing human embryo as nothing more than 'mere cells': misunderstands the meaning of potentiality – and, specifically, the difference between a 'being-on-the-way' (such as a developing human embryo) and a 'pile of raw materials,' which has no definite potential and which might become anything at all. The suggestion that extracorporeal embryos are not yet individual human organisms-on-the-way, but rather special human cells that acquire only through implantation the potential to become individual human organisms-on-the-way, rests on a misunderstanding of the meaning and significance of potentiality. An embryo is, by definition and by its nature, potentially a fully developed human person; its potential for maturation is a characteristic it *actually* has, and from the start. The fact that embryos

have been created outside their natural environment – which is to say, outside the woman's body – and are therefore limited in their ability to realize their natural capacities, does not affect either the potential or the moral status of the beings themselves. A bird forced to live in a cage its entire life may never learn to fly. But this does not mean that it is less of a bird, or that it lacks the imma-nent potentiality to fly on feathered wings. It means only that a caged bird – like an in vitro human embryo – has been deprived of its proper environment. There may, of course, be good human reasons to create embryos outside their natural environments – most obviously, to aid infertile couples. But doing so does not obliterate the moral status of the embryos themselves.[1]

Although the Council never provides a more detailed and developed explanation of potentiality, it is clear that the majority holds that potentiality is not affected by external conditions that may prevent the potential from ever being realized and that what potentialities something has is true by definition of the kind of thing that it is. In this paper, I propose an alternative account of potentiality that chal-lenges these two claims. I start with what I believe is a common ordinary assumption about potentiality, i.e. if X is potentially Y, then it must be possible for X to be Y. Building on Edward Covey's analysis of the sense of possibility implied in such a notion of potentiality, I argue that whether an embryo has the potential to be a person is not determined by definition. Instead, it is an empirical matter involving an assessment of the actual physical conditions that may restrict the embryo's possibilities. In this view, some human embryos may lack the potential to become a person that others have. Assuming for the sake of argument that the potential to be a person gives a being special moral status, it would follow that some human embryos lack this status.

My strategy in this paper is thus to distinguish potentialities that are realistic and ethically relevant from those that are not on the basis of empirical probability. While I do not propose a fine line of *how much* probability is necessary for some potentiality to be realistic and ethically relevant, this does not preclude recognition of the distinction. The distinction, I believe, explains why the concept of potentiality is not useless. Indeed, if we do not take into account the probabilities of whether a potentiality may be realized, our notion of potentiality would be 'too promiscuous' to be of any use in our ethical deliberations. This is the problem with the concept of potentiality invoked by the Council. Insofar as I am rejecting the idea that an embryo's potentiality for personhood is true by definition of the kind of thing it is, my account of potentiality is revisionist or at least proposes an alternative to the account of potentiality relied on by the President's Council. However, since my case is grounded on the common assumption that if X is potentially Y, then it must be possible for X to be Y, I am, in part, trying to elucidate what I believe is part of the meaning of our ordinary concept of potentiality. Moreover, consideration of empirical conditions that restrict what is physically possible should lead us to distinguish

degrees of potentiality among human embryos and therewith their moral status.

One of the problems with invoking any notion of potentiality is that it initially appears to be, as Joel Feinberg observes, too 'promiscuous' to be of much use.[2] Feinberg asks us to consider the potential of a pile of dehydrated orange juice. He states that the powder is potentially orange juice (just add water). However, it is potentially many other things as well. All that we have to do is adjust the conditions. For example, since we could add water and arsenic, it is potentially poison. Since we could add flour, eggs, and yeast, and then bake, it is potentially orange cake. Feinberg correctly observes that this sense of 'potentiality' is so 'promiscuous' as to be practically useless. Aristotle also seems to recognize this problem when he observes that any matter is potentially anything.[3]

In the context of the abortion debate, this problem takes the form of what David Annis has called 'the extension argument'.[4] The extension argument is a *reductio ad absurdum* criticism of the claim that the human embryo has the potential for personhood in a morally relevant sense. Critics note that an embryo requires many external factors, such as implantation in a hospitable womb and nutrition, in order for it to develop intellect, will, and other characteristics of personhood. However, the same could be said about the sperm and egg before they are conjoined. Given the appropriate conditions, these entities have the potential to develop into persons and so we would have to accord them the same moral status as the embryo. Moreover, if every cell in our body could be used for reproductive cloning, every cell would then have the potential for personhood. If we wish to deny the potential for personhood to gametes and these other cells, we must deny it to the embryo. In all these cases, external factors must be added or assumed in order to attribute the potential for personhood to the entity.

To address the extension argument and begin to formulate a useful notion of potentiality, Edward Covey observes that it is generally agreed that 'X is potentially a person' is to be understood as 'it is possible that X become a person, where "possible" is taken in the sense of "physically possible" rather than "logically possible".'[5] Covey contrasts physical possibility with logical possibility and analyzes the former in terms of nomic regularities, i.e. 'an event or state of affairs is nomically possible just in case its coming about is in accordance with the laws of nature, given the initial state of affairs which actually obtains in the world'.[6] To illustrate the distinction, Covey cites the example from Nicholas Rescher that it is logically possible or conceivable that an acorn could develop into a tree that produces pears, but it is not nomically possible that it can.

Covey, however, goes on to draw another distinction within the category of nomic possibility, which qualifies his treatment of the example cited from Rescher. He observes that sometimes we make conjectures about what might be nomically possible but which fall short of the lawlessness of logical possibility. For example, although it is absolutely nomically impossible for two masses to be near one another without being affected by gravity, as this would be inconsistent with the most basic physical laws in the actual world, it is not absolutely nomically impossible for an acorn to grow into a tree that produces pears. While respecting the laws of nature, it is possible that people could exploit them by developing techniques of genetic alteration, which reprogram acorns with pear genes.[7] However, Covey observes that given our existing state of knowledge about genetics, it is actually nomically impossible for a particular acorn starting to germinate at this time to produce pears. In these cases, consideration of certain absolute physical possibilities, e.g. that we could manipulate the genes of the acorn to get pears, fail to generate actual possibilities about, e.g. currently existing acorns. In other words, the possibility of a possibility does not yield an actual possibility.

Covey uses this distinction to generate two principles that he believes are helpful to understanding whether a possibility is actual, given that certain intervening steps are required for that possibility to be actualized. The first principle is:

1 Where the transition from S1 to S2 is possible only given an intermediate condition S3, and where S3 is merely absolutely possible, but not actually possible, the (absolute) possibility of S3's obtaining, together with the (absolute) possibility of S2's obtaining if S3 obtains, will not yield the (actual) possibility of S2's obtaining . . .[8]

This principle is a rule that excludes attributing a potentiality to X, rather than a rule that tells us when such a potentiality exists. For example, it is not absolutely nomically impossible that an acorn could grow into a tree that produced pears. However, given the actual conditions in the world that there is no way to change the genetic structure of the acorn, it is actually impossible that a given acorn that germinates today could grow into a tree that produced pears. We are thus justified in denying that the acorn has the potential to grow into a tree that produces pears.

Covey claims that this principle applies when our own decisions do not form part of the conditions having the role of S3. He claims that a second principle applies when human choice is involved:

2 Where human choice and rational agency is involved, it is never correct for a person or group to consider their own choices of possible actions as part of the given background

conditions which would imply that certain outcomes are physically impossible.[9]

Covey justifies this second principle by appeal to the notion of free agency, which he claims entails that 'the dispositional states which may determine our actual choices never count as criteria of physical necessity in our calculations of what we can and cannot do.'[10] To illustrate this principle, he offers the following example:

Melissa, who has removed Stan's carburetor, might say, 'It is impossible for him to drive his car today because I've decided to keep the carburetor off his engine until tomorrow, even though if I reinstall it today he will be able to drive the car, and I can (but I won't) reinstall it.' This has the apparent form of a judgment that *possibly possibly p*, but *not possibly p*. In ordinary discourse this is fine, but the more precise way for Melissa to characterize the situation would be to say that he could, indeed, drive the car today, but *will* not because she chooses one of the particular possibilities open to her. Stan, though, may rightly judge that given her actual decision, *he* is *personally* powerless to bring about the desired result (though he may not be – he might be able to achieve the result through persuasion or some mechanical work of his own).[11]

I will ultimately reject Covey's second principle and argue that decisional factors may be part of the conditions that determine actual possibilities. However, I accept his first principle and shall show how this principle applies to the evaluation of whether an embryo is potentially a person. Consider, for example, an abnormal human embryo that lacks the genes to develop a brain or heart. Also, assume that we have no way of altering its genetic structure to allow for such development. Such an embryo would not be potentially a person, since attributing such a potential to it would entail relying on absolute possibilities, rather than actual ones. Such absolutely possible conditions, however, do not generate actual, physical possibilities, according to Covey's first principle. Moreover, because the abnormal embryo lacks the actual possibility of becoming a person, it lacks the potential for being a person. Finally, if the potential to be a person were what garners an embryo special moral status, then it would follow that this abnormal embryo would fail to have this status.

The potentiality of a human embryo to develop intellect and will may also be restricted by physical factors external to the embryo that cannot be rectified in any realistic way. For example, an embryo may have the requisite genes for further development, but it may be in an environment that will prevent those genes from directing that development, e.g. the embryo may be situated in a uterus that is so scarred that implantation is physically impossible. If it is only absolutely possible but not actually possible to change the external conditions, then the embryo would lack the actual possibility and hence the potential to be a person.[12]

However, suppose the technology, e.g. uterine flushing, is readily available to harvest an embryo from a woman with a uterus that is unable to sustain the embryo, but she refuses to undergo the procedure. Does her decision imply that the embryo lacks the actual possibility and hence the potential to become a person? At this point, we need to consider Covey's second principle and his example involving Melissa who has removed the carburetor from Stan's car. I disagree with Covey's analysis of this particular case and do not think that it supports his general principle (2). To see this, let us assume that the reinstallation of the carburetor is a necessary condition of Stan's starting the car. Nothing else he or anyone else could do would enable him to start the car. This enables us to better isolate the role that Melissa's decision to keep the carburetor has on the possibility of Stan's starting his car today. What makes Covey's interpretation initially plausible is that we can easily imagine Melissa changing her mind and the carburetor being subsequently installed. However, we can also easily imagine cases in which Melissa's decision has led to actions that make it impossible for Stan to drive his car today. For example, if Stan is in New York and Melissa is on a plane to China with the carburetor, then it is not actually possible for Stan to drive his car today. We can speculate about how it is absolutely possible for the carburetor to be reinstalled in time to enable him to drive the car today. For example, the commercial airliner could turn around in mid-flight and by some sequence of events that do not violate basic physical laws, the carburetor could be reinstalled in time for Stan to drive the car. However, if the likelihood of this sequence of events happening is about as likely as someone discovering tomorrow how to alter the genes of an oak tree so that it produces pears, then we should conclude that it is not actually possible for Stan to drive his car today.

Thus, Covey's example is not adequately described. To know whether a decision makes some event physically impossible, we need to consider other conditions that may obtain in conjunction with the decision. Indeed, the actual decisions may lead to conditions that cannot be changed within a timeframe that would allow for the physical possibility of a certain outcome. In addition, we need to consider Melissa's resolve of her intention not to return the carburetor. If the possibility of her changing her mind is absolutely possible but not actually possible, then it may not be actually possible for Stan to drive his car today.

To hold that choices or decisions can never be part of the background conditions that restrict what is physically possible is to deny the causal

efficacy of those choices and decisions. In addition, once decisions are made that put realistic restrictions on possible outcomes, the possibility of those outcomes occurring is dependent on the possibility of those choices being changed or overridden. If there is no reason to think that the decisions will be changed or overridden, then there is no reason to think that the outcomes restricted by those choices are actually possible.

These observations are relevant to evaluating the potential of certain embryos. For example, contraceptive measures such as an intrauterine device (IUD) or the 'morning after pill' may work by preventing the fertilized ovum from implanting in the womb. Indeed, these measures may be in place before the fertilization of the ovum. In these cases, there may be some possibility of the embryo developing, as these methods of intervention are not 100 percent effective. However, the embryo's potential for development will be much more remote than that of an embryo in an environment in which these measures are not present. In terms of potentiality, the potentiality of these embryos is similar to that of embryos that are in women who, because of some physical condition such as significant scarring of the uterine wall, are unable to sustain the life of the embryo. Unless it is realistically possible to correct the physical conditions before the embryo is expelled from the womb, this embryo lacks the actual possibility of further development. However, the embryo does have an absolute possibility of developing, since we can speculate about conditions that could change to allow for the development. Similarly, unless it is realistically possible that the physical conditions resulting from the implantation of the IUD or the ingestion of the oral contraceptive can be reversed in a timeframe that would allow the embryo to develop, the embryo lacks the actual possibility of further development. We could speculate on the possibility of the woman, for example, deciding to remove the IUD or consenting to have her uterus flushed to remove the embryo and implant it in another womb. However, if the possibility of her removing the IUD or flushing the uterus in the necessary timeframe is as unrealistic as speculating about the possibility of discovering some way to surgically intervene to correct an abnormal womb, then the potential of the embryos in these situations is similar. Since potentiality is determined by the actual conditions in the world, it is irrelevant whether those conditions are set by physical conditions beyond our control or by physical conditions within our control. It may be as improbable to move a mind as it is to move a mountain.

Recognition of the role that decisions may have in fixing conditions in the world that in turn restrict what is realistically possible implies that in many cases there may not be a value-neutral way of determining potentiality. For example, one could claim that an embryo in a woman with an abnormal uterus that cannot sustain an embryo has the potential to be a person, because we can imagine conditions that would make it physically possible for the embryo to develop. For example, we could compel the woman to undergo regular testing for the presence of an embryo, harvest the embryo, and implant it in a surrogate mother, regardless of whether we had consent from the biological or surrogate mother. However, if the biological mother or surrogate has refused this type of intervention and we respect and legally enforce such refusals, then the claim that this embryo is potentially a person would entail invoking absolute possibilities, rather than actual possibilities. Thus, unless one is willing to challenge the moral and legal rationale for allowing the mother and surrogate choices in this matter, there is no basis for attributing the potential to be a person to the embryo. Since their decisions not to undergo the procedures are incompatible with the embryo becoming a person, the

embryo lacks the possibility and therefore the potentiality to be a person. Thus, the determination of whether a potentiality exists must be done in a context that assesses the actual causal factors, including decisions and laws that are effective in that context. If the context were altered so that the women consented to the procedures or if the laws were different, the actual possibilities and potentiality for the embryo would change.

This view supports Gene Outka's suggestion that it would be permissible to experiment on 'spare' frozen embryos that are destined to be destroyed.[13] Outka attempted to distinguish using embryos created for the purpose of research and using embryos that were originally created for reproduction but that were no longer needed and would be destroyed. He tried to ground this moral distinction by appealing to the intention behind the acts. However, members of the President's Council criticized his position on grounds that spare embryos still have the potential for intellect and will and therefore deserve special respect. The intention behind their creation was viewed as irrelevant to their moral standing. However, it is important to recognize that whether these embryos have this potential depends on whether one accepts the legitimacy of parental discretion in determining their fate. If that legitimacy is respected, then the potential of those embryos destined to be destroyed is different than that of other frozen or living embryos. Failure to recognize this difference would be to ignore the actual conditions in the world that restrict which possibilities are realistic and ethically relevant. However, this is not to say that the members of the Council were necessarily wrong in their claim about the potentiality of these embryos. Rather, it is to simply point out that it would be inconsistent to recognize the legitimacy of treatment for infertility that leads to the creation of extra embryos whose disposition rests with the parents and hold that all of the embryos created by this process have the potential to develop intellect and will. The claim that these 'spare' embryos have the potential for intellect and will stands or falls with acceptance or rejection of the procedures that lead to their creation. If there are good moral and legal reasons for allowing spare embryos to be created to treat infertility and allowing parental discretion of their fate, then there are good moral and legal reasons for denying the potentiality for intellect and will to embryos that parents have decided to destroy. The objection that it would be wrong to experiment on these embryos because their potentiality garners them special respect would thus be removed.

If the logic of my argument in this paper is correct, rejection of the conclusion would have to take the form of rejecting one of its premises. Before closing, I wish to consider one reason for rejecting the premise that a necessary condition for X being potentially Y is that it is possible that X become Y.[14] Consider the following statements that appear to violate this condition but in some sense are undeniably true:

(1) Marcia made Law Review at Harvard and clearly had the potential to become a brilliant corporate attorney, but that was not possible for her because she had to care for her mother throughout a long-term illness.

(2) Charles had a terrific voice and had the potential to sing leading roles at the Metropolitan Opera. However, Charles was African-American and lived at a time when African-Americans were not allowed to sing at the Metropolitan Opera, so such an achievement was not possible for him.

Claims like (1) and (2) do not appear to be self-contradictory yet they appear to violate the premise that potentiality assumes possibility. Such retrospective claims, however, are not counterexamples. Consider when Marcia lost the potential to become a brilliant corporate attorney and why she lost that potential. Presumably, her decision to care for her mother led to conditions that made it no longer realistically possible for her to become a corporate attorney. Until that point was reached, Marcia did indeed have the potential to become a brilliant attorney, but after that point, she no longer had the potential. The retrospective claim should not lead us to think that at any point Marcia had the potential but not the possibility to become a brilliant attorney. (2) is more complicated. Consider the narrower claim (3), uttered when Charles was at the height of his singing ability:

(3) Charles has the potential but no possibility to sing leading roles at the Metropolitan Opera.

(3) might be used to distinguish Charles from, say, Jane, who is also African-American but cannot sing well at all. What is meant by (3) is that Charles has the ability and talent to sing at the Met but has no realistic opportunity to do so. However, ascribing to Charles the 'potential' to sing at the Met is made in the context of assuming counterfactually that the barrier preventing him from doing so is or should be removed. It would make no practical sense to ascribe such a 'potential' to someone under the assumption that the barrier should not be removed. For example, suppose Charles was not a US citizen. One would not ascribe to Charles the potential to vote in an imminent US election, since there are laws prohibiting non-US citizens from voting and presumably those laws should not be changed. If one thought that such laws should be changed, one might then sensibly ascribe the potential to vote to Charles. Indeed, one might refer to Charles's intelligence, age, and other characteristics commonly assumed as qualifications for voting as reasons for thinking

that, since people like Charles have the potential to vote in US elections, the voting laws in the US should be changed. If the ascription of these potentials makes sense only under the assumption that barriers that effectively prevent the possibility of these potentials from being realized should be removed, then statements like (2) and (3) do not challenge the assumption that potentiality presupposes a realistic possibility for the potential to be realized. Indeed, whatever sense there is to attributing to Charles the potential to sing at the Met stands or falls with whatever sense there is to thinking that the barriers that make it currently impossible for him to do so should be removed.

Notes

1   President's Council on Bioethics. 2002, *Human Cloning and Human Dignity: An Ethical Inquiry.* Washington, D.C. U. S. Government Printing: 175.

2   Joel Feinberg. 1974. The Rights of Animals and Unborn Generations (Appendix: The Paradoxes of Potentiality). In *Philosophy and Environmental Crisis.* W. T. Blackstone ed. Athens, GA: U of Georgia Press: 67–68.

3   Aristotle. *Metaphysics* IX 7, 1048b35–1049bl.

4   David B. Annis. Abortion and the Potentiality Principle. *South J Philos* 1984; 22: 155–163.

5   Edward Covey. Physical Possibility and Potentiality in Ethics. *Am Philos Q* 1991; 28(3): 237.

6   Ibid.

7   Covey, *op. cit.* note 2, 240.

8   Ibid.

9   Ibid.

10   Ibid: 241.

11   Ibid.

12   In some cases, we may have the knowledge and technology to save the embryo. For example, we could flush the scarred uterus, harvest the embryo, and implant it into a surrogate. However, the fact that we simply have the knowledge and technology to do this does not imply that all such embryos have the potential for further development. The knowledge and technology would have to be available, not simply hypothetically, but in a realistic way and timeframe. If it were not, the embryo would lack the realistic potential for further development. Consider, for example, the difference in realistic potential between an embryo in a woman with a damaged uterus who is about to undergo uterine flushing and embryo harvesting at a reputable clinic and an embryo in a woman with the same uterine condition but who is so far from medical care that it is practically impossible to save the embryo. Consider also the difference in potentiality between a frozen embryo preserved in a facility with power and one preserved in a facility that has lost its power and where there is no possibility of restoring the power to save the embryo. In these cases, features of the embryo's physical environment and external factors combine to limit the potentiality of the embryo. To claim that the embryos in these imperiled circumstances still have the potential to develop would be to ignore the actual conditions that realistically restrict their possibilities.

13   Gene Outka. Transcript of Testimony before President's Council on Bioethics, April 25, 2002. Available at: www.bioethics.gov/transcripts/apr02/apr25session3.html [Accessed June 10 2004].

14   I thank a referee of this journal for raising this objection and formulating the counterexamples (1) and (2) that follow.

Chapter 4

John Harris

STEM CELLS, SEX, AND PROCREATION

The ethics of stem cell research

Stem cell research is of ethical significance for three major reasons:

1   It will for the foreseeable future involve the use and sacrifice of human embryos.
2   Because of the regenerative properties of stem cells, stem cell therapy may always be more than therapeutic – it may involve the enhancement of human functioning and indeed the extension of the human lifespan.
3   So-called therapeutic cloning – the use of cell nuclear replacement to make the stem cells clones of the genome of their intended recipient – involves the creation of cloned pluripotent and possibly totipotent cells, which some people find objectionable.

Elsewhere I have discussed in detail the ethics of genetic enhancement[41] and the ethics of cloning,[42] and I noted above the immortalizing potential of stem cell research. In this essay, I concentrate on objections to the use of embryos and fetuses as sources of stem cells.

Given that, currently, the most promising source of stem cells for research and therapeutic purposes is either aborted fetuses or preimplantation embryos, their recovery and use for current practical purposes seems to turn crucially on the moral status of the embryo and the fetus. A number of recent indications are showing promise for the recovery and use of adult stem cells. It was reported recently that Catherine Verfaillie and her group at the University of Minnesota had successfully isolated adult stem cells from bone marrow and that these seemed to have pluripotent properties (i.e., capable of development in many ways but not in all, and not capable of becoming a new separate creature) like most HES cells.[43] Simultaneously, Nature Online published a paper from Ron McKay at NIH showing the promise of embryo derived cells in the treatment of Parkinson's disease.[44]

This indicates the importance of pursuing both lines of research in parallel. The dangers of abjuring embryo research in the hope that adult stem cells will be found to do the job adequately is highly dangerous and problematic for a number of reasons. The first is that we do not yet know whether adult cells will prove as good as embryonic cells for therapeutic purposes. At the moment there is simply much more accumulated data and much more therapeutic promise from human embryonic stem cells. The second is that it might turn out that adult cells will be good for some therapeutic purposes and human embryonic stem cells for others. Third, we already know that it is possible to modify or replace virtually any gene in human embryonic stem cells; whether this will also be true of adult stem cells has yet to be established. Finally, it would be an irresponsible gamble with human lives to back one source of cells rather than another and to make people wait and possibly die while what is still the less favored source of stem cells is further developed. This means that the ethics of HESC research is still a vital and pressing problem and cannot for the foreseeable future be bypassed by concentration on adult stem cells.

Stem cells from early embryos

It is possible to remove cells from early preimplantation embryos without damage to the original embryo. This may be one solution to the problem of obtaining embryonic stem cells. However, if the cells removed are totipotent (i.e., capable of becoming literally any part of the creature, including the whole creature), and if moreover they are capable of dividing until the cell mass achieves sufficient cells for autonomy (i.e., the ability to implant successfully and continue to grow to maturity),[45] then they are in effect separate zygotes, they are themselves "embryos," and so they must be protected to whatever extent embryos are protected. If however, such cells are merely pluripotent, then they could not be regarded as embryos and the use of them would, presumably, not offend those who regard the embryo as sacrosanct. Unfortunately, it is not at present possible to tell in advance whether a particular cell is totipotent or simply pluripotent. This can only be discovered for sure retrospectively by observing the cell's capabilities.

I will now set out one ethical principle that I believe must be added to the central principles cited in guiding our approach to HESC research

and raise two issues of the consistency of attitudes and judgments about stem cell research with other practices and treatments used and considered acceptable (albeit with qualifications) not only in the European Union but indeed in the world at large. The two issues of consistency are:

1   consistency of stem cell research with what is regarded as acceptable and ethical with respect to normal sexual reproduction
2   consistency with attitudes to and moral beliefs about abortion and assisted reproduction.

The ethical principle that I believe we all share and that applies to the use of embryos in stem cell research is the Principle of Waste Avoidance.

The Principle of Waste Avoidance

This widely shared principle assumes that it is right to benefit people, if we can, and wrong to harm them, and it states that, faced with the opportunity to use resources for a beneficial purpose when the alternative is that those resources are wasted, we have powerful moral reasons to avoid waste and do good instead. I will start with consideration of the first requirement of consistency.

Lessons from sexual reproduction. Let us start with the free and completely unfettered liberty to establish a pregnancy by sexual reproduction without any "medical" assistance. What are people and societies who accept this free and unfettered liberty committing themselves to? What has a God who has ordained natural procreation committed herself to?

We now know that for every successful pregnancy that results in a live birth many, perhaps as many as five,[46] early embryos will be lost or will "miscarry" (although these are not perhaps "miscarriages" as the term is normally used because this sort of very early embryo loss is almost always entirely unnoticed). Many of these embryos will be lost because of genetic abnormalities, but some would have been viable. Many people believe that the fact that perhaps a large proportion of these embryos are not viable somehow makes their sacrifice irrelevant. But those who believe that the embryo is morally important do not usually believe that this importance applies only to healthy embryos. Those who accept the moral importance of the embryo would be no more justified in discounting the lives of unhealthy embryos than those who accept the moral importance of adult humans would be in discounting the lives of the sick or of persons with disability.

How are we to think of the decision to attempt to have a child in the light of these facts? One obvious and inescapable conclusion is that God and/or nature has ordained that "spare" embryos be produced for almost every pregnancy and that most of these will have to die in order that a sibling embryo can come to birth. Thus, the sacrifice of embryos seems to be an inescapable and inevitable part of the process of procreation. It may not be intentional sacrifice, and it may not attend every pregnancy, but the loss of many embryos is the inevitable consequence of the vast majority (perhaps all) pregnancies. For everyone who knows the facts, it is a conscious, knowing, and therefore deliberate sacrifice; and for everyone, regardless of "guilty" knowledge, it is part of the true description of what they do in having or attempting to have children.

We may conclude that the production of spare embryos, some of which will be sacrificed, is not unique to assisted reproduction technologies (ART); it is an inevitable (and presumably acceptable, or at least tolerable?) part of all reproduction.

Both natural procreation and ART involve a process in which embryos, additional to those

that will actually become children, are created only to die. I will continue to call these "spare" embryos in each case. If either of these processes is justified it is because the objective of producing a live healthy child is judged worth this particular cost. The intentions of the actors, appealed to in the frequently deployed but fallacious doctrine of double effect,[47] are not relevant here. What matters is what the agents knowingly and voluntarily bring about. That this is true can be seen by considering the following example.

Suppose we discovered that the use of mobile telephones within 50 meters of a pregnant woman resulted in a high probability, near certainty, of early miscarriage. No one would suggest that, once this is known, it would be legitimate to continue use of mobile telephones in such circumstances on the grounds that phone owners did not intend to cause miscarriages. Any claim by phone users that they were merely intent on causing a public nuisance or, less probably, making telephonic communication with another person and therefore not responsible for the miscarriages would be rightly dismissed. It might, of course, be the case that we would decide that mobile communications were so important that the price of early miscarriage and the consequent sacrifice of embryos was a price well worth paying for the freedom to use mobile telephones. And this is, presumably, what we feel about the importance of establishing pregnancies and having children. Mobile telephone users, of course, usually have an alternative method of communication available, but let us suppose they do not.

This example shows the incoherence of the so-called doctrine of double effect. The motives or primary purposes of the phone user are clearly irrelevant to the issue of their responsibility for the consequences of their actions. They are responsible for what they knowingly bring about. The only remaining question is whether, given the moral importance of what they are trying to achieve (phoning their friends), the consequent miscarriages are a price it is morally justifiable to exact to achieve that end. Here the answer is clearly "no." Sometimes proponents of the doctrine of double effect attempt to make proportionality central to the argument. It is not, so it is claimed, the fact that causing miscarriage is not the primary or first intention or effect that matters but the fact that miscarriage is a serious wrong compared with the benefit of using a mobile telephone. However, this is to miss the point of the doctrine of double effect. Proportionality cannot be the issue because the doctrine of double effect was designed to exculpate people from the wrong of intending a forbidden act. The proportionality of the various outcomes cannot speak to the issue of primary or second effects. Only the true account of what the agents wanted to achieve or were "trying to do," of what the main intention or purpose actually was or is, can do that.

However, when we pose the same question about the moral acceptability of sacrificing embryos in pursuance of establishing a successful pregnancy, the answer seems different. My point is that the same issues arise when considering the use of embryos to obtain embryonic stem cells. Given the possible therapeutic uses I have reviewed, it would be difficult, I suggest, to regard such uses as other than morally highly significant. Given that decisions to attempt to have children using sexual reproduction as the method (or even decisions to have unprotected intercourse) inevitably create embryos that must die, those who believe having children or even running the risk of conception is legitimate cannot consistently object to the creation of embryos for comparably important moral reasons. The only remaining question is whether the use of human embryonic stem cells for therapies designed to save lives and ameliorate suffering are purposes of moral importance comparable to those of attempting to have (or

risking the conception of) children by sexual reproduction.

The conscious voluntary production of embryos for research, not as the by-product of attempts (assisted or not) at reproduction, is a marginally different case, although some will think the differences important. However, if the analysis so far is correct, then this case is analogous in that it involves the production and destruction of embryos for an important moral purpose. All that remains is to decide what sorts of moral objectives are comparable in importance to that of producing a child. Although some would defend such a position,[48] it would seem more than a little perverse to imagine that saving an existing life could rank lower in moral importance to creating a new life. Assisted reproduction is, for example, given relatively low priority in the provision of healthcare services. Equally, saving a life that will exist in the future seems morally comparable to creating a future life. In either case, the moral quality and importance of the actions and decisions involved and of their consequences seem comparable.

Instrumentalization. It is important to note that prolife advocates or Catholics are necessarily acting instrumentally when they attempt to procreate. They are treating the 1–4 embryos that must be sacrificed in natural reproduction as a conscious (though not intended) means to have a live birth. This is something Catholics certainly and probably most others who hold a "prolife" position should not do.

However, the issue is not whether Catholics or those who take a prolife position may or may not be permitted to create embryos, which certainly or highly probably will die prematurely, and whether this constitutes reckless endangerment of embryos or even the unjustifiable killing of embryos. Rather, the facts of life, the facts of natural reproduction, show that the creation and destruction of embryos is something that all those who indulge in unprotected intercourse and certainly all those who have children are engaged in. It is not something that only those who use assisted reproduction or those who accept experimentation on embryos are "guilty" of. It is a practice in which we are all, if not willing, at least consenting participants, and it shows that a certain reverence for or preciousness about embryos is misplaced.

Embryo-sparing ART. It might be said that there is a difference – those who engage in assisted reproduction engage in the destruction of embryos at a greater rate than need be. Those who engage in sex are not engaged in the destruction of embryos at a greater rate than is required for the outcome they seek. It would be interesting to know whether, if a creating a single embryo by IVF became a reliable technique, prolife supporters would feel obliged to use this method rather than sexual reproduction because of its embryo-sparing advantages. It looks as though there would indeed be a strong moral obligation to abandon natural procreation and use only embryo-sparing ART.

Consider a fictional IVF scenario. A woman has two fertilized eggs and is told it is certain that if she implants both only one will survive but that if she implants only one it will not survive. Would she be wrong to implant two embryos to ensure a successful singleton pregnancy? This example is, of course, fictional only in terms of the degree of certainty supposed. It is good practice in IVF to implant two or three embryos in the hope of achieving the successful birth of one child. Thus, in normal IVF as in normal sexual reproduction, the creation and "sacrifice" of embryos in pursuit of a live child is not only accepted as necessary but is part of the chosen means for achieving the objective. Most people would, I believe, judge this to be permissible, and indeed it is what often happens in successful IVF pregnancies, where up to three embryos are implanted in the hope of one live birth. Even in Germany, where stem cell research

using embryos is currently banned and where legal protections for the embryo are enshrined in the constitution, IVF is permitted, and it is usual to implant three embryos in the hope and expectation of achieving no more than a single live birth.

Even if we could accurately predict in advance which embryos would survive and which would not, the ethics would not change. Suppose that for some biological reason there was a condition that required that, for one embryo to implant, it was necessary to introduce a companion embryo that would not, and we could tell in advance which would be which. It is difficult to imagine how or why this fact would alter the ethics of the procedure; it would remain the case that one must die in order that the other would survive. If people in this condition wanted ART, would we judge it unethical to provide it to them but not to "normal" IVF candidates when the "costs" were the same in each case – namely, the loss of one embryo in pursuit of a healthy birth?

It might be objected that the parallel with sexual reproduction is like saying that, because we know that road traffic causes thousands of deaths per year, to drive a car is to accept that the sacrifice of thousands of lives in almost every country, for example, is a price worth paying for the institution of motor transport. This might seem a telling analogy showing that we do not willingly accept the inevitable consequences of what we do. There are, of course, many disanalogous features of the purported reductio ad absurdum comparison with road deaths. The vast majority of drivers will go all their lives without having an injury-causing incident, let alone a fatality, and the probability of any individual causing a death once exacerbating factors such as alcohol use and reckless fatigue are taken into account is vanishingly small by any standards and insignificant when compared with the high risk of production of embryos in unprotected sex between fertile partners. However, suppose an individual knew that, despite a long driving career without accidents, today is the day that either they will surely be involved in a fatal accident and cause someone's death or that the probability of this happening is very high indeed. Would it be conceivable that it might be permissible, let alone ethical, to drive today? And yet that is the situation with normal sexual intercourse, at least for those who regard the embryo as protected.

The natural is not connected to the moral. It is important to be clear about the form of this argument. I am not, of course, suggesting that because something happens in nature it must be morally permissible for humans to choose to do it. I am not suggesting that, because embryos are produced only to die in natural procreation, that the killing of embryos must be morally sound. I am saying, rather, that if something happens in nature and we find it acceptable in nature given all the circumstances of the case, then if the circumstances are relevantly similar it will for the same reasons be morally permissible to achieve the same result as a consequence of deliberate human choice. I am saying that we do as a matter of fact and of sound moral judgment accept the sacrifice of embryos in natural reproduction, because although we might rather not have to sacrifice embryos to achieve a live healthy birth, we judge it to be defensible to continue natural reproduction in the light of the balance between the moral costs and the benefits. And if we make this calculation in the case of normal sexual reproduction we should, for the same reasons, make a similar judgment in the case of the sacrifice of embryos in stem cell research.

To take a different but analogous case: if we say that God and/or nature "approves" of cloning by cell division because of the high rate of natural monozygotic twinning[49] and that therefore the duplication of the human genome is not per se unethical we are not saying that cloning by cell division is ethically unproblematic because it occurs naturally. The point of the analogy is rather that, because the birth of natural identical twins is generally not considered regrettable, we are reminding ourselves that there is nothing here to regret. Indeed, it is the occasion for unmitigated joy or at least moral neutrality. We should, therefore, unless we can find a difference, feel the same about choosing deliberately to create twins by duplication of the human genome.[50] If we then object to cloning by a different method, cell nuclear transfer objections must obviously be to features that arise uniquely in cell nuclear transfer and cannot simply be to such features as duplication of the human genome. Our acceptance of the natural does not, of course, apply to naturally occurring premature death; here we do think there is something to regret, even if it is natural and inevitable.

Instrumentalization revisited. Another possible concern involves a version of the instrumentalization objection that demands that embryos not be produced only to be used for the benefit of others but that, as in sexual reproduction, they should all have some chance of benefiting from a full normal lifespan.[51] In normal sexual reproduction, embryos must be created only to die so that a sibling embryo can come to birth. But, arguably, it is in each embryo's interest that reproduction continues because it is the embryo's only chance to be the one that survives. Embryos (if they had rationality) would have a rational motive to participate (albeit passively) in sexual reproduction. By contrast, so it might be claimed, embryos produced specifically for research would not rationally choose to participate for they stand to gain nothing. All research embryos will die, and none have a chance of survival. If this argument is persuasive against the production of research embryos, it is easily answered by ensuring that the embryos produced for research have to some appropriate extent a real chance of survival. One would simply have to produce more embryos than are required for research, randomize allocation to research, and ensure that the remainder are implanted with a chance to become persons. To ensure that it would be in every embryo's interest to be "a research embryo," all research protocols permitting the production of research embryos would have to produce extra embryos for implantation. To take a figure at random but one that, as it happens, mirrors natural reproduction and gives a real chance of survival to all embryos, we could ensure that for every, say, 100 embryos needed for research another 10 would be produced for implantation. The 100 embryos would be randomized 90 for research, 10 for implantation, and all would have a chance of survival and an interest in the maintenance of a process that gave them this chance.

The third case concerns spare embryos that become available for research as a result of an ART program in which they have been produced and to which they are now superfluous because their "mother" has now declined for whatever reason to accept more embryos for implantation and has refused consent for their implantation into others. Here it might be suggested that these embryos are also like the research embryos just considered. However, this is not the case. These embryos have had their chance of implantation, but unfortunately for them, they have missed out. The fact that now they are irredeemably surplus to requirements for implantation does not show that they always were. These embryos have had their chance of life, their "motive" for participating in the program is as strong as in sexual reproduction or randomized research embryos.

Born to die. The force of the sexual reproduction analogy may seem vulnerable to the following claim.[52] It can be said that, just as parents are responsible for the deaths of the embryos inevitably produced as a consequence of

unprotected intercourse, so also and to the same extent are they responsible for the deaths of the children they actually produce when these children eventually die of old age. This is because in each case the parents have produced a life, which will end at a particular point and that point is in each case out of the parents' control. So, if parents are responsible for the deaths of the embryos lost as a result of unprotected intercourse, they are also responsible for the deaths of their children lost in old age. In neither case, however, have the parents been the proximate cause of death, but they have caused the life and death cycle. This objection, like the objection from the acceptability of motorized road transport, purports to constitute a reductio ad absurdum.

This is a puzzling objection. As I have argued, people accept the necessity of and the justification for producing surplus embryos because they wish to have a baby. Those who judge the embryo to have moral importance comparable to adults or children will have to justify their instrumentalization of the embryos that are sacrificed to this end.

On the other hand, those who think that dying of old age or being given a worthwhile life is a good will see nothing to justify. The parents are responsible for that life, to be sure, but they are morally justified in that responsibility, and in that the life for which they are responsible has been or is reasonably likely to be a worthwhile life, then, unless they have also arranged the death, their responsibilities have been exercised in a way that is both morally and socially appropriate.

The life of their child was in this case neither created nor ended to be a means to the interests of others. It is a good life, the creation of which requires no justification and the end of which was neither caused by the parents nor was its timing predictable by them. They therefore have no excuses to make. By contrast, the lives of the embryos that must die early are, if those lives are morally important at all, not lives the ending of which is a reasonable price to pay for the life lived.

The United States condemns human reproduction! Shocked by the idea of any activity that threatens the embryo, the U.S. government has adopted the revolutionary strategy of attempting to condemn human reproduction and, for good measure, has included all unprotected intercourse in the condemnation and to ban all federal support for such activities.

How have our cousins in the United States arrived at this daring and groundbreaking social policy? In the United States, current federal law prohibits the use of federal funds for "the creation of a human embryo" explicitly for research purposes or, more crucially, for "research in which a human embryo or embryos are destroyed, discarded or knowingly subjected to the risk of injury or death." [53] Such law is presumably animated by concern about the morally problematic nature of such actions and also by the idea that federal support in the form (among others) of "tax dollars" should not be given to activities that a significant number of people find offensive or objectionable. As I have noted, normal sexual reproduction inevitably involves a process in which a human embryo or embryos are destroyed or discarded. It is also incontrovertibly an activity in which a human embryo or embryos are "knowingly subjected to the risk of injury or death," at least for anyone who knows the facts of life. The perpetuation of this position seems likely, as President George W. Bush had made an election promise never to provide federal support for research that involves living human embryos. Those who can read his lips may have less confidence that this promise will be kept.

Consistency with attitudes to and moral beliefs about abortion and assisted reproduction. In most countries of the European Union and indeed in most countries of the world, abortion is permissible under some circumstances. Usually,

permissibility is considered greater at very early stages of pregnancy, permissibility waning with embryonic and fetal development. The most commonly accepted ground for abortion (where it is acceptable) is to protect the life and the health of the mother. Sometimes the idea of protection of the life and health of the mother is very broadly and liberally interpreted, as it is in the United Kingdom; sometimes the requirement is very strict, demanding real and present danger to the life and health of the mother (Northern Ireland, for example). Given that the therapies initially posited for stem cell research – the treatment of Parkinson's disease and the development of tailor-made transplant organs – are all for serious diseases that threaten life and dramatically compromise health, it is difficult to see how those who think the sacrifice of early embryos for these purposes is or could be justified could find principled objections to the use of embryos in other lifesaving therapies. [54]

The same is, of course, true, as I have already noted of ART. All IVF involves the creation of spare embryos, and all IVF now practiced is built on research done on many thousands of embryos. Most countries and most religions accept IVF and its benefits and in doing so accept that spare embryos will be produced only to die. Even Germany, which has, as I noted, an Embryo Protection Act, accepts the practice of implanting up to three embryos in the hope and expectation that at least one will survive. The acceptance of the practice of IVF is necessarily an acceptance that embryos may be created and destroyed for a suitably important moral purpose.

The principle of waste avoidance. As I stated previously, this widely shared principle states that it is right to benefit people if we can and wrong to harm them, and that, faced with the opportunity to use resources for a beneficial purpose when the alternative is that those resources are wasted, we have powerful moral reasons to avoid waste and do good instead.

That it is surely better to do something good than to do nothing good should be reemphasized. It is difficult to find arguments in support of the idea that it could be better (more ethical) to allow embryonic or fetal material to go to waste than to use it for some good purpose. It must, logically, be better to do something good than to do nothing good; it must be better to make good use of something than to allow it to be wasted. It must surely be more ethical to help people than to help no one. This principle – that, other things being equal, it is better to do some good than no good – implies that tissue and cells from aborted fetuses should be available for beneficial purposes in the same way that it is ethical to use organs and tissue from cadavers in transplantation.

It does not follow, though, that it is ethical to create embryos specifically for the purposes of deriving stem cells from them. However, as I discussed, there may be problems in objecting to creating embryos for this purpose from people who do not object to the sacrifice of embryos in pursuit of another supposedly beneficial objective – namely, the creation of a new human being. Only those who think that it is more important to create new humans than to save existing ones will be attracted to the idea that sexual reproduction is permissible whereas the creation of embryos for therapy is not.

Notes

This paper draws on, but also answers objections to, my earlier paper "The ethical use of human embryonic stem cells in research and therapy" published in: Burley JC, Harris J, eds. *A Companion to Genethics: Philosophy and the Genetic Revolution*. Oxford: Blackwell; 2001. I am indebted to my colleague Louise Irving for the data on social policy in Europe and to Julian Savulescu for many stimulating conversations and exchanges. Work on this paper was supported by a project grant from the European Commission for EUROSTEM under its

Quality of Life and Management of Living Resources Programme, 2002.

[. . .]

41 See: Harris J. *Wonderwoman and Superman: The Ethics of Human Biotechnology*. New York: Oxford University Press; 1992; see also Harris, J. *Clones, Genes, and Immortality*. New York: Oxford University Press; 1998.

42 See note 19, Harris 1997, Harris 1998, Harris 1999.

43 International Herald Tribune 2002 Jun 22–23:1. In a paper presented at FENS Forum Workshop, Paris, 13 Jul 2002, Austin Smith emphasized the importance of pursuing research on all sources of stem cells simultaneously.

44 McKay R. 'Stem cells – hype and hope'. *Nature* 2000;406:361–4.

45 Pedersen R. Embryonic steps towards stem cell medicine. Paper presented at the EUROSTEM conference Regulation and Legislation under Conditions of Scientific Uncertainty. 2002 Mar 6–9; Bilbao, Spain. The conference was supported by a Project Grant from The European Commission Directorate-General for Research, "Quality of Life."

46 Robert Winston gave the figure of five embryos for every live birth some years ago in a personal communication. Anecdotal evidence to me from a number of sources confirms this high figure, but the literature is rather more conservative, making more probable a figure of three embryos lost for every live birth. See: Boklage CE. Survival probability of human conceptions from fertilization to term. *International Journal of Fertility* 1990;35(2)75–94. See also: Leridon H. *Human Fertility: The Basic Components*. Chicago: University of Chicago Press; 1977. Again, in a recent personal communication, Henri Leridon confirmed that a figure of three lost embryos for every live birth is a reasonable conservative figure.

47 For a conclusive refutation of that doctrine, see: Harris J. *Violence and Responsibility*. London: Routledge and Kegan Paul; 1980. For a more recent discussion of these broad issues, see: Kamm FH. The doctrine of triple effect and why a rational agent need not intend the means to his end. *Proceedings of the Aristotelian Society* (Supplementary volume) 2000;74:21–39;

and Harris J. The moral difference between throwing a trolley at a person and a throwing a person at a trolley: a reply to Francis Kamm. *Proceedings of the Aristotelian Society* (Supplementary volume) 2000;S74:40–57.

48 Some hedonistic utilitarians, for example.

49 Human monozygotic twinning occurs in roughly one per 270 births (three per 1,000). I take this figure of the rate of natural twinning from: Moore KL, Persaud TVN. *The Developing Human*, 5th ed. Philadelphia: W. B. Saunders; 1993.

50 Or indeed cloning by cell nuclear substitution, but that is another story. For the full story, see note 18, Harris 1997.

51 The possible objection was put to me by Julian Savulescu – the response to it with all its defects is mine.

52 A point made to me by Louis G Aldrich at the Third International Conference of Bioethics, National Central University, Shungli, Taiwan. 2002 Jun 24–29.

53 U.S. Public Law 105-277, sect. 511, 1998 Oct 21, slip copy. H.R. 4328.

54 See Harris J. 'Should we experiment on embryos?' In: Lee R, Morgan D, eds. *Birthrights: Law and Ethics at the Beginnings of Life*. London: Routledge; 1988: 85–95.

On what grounds should we select and enhance our offspring?

INTRODUCTION TO PART TWO

BIOMEDICAL INNOVATIONS HAVE increased our control of reproductive processes. First, we have means to select offspring: for example, pre-implantation genetic diagnosis (PGD) involves testing the genetic make-up of *in vitro* embryos before deciding which to implant in the womb; and prenatal testing provides information on the developing foetus for parents to use in deciding whether to continue with a pregnancy. Second, although not currently practicable, genetic enhancement of our offspring has been made feasible by advances in genetic science. On what grounds should we exercise these powers of selection and enhancement?

An established rationale for selection decisions is to avoid disease and disability. Parens and Asch object to this in the extract from their disability rights critique of prenatal testing for genetic disorders. Keenly aware that 'prospective parents will use positive prenatal test results primarily as the basis of a decision to abort foetuses that carry mutations associated with disease and/or disability', Parens and Asch present two rejoinders. First, prenatal testing is morally problematic: to select out disabilities expresses and thereby reinforces a hostile attitude to the disabled (the 'expressivist argument'); and testing evinces an inappropriate attitude towards parenting (the 'parental attitude argument'). Second, prenatal testing is supported by the failure – based on the tendency to allow a person's disability to 'stand in' for the whole of them – to appreciate that being disabled does not preclude living a good life, and parenting a disabled child can be rewarding and enriching.

By contrast, Savulescu defends what he calls the principle of procreative beneficence: 'couples (or single reproducers) should select the child, of the possible children they could have, who is expected to have the best life, or at least as good a life as the others, based on the relevant, available information'. Savulescu modulates this principle in two ways: 'persuasion is justified, but not coercion'; and the principle 'must be balanced against others'. Hence, procreative beneficence should inform all selection decisions and can justify persuasion but 'ultimately, we should allow couples to make their own decisions about which child to have'. Nonetheless, the principle of procreative beneficence supports selection decisions aimed at avoiding disease and disability, and promoting desirable non-disease traits.

Avoiding disease and disability is also one rationale for selecting offspring on the basis of their sex, i.e. to avoid X-linked diseases by ensuring a female child. But there are non-disease related motives for choosing the sex of one's child, such as to balance up one's family. By focusing on the technique of flow cytometric separation of X- and Y-bearing sperm, Robertson avoids the ethically problematic features of other methods of ensuring the sex of a child – such as abortion – thereby bringing the ethics of gender selection *per se* into focus. He considers various motives to sex select, including gender variety and first-born gender preference, focusing on the charge that all gender selection is inevitably sexist. Circumspectly, he concludes that the strength of the 'claim of a right to choose offspring gender' varies between cases, and more evidence on the harmful effects of gender selection is required.

Moving from selection to enhancement, familiar ways of making a child as accomplished as possible involve the behaviour of the mother when pregnant (taking folic acid and avoiding smoking and alcohol, for example), formal and informal education, lifestyle choices concerning diet and exercise, and promoting hobbies and good habits. But the bioethics of enhancement focuses on something more radical, namely attempts to improve the human species by means of genetic manipulation. Genetic enhancement avoids the genetic lottery of natural reproduction by designing offspring according to a wish-list of desirable non-disease traits such as athleticism, intelligence and longevity. Some distinctions help to focus the topic. First, gene therapies have what is sometimes called the 'normalising intent' of helping victims of genetic diseases to achieve normal functioning; by contrast, the intention behind genetic enhancement is to raise people above normal or 'species-typical' functioning by making them more intelligent, athletic, beautiful, and so on. Second, whilst it is feasible to enhance individuals, the main bioethical worries are about germ-line genetic modifications because these would be inherited by future generations.

Sandel's piece addresses the puzzle that genetic enhancement is morally troubling despite the weakness of the obvious objections based on autonomy (i.e., genetically engineered people would not be fully free) and fairness (i.e., the genetically enhanced would enjoy an unfair advantage over others). Sandel's reliance on metaphors – the 'deeper danger' of enhancement, he suggests, is in its 'Promethean aspiration to remake nature', 'the drive to mastery' and the failure to 'acknowledge the giftedness of life' – lends his paper a fanciful air. But his case is coherent: genetic enhancement is morally troubling because it represents a wilful determination to achieve control and mastery over, as opposed to quiescent acceptance of, the given. Sandel insists that this is morally troubling even outside of a religious standpoint because it 'would transform three key features of our moral landscape: humility, responsibility, and solidarity'.

Arguably, Sandel's response to genetic enhancement equivocates between presenting the failure to 'acknowledge the giftedness of life' as intrinsically wrong (because it epitomises 'dominion over reverence', for example), and as wrong because of its consequences (such as the erosion of solidarity). By contrast, Bostrom is univocal. He dismisses 'customary injunctions against playing God, messing with nature, tampering with our human essence, or displaying punishable hubris' in order to pursue a consequentialist approach. As a 'transhumanist', Bostrom takes 'human nature as a work-in-progress, a half-baked beginning that we can learn to remold in desirable ways', including in ways that would usher in 'post-humans', i.e. 'beings with vastly greater capacities than present human beings'. Despite this robust approach, Bostrom is circumspect in acknowledging that not all modifications would be for the best, and he is as concerned as more conservative bioethicists to avoid dystopia. For example, Bostrom considers whether worries about the consequences of modifying the germ-line are well placed, concluding that opponents have not yet met the challenge of showing that 'the balance of reason' tilts in their favour.

Selections

Parens, E. and Asch, A. 1999. 'The Disability Rights Critique of Prenatal Testing: Reflections and Recommendations', *Hastings Center Report*, 29 (5) (Supplement), S1–S16.

Savulescu, J. 2001. 'Procreative Beneficence: Why we Should Select the Best Children', *Bioethics*, 15 (5&6), 413–26.

Robertson, J.A. 2001. 'Preconception Gender Selection', *American Journal of Bioethics*, 1 (1), 2–9.

Sandel, M.J. 2004. 'The Case Against Perfection', *The Atlantic Monthly*, 293 (3), 51–62.

Bostrom, N. 2003. 'Human Genetic Enhancements: A Transhumanist Perspective', *The Journal of Value Inquiry* 37 (4), 493–506.

Chapter 5

E. Parens and A. Asch

THE DISABILITY RIGHTS CRITIQUE OF PRENATAL GENETIC TESTING: REFLECTIONS AND RECOMMENDATIONS

The international project to sequence the human genome was undertaken in the expectation that knowing the sequence will offer new ways to understand and treat disease and disability. If researchers can identify the sequences of genes that code for the body's building blocks, then, it is hoped, they can identify and correct the sequences associated with disease and disability.

So far, researchers have enjoyed only minimal success in using gene therapy to correct such conditions, and no researcher has yet even attempted to use gene therapy to correct genetic impairments in a fetus. Rather, the discovery of abnormal or incorrect sequences has led primarily to the development of genetic tests that can reveal whether a person, embryo, or (in the usual case) a fetus carries an abnormality or "mutation" associated with disease or disability. It is now possible to test for gene mutations associated with some 400 conditions, from those universally viewed as severe, such as Tay Sachs, to those that many might describe as relatively minor, such as Polydactyly (a trait involving an extra little finger). The number and variety of conditions for which tests are available grows almost daily.[1]

Today we test for one trait at a time. In the future, however, with advances in biochip technology, it will be possible to test simultaneously for as many traits as one would like. In principle, we will be able to test for any trait we wish that has been associated with any given allele. Not only will the cost of such testing likely decrease as the diagnostic technology advances, but advances in the technology will make it possible to do the testing earlier in the pregnancy. One such technology will isolate the very small number of fetal cells that circulate in the maternal blood. Insofar as these earlier tests will be performed on fetal cells obtained from the mother's blood (rather than from the amniotic sac or chorionic villi) they will be noninvasive. Thus it will be possible to do many more tests, at once, and with less cost to the pregnant woman in time, inconvenience, risk, or dollars, than is now the case.[2]

As the ease of testing increases, so does the perception within both the medical and broader communities that prenatal testing is a logical extension of good prenatal care: the idea is that prenatal testing helps prospective parents have healthy babies. On the one hand, this perception is quite reasonable. Though no researcher has yet even attempted to correct a genetic impairment with in-utero gene therapy, increasingly there are nongenetic approaches to such impairments. At the time of this writing, more than fifty fetuses have undergone in-utero surgery to repair neural tube impairments (myleomeningoceles).[3] Moreover, negative (or reassuring) prenatal test results will reduce the anxiety felt by many prospective parents, and this in itself can be construed as part of good prenatal care. On the other hand, as long as in-utero interventions remain relatively rare, and as long as the number of people seeking prenatal genetic information to prepare for the birth of a child with a disability remains small, prospective parents will use positive prenatal test results primarily as the basis of a decision to abort fetuses that carry mutations associated with disease and/or disability. Thus there is a sense in which prenatal testing is not simply a logical extension of the idea of good prenatal care.

Logical extension or no, using prenatal tests to prevent the birth of babies with disabilities seems to be self-evidently good to many people. Even if the testing will not help bring a healthy baby to term this time, it gives prospective parents a chance to try again to conceive. To others, however, prenatal testing looks rather different. If one thinks for even a moment about the history of our society's treatment of people with disabilities, it is not difficult to appreciate why people identified with the disability rights movement might regard such testing as dangerous. For the members of this movement, including people with and without disabilities and both issue-focused and disability-focused groups, living with disabling traits need not be detrimental either to an individual's prospects of leading a worthwhile life, or to the families in which they grow up, or to society at large. Although the movement has no one position on prenatal diagnosis, many adherents of the disability rights movement believe that public support for prenatal diagnosis and abortion based on disability contravenes the movement's basic philosophy and goals. Critics contend that:

(1) Continuing, persistent, pervasive discrimination constitutes the major problem of having a disability for people themselves and for their families and communities. Rather than improving the medical or social situation of today's or tomorrow's disabled citizens, prenatal diagnosis reinforces the medical model that disability itself, not societal discrimination against people with disabilities, is the problem to be solved.

(2) In rejecting an otherwise desired child because they believe that the child's disability will diminish their parental experience, parents suggest that they are unwilling to accept any significant departure from the parental dreams that a child's characteristics might occasion.

(3) When prospective parents select against a fetus because of predicted disability, they are making an unfortunate, often misinformed decision that a disabled child will not fulfill what most people seek in child rearing, namely, "to give ourselves to a new being who starts out with the best we can give, and who will enrich us, gladden others, contribute to the world, and make us proud."[4]

This document, the product of two years of discussions by a diverse group drawn from within and outside the disability rights movement, reshuffles what is contained in these criticisms and discerns in them two broad claims: simply put, that prenatal genetic testing followed by selective abortion is morally problematic and that it is driven by misinformation. The document elaborates and evaluates these two claims, turns to explore the prospects for distinguishing between acceptable and unacceptable testing, and draws out of the ongoing debate that it seeks to focus – not to put to rest – recommendations to guide professional providers of genetic testing through this difficult terrain.

Understanding and evaluating the disability rights critique

Prenatal testing is morally problematic

The disability critique holds that selective abortion after prenatal diagnosis is morally problematic, and for two reasons. First, selective abortion expresses negative or discriminatory attitudes not merely about a disabling trait, but about those who carry it. Second, it signals an intolerance of diversity not merely in the society but in the family, and ultimately it could harm parental attitudes toward children.

The expressivist argument

The argument that selective abortion expresses discriminatory attitudes has been called the expressivist argument.[5] Its central claim is that prenatal tests to select against disabling traits express a hurtful attitude about and send a hurtful message to people who live with those same traits. In the late 1980s, Adrienne Asch put the concern this way: "Do not disparage the lives of existing and future disabled people by trying to screen for and prevent the birth of babies with their characteristics."[6] More recently, she has clarified what the hurtful or disparaging message is:

> As with discrimination more generally, with prenatal diagnosis, a single trait stands in for the whole, the trait obliterates the whole. With both discrimination and prenatal diagnosis, nobody finds out about the rest. The tests send the message that there's no need to find out about the rest.[7]

Indeed, many people with disabilities, who daily experience being seen past because of some single trait they bear, worry that prenatal testing repeats and reinforces that same tendency toward letting the part stand in for the whole.

Prenatal testing seems to be more of the discriminatory same: a single trait stands in for the whole (potential) person. Knowledge of the single trait is enough to warrant the abortion of an otherwise wanted fetus. On Asch's more recent formulation, the test sends the hurtful message that people are reducible to a single, perceived-to-be-undesirable trait.

This observation about letting the part stand in for the whole is surely enormously important. In everyday life, traits do often stand in for the whole, people do get looked past because of them. Indeed, one form of the expressivist argument has been regarded rather highly in another context. Many people who are concerned to support women's rights, have argued that prenatal sex selection is morally problematic because it embodies and reinforces discriminatory attitudes toward women.[8] The sex trait is allowed to obliterate the whole, as if the parents were saying, "We don't want to find out about 'the rest' of this fetus; we don't want a girl."

Marsha Saxton has put the expressivist argument this way:

> The message at the heart of widespread selective abortion on the basis of prenatal diagnosis is the greatest insult: some of us are "too flawed" in our very DNA to exist; we are unworthy of being born. . . . fighting for this issue, our right and worthiness to be born, is the fundamental challenge to disability oppression; it underpins our most basic claim to justice and equality – we are indeed worthy of being born, worth the help and expense, and we know it![9]

And as Nancy Press has argued, by developing and offering tests to detect some characteristics and not others, the professional community is expressing the view that some characteristics, but not all, warrant the attention of prospective parents.[10]

For several reasons, however, there is disagreement about the merit of the expressivist argument as a basis for any public policy regarding prenatal diagnosis of disability. Individual women and families have a host of motives and reasons for seeking out genetic information, and as James Lindemann Nelson and Eva Feder Kittay argue, it is impossible to conclude just what "message" is being sent by any one decision to obtain prenatal testing.[11] Acts (and the messages they convey) rarely have either a single motivation or meaning.

Some prospective parents no doubt have wholly negative attitudes toward what they imagine a life with a disability would be like for them and their child; others may believe that life could be rich for the child, but suspect that their own lives would be compromised. Others who have disabilities perhaps see passing on their disabling trait as passing on a part of life that for them has been negative. Parents of one child with a disability may believe that they don't have the emotional or financial resources for another. The point is that the meaning of prenatal testing for would-be parents is not clear or singular. In any case, those sympathetic to at least some forms of prenatal testing point out that prospective parents do not decide about testing to hurt existing disabled people but to implement their own familial goals. In that sense, there is no "message" being sent at all.

To many in the disability rights movement, however, regardless of the parental motive to avoid the birth of a child who will have a disability, the parent may still be letting a part stand in for the whole. That prospective parents do not intend to send a hurtful message does not speak to the fact that many people with disabilities receive such a message and are pained by it.

A second criticism of the expressivist argument is that it calls into question the morality of virtually all abortions. The argument presumes that we can distinguish between aborting "any" fetus and a "particular" fetus that has a disability – what Adrienne Asch has called the any-particular distinction. According to Asch, most abortions reflect a decision not to bring any fetus to term at this time; selective abortions involve a decision not to bring this particular fetus to term because of its traits. Prochoice individuals within and outside the disability community agree that it is morally defensible for a woman to decide, for example, that she doesn't want any child at a given time because she thinks she's too young to mother well, or because it would thwart her life plan, or because she has all the children she wants to raise. The question is whether that decision is morally different from a decision to abort an otherwise-wanted fetus.

But it is not clear that the distinction is adequate. Sometimes the decision to abort "any" fetus can be recast as a decision to abort a "particular" fetus. James Lindemann Nelson, for example, argues that if parents of three children chose to end a pregnancy that would have produced a fourth child, such parents would not be making a statement about the worthwhileness of other families with four children, or about the worth of fourth-born children as human beings.[12] Rather, they would be deciding what would be right for their particular situation. If, as Asch and others have argued, prenatal testing is morally suspect because it lets a trait stand in for the whole potential person, precisely the same argument would apply to aborting a fetus because it was the fourth child. The trait of being fourth-born makes the prospective parents ignore every other respect in which that fetus could become a child that would be a blessing to its family and community. Nelson's example of the potential fourth-born child suggests one reason to doubt the merit of the any-particular distinction; he thinks that the disability critics have failed to explain why traits like being fourth-born could be a legitimate basis for an abortion while disabling traits could not.

A third criticism of the expressivist argument is that it presumes that selective abortion based on prenatal testing is morally problematic in a way that other means of preventing disability are not. Such other means include, for example, taking folic acid to reduce the likelihood of spina bifida, or eschewing medication that is known to stunt the growth, or harm the organs or limbs of a developing fetus. Such acts (or refraining from such acts) on the part of the pregnant woman are designed to protect the health of the developing fetus.

Disability critics hold, however, that abortion does not protect the developing fetus from anything. It prevents disability by simply killing the fetus. Proponents of this disability critique hold a strong prochoice position. Their objection is only to a certain way of using abortion.

But those from the mainstream prochoice community think of selective abortion in different terms. They do not see an important moral difference between selective abortion and other modes of preventing disability in large part because they do see an important moral distinction between a born child with a disabling trait and an embryo or fetus with a disabling trait. They argue that parents of all born children have an obligation to love and care for those children – regardless of their traits. They also argue, however, that the pregnant woman (and her partner) are not "parents" before the child is born. Just as a woman or couple may decide during the first two trimesters of any pregnancy that becoming a parent to a first child, or to any child, is not in accord with their life plans, so may they make the same decision on the grounds that the fetus has disabling traits. The woman may terminate the pregnancy and try again to become pregnant with a fetus that has not been identified as carrying a disabling trait. On this view, if it is reasonable to prevent disability in a developing child by adhering to a particular lifestyle, taking specified medications

or refraining from taking others, it is equally acceptable to opt for abortion to prevent the birth of a child with a significant disability.[13]

Even if expressivist arguments will not dissuade all people from using tests in making reproductive decisions for their own lives, policies that would in any way penalize those who continue pregnancies in spite of knowing that their child will live with a disabling trait must be avoided. Those prospective parents who either forgo prenatal testing or decide that they want to continue a pregnancy despite the detection of a disabling trait should not have to contend with losing medical services or benefits for their child, nor feel obliged to justify their decisions. Further, the availability of prenatal testing in no way reduces our societal obligations to those people who are born with or acquire disabilities. Even if prenatal diagnosis says nothing to or about existing or future disabled people, we should as a society vigorously enforce antidiscrimination laws and improve services and supports for disabled people and their families.

The parental attitude argument

The second argument that prenatal testing is morally problematic we call the *parental attitude argument*. According to it, using prenatal tests to select against some traits indicates a problematic conception of and attitude toward parenthood. Part of the argument is that prenatal testing is rooted in a "fantasy and fallacy" that "parents can guarantee or create perfection" for their children.[14] If parents were to understand what they really should seek in parenting, then they would see how relatively unimportant are the particular traits of their children.

The parental attitude argument also involves the thought that in the context of prenatal testing, a part, a disability, stands in for the whole, a person. The prospective parent who wants to avoid raising a child with a diagnosable disability forgets that along with the disabling trait come other traits, many of which are likely to be as enjoyable, pride-giving, positive (and as problematic, annoying, and complicated) as any other child's traits. If prospective parents imagine that disability precludes everything else that could be wonderful about the child, they are likely acting on misinformation and stereotype. The prospective parent has made biology destiny in the way that critics of the medical model of disability consistently resist.

According to the parental attitude argument, prospective parents should keep in mind that the disabling trait is only one of a fetus's characteristics. The activity of appreciating and nurturing the particular child one has is what the critics of selection view as the essence of good parenting. Loving and nurturing a child entails appreciating, enjoying, and developing as best one can the characteristics of the child one has, not turning the child into someone she is not or lamenting what she is not. If we were to notice that it is a fantasy and fallacy to think that parents can guarantee or create perfection for their child, if we were to recognize what is really important about the experience of parenting, we would see that we should be concerned with certain attitudes toward parenting, not with "disabling" traits in our children. Good parents will care about raising whatever child they receive and about the relationship they will develop, not about the traits the child bears. In short, what bothers those wary of prenatal diagnosis is what might be called "the selective mentality." The attention to particular traits indicates a morally troubling conception of parenthood, a preoccupation with what is trivial and an ignorance of what is profound.

Those who connect acceptance of disability to what is desirable in any parent-child relationship will worry that our attitudes toward parenthood and ultimately toward each other are changing as a result of technologies like prenatal diagnosis.[15] Do these technologies lead us, one might ask, toward the commodification of children, toward thinking about them and treating them as products rather than as "gifts" or "ends in themselves"? Is it making us as a society less resilient in the face of the inevitable risks that our children face, and less willing to acknowledge the essential fragility of our species? When members of our society are confronted with, for example, sex selection or with the possibility of selecting for non-health-related traits like sexual orientation, concerns about the selective mentality come quickly to our lips. Indeed, those who want to reject the parental attitude argument in the context of disabling traits must recognize that they are criticizing an argument that they themselves may well want to use in the context of non-health-related traits. Certainly many worry about the cumulative effect of individual choices, about the technologization of reproduction, and about a decreasing cultural ability or willingness to accept the reality of uncontrollable events. These concerns trouble even those who profess to be comfortable with genetic testing and selective abortion.

Nonetheless, many find significant problems with the parental attitude argument. One of the most important is that it makes what William Ruddick calls the "maternalist assumption," namely, that "a woman who wants a child should want any child she gets."[16] Ruddick acknowledges that many women do hold "maternalist" conceptions of pregnancy and motherhood, out of which that assumption grows. But he points out that there are other legitimate conceptions of pregnancy and motherhood that do not depend on or give rise to the same assumption. He suggests that some prospective parents may legitimately adopt a "projectivist" or "familial" conception of parenthood, and that either of these views is compatible with trying to assure that any child they raise has characteristics that accord with these parental goals. In the projectivist parent's understanding of child rearing, the child is a part of her parental projects, and, within limits, parents may legitimately undertake to ensure that a child starts out with the requisites for fulfilling these parental hopes and aims. Ruddick is not claiming that projectivist parents could ignore a child's manifested commitments to things beyond the parents' life plans, but he is saying that, for example, the parent passionate about music may legitimately select against a future child whose deafness would make a love of some forms of music impossible. If a hearing child turns out to be tone deaf and enthusiastic about rock collecting and bird watching but not music, and if the parent views these activities as inimical to her parental values or projects, she need not support them, or (within limits) allow other people to do so.

According to Ruddick, the "familial" conception of parenthood highlights a parent's vision of her child as herself a parent, sibling – a participant in a nuclear and extended family that gives central meaning to life. For example, parents whose dreams of child rearing include envisioning their own child as a parent would be acting consistently with their conception of parenthood if they decided not to raise a boy with cystic fibrosis, whose sterility and shortened life span might preclude either biological or adoptive parenthood. A child of such a parent might, of course, reject family life in favor of solitude or communal adult companionship, but in using available technology to avoid raising a child who would never be able to fulfill a deeply cherished parental dream, the parent is acting in accordance with a legitimate conception of parenthood.

Although Ruddick is not alone in thinking that a selective mentality may be compatible with praiseworthy parenting, many share the disability community's worry that prenatal testing threatens our attitudes toward children,

parenthood, and ultimately ourselves. Certainly, it would be to the good if we would think more deeply about our attitudes. If we want to be parents, why do we want to be parents? What do we hope it will bring for our children-to-be and for ourselves? And prospective parents would benefit from grappling with those questions in the context of prenatal diagnosis. However, such concerns could not undergird specific policies regarding prenatal testing for disabling traits.

Prenatal testing is based on misinformation

The second major claim of the disability critique is that prenatal testing depends on a misunderstanding of what life with disability is like for children with disabilities and their families. Connected with this claim is the question whether disability is one more form of "neutral" human variation, or whether it is different from variations usually thought of as nondisabling traits, such as eye color, skin color, or musicality.

There are many widely accepted beliefs about what life with disability is like for children and their families. Most of these beliefs are not based on data. They include assumptions that people with disabilities lead lives of relentless agony and frustration and that most marriages break up under the strain of having a child with a disability. Recent studies suggest, for example, that many members of the health professions view childhood disability as predominantly negative for children and their families, in contrast to what research on the life satisfaction of people with disabilities and their families has actually shown.[17] One strand of this project, then, involved wrestling with what to make of conflicting perceptions about how people with disabilities and their families experience life. Three disability researchers in the Hastings Center group – Philip Ferguson, Alan Gartner, and Dorothy Lipsky – analyzed empirical data on the impact of children with disabilities on families.[18] Their review, surprising to many, concludes that the adaptational profiles of families that have a child with a disability basically resemble those of families that do not.

According to Ferguson, Gartner, and Lipsky's reading of the data, families that include disabled children fare on average no better or worse than families in general. Some families founder, others flourish. Ferguson, Gartner, and Lipsky do not deny that families are often distressed upon first learning that their child has a disability. And they acknowledge that families with children who evince significantly challenging behavior experience more disruption than do other families. But recent research on raising a child with a disability offers happier news for families than many in our society have been led to expect. In the words of one leading family researcher, "The most recent literature suggests that families of children with handicaps [sic] exhibit variability comparable to the general population with respect to important outcomes such as parent stress, . . . family functioning, . . . and marital satisfaction."[19] Studies of family adaptation have begun to recognize the prevalence of positive outcomes in many families.[20] Indeed, one recent study found that parents of disabled adolescents reported more positive perceptions of their children than do parents of nondisabled adolescents.[21]

In a 1995 study intended to learn how a child's disability affected the work lives of dual career families, the authors found that the needs and concerns of families with and without children with disabilities were "strikingly similar." They did, however, observe:

What seems to distinguish families of children with disabilities from other working families is the intensity and complexity of the arrangements required to balance work and home responsibilities successfully. For example, parents of children with disabilities, particularly those with serious medical or behavioral problems, find it more difficult to locate appropriate, affordable child care. . . . Similarly, these families are more dependent upon health insurance policies with comprehensive coverage.[22]

This same study reminds us of a point that both Ruddick and Kittay made: a child's disability may sometimes alter the customary parent-child life cycle, in which parents gradually relinquish daily guidance and caretaking and – if they are fortunate – see their children take on adult productive and caretaking roles. Depending on the impairment and on the social arrangements that parents help a growing child construct, some people with disabilities may require their parents' help through adulthood in securing shelter, social support, and safety. Increasingly, adults with disabilities such as muscular dystrophy, spina bifida, cystic fibrosis, Down syndrome, and other conditions do not stay "eternal children," as they were once thought to do. Nonetheless, some, albeit small, portion of the population of disabled people will be more vulnerable for longer than others, and more in need of what Kittay (borrowing from Sara Ruddick) described as "attentive, protective love."[23]

While it is important to demolish the myth that disability entails relentless agony for the child and family, there is still considerable disagreement about what conclusions to draw from the literature on the family impact of a child with disability. In the view of the disability community, this literature suggests that prenatal testing to select against disabling traits is misguided in the sense that it is based on misinformation. That is, if prospective parents could see that families with children who have disabilities fare much better than the myth would have it, then parents would be less enthusiastic about the technology.

However, recognizing that there are erroneous beliefs that need to be dispelled may not show that the desire for prenatal testing stems from misinformation alone. The first problem with the argument from misinformation has to do with the difference between retrospective and prospective judgments. It is one thing to look back on a stressful but ultimately rewarding experience and say, I'm glad I did that. It is another to look forward to the possibility of a stressful and perhaps ultimately rewarding experience and say, I'm glad to give it a try. To appreciate that many families respond well to stress does not commit one to thinking that it would be a mistake for families to try to avoid it. It may be true that, as one of the studies of working families points out, the concerns of working parents with disabled children very much resemble the concerns of any working parent – ensuring that children are safe, happy, stimulated, and well cared for at home, at school, and in after-school activities. But that study also acknowledges that working parents of children with special medical or behavioral needs find that meeting those needs takes more time, ingenuity, and energy than they think would have to be spent on the needs of nondisabled children. To appreciate that many families emerge stronger, wiser, and even better as a result of such an experience may not suggest that it is unreasonable or morally problematic to try to avert it. As Mary Ann Baily put it, child rearing is already like mountain climbing. That I want to climb Mount Rainier doesn't commit me to wanting to climb Everest. I appreciate that the rewards of climbing Everest might be extraordinary, beyond my wildest dreams, but I'd settle for Rainier.[24]

The disability researchers and theorists did not persuade everyone in the project group that raising a child with a disability is not more demanding than raising a child without this condition. As a specific type of life challenge, raising a child who has a disability may provide one individual of a particular aptitude or orientation with a life experience of great reward and fulfillment, perhaps with a positive transformation. For a different individual, who possesses a different character or aptitude, the overall experience may be negative. Parents may examine themselves and conclude that they are not choosing against a child's specific traits; they may be making an honest and informed acceptance of their own character and goals.[25]

Disability in society

Perhaps the most fundamental and irreconcilable disagreement over the argument from misinformation has to do with just what having a disability is "really" like for people themselves and for their families. Just how much of the problem of disability is socially constructed? Is it reasonable to say that in a differently constructed social environment, what are now disabling traits would become "neutral" characteristics?

Undoubtedly, more of the problem of disability is socially constructed than many people generally believe. But does that imply that having a characteristic like cystic fibrosis or spina bifida is of no more consequence than being left-handed or being a man who is five feet, three inches tall? According to the disability rights critique of prenatal testing, if people with disabilities were fully integrated into society, then there would be no need for the testing. In the world they seek to create, if a given health status turned out to be a handicap, that would be because of societal, not personal, deficits; the appropriate response would be to change society so that the person could live a full life with a range of talents, capacities, and difficulties that exist for everyone. In a society that welcomed the disabled as well as the nondisabled, there would be no reason to prevent the births of people with traits now called disabling.

In this project, those sympathetic to at least some forms of prenatal testing were struck by the fact that, for reasons that seem to be complex, members of the disability community speak at different times in different modes about the nature of disability. Sometimes, members of that community are clear about the fact that disabling traits have a "biological reality" or are not neutral. Adrienne Asch writes, "The inability to move without mechanical aid, to see, to hear, or to learn is not inherently neutral. Disability itself limits some options."[26] At other times, however, and this is the mode usually emphasized in critiques of prenatal testing, those in the disability rights movement speak as if those traits indeed are inherently neutral. Thus Deborah Kent writes: "I premised my life on the conviction that blindness was a neutral characteristic."[27] In this other mode, the disability community argument often is that, different from what prospective parents imagine, these so-called disabling traits are not, to coin a term, "disvaluable" in themselves; they are disvaluable because of the way they are socially constructed.

Nora Groce's work illustrates the point about how social arrangements shape whether a characteristic is disabling.[28] In Martha's Vineyard in the 19th century, Groce argues, being unable to hear was not disabling because everyone spoke sign language. Groce's work establishes that much of what is difficult about having a disability stems from manifold facets of society, from architecture to education to aesthetic preferences. In choosing how to construct our societies, we do, as Allen Buchanan puts it, "choose who will be disabled."[29] We could choose differently than we have, and if we were to choose differently, what's disabling about what we now call disabilities would be largely eliminated. Plainly, then, the social constructionist argument is powerful. The objection concerns, rather, what appears to be a correlative claim of the disability position: that so-called disabling traits

are neither disabling nor "disvaluable," but neutral.

Trying to delineate, understand, and come to consensus over this claim is perhaps the most contentious and difficult part of thinking about prenatal testing in the context of the disability critique. It is worth restating what Asch, Saxton, Lipsky, and others do and do not mean by the "neutrality" of disability. Adherents of the disability critique acknowledge that some characteristics now labeled disabilities are easier to incorporate into today's society, or into a reconstructed society, than are others. Thus, no one would deny that disabling traits – departures from species-typical functioning – foreclose some options, or that some disabilities foreclose more options than others. A child with Down syndrome may never climb Mount Rainier because his strength, agility, and stamina may preclude it; he may also never read philosophy because he does not have the skills to decipher abstract material. Granting that people who can climb mountains and read abstract papers derive enjoyment and meaning from such activities, then being foreclosed from them, not by one's own choice, is regrettable. The lack of possibility is widely seen as disvaluable. In addition, these lacks of capacity stem from the characteristics of the individual who is not strong enough or agile enough to climb, or who is unable by any teaching now known to us to grasp complex abstract discourse. In that sense, disability community critics acknowledge that these facets of some disabilities are "real," inherent in the characteristic itself and not an artifact of any interaction with the environment. Even if all traits are to some extent "socially constructed," that is irrelevant to the fact that the existence of these traits forecloses for those who have them the opportunity to engage in some highly desirable and valuable activities; not being able to engage in those activities is disvaluable.

Disability community critics of the medical model of disability acknowledge that they would be going too far if they claimed that society should not value activities that some of its members cannot engage in; it is harmless to value the capacity of sight that permits people to behold Rembrandt's masterpieces, sunsets, or the faces of family members and friends. It is not offensive to prize intellectual accomplishment, athletic prowess, or the ability to appreciate visual beauty and to regret that not everyone we know can enjoy them. To the extent that spina bifida, Down syndrome, blindness, or cystic fibrosis currently preclude people from undertaking some parts of life that people who do not have those traits might experience, the disability critique acknowledges that disability puts some limits on the "open future"[30] people seek for themselves and their children.

As Bonnie Steinbock argues, if we really thought disability "neutral," we would not work as we do to maintain, restore, and promote health in ourselves and others. We use medicine in the hope that it will cure or ameliorate illness and disability. We urge pregnant women to refrain from activities that risk harming the fetus. If we thought that disabilities were "neutral," then we could tell women who smoke or drink during pregnancy to rest easy, for developmental delay, low birth weight, and fetal alcohol syndrome would all be just "neutral variations," of no consequence to the future child.[31]

While disability community critics acknowledge that some disabilities foreclose some opportunities, they also hold that calling attention to the foreclosure obscures two important points. The first is that rather than dwell on the extent to which opportunities to engage in some activities are truncated, we should concentrate on finding ways for people with disabilities to enjoy alternative modes of those same activities. Philip Ferguson puts it this way:

The point is not so much whether . . . a blind person cannot enjoy a Rembrandt . . . but whether social arrangements can be imagined that allow blind people to have intense aesthetic experiences. . . . People in wheelchairs may not be able to climb mountains, but how hard is it to create a society where the barriers are removed to their experiences of physical exhilaration? . . . Someone with Down syndrome may not be able to experience the exquisite joy of reading bioethics papers and debating ethical theory, but . . . that person can experience the joy of thinking hard about something and reflecting on what he or she really believes. . . . The challenge is to create the society that will allow as many different paths as possible to the qualities of life that make us all part of the human community.[32]

The second fundamental point is that rather than concentrate on the truncation or loss of some opportunities, our society generally – and prospective parents in particular – should concentrate on the nearly infinite range of remaining opportunities. Indeed, every life course necessarily closes off some opportunities in the pursuit of others. Thus while the disability critics of prenatal diagnosis acknowledge that disability is likely to entail some amount of physical, psychological, social, and economic hardship, they hold that when viewed alongside any other life, on balance, life is no worse for people who have disabilities than it is for people who do not. No parent should assume that disability assures a worse life for a child, one with more suffering and less quality, than will be had by those children with whom she or he will grow up.

The claim then is that overall, there is no more stress in raising a child with a disability than in raising any other child, even if at some times there is more stress, or different stress. In that sense the disability community claims that disability is on balance neutral. Even here, however, many find that the terms "neutral" and "normal" are either inaccurate characterizations of disability or are being used in confusing ways. Specifically, some worry that these terms are used sometimes only to describe or evaluate traits and at other times to describe or evaluate persons.

Evaluations of traits versus evaluations of persons

As already mentioned, the disability community itself sometimes speaks about the descriptive and evaluative senses in which disabling traits are not neutral, not normal. Legislation like the ADA could not exist without a recognition that in some sense disabling traits are neither neutral nor normal. Indeed, the societal provision of special resources and services to people with disabilities depends on noticing the descriptive and evaluative senses in which disabling traits are not neutral, and how the needs of the people who live with them are, descriptively speaking, not normal. Yet the recognition of the obligation to provide those special resources is rooted in a commitment to the fundamental idea that the people living with those traits are, morally speaking, "normal"; the people bearing the traits are evaluatively normal in the sense of deserving the normal respect due equally to all persons. Unequal or special funding expresses a commitment to moral equality. Recognizing the non-neutrality of the trait and the "ab-normality" of the person's needs is necessary for expressing the commitment to moral equality and equal opportunity. There is nothing paradoxical about appreciating the descriptive sense in which people with disabling traits are abnormal while also appreciating the evaluative or moral sense in which they are normal.

Some who are sympathetic to prenatal testing worry that people in the disability community (as well as others) often conflate descriptive

claims about traits and evaluative or moral claims about persons, as for example when Deborah Kent, who is blind, writes:

> When I was growing up people called my parents "wonderful." They were praised for raising me "like a normal child." As far as I could tell, they were like most of the other parents in my neighborhood, sometimes wonderful and sometimes very annoying. And from my point of view I wasn't like a normal child – I was normal.[33]

What does Kent mean when she says that she "was normal"? As a descriptive claim, it is not reasonable to say that the trait of blindness is normal. Statistically speaking, it is not. Also, as an evaluative claim, insofar as the trait can make it impossible to enjoy some wonderful opportunities, it does not seem reasonable to say that the trait is neutral. The trait may indeed seem neutral and insignificant when viewed in the context of the whole person; but that is a claim about the person, not the trait. On the view of those sympathetic to testing, the descriptive and evaluative claims about the trait do not bear a necessary logical relation to evaluative claims about the person who bears it. As an evaluative or moral claim about the person, it makes perfect sense to say that a person who is blind is normal; she is normal in the sense that she deserves the normal, usual, equal respect that all human beings deserve.

But if it is easy to notice the difference between the descriptive and evaluative claims about traits and the evaluative claims about persons, why do people in the disability community (and others) keep slipping between the two? Erik Parens suggests that there may be an important reason for this seemingly imprecise slipping. Discrimination against people with disabilities often involves a tendency to allow the part to stand in for the whole; Parens's suggestion is that members of the disability community sometimes succumb to a similar, equally problematic error. The majority community sometimes uses the trait to deny the moral significance of the person; the disability community sometimes uses the moral significance of the person to deny the significance of the trait. The majority community slips from an observation about a trait to a claim about a person; the disability community slips from an observation about a person to a claim about a trait. At important moments, both groups fail to distinguish evaluations of traits from evaluations of persons. While such slippage may be easily committed in both communities, and particularly understandable on the part of the disability community, it may be equally counterproductive in both.

In the end, for all of the project group's disagreements about the appropriateness of employing selective abortion to avoid raising a child with a so-called disabling trait, and about the aptness of the distinction between aborting any fetus versus aborting a particular fetus with a disability, at least these disagreements forced the group to grapple with what many think is disvaluable or undesirable about these traits. Albeit uneasily, the majority of the working group seems to think that disabling traits are disvaluable insofar as they constrain or limit some opportunities. To say that a disability is disvaluable is only to say that, in the world we now inhabit and in the world we can imagine living in any time soon, to have a given trait would make it impossible or very difficult to engage in some activities that most people would want themselves or their children to have the option of engaging in. For this reason, then, the majority seems uneasily to think that traits are disvaluable insofar as they preclude what many find precious. This view was held "uneasily" because many are keenly aware of how limited our ability is to imagine alternative

social constructions – as well as of the extent to which traits once thought unreconstructable are now thought to be nearly infinitely plastic. We are keenly aware of the extent to which the trait that is sex was constructed in the past in arbitrary and pernicious ways, as well as of past arguments that sex could not be constructed much differently. And we recognize how paltry our ability is to imagine what the experience of others is like. Few of us would have believed before the project meetings began that conjoined twins would report feeling about their lives pretty much like people with "normal" bodies report feeling about theirs.[34]

It is important to remember that the disability community arguments are not intended to justify wholesale restrictions on prenatal testing for genetic disability. Rather, they are intended to make prospective parents pause and think about what they are doing, and to challenge professionals to help parents better examine their decisions. They are intended to help make our decisions thoughtful and informed, not thoughtless and automatic. In his book about his son who has Down syndrome, Michael Bérubé attempts to steer a path much like the one ultimately adopted here. He writes:

> I'm . . . not sure whether I can have any advice for prospective parents who are contemplating what course of action to take when they discover they will bear a "disabled" child. Obviously I can't and don't advocate abortion of fetuses with Down syndrome; indeed, the only argument I have is that such decisions should not be automatic.[35]

To some, the advice that such decisions shouldn't be automatic may seem wishy-washy and disheartening. But to those who, like Hannah Arendt, think that evil can arise from thoughtlessness, it seems neither.

[. . .]

Notes

1 Cynthia M. Powell, "The Current State of Prenatal Genetic Testing in the U.S.," in *Prenatal Genetic Testing and the Disability Rights Critique*, ed. Erik Parens and Adrienne Asch (Washington, D.C.: Georgetown University Press, forthcoming).

2 Thomas H. Murray, *The Worth of a Child* (Berkeley and Los Angeles: University of California Press, 1996), pp. 116-17.

3 Diana W. Bianchi, Timothy M. Crombleholme, Mary D'Alton, *Fetology: Diagnosis and Management of the Fetal Patient* (Blacklick, Ohio: McGraw Hill, forthcoming).

4 Adrienne Asch "Reproductive Technology and Disability," in *Reproductive Laws for the 1990s*, ed. Sherrill Cohen and Nadine Taub (Clifton, N.J.: Humana Press, 1989), pp. 69–124, at 86.

5 Allen E. Buchanan, "Choosing Who Will Be Disabled: Genetic Intervention and the Morality of Inclusion," *Social Philosophy and Policy* 13 (1996): 18–46.

6 Adrienne Asch, "Reproductive Liberty," p. 81.

7 Adrienne Asch, "Why I Haven't Changed My Mind," in Parens and Asch, *Prenatal Genetic Testing*.

8 Dorothy C. Wertz and John C. Fletcher, "Sex Selection through Prenatal Diagnosis: A Feminist Critique," in *Feminist Perspectives in Medical Ethics*, ed. Helen Bequaert Holms and Laura M. Purdy (Bloomington, Ind.: Indiana University Press, 1992), pp. 240–53.

9 Marsha Saxton, "Disability Rights and Selective Abortion," in *Abortion Wars: A Half Century of Struggle, 1950–2000*, ed. Rickie Solinger (Berkeley and Los Angeles: University of California Press, 1997), pp. 374–95, at 391.

10 Nancy Press, "Assessing the Expressive Character of Prenatal Testing: The Choices Made or the Choices Made Available?" in Parens and Asch, *Prenatal Genetic Testing*.

11 Eva Feder Kittay with Leo Kittay, "On the Expressivity and Ethics of Selective Abortion for Disability: Conversations with My Son," and James Lindemann Nelson, "The Meaning of the Act: Reflections on the Expressive Force of Reproductive Decision-making and Policies," both in Parens and Asch, *Prenatal Genetic Testing*.

12 Nelson, "Meaning of the Act."

13 Bonnie Steinbock, "Disability, Prenatal Testing, and Selective Abortion," in Parens and Asch, *Prenatal Genetic Testing*.

14 Asch, "Reproductive Technology and Disability," p. 88.

15 Murray, "Worth of a Child," pp. 115–41; and Adrienne Asch and Gail Geller, "Feminism, Bioethics, and Genetics," in *Feminism and Bioethics: Beyond Reproduction*, ed. Susan M. Wolf (New York: Oxford University Press, 1996).

16 William Ruddick, "Pregnancies and Prenatal Tests," in Parens and Asch, *Prenatal Genetic Testing*.

17 J. A. Blier Blaymore, J. A. Liebling, Y. Morales, M. Carlucci, "Parents' and Pediatricians' Views of Individuals with Meningomyelocile," *Clinical Pediatrics* 35, no. 3 (1996): 113–17; M. L. Wollraich, G. N. Siperstein, P. O'Keefe, "Pediatricians' Perceptions of Mentally Retarded Individuals," *Pediatrics* 80, no. 5 (1987): 643–49.

18 Philip Ferguson, Alan Gartner, Dorothy Lipsky, "The Experience of Disability in Families: A Synthesis of Research and Parent Narratives," in Parens and Asch, *Prenatal Genetic Testing*.

19 M. W. Krauss, "Child-Related and Parenting Stress: Similarities and Differences Between Mothers and Fathers of Children with Disabilities," *American Journal of Mental Retardation* 97 (1993): 393–404.

20 D. A. Abbott and W. H. Meredith, "Strengths of Parents with Retarded Children," *Family Relations* 35 (1986): 371–75; A. P. Turnbull, J. M. Patterson, S. K. Behr et al., eds., *Cognitive Coping: Families and Disability* (Baltimore, Md.: Paul H. Brookes, 1993).

21 J. P. Lehman and K. Roberto, "Comparison of Factors Influencing Mothers' Perceptions about the Future of Their Adolescent Children with and without Disabilities," *Mental Retardation* 34 (1996): 27–38.

22 Ruth I. Freedman, Leon Litchfield, Marjl Erickson Warfield, "Balancing Work and Family: Perspectives of Parents of Children with Developmental Disabilities," *Families in Society: The Journal of Contemporary Human Services* (October 1995): 50714, at 511.

23 Sara Ruddick, *Maternal Thinking: Toward a Politics of Peace* (Boston: Beacon Press, 1989).

24 Mary Ann Baily, personal communication.

25 Caroline Moon, unpublished paper on file with Luce Program at Wellesley College.

26 Asch, "Reproductive Technology," p. 73.

27 Deborah Kent, "Somewhere a Mockingbird," in Patens and Asch, *Prenatal Genetic Testing*.

28 Nora Ellen Groce, *Everyone Here Spoke Sign Language: Hereditary Deafness on Martha's Vineyard* (Cambridge, Mass.: Harvard University Press, 1985).

29 Allen Buchanan, "Choosing Who Will Be Disabled."

30 Dena S. Davis, "Genetic Dilemmas and the Child's Right to an Open Future," *Hastings Center Report* 27, no. 2 (1997): 7–15; Bonnie Steinbock and Ronald McClamrock, "When Is Birth Unfair to the Child?" *Hastings Center Report* 24, no. 6 (1994): 15–21; Ronald Green, "Parental Autonomy and the Obligation Not to Harm One's Child Genetically," *Journal of Law, Medicine & Ethics* 25, no. 1 (1997): 5–16.

31 Steinbock, "Disability Prenatal Testing."

32 Philip Ferguson, personal communication.

33 Kent, "Somewhere a Mockingbird."

34 Alice Domurat Dreger, "The Limits of Individuality: Ritual and Sacrifice in the Lives and Medical Treatment of Conjoined Twins," *Studies in the History and Philosophy of Science* 29, no. 1 (1998): 1–29.

35 Michael Bérubé, *Life As We Know It: A Father, a Family, and an Exceptional Child* (New York: Pantheon, 1996).

Chapter 6

Julian Savulescu

PROCREATIVE BENEFICENCE: WHY WE SHOULD SELECT THE BEST CHILDREN

Introduction

Imagine you are having in vitro fertilisation (IVF) and you produce four embryos. One is to be implanted. You are told that there is a genetic test for predisposition to scoring well on IQ tests (let's call this intelligence). If an embryo has gene subtypes (alleles) A, B there is a greater than 50 per cent chance it will score more than 140 if given an ordinary education and upbringing. If it has subtypes C, D there is a much lower chance it will score over 140. Would you test the four embryos for these gene subtypes and use this information in selecting which embryo to implant?

Many people believe intelligence is a purely social construct and so it is unlikely to have a significant genetic cause. Others believe there are different sorts of intelligence, such as verbal intelligence, mathematical intelligence, musical ability and no such thing as general intelligence. Time will tell. There are several genetic research programs currently in place which seek to elucidate the genetic contribution to intelligence. This paper pertains to any results of this research even if it only describes a weak probabilistic relation between genes and intelligence, or a particular kind of intelligence.

Many people believe that research into the genetic contribution to intelligence should not be performed, and that if genetic tests which predict intelligence, or a range of intelligence, are ever developed, they should not be employed in reproductive decision-making. I will argue that we have a moral obligation to test for genetic contribution to non-disease states such as intelligence and to use this information in reproductive decision-making.

Imagine now you are invited to play the Wheel of Fortune. A giant wheel exists with marks on it from 0–$1,000,000, in $100 increments. The wheel is spun in a secret room. It stops randomly on an amount. That amount is put into Box A. The wheel is spun again. The amount which comes up is put into Box B. You can choose Box A or B. You are also told that, in addition to the sum already put in the boxes, if you choose B, a dice will be thrown and you will lose $100 if it comes up 6.

Which box should you choose?

The rational answer is Box A. Choosing genes for non-disease states is like playing the Wheel of Fortune. You should use all the available information and choose the option most likely to bring about the best outcome.

Procreative beneficence: the moral obligation to have the best children

I will argue for a principle which I call Procreative Beneficence:

> couples (or single reproducers) should select the child, of the possible children they could have, who is expected to have the best life, or at least as good a life as the others, based on the relevant, available information.

I will argue that Procreative Beneficence implies couples should employ genetic tests for non-disease traits in selecting which child to bring into existence and that we should allow selection for non-disease genes in some cases even if this maintains or increases social inequality.

By 'should' in 'should choose', I mean 'have good reason to.' I will understand morality to require us to do what we have most reason to do. In the absence of some other reason for action, a person who has good reason to have the best child is morally required to have the best child.

Consider the following three situations involving normative judgements.

(1) 'You are 31. You will be at a higher risk of infertility and having a child with an abnormality if you delay child-bearing. But that has to be balanced against taking time out of your career now. That's only something you can weigh up.'
(2) 'You should stop smoking.'
(3) 'You must inform your partner that you are HIV positive or practise safe sex.'

The 'should' in 'should choose the best child' is that present in the second example. It implies that persuasion is justified, but not coercion, which would be justified in the third case. Yet

the situation is different to the more morally neutral (1).

Definitions

A disease gene is a gene which causes a genetic disorder (e.g. cystic fibrosis) or predisposes to the development of disease (e.g. the genetic contribution to cancer or dementia). A non-disease gene is a gene which causes or predisposes to some physical or psychological state of the person which is not itself a disease state, e.g. height, intelligence, character (not in the subnormal range).

Selection

It is currently possible to select from a range of possible children we could have. This is most frequently done by employing fetal selection through prenatal testing and termination of pregnancy. Selection of embryos is now possible by employing in vitro fertilisation and preimplantation genetic diagnosis (PGD). There are currently no genetic tests available for non-disease states except sex. However, if such tests become available in the future, both PGD and prenatal testing could be used to select offspring on the basis of non-disease genes. Selection of sex by PGD is now undertaken in Sydney, Australia.[1] PGD will also lower the threshold for couples to engage in selection since it has fewer psychological sequelae than prenatal testing and abortion.

In the future, it may be possible to select gametes according to their genetic characteristics. This is currently possible for sex, where methods have been developed to sort X and Y bearing sperm.[2]

Behavioural genetics

Behavioural Genetics is a branch of genetics which seeks to understand the contribution of genes to complex behaviour. The scope of behavioural genetics is illustrated in Table 6.1.

Consider the *Simple Case of Selection for Disease Genes*. A couple is having IVF in an attempt to have a child. It produces two embryos. A battery of tests for common diseases is performed. Embryo A has no abnormalities on the tests performed. Embryo B has no abnormalities on the tests performed except its genetic profile reveals it has a predisposition to developing asthma. Which embryo should be implanted?

Embryo B has nothing to be said in its favour over A and something against it. Embryo A should (on pain of irrationality) be implanted. This is like choosing Box A in the Wheel of Fortune analogy.

Why shouldn't we select the embryo with a predisposition to asthma? What is relevant about asthma is that it reduces quality of life. Attacks cause severe breathlessness and in extreme cases, death. Steroids may be required to treat it. These are among the most dangerous drugs which exist if taken long term. Asthma can be lifelong and require lifelong drug treatment. Ultimately it can leave the sufferer wheel chair bound with chronic obstructive airways disease. The morally relevant property of 'asthma' is that it is a state which reduces the well-being a person experiences.

Table 6.1 Behavioural Genetics

Aggression and criminal behaviour
Alcoholism
Anxiety and anxiety disorders
Attention deficit hyperactivity disorder (ADHD)
Antisocial personality disorder
Bipolar disorder
Homosexuality
Maternal behaviour
Memory and intelligence
Neuroticism
Novelty seeking
Schizophrenia
Substance addiction

Parfitian defence of voluntary procreative beneficence in the Simple Case

The following example, after Parfit,[3] supports Procreative Beneficence. A woman has rubella. If she conceives now, she will have a blind and deaf child. If she waits three months, she will conceive another different but healthy child. She should choose to wait until her rubella is passed.

Or consider the Nuclear Accident. A poor country does not have enough power to provide power to its citizens during an extremely cold winter. The government decides to open an old and unsafe nuclear reactor. Ample light and heating are then available. Citizens stay up later, and enjoy their lives much more. Several months later, the nuclear reactor melts down and large amounts of radiation are released into the environment. The only effect is that a large number of children are subsequently born with predispositions to early childhood malignancy.

The supply of heating and light has changed the lifestyle of this population. As a result of this change in lifestyle, people have conceived children at different times than they would have if there had been no heat or light, and their parents went to bed earlier. Thus, the children born after the nuclear accident would not have existed if the government had not switched to nuclear power. They have not been harmed by the switch to nuclear power and the subsequent accident (unless their lives are so bad they are worse than death). If we object to the Nuclear Accident (which most of us would), then we must appeal to some form of harmless wrong-doing. That is, we must claim that a wrong was done, but no one was harmed. We must appeal to something like the Principle of Procreative Beneficence.

An objection to Procreative Beneficence in the simple case

The following objection to Procreative Beneficence is common.

If you choose Embryo A (without a predisposition to asthma), you could be discarding someone like Mozart or an Olympic swimmer. So there is no good reason to select A.

It is true that by choosing A, you could be discarding a person like Mozart. But it is equally true that if you choose B, you could be discarding someone like Mozart without asthma. A and B are equally likely (on the information available) to be someone like Mozart (and B is more likely to have asthma).

Other principles of reproductive decision-making applied to the simple case

The Principle of Procreative Beneficence supports selecting the embryo without the genetic predisposition to asthma. That seems intuitively correct. How do other principles of reproductive decision-making apply to this example?

1 *Procreative Autonomy*: This principle claims that couples should be free to decide when and how to procreate, and what kind of children to have.[4] If this were the only decision-guiding principle, it would imply couples might have reason to choose the embryo with a predisposition to asthma, if for some reason they wanted that.
2 *Principle of Non-Directive Counselling*: According to this principle, doctors and genetic counsellors should only provide information about risk and options available to reduce that risk.[5] They should not give advice or other direction. Thus, if a couple wanted to transfer Embryo B, and they knew that it would have a predisposition to asthma, nothing more is to be said according to Non-Directive Counselling.
3 *The 'Best Interests of the Child' Principle*: Legislation in Australia and the United

Kingdom related to reproduction gives great weight to consideration of the best interests of the child. For example, the Victorian Infertility Treatment Act 1995 states '*the welfare and interests of any person born or to be born as a result of a treatment procedure are paramount*.'[6] This principle is irrelevant to this choice. The couple could choose the embryo with the predisposition to asthma and still be doing everything possible in the interests of that child.

None of the alternative principles give appropriate direction in the Simple Case.

Moving from disease genes to non-disease genes: what is the 'best life'?

It is not asthma (or disease) which is important, but its impact on a life in ways that matter which is important. People often trade length of life for non-health related well-being. Non-disease genes may prevent us from leading the best life.

By 'best life', I will understand the life with the most well-being. There are various theories of well-being: hedonistic, desire-fulfilment, objective list theories.[7] According to hedonistic theories, what matters is the quality of our experiences, for example, that we experience pleasure. According to desire-fulfilment theories, what matters is the degree to which our desires are satisfied. According to objective list theories, certain activities are good for people, such as achieving worthwhile things with your life, having dignity, having children and raising them, gaining knowledge of the world, developing one's talents, appreciating beautiful things, and so on.

On any of these theories, some non-disease genes will affect the likelihood that we will lead the best life. Imagine there is a gene which contributes significantly to a violent, explosive, uncontrollable temper, and that state causes people significant suffering. Violent outbursts lead a person to come in conflict with the law and fall out of important social relations. The loss of independence, dignity and important social relations are bad on any of the three accounts.

Buchanan et al. argue that what is important in a liberal democracy is providing people with general purpose means, i.e. those useful to any plan of life.[8] In this way we can allow people to form and act on their own conception of the good life. Examples of general purpose means are the ability to hear and see. But similarly the ability to concentrate, to engage with and be empathetic towards other human beings may be all purpose means. To the degree that genes contribute to these, we have reason to select those genes.

Consider another example. Memory (M) is the ability to remember important things when you want to. Imagine there is some genetic contribution to M: Six alleles (genes) contribute to M. IVF produces four embryos. Should we test for M profiles?

Does M relate to well-being? Having to go to the supermarket twice because you forgot the baby formula prevents you doing more worthwhile things. Failing to remember can have disastrous consequences. Indeed, forgetting the compass on a long bush walk can be fatal. There is, then, a positive obligation to test for M and select the embryo (other things being equal) with the best M profile.

Does being intelligent mean one is more likely to have a better life? At a folk intuitive level, it seems plausible that intelligence would promote well-being on any plausible account of well-being. On a hedonistic account, the capacity to imagine alternative pleasures and remember the salient features of past experiences is important in choosing the best life. On a desire-fulfilment theory, intelligence is important to choosing means which will best satisfy one's

ends. On an objective list account, intelligence would be important to gaining knowledge of the world, and developing rich social relations. Newson has reviewed the empirical literature relating intelligence to quality of life. Her synthesis of the empirical literature is that 'intelligence has a high instrumental value for persons in giving them a large amount of complexity with which to approach their everyday lives, and that it equips them with a tool which can lead to the provision of many other personal and social goods.'[9]

Socrates, in Plato's Philebus, concludes that the best life is a mixture of wisdom and pleasure. Wisdom includes thought, intelligence, knowledge and memory.[10] Intelligence is clearly a part of Plato's conception of the good life:

> without the power of calculation you could not even calculate that you will get enjoyment in the future; your life would be that not of a man, but of a sea-lung or one of those marine creatures whose bodies are confined by a shell.[11]

Choice of means of selecting

This argument extends in principle to selection of fetuses using prenatal testing and termination of affected pregnancy. However, selection by abortion has greater psychological harms than selection by PGD and these need to be considered. Gametic selection, if it is ever possible, will have the lowest psychological cost.

Objections to the Principle of Procreative Beneficence applied to non-disease genes

1 *Harm to the child:* One common objection to genetic selection for non-disease traits is that it results in harm to the child. There are various versions of this objection, which include the harm which arises from excessive and

overbearing parental expectations, using the child as a means, and not treating it as an end, and closing off possible future options on the basis of the information provided (failing to respect the child's 'right to an open future').

There are a number of responses. Firstly, in some cases, it is possible to deny that the harms will be significant. Parents come to love the child whom they have (even a child with a serious disability). Moreover, some have argued that counselling can reduce excessive expectations.[12]

Secondly, we can accept some risk of a child experiencing some state of reduced well-being in cases of selection. One variant of the harm to child objection is: 'If you select embryo A, it might still get asthma, or worse, cancer, or have a much worse life than B, and you would be responsible.' Yet selection is immune to this objection (in a way which genetic manipulation is not).

Imagine you select Embryo A and it develops cancer (or severe asthma) in later life. You have not harmed A unless A's life is not worth living (hardly plausible) because A would not have existed if you had acted otherwise. A is not made worse off than A would otherwise have been, since without the selection, A would not have existed. Thus we can accept the possibility of a bad outcome, but not the probability of a very bad outcome. (Clearly, Procreative Beneficence demands that we not choose a child with a low predisposition to asthma but who is likely to have a high predisposition to cancer.)

This is different to genetic manipulation. Imagine you perform gene therapy to correct a predisposition to asthma and you cause a mutation which results in cancer later in life. You have harmed A: A is worse off in virtue of the genetic manipulation than A would have been if the manipulation had not been performed (assuming cancer is worse than asthma).

There is, then, an important distinction between:

- interventions which are genetic manipulations of a single gamete, embryo or fetus
- selection procedures (e.g. sex selection) which select from among a range of different gametes, embryos and fetuses.

2 Inequality: One objection to Procreative Beneficence is that it will maintain or increase inequality. For example, it is often argued that selection for sex, intelligence, favourable physical or psychological traits, etc. all contribute to inequality in society, and this is a reason not to attempt to select the best.

In the case of selection against disease genes, similar claims are made. For example, one version of the Disability Discrimination Claim maintains that prenatal testing for disabilities such as Down syndrome results in discrimination against those with those disabilities both by:

- the statement it makes about the worth of such lives
- the reduction in the numbers of people with this condition.

Even if the Disability Discrimination Claim were true, it would be a drastic step in favour of equality to inflict a higher risk of having a child with a disability on a couple (who do not want a child with a disability) to promote social equality.

Consider a hypothetical rubella epidemic. A rubella epidemic hits an isolated population. Embryos produced prior to the epidemic are not at an elevated risk of any abnormality but those produced during the epidemic are at an increased risk of deafness and blindness. Doctors should encourage women to use embryos which they have produced prior to the epidemic in preference to ones produced during the epidemic. The reason is that it is bad that blind and deaf children are born when sighted and hearing children could have been born in their place.

This does not necessarily imply that the lives of those who now live with disability are less deserving of respect and are less valuable. To attempt to prevent accidents which cause paraplegia is not to say that paraplegics are less deserving of respect. It is important to distinguish between disability and persons with disability. Selection reduces the former, but is silent on the value of the latter. There are better ways to make statements about the equality of people with disability (e.g., we could direct savings from selection against embryos/fetuses with genetic abnormalities to improving well-being of existing people with disabilities).

These arguments extend to selection for non-disease genes. It is not disease which is important but its impact on well-being. In so far as a non-disease gene such as a gene for intelligence impacts on a person's well-being, parents have a reason to select for it, even if inequality results.

This claim can have counter-intuitive implications. Imagine in a country women are severely discriminated against. They are abandoned as children, refused paid employment and serve as slaves to men. Procreative Beneficence implies that couples should test for sex, and should choose males as they are expected to have better lives in this society, even if this reinforces the discrimination against women.

There are several responses. Firstly, it is unlikely selection on a scale that contributes to inequality would promote well-being. Imagine that 50 per cent of the population choose to select boys. This would result in three boys to every one girl. The life of a male in such a society would be intolerable.

Secondly, it is social institutional reform, not interference in reproduction, which should be promoted. What is wrong in such a society is the treatment of women, which should be addressed separately to reproductive decision-making. Reproduction should not become an instrument

of social change, at least not mediated or motivated at a social level.

This also illustrates why Procreative Beneficence is different to eugenics. Eugenics is selective breeding to produce a better population. A public interest justification for interfering in reproduction is different from Procreative Beneficence which aims at producing the best child, of the possible children, a couple could have. That is an essentially private enterprise. It was the eugenics movement itself which sought to influence reproduction, through involuntary sterilisation, to promote social goods.

Thirdly, consider the case of blackmail. A company says it will only develop an encouraging drug for cystic fibrosis (CF) if there are more than 100,000 people with CF. This would require stopping carrier testing for CF. Should the government stop carrier testing?

If there are other ways to fund this research (e.g., government funding), this should have priority. In virtually all cases of social inequality, there are other avenues to correct inequality than encouraging or forcing people to have children with disabilities or lives of restricted genetic opportunity.

Limits on procreative beneficence: personal concern for equality or self interest

Consider the following cases. David and Dianne are dwarfs. They wish to use IVF and PGD to select a child with dwarfism because their house is set up for dwarfs. Sam and Susie live a society where discrimination against women is prevalent. They wish to have a girl to reduce this discrimination. These choices would not harm the child produced if selection is employed. Yet they conflict with the Principle of Procreative Beneficence.

We have here an irresolvable conflict of principles:

- personal commitment to equality, personal interests and Procreative Autonomy
- Procreative Beneficence.

Just as there are no simple answers to what should be done (from the perspective of ethics) when respect for personal autonomy conflicts with other principles such as beneficence or distributive justice, so too there are no simple answers to conflict between Procreative Autonomy and Procreative Beneficence.

For the purposes of public policy, there should be a presumption in favour of liberty in liberal democracies. So, ultimately, we should allow couples to make their own decisions about which child to have. Yet this does not imply that there are no normative principles to guide those choices. Procreative Beneficence is a valid principle, albeit one which must be balanced against others.

The implication of this is that those with disabilities should be allowed to select a child with disability, if they have a good reason. But the best option is that we correct discrimination in other ways, by correcting discriminatory social institutions. In this way, we can achieve both equality and a population whose members are living the best lives possible.

Conclusions

With respect to non-disease genes, we should provide:

- information (through PGD and prenatal testing)
- free choice of which child to have
- non-coercive advice as to which child will be expected to enter life with the best opportunity of having the best life.

Selection for non-disease genes which significantly impact on well-being is *morally required* (Procreative Beneficence). 'Morally required' implies moral persuasion but not coercion is justified.

If, in the end, couples wish to select a child who will have a lower chance of having the best life, they should be free to make such a choice. That should not prevent doctors from attempting to persuade them to have the best child they can. In some cases, persuasion will not be justified. If self-interest or concern to promote equality motivate a choice to select less than the best, then there may be no overall reason to attempt to dissuade a couple. But in cases in which couples do not want to use or obtain available information about genes which will affect well-being, and their desires are based on irrational fears (e.g., about interfering with nature or playing God), then doctors should try to persuade them to access and use such information in their reproductive decision-making.

Notes

1 J. Savulescu. 'Sex Selection – the case for.' *Medical Journal of Australia* 1999; 171: 373–5.

2 E.F. Fugger, S.H. Black, K. Keyvanfar, J.D. Schulman. 'Births of normal daughters after Microsort sperm separation and intrauterine insemination, invitro fertilization, or intracytoplasmic sperm injection.' *Hum Reprod* 1998; 13: 2367–70.

3 D. Parfit. 1976. 'Rights, Interests and Possible People,' in *Moral Problems in Medicine*, S. Gorovitz, et al., eds. Englewood Cliffs. Prentice Hall. D. Parfit. 1984. *Reasons and Persons*, Oxford. Clarendon Press: Part IV.

4 R. Dworkin. 1993. *Life's Dominion: An Argument about Abortion and Euthanasia*. London. Harper Collins; J. Harris. 'Goodbye Dolly? The ethics of human cloning.' *Journal of Medical Ethics* 1997; 23: 353–60; J. Harris. 1998. 'Rights and Reproductive Choice,' in *The Future of Reproduction*, J. Harris and S. Holm, eds. Oxford. Clarendon Press; J.A. Robertson. 1994. *Children of Choice: Freedom and the New Reproductive Technologies*. Princeton. Princeton University Press;

C. Strong. 1997. *Ethics in reproductive and perinatal medicine*. New Haven. Yale University Press.

5   J.A.F. Roberts. 1959. *An Introduction to Human Genetics*. Oxford. OUP.

6   The *Human Fertilization and Embryology Act* 1990 in England requires that account be taken of the welfare of any child who will be born by assisted reproduction before issuing a licence for assistance (S.13(5)).

7   Parfit, *op. cit.*, Appendix I, pp. 493–502; J. Griffin. 1986. *Well-Being*. Oxford. Clarendon Press.

8   A. Buchanan, D.W. Brock, N. Daniels, D. Wikler. 2000. *From Chance to Choice*. Cambridge. CUP: 167. Buchanan and colleagues argue in a parallel way for the permissibility of genetic manipulation (enhancement) to allow children to live the best life possible (Chapter Five). They do not consider selection in this context.

9   A. Newson. The value of intelligence and its implications for genetic research. *Fifth World Congress of Bioethics*, Imperial College, London, 21–24 September 2000.

10   *Philebus* 21 C 1–12. A.E. Taylor's translation. 1972. Folkstone. Dawsons of Pall Mall: 21 D 11–3, E 1–3.

11   *Philebus* 21 C 1–12.

12   J. Robertson. Preconception Sex Selection. *American Journal of Bioethics* 1:1 (Winter 2001).

Chapter 7

John A. Robertson

PRECONCEPTION GENDER SELECTION[1]

Advances in genetics and reproductive technology present prospective parents with an increasing number of choices about the genetic makeup of their children. Those choices now involve the use of carrier and prenatal screening techniques to avoid the birth of children with serious genetic disease, but techniques to choose nonmedical characteristics will eventually be available. One nonmedical characteristic that may soon be within reach is the selection of offspring gender by preconception gender selection (PGS).

Gender selection through prenatal diagnosis and abortion has existed since the 1970s. More recently, preimplantation sexing of embryos for transfer has been developed (Tarin and Handyside 1993; The Ethics Committee of the American Society of Reproductive Medicine 1999). Yet prenatal or preimplantation methods of gender selection are unattractive because they require abortion or a costly, intrusive cycle of in vitro fertilization (IVF) and embryo discard. Attempts to separate X- and Y-bearing sperm for preconception gender selection by sperm swim-up or swim-through techniques have not shown consistent X- and Y-sperm cell separation or success in producing offspring of the desired gender.

The use of flow cytometry to separate X- and Y-bearing sperm may turn out to be a much more reliable method of enriching sperm populations for insemination. Laser beams passed across a flowing array of specially dyed sperm can separate most of the 2.8 percent heavier X- from Y-bearing sperm, thus producing an X-enriched sperm sample for insemination.[2] Flow cytometry has been used successfully in over 400 sex selections in rabbit, swine, ovine, and bovine species, including successive generations in swine and rabbit (Fugger et al. 1998). A human pregnancy was reported in 1995 (Levinson, Keyvanfar, and Wu 1995).

The United States Department of Agriculture (USDA), which holds a patent on the flow cytometry separation process, has licensed the Genetics and IVF Institute in Fairfax, Virginia, to study the safety and efficacy of the technique for medical and "family balancing" reasons in an institutional review board-approved clinical trial.[3] In 1998 researchers at the Institute reported a 92.9 percent success rate for selection of females in 27 patients, with most fertilizations occurring after intrauterine insemination (Fugger et al. 1998). A lower success rate (72 percent) was reported for male selection.[4]

At this early stage of development much more research is needed to establish the high degree of safety and efficacy of flow cytometry methods of PGS that would justify widespread use. With only one published study of outcomes to date, it is too soon to say whether the 92 percent success rate in determining female gender will hold for other patients, much less that male selection will reach that level of efficacy. Animal safety data have shown no adverse effect of the dye or laser used in the technique on offspring, but that is no substitute for more extensive human studies (Vidal et al. 1999). In addition, if flow cytometry instruments are to be used for sperm separation purposes, they may be classified as medical devices that require U.S. Food and Drug Administration (FDA) approval. Finally, the holder of the process patent – the USDA – will have to agree to license the process for human uses.

If further research establishes that flow cytometry is a safe and effective technique for both male and female PGS, and regulatory and licensing barriers are overcome, then a couple wishing to choose the gender of their child would need only provide a sperm sample and undergo one or more cycles of intrauterine insemination with separated sperm.[5] A clinic or physician that offers assisted reproductive technologies (ART) and invests in the flow cytometry equipment could run the separation and prepare the X- or Y-enriched sperm for insemination, or it could have the sperm processed by a clinic or firm that has made that investment. Flow cytometry separation would not be as cheap and easy as determining gender by taking a pill before intercourse, but it would be within reach of most couples who have gender preferences in offspring.[6]

Demand for preconception gender selection

Unknown at present is the number of people who have offspring gender preferences robust enough to incur the costs and inconvenience of PGS. Although polls have often shown a preference for firstborn males, they have not shown that a large number of couples would be willing to forego coital conception in order to select the gender of their children. If PGS proves to be safe and effective, however, it may be sought by two groups of persons with gender preferences.

One group would seek PGS in order to have a child of a gender different from that of a previous child or children. A preference for gender variety in offspring would be strongest in families that have already had several children of one gender. They may want an additional child only if they can be sure that it will be of the gender opposite to their existing children.[7] Couples who wish to have only two children might use PGS for the second child to ensure that they have one child of each gender. If social preferences for two-child families remain strong, some families may use PGS to choose the gender of the second child.

A second group of PGS users would be those persons who have strong preferences for the gender of the first child. The most likely candidates here are persons with strong religious or cultural beliefs about the role or importance of children with a particular gender. Some Asian cultures have belief systems that strongly prefer that the firstborn child be a male. In some cases the preference reflects religious beliefs or traditions that require a firstborn son to perform funeral rituals to assure his parents' entrance into heaven (for a discussion of son preferences in India and China, see Macklin 1999, 148–51). In others it simply reflects a deeply embedded social preference for males over females. The first-child preference will be all the stronger if a one-child-per-family policy is in effect, as occurred for a while in China (Greenlagh and Li 1995, 627). While the demand for PGS for firstborn children is likely to be strongest in those countries, there has been a sizable migration of those groups to the United States, Canada, and Europe.[8] Until they are more fully assimilated, immigrant groups in Western countries may retain the same gender preferences that they would have held in their homelands.

Other persons with strong gender preferences for firstborn children would be those who prize the different rearing or relational experiences they think they would have with children of a particular gender. They may place special value on having their firstborn be male or female because of personal experiences or beliefs. Numerous scenarios are likely here, from the father who very much wants a son because of a desire to provide his child with what he lacked growing up, to the woman who wants a girl because of the special closeness that she thinks she will have with a daughter (Belkin 1999).

The ethical dilemma of preconception gender selection

The prospect of preconception gender selection appears to pose the conflict – long present in other bioethical issues – between individual desires and the larger common good. Acceding to individual desires about the makeup of children seems to be required by individual autonomy. Yet doing so leads to the risk that children will be treated as vehicles of parental satisfaction rather than as ends in themselves, and could accelerate the trend toward negative and even positive selection of offspring characteristics. The dilemma of reconciling procreative liberty with the welfare of offspring and families will only intensify as genetic technology is further integrated with assisted reproduction and couples seek greater control over the genes of offspring.

Arguments for preconception gender selection

The strongest argument for preconception gender selection is that it serves the needs of

couples who have strong preferences about the gender of their offspring and would not reproduce unless they could realize those preferences. Because of the importance of reproduction in an individual's life, the freedom to make reproductive decisions has long been recognized as a fundamental moral and legal right that should not be denied to a person unless exercise of that right would cause significant harm to others (Robertson 1994, 22–42). A corollary of this right, which is now reflected in carrier and prenatal screening practices to prevent the birth of children with genetic disease, is that prospective parents have the right to obtain preconception or prenatal information about the genetic characteristics of offspring, so that they may decide in a particular case whether or not to reproduce (Robertson 1996, 424–35).[9]

Although offspring gender is not a genetic disease, a couple's willingness to reproduce might well depend on the gender of expected offspring. Some couples with one or more children of a particular gender might refuse to reproduce if they cannot use PGS to provide gender variety in their offspring or to have additional children of the same gender (E. F. Fugger, personal communication to author). In other cases they might have such strong rearing preferences for their firstborn child that they might choose not to reproduce at all if they cannot choose that child's gender. Few persons contemplating reproduction may fall into either group; but for persons who strongly hold those preferences, the ability to choose gender may determine whether they reproduce.

In cases where the gender of offspring is essential to a couple's decision to reproduce, the freedom to choose offspring gender would arguably be part of their procreative liberty (Robertson 1996, 434). Since respect for a right is not dependent on the number of persons asserting the right, they should be free to use a technique essential to their reproductive decision unless the technique would cause the serious harm to others that overcomes the strong presumption that exists against government interference in reproductive choice. Until there is a substantial basis for thinking that a particular use of PGS would cause such harms, couples should be free to use the technique in constituting their families. The right they claim is a right against government restriction or prohibition of PGS. It is not a claim that society or insurers are obligated to fund PGS or that particular physicians must provide it.

Arguments against preconception gender selection

There are several arguments against preconception gender selection. Although such methods do not harm embryos and fetuses or intrude on a woman's body as *prenatal* gender selection does, they do raise other important issues. One concern is the potential of such techniques to increase or reinforce sexism, either by allowing more males to be produced as first or later children, or by paying greater attention to gender itself. A second concern is the welfare of children born as a result of PGS whose parents may expect them to act in certain gender specific ways when the technique succeeds, but who may be disappointed if the technique fails. A third concern is societal. Widely practiced, PGS could lead to sex-ratio imbalances, as have occurred in some parts of India and China due to female infanticide, gender-driven abortions, and a one-child-per-family policy (Sen 1990). Finally, the spread of PGS would be another incremental step in the growing technologization of reproduction and genetic control of offspring. While each step alone may appear to be justified, together they could constitute a threat to the values of care and concern that have traditionally informed norms of parenting and the rearing of children.

Evaluation of ethical and social issues

Concerns about sex-ratio imbalances, welfare of offspring, and technologizing reproduction may be less central to debates over PGS than whether such practices would be sexist or contribute to sexism. If the number of persons choosing PGS is small, or the technique is used solely for offspring gender variety, sex-ratio imbalances should not be a problem. If use patterns did produce drastic changes in sex ratios, self-correcting or regulatory mechanisms might come into play. For example, an overabundance of males would mean fewer females to marry, which would make being male less desirable, and provide incentives to increase the number of female births. Alternatively, laws or policies that required providers of PGS to select for males and females in equal numbers would prevent such imbalances.[10] A serious threat of a sex-ratio imbalance would surely constitute the compelling harm necessary to justify limits on reproductive choice.

It may also be difficult to show that children born after PGS were harmed by use of the technique. Parents who use PGS may indeed have specific gender role expectations of their children, but so will parents who have a child of a preferred gender through coitus. Children born with the desired gender after PGS will presumably be wanted and loved by the parents who sought this technique. Parents who choose PGS should be informed of the risk that the technique will not succeed, and counseled about what steps they will take if a child of the undesired gender is born.[11] If they commit themselves in advance to the well-being of the child, whatever its gender, the risk to children should be slight. However, it is possible that some couples will abort if the fetus is of the undesired gender. PGS might thus inadvertently increase the number of gender-selection abortions.

Finally, technological assistance in reproduction is now so prevalent and entrenched that a ban on PGS would probably have little effect on the use of genetic and reproductive technologies in other situations. With some form of prenatal screening of fetuses occurring in over 80 percent of United States pregnancies, genetic selection by negative exclusion is already well-installed in contemporary reproductive practice. Although there are valid concerns about whether positive forms of selection, including nonmedical genetic alteration of offspring genes, should also occur, drawing the line at all uses of PGS will not stop the larger social and technological forces that lead parents to use genetic knowledge to have healthy, wanted offspring. If a particular technique can be justified on its own terms, it should not be barred because of speculation of a slippery slope toward genetic engineering of offspring traits (for an analysis of the slippery-slope problem with genetic selection, see Robertson 1994, 162–65).

Is gender selection inherently sexist?

A central ethical concern with PGS is the effect of such practices on women, who in most societies have been subject to disadvantage and discrimination because of their gender. Some ethicists have argued that any attention to gender, male or female, is per se sexist, and should be discouraged, regardless of whether one can show actual harmful consequences for women (see Grubb and Walsh 1994; and Wertz and Fletcher 1989). Others have argued that there are real differences between male and female children that affect parental rearing experiences and thus legitimate nonsexist reasons for some couples to prefer to rear a girl rather than a boy or vice versa, either as a single child or after they have had a child of the opposite gender.

To assess whether PGS is sexist we must first be clear about what we mean by sexism. *The Compact OED* (1991, 1727) defines sexism as "the

assumption that one sex is superior to the other and the resultant discrimination practised against members of the supposed inferior sex, especially by men against women." By this definition, sexism is wrong because it denies the essential moral, legal, and political equality between men and women. Under this definition, if a practice is not motivated by judgments or evaluations that one gender is superior to the other, or does not lead to discrimination against one gender, it is not sexist.

Professor Mary Mahowald, an American bioethicist writing from an egalitarian feminist perspective, makes the same point with a consequentialist twist:

> Selection of either males or females is justifiable on medical grounds and *morally defensible in other situations* [emphasis added] so long as the intention and the consequences of the practice are not sexist. Sexist intentions are those based on the notion that one sex is inferior to the other; sexist consequences are those that disadvantage or advantage one sex vis-à-vis the other.
>
> (2000, 121)

In my view, the OED definition, modified by Mahowald's attention to consequences, is a persuasive account of the concept of sexism. If that account is correct, then not all attention to the biologic, social, cultural, or psychological differences between the sexes would necessarily be sexist or disadvantage females. That is, one could recognize that males and females have different experiences and identities because of their gender, and have a preference for rearing a child of one gender over another, without disadvantaging the dispreferred gender or denying it the equal rights, opportunities, or value as a person that constitutes sexism.

If this conjecture is correct, it would follow that some uses of PGS would clearly be sexist, while others would clearly not be. It would be sexist to use PGS to produce males because of a parental belief that males are superior to females. It would be nonsexist to use PGS to produce a girl because of a parental recognition that the experience of having and rearing a girl will be different than having a boy. In the latter case, PGS would not rest on a notion of the greater superiority of one gender over another, nor, if it occurred in countries that legally recognize the equal rights of women, would it likely contribute to sexism or further disadvantage women. As Christine Overall, a British feminist bioethicist, has put it, "sexual similarity or sexual complementarity are morally acceptable reasons for wanting a child of a certain sex" (1987, 27; quoted in Mahowald 2000, 117).

Psychological research seems to support this position. It has long been established that there are differences between boys and girls in a variety of domains, such as (but not limited to) aggression, activity, toy preference, psychopathology, and spatial ability (Maccoby and Jacklin 1974; Gilligan 1980; Kimura and Hampson 1994; Feingold 1994; Collaer and Hines 1995; and Halpern 1997). Whether these differences are primarily inborn or learned, they are facts that might rationally lead people to prefer rearing a child of one gender rather than another, particularly if one has already had one or more children of a particular gender. Indeed, Supreme Court Justice Ruth Bader Ginsburg, a noted activist for women's rights before her appointment to the Supreme Court, in her opinion striking down a male-only admissions policy at the Virginia Military Institute United States v. Virginia, (116 S. Ct. 2264 [1996]), noted that:

Physical differences between men and women . . . are enduring: "[T]he two sexes are not fungible; a community made up exclusively of one [sex] is different from a community composed of both." . . . "Inherent differences" between men and women, we have come to appreciate, remain cause for celebration.

Some persons will strongly disagree with this account of sexism and argue that any attention to gender difference is inherently sexist because perceptions of gender difference are themselves rooted in sexist stereotypes. They would argue that any offspring gender preference is necessarily sexist because it values gender difference and thus reinforces sexism by accepting the gendered stereotypes that have systematically harmed women (Grubb and Walsh 1994; and Wertz and Fletcher 1989, 21[12]). According to them, a couple with three boys who use PGS to have a girl are likely to be acting on the basis of deeply engrained stereotypes that harm women. Similarly, a couple's wish to have only a girl might contribute to unjustified gender discrimination against both men and women, even if the couple especially valued females and would insist that their daughter receive every benefit and opportunity accorded males.

Resolution of this controversy depends ultimately on one's view of what constitutes sexism and what actions are likely to harm women. Although any recognition of gender difference must be treated cautiously, I submit that recognizing and preferring one type of childrearing experience over the other can occur without disadvantaging women generally or denying them equal rights and respect. On this view, sexism arises not from the recognition or acceptance of difference, but from unjustified reactions to it. Given the biological and psychological differences between male and female children, parents with a child of one gender might without being sexist prefer that their next child be of the opposite gender. Similarly, some parents might also prefer that their firstborn or only child be of a particular gender because they desire a specific rearing and companionship experience.

If it is correct that using PGS for offspring diversity is sexist, then those who deny that biological gender differences exist, or who assume that any recognition of them always reinforces sexism or disadvantages women, will not have carried the burden of showing that a couple's use of PGS for offspring gender variety or other nonintentionally sexist uses is so harmful to women that it justifies restricting procreative choice. Until a clearer ethical argument emerges, or there is stronger empirical evidence that most choices to select the gender of offspring would be harmful, policies to prohibit or condemn as unethical all uses of nonmedically indicated PGS would not be justified.

The matter is further complicated by the need to respect a woman's autonomy in determining whether a practice is sexist. If a woman is freely choosing to engage in gender selection, even gender-selection abortion, she is exercising procreative autonomy. One might argue in response that the woman choosing PGS or abortion for gender selection is not freely choosing if her actions are influenced by strong cultural mores that prefer males over females. Others, however, would argue that the straighter path to equal rights is to respect female reproductive autonomy whenever it is exercised, even if particular exercises of autonomy are strongly influenced by the sexist norms of her community (Mahowald 2000, 188).

Public policy and preconception gender selection

Because of the newness of PGS and uncertainties about its effects, the best societal approach would, of course, be to proceed slowly, first requiring extensive studies of safety and efficacy, and then at first only permitting PGS for increasing the gender variety of offspring in particular families.[13] Only after the demographic and other effects of PGS for gender variety have been found acceptable should PGS be available for firstborn children.

However, given the close connection between parental gender preferences for offspring and reproductive choice, public policies that bar all nonmedical uses of PGS or that restrict it to choosing gender variety in offspring alone could be found unconstitutional or illegal. If there are physical, social, and cultural differences between girls and boys that affect the rearing or relational experiences of parents, individuals and couples would have the right to implement those preferences as part of their fundamental procreative liberty. The risk that exercising rights of procreative liberty would hurt offspring or women – or contribute to sexism generally – is too speculative and uncertain to justify infringement of those rights.

The claim of a right to choose offspring gender is clearest in the case of PGS for gender variety. If flow cytometry or other methods of PGS are found to be safe and effective, there would be no compelling reason to ban or restrict their nonmedical use by persons seeking gender variety in the children they rear. Couples with one child or several children of a particular gender might, without being sexist or disadvantaging a particular gender, prefer to have an additional child of the opposite gender. ART clinics should be free to proceed with PGS for offspring variety in cases where couples are aware of the risk of failure, and have undergone counseling that indicates that they will accept and love children of the dispreferred gender if PGS fails. Clinics providing PGS should also ask couples to participate in research to track and assess the effects of PGS on children and families.

The use of PGS to determine the gender of firstborn children is a more complicated question. The choice to have one's first or only child be female has the least risk of being sexist, because it is privileging or giving first place to females, who have traditionally been disfavored.[14] The use of PGS to select firstborn males is more problematic because of the greater risk that this choice reflects sexist notions that males are more highly valued. It is also more likely to entrench male dominance. The danger of sexism is probably highest in those ethnic communities that place a high premium on male offspring, but it could exist independently of those settings.

Yet restricting PGS to offspring gender variety and firstborn females may be difficult to justify. Given that individuals could prefer to have a boy rather than a girl because of the relational and rearing experiences he will provide, just as they might prefer a girl for those reasons, it might be difficult to show that all preferences for firstborn males are sexist. Nor could one easily distinguish firstborn male preferences when the couple demanding them is of a particular ethnic origin. Although the risk that firstborn male preferences would be sexist is greatest if the PGS occurred in a country in which those beliefs prevailed, the chance that PGS would contribute to societal sexism lessens greatly if the child is reared in a country that legally protects the equal status of women and men.

If prohibitions on some or all nonmedical uses of PGS could not be justified and might even be unconstitutional, regulation would have to take different forms. One form would be to deny public or private insurance funding of PGS procedures, which would mean that only those willing to pay out-of-pocket would utilize them. Another form would be for the physicians who control access to PGS techniques to take steps to assure that it is used wisely. If they comply with laws banning discrimination, physician organizations or ART clinics could set guidelines concerning access to PGS. They might, for example, limit its use to offspring gender variety or firstborn female preferences only. As a

condition of providing services, they might also require that any couple or individual seeking PGS receive counseling about the risks of failure and commit to rear a child even if its gender is other than that sought through PGS.[15] Although such guidelines would not have the force of statutory law, they could affect the eligibility of ART clinics to list their ART success rates in national registries and could help define the standard of care in malpractice cases.

Conclusion

The successful development of flow cytometry separation of X- and Y-bearing sperm would make safe, effective, and relatively inexpensive means of nonmedical preconception gender selection available for selecting female, if not also for male, offspring. The nonmedical use of PGS raises important ethical, legal, and social issues, including the charge that any or most uses of PGS would be sexist and should therefore be banned or discouraged. Assessment of this charge, however, shows that the use of PGS to achieve offspring gender variety and (in some cases) even firstborn gender preference, may not be inherently sexist or disadvantaging of women. Although it would be desirable to have extensive experience using PGS to increase the variety of offspring gender before extending it to firstborn gender preferences, it may not be legally possible to restrict the technique in this way. However, practitioners offering PGS should restrict their PGS practice to offspring gender variety until further debate and analysis of the issues has occurred. In any event, physicians offering PGS should screen and counsel prospective users to assure that persons using PGS are committed to the well-being of their children, whatever their gender.

A policy solution that gives practitioners and patients primary control without direct legal or social oversight, although not ideal, may be the best way to deal with new reprogenetic techniques. Society should not prohibit or substantially burden reproductive decisions without stronger evidence of harm than PGS now appears to present. Ultimately, the use of PGS and other reprogenetic procedures will depend on whether they satisfy ethical norms of care and concern for children while meeting the needs of prospective parents.

Notes

1 "Preconception gender selection" (PGS) rather than "preconception sex selection" (PSS) is used throughout this article to convey the importance of the social and psychological meanings with which biologic sex is invested for prospective parents and society generally. Because earlier versions of this article used "sex" rather than "gender," commentators may not have had the opportunity to revise their comments in response to the change. For the discussion at hand, either "sex" or "gender," "PSS" or "PGS" is acceptable.

2 When combined with the X-chromosomes of oocytes, X-bearing sperm can produce only XX or female offspring. Similarly, Y-bearing sperm combined with the X-chromosome of oocytes can produce only XY or male offspring.

3 See the study's web page, www.microsort.net. This article uses the term "gender variety" rather than "family balancing" to avoid the misconception that a family is "unbalanced" if it has many or only children of one gender. (I am grateful to George Annas for this suggestion.)

4 See the study's web page, www.microsort.net. Because Y-bearing sperm are smaller and contain less DNA, there is more chance that the sorting machine will fail to distinguish X's and Y's, and thus provide samples that are insufficiently enriched with Y-sperm to give a high chance of having a male child.

5 Presumably flow cytometry separation of sperm could occur with donor as well as couple sperm. It could also be requested by couples undergoing IVF or intracytoplasmic sperm injection (ICSI) who request that the sperm provided be enriched or chosen to effect the gender of choice.

6 The current cost of $1,500 per insemination cycle should decrease as further progress in the field occurs.

7 Persons requesting PGS for gender balancing in the Fairfax study had an average of 3.4 children of the same gender, and sought boys and girls in roughly equal numbers (Edward Fugger, personal communication to author, 23 February 2000).

8 See Chen (1999). The article describes immigration of a middle class family from the Indian state of Gujarat to Bridgewater, New Jersey, a suburb 40 miles from New York City.

9 It should be emphasized that the right claimed here is a negative right against government interference, not an obligation of a particular provider or public or private insurers to provide those services.

10 See Glover (1994). Professor Glover has apparently changed his position from the more negative one he took in his earlier *Ethics of New Reproductive Technologies: The Glover Report to the European Commission* (1989, 141–44). See also Jones (1992).

11 The risk arises because flow cytometry separation can only provide a greatly enriched sample of X- or Y-bearing sperm for insemination. It cannot guarantee that every sperm in the sample is either X or Y.

12 Wertz and Fletcher overlook how one could have gender preferences based on perceptions of experiential and rearing differences, rather than on differences in the worth or rights of women, when they assert that any form of gender selection violates the principle of equality between the genders "because it is premised upon a belief in sexual inequality."

13 FDA approval of the safety and efficacy of flow cytometry methods of PGS would also be required before widespread use.

14 Persons taking a more purist approach to sexism would, of course, differ with this assessment.

15 They might also require that consumers agree to participate in research so that policymakers will have reliable information about the uses of PGS and the problems it presents.

References

Belkin, L. 1999. "Getting the girl." *New York Times Magazine*, 25, 26.

Chen, D. W. 1999. "Asian middle class alters a rural enclave." *New York Times*, 28 December, A1.

Collaer, M. L., and M. Hines. 1995. "Human behavioral sex differences: A role for gonadal hormones during early development?" *Psychological Bulletin* 118(1): 55–107.

Compact OED, The, new ed. 1991. New York: Oxford University Press.

Ethics Committee of the American Society of Reproductive Medicine, The 1999. "Sex selection and pre-implantation genetic diagnosis." *Fertility and Sterility* 72(4): 595.

Feingold, A. 1994. "Gender differences in personality: A meta-analysis." *Psychological Bulletin* 116(3): 429–56.

Fugger, E. F., S. H. Black, K. Keyvanfar, and J. D. Schulman. 1998. "Births of normal daughters after microsort sperm separation and intrauterine insemination, in-vitro fertilization, or intracytoplasmic sperm injection." *Human Reproduction* 13:2367.

Gilligan, C. 1980. *In a different voice*. Cambridge: Harvard University Press.

Glover, J. 1989. *Ethics of new reproductive technologies: The Glover report to the European Commission*. DeKalb, IL: Northern Illinois University Press.

———. 1994. "Comments on some ethical issues in sex selection" 6. Paper presented at International Symposium on Ethics in Medicine and Reproductive Biology, July.

Greenhalgh, S., and J. Li 1995. "Engendering reproductive policy and practice in peasant china: For a feminist demography of reproduction." *Signs* 20:601.

Grubb, A., and P. Walsh. 1994. "Gender-vending II." *Dispatches* 1 (summer).

Halpern, D. F. 1997. "Sex differences in intelligence: Implications for education." *American Psychologist* 52(10): 1091–1102.

Jones, O. D. 1992. "Sex selection: Regulating technology enabling predetermination of a child's gender." *Harvard Journal of Law and Technology* 6:51.

Kimura, D., and E. Hampson. 1994. "Cognitive pattern in men and women is influenced by fluctuations in sex hormones." *Current Directions in Psychological Science* 3(2): 57–61.

Levinson, G., K. Keyvanfar, and J. C. Wu. 1995. "DNA based X-enriched sperm separation as an adjunct to preimplantation genetic testing for the prevention of X-linked disease." *Human Reproduction* 10:979–82.

Maccoby, E. E., and C. N. Jacklin. 1974. *The psychology of sex differences*. Palo Alto: Stanford University Press.

Macklin, R. 1999. *Against relativism*. New York: Oxford University Press.

Mahowald, M. B. 2000. *Genes, women, equality*. New York: Oxford University Press.

Overall, C. 1987. *Ethics and human reproduction*. Boston: Allen and Unwin.

Robertson, J. A. 1994. *Children of choice: Freedom and the new reproductive technologies*. Princeton: Princeton University Press.

———1996. "Genetic selection of offspring characteristics." *Boston University Law Review*. 76:421.

Sen, A. 1990. "More than 100 million women are missing." *New York Review of Books*, 20 December, 61–66.

Tarin, J. J., and A. H. Handyside. 1993. "Embryo biopsy strategies for preimplantation diagnosis." *Fertility and Sterility* 59:943.

Vidal, F., J. Blanco, E. F. Fugger, et al. 1999. "Preliminary study of the incidence of disomy in sperm fractions after microsort flow cytometry." *Human Reproduction* 14:2987.

Wertz, D. C., and J. C. Fletcher. 1989. "Fatal knowledge: Prenatal diagnosis and sex selection." *Hastings Center Report* 19:21.

Chapter 8

Michael J. Sandel

THE CASE AGAINST PERFECTION: WHAT'S WRONG WITH DESIGNER CHILDREN, BIONIC ATHLETES, AND GENETIC ENGINEERING?

Breakthroughs in genetics present us with a promise and a predicament. The promise is that we may soon be able to treat and prevent a host of debilitating diseases. The predicament is that our newfound genetic knowledge may also enable us to manipulate our own nature – to enhance our muscles, memories, and moods; to choose the sex, height, and other genetic traits of our children; to make ourselves "better than well." When science moves faster than moral understanding, as it does today, men and women struggle to articulate their unease. In liberal societies they reach first for the language of autonomy, fairness, and individual rights. But this part of our moral vocabulary is ill equipped to address the hardest questions posed by genetic engineering. The genomic revolution has induced a kind of moral vertigo.

Consider cloning. The birth of Dolly the cloned sheep, in 1997, brought a torrent of concern about the prospect of cloned human beings. There are good medical reasons to worry. Most scientists agree that cloning is unsafe, likely to produce offspring with serious abnormalities. (Dolly recently died a premature death.) But suppose technology improved to the point where clones were at no greater risk than naturally conceived offspring. Would human cloning still be objectionable? Should our hesitation be moral as well as medical? What, exactly, is wrong

with creating a child who is a genetic twin of one parent, or of an older sibling who has tragically died – or, for that matter, of an admired scientist, sports star, or celebrity?

Some say cloning is wrong because it violates the right to autonomy: by choosing a child's genetic makeup in advance, parents deny the child's right to an open future. A similar objection can be raised against any form of bioengineering that allows parents to select or reject genetic characteristics. According to this argument, genetic enhancements for musical talent, say, or athletic prowess, would point children toward particular choices, and so designer children would never be fully free.

At first glance the autonomy argument seems to capture what is troubling about human cloning and other forms of genetic engineering. It is not persuasive, for two reasons. First, it wrongly implies that absent a designing parent, children are free to choose their characteristics for themselves. But none of us chooses his genetic inheritance. The alternative to a cloned or genetically enhanced child is not one whose future is unbound by particular talents but one at the mercy of the genetic lottery.

Second, even if a concern for autonomy explains some of our worries about made-to-order children, it cannot explain our moral hesitation about people who seek genetic remedies or enhancements for themselves. Gene therapy on somatic (that is, nonreproductive) cells, such as muscle cells and brain cells, repairs or replaces defective genes. The moral quandary arises when people use such therapy not to cure a disease but to reach beyond health, to enhance their physical or cognitive capacities, to lift themselves above the norm.

Like cosmetic surgery, genetic enhancement employs medical means for nonmedical ends – ends unrelated to curing or preventing disease or repairing injury. But unlike cosmetic surgery, genetic enhancement is more than skin-deep. If we are ambivalent about surgery or Botox injections for sagging chins and furrowed brows, we are all the more troubled by genetic engineering for stronger bodies, sharper memories, greater intelligence, and happier moods. The question is whether we are right to be troubled, and if so, on what grounds.

In order to grapple with the ethics of enhancement, we need to confront questions largely lost from view – questions about the moral status of nature, and about the proper stance of human beings toward the given world. Since these questions verge on theology, modern philosophers and political theorists tend to shrink from them. But our new powers of biotechnology make them unavoidable. To see why this is so, consider four examples already on the horizon: muscle enhancement, memory enhancement, growth-hormone treatment, and reproductive technologies that enable parents to choose the sex and some genetic traits of their children. In each case what began as an attempt to treat a disease or prevent a genetic disorder now beckons as an instrument of improvement and consumer choice.

Muscles. Everyone would welcome a gene therapy to alleviate muscular dystrophy and to reverse the debilitating muscle loss that comes with old age. But what if the same therapy were used to improve athletic performance? Researchers have developed a synthetic gene that, when injected into the muscle cells of mice, prevents and even reverses natural muscle deterioration. The gene not only repairs wasted or injured muscles but also strengthens healthy ones. This success bodes well for human applications. H. Lee Sweeney, of the University of Pennsylvania, who leads the research, hopes his discovery will cure the immobility that afflicts the elderly. But Sweeney's bulked-up mice have already attracted the attention of athletes seeking a competitive edge. Although the therapy is not yet approved for human use, the prospect of genetically enhanced weight lifters, home-run sluggers, linebackers, and sprinters is easy to imagine. The widespread use of steroids and other performance-improving drugs in professional sports suggests that many athletes will be eager to avail themselves of genetic enhancement.

Suppose for the sake of argument that muscle-enhancing gene therapy, unlike steroids, turned out to be safe – or at least no riskier than a rigorous weight-training regimen. Would there be a reason to ban its use in sports? There is something unsettling about the image of genetically altered athletes lifting SUVs or hitting 650-foot home runs or running a three-minute mile. But what, exactly, is troubling about it? Is it simply that we find such superhuman spectacles too bizarre to contemplate? Or does our unease point to something of ethical significance?

It might be argued that a genetically enhanced athlete, like a drug-enhanced athlete, would have an unfair advantage over his unenhanced competitors. But the fairness argument against enhancement has a fatal flaw: it has always been the case that some athletes are better endowed genetically than others, and yet we do not consider this to undermine the fairness of competitive sports. From the standpoint of fairness, enhanced genetic differences would be no worse than natural ones, assuming they were safe and made available to all. If genetic enhancement in sports is morally objectionable, it must be for reasons other than fairness.

Memory. Genetic enhancement is possible for brains as well as brawn. In the mid-1990s scientists managed to manipulate a memory-linked gene in fruit flies, creating flies with photographic memories. More recently researchers have produced smart mice by inserting extra copies of a memory-related gene into mouse embryos. The altered mice learn more quickly and remember things longer than normal mice. The extra copies were programmed to remain active even in old age, and the improvement was passed on to offspring.

Human memory is more complicated, but biotech companies, including Memory Pharmaceuticals, are in hot pursuit of memory-enhancing drugs, or "cognition enhancers," for human beings. The obvious market for such drugs consists of those who suffer from Alzheimer's and other serious memory disorders. The companies also have their sights on a bigger market: the 81 million Americans over fifty, who are beginning to encounter the memory loss that comes naturally with age. A drug that reversed age-related memory loss would be a bonanza for the pharmaceutical industry: a Viagra for the brain. Such use would straddle the line between remedy and enhancement. Unlike a treatment for Alzheimer's, it would cure no disease; but insofar as it restored capacities a person once possessed, it would have a remedial aspect. It could also have purely nonmedical uses: for example, by a lawyer cramming to memorize facts for an upcoming trial, or by a business executive eager to learn Mandarin on the eve of his departure for Shanghai.

Some who worry about the ethics of cognitive enhancement point to the danger of creating two classes of human beings: those with access to enhancement technologies, and those who must make do with their natural capacities. And if the enhancements could be passed down the generations, the two classes might eventually become subspecies – the enhanced and the merely natural. But worry about access ignores the moral status of enhancement itself. Is the scenario troubling because the unenhanced poor would be denied the benefits of bioengineering, or because the enhanced affluent would somehow be dehumanized? As with muscles, so with memory: the fundamental question is not how to ensure equal access to enhancement but whether we should aspire to it in the first place.

Height. Pediatricians already struggle with the ethics of enhancement when confronted by parents who want to make their children taller. Since the 1980s human growth hormone has been approved for children with a hormone deficiency that makes them much shorter than average. But the treatment also increases the height of healthy children. Some parents of healthy children who are unhappy with their stature (typically boys) ask why it should make a difference whether a child is short because of a hormone deficiency or because his parents happen to be short. Whatever the cause, the social consequences are the same.

In the face of this argument some doctors began prescribing hormone treatments for children whose short stature was unrelated to any medical problem. By 1996 such "off-label" use accounted for 40 percent of human-growth-hormone prescriptions. Although it is legal to prescribe drugs for purposes not approved by the Food and Drug Administration, pharmaceutical companies cannot promote such use. Seeking to expand its market, Eli Lilly & Co. recently persuaded the FDA to approve its human growth hormone for healthy children whose projected adult height is in the bottom one percentile – under five feet three inches for boys and four feet eleven inches for girls. This concession raises a large question about the ethics of enhancement: If hormone treatments need not be limited to those with hormone deficiencies, why should they be available only to very short children? Why shouldn't all shorter-than-average children be able to seek treatment? And what about a child of average height who wants to be taller so that he can make the basketball team?

Some oppose height enhancement on the grounds that it is collectively self-defeating; as some become taller, others become shorter relative to the norm. Except in Lake Wobegon, not every child can be above average. As the unenhanced began to feel shorter, they, too, might seek treatment, leading to a hormonal arms race that left everyone worse off, especially those who couldn't afford to buy their way up from shortness.

But the arms-race objection is not decisive on its own. Like the fairness objection to bioengineered muscles and memory, it leaves unexamined the attitudes and dispositions that prompt the drive for enhancement. If we were bothered only by the injustice of adding shortness to the problems of the poor, we could remedy that unfairness by publicly subsidizing height enhancements. As for the relative height deprivation suffered by innocent bystanders, we could compensate them by taxing those who buy their way to greater height. The real question is whether we want to live in a society where parents feel compelled to spend a fortune to make perfectly healthy kids a few inches taller.

Sex selection. Perhaps the most inevitable nonmedical use of bioengineering is sex selection. For centuries parents have been trying to choose the sex of their children. Today biotech succeeds where folk remedies failed.

One technique for sex selection arose with prenatal tests using amniocentesis and ultrasound. These medical technologies were developed to detect genetic abnormalities such as spina bifida and Down syndrome. But they can also reveal the sex of the fetus – allowing for the abortion of a fetus of an undesired sex. Even among those who favor abortion rights, few advocate abortion simply because the parents do not want a girl. Nevertheless, in traditional societies with a powerful cultural preference for boys, this practice has become widespread.

Sex selection need not involve abortion, however. For couples undergoing in vitro fertilization (IVF), it is possible to choose the sex of the child before the fertilized egg is implanted in the womb. One method makes use of preimplantation genetic diagnosis (PGD), a procedure developed to screen for genetic diseases. Several eggs are fertilized in a petri dish and grown to the eight-cell stage (about three days). At that point the embryos are tested to determine their sex. Those of the desired sex are implanted; the others are typically discarded. Although few couples are likely to undergo the difficulty and expense of IVF simply to choose the sex of their child, embryo screening is a highly reliable means of sex selection. And as our genetic knowledge increases, it may be possible to use PGD to cull embryos carrying undesired genes, such as those associated with obesity, height, and skin color. The science-fiction movie *Gattaca* depicts a future in which parents routinely screen embryos for sex, height, immunity to disease, and even IQ. There is something troubling about the *Gattaca* scenario, but it is not easy to identify what exactly is wrong with screening embryos to choose the sex of our children.

One line of objection draws on arguments familiar from the abortion debate. Those who believe that an embryo is a person reject embryo screening for the same reasons they reject abortion. If an eight-cell embryo growing in a petri dish is morally equivalent to a fully developed human being, then discarding it is no better than aborting a fetus, and both practices are equivalent to infanticide. Whatever its merits, however, this "pro-life" objection is not an argument against sex selection as such.

The latest technology poses the question of sex selection unclouded by the matter of an embryo's moral status. The Genetics & IVF Institute, a for-profit infertility clinic in Fairfax, Virginia, now offers a sperm-sorting technique that makes it possible to choose the sex of one's child before it is conceived. X-bearing sperm, which produce girls, carry more DNA than Y-bearing sperm, which produce boys; a device called a flow cytometer can separate them. The process, called MicroSort, has a high rate of success.

If sex selection by sperm sorting is objectionable, it must be for reasons that go beyond the debate about the moral status of the embryo. One such reason is that sex selection is an instrument of sex discrimination – typically against girls, as illustrated by the chilling sex ratios in India and China. Some speculate that societies with substantially more men than women will be less stable, more violent, and more prone to crime or war. These are legitimate worries – but the sperm-sorting company has a clever way of addressing them. It offers MicroSort only to couples who want to choose the sex of a child for purposes of "family balancing." Those with more sons than daughters may choose a girl, and vice versa. But customers may not use the technology to stock up on children of the same sex, or even to choose the sex of their firstborn child. (So far the majority of MicroSort clients have chosen girls.) Under restrictions of this kind, do any ethical issues remain that should give us pause?

The case of MicroSort helps us isolate the moral objections that would persist if muscle-enhancement, memory-enhancement, and height-enhancement technologies were safe and available to all.

It is commonly said that genetic enhancements undermine our humanity by threatening our capacity to act freely, to succeed by our own efforts, and to consider ourselves responsible – worthy of praise or blame – for the things we do and for the way we are. It is one thing to hit seventy home runs as the result of disciplined training and effort, and something else, something less, to hit them with the help of steroids or genetically enhanced muscles. Of course, the roles of effort and enhancement will be a matter of degree. But as the role of enhancement increases, our admiration for the achievement fades – or, rather, our admiration for the achievement shifts from the player to his pharmacist. This suggests that our moral response to enhancement is a response to the diminished agency of the person whose achievement is enhanced.

Though there is much to be said for this argument, I do not think the main problem with enhancement and genetic engineering is that they undermine effort and erode human agency. The deeper danger is that they represent a kind of hyperagency – a Promethean aspiration to remake nature, including human nature, to serve our purposes and satisfy our desires. The problem is not the drift to mechanism but the drive to mastery. And what the drive to mastery misses and may even destroy is an appreciation of the gifted character of human powers and achievements.

To acknowledge the giftedness of life is to recognize that our talents and powers are not wholly our own doing, despite the effort we expend to develop and to exercise them. It is also to recognize that not everything in the world is open to whatever use we may desire or devise. Appreciating the gifted quality of life constrains the Promethean project and conduces to a certain humility. It is in part a religious sensibility. But its resonance reaches beyond religion.

It is difficult to account for what we admire about human activity and achievement without drawing upon some version of this idea. Consider two types of athletic achievement. We appreciate players like Pete Rose, who are not blessed with great natural gifts but who manage, through striving, grit, and determination, to excel in their sport. But we also admire players like Joe DiMaggio, who display natural gifts with grace and effortlessness. Now, suppose we learned that both players took performance-enhancing drugs. Whose turn to drugs would we find more deeply disillusioning? Which aspect of the athletic ideal – effort or gift – would be more deeply offended?

Some might say effort: the problem with drugs is that they provide a shortcut, a way to win without striving. But striving is not the point of sports; excellence is. And excellence consists at least partly in the display of natural talents and gifts that are no doing of the athlete who possesses them. This is an uncomfortable fact for democratic societies. We want to believe that success, in sports and in life, is something we earn, not something we inherit. Natural gifts, and the admiration they inspire, embarrass the meritocratic faith; they cast doubt on the conviction that praise and rewards flow from effort alone. In the face of this embarrassment we inflate the moral significance of striving, and depreciate giftedness. This distortion can be seen, for example, in network-television coverage of the Olympics, which focuses less on the feats the athletes perform than on heartrending stories of the hardships they have overcome and the struggles they have waged to triumph over an injury or a difficult upbringing or political turmoil in their native land.

But effort isn't everything. No one believes that a mediocre basketball player who works and trains even harder than Michael Jordan deserves greater acclaim or a bigger contract. The real problem with genetically altered athletes is that they corrupt athletic competition as a human activity that honors the cultivation and display of natural talents. From this standpoint, enhancement can be seen as the ultimate expression of the ethic of effort and willfulness – a kind of high-tech striving. The ethic of willfulness and the biotechnological powers it now enlists are arrayed against the claims of giftedness.

The ethic of giftedness, under siege in sports, persists in the practice of parenting. But here, too, bioengineering and genetic enhancement threaten to dislodge it. To appreciate children as gifts is to accept them as they come, not as objects of our design or products of our will or instruments of our ambition. Parental love is not contingent on the talents and attributes a child happens to have. We choose our friends and spouses at least partly on the basis of qualities we find attractive. But we do not choose our children. Their qualities are unpredictable, and even the most conscientious parents cannot be held wholly responsible for the kind of children they have. That is why parenthood, more than other human relationships, teaches what the theologian William F. May calls an "openness to the unbidden."

May's resonant phrase helps us see that the deepest moral objection to enhancement lies less in the perfection it seeks than in the human disposition it expresses and promotes. The problem is not that parents usurp the autonomy of a child they design. The problem lies in the hubris of the designing parents, in their drive to master the mystery of birth. Even if this disposition did not make parents tyrants to their children, it would disfigure the relation between parent and child, and deprive the parent of the humility and enlarged human sympathies that an openness to the unbidden can cultivate.

To appreciate children as gifts or blessings is not, of course, to be passive in the face of illness or disease. Medical intervention to cure or prevent illness or restore the injured to health does not desecrate nature but honors it. Healing sickness or injury does not override a child's natural capacities but permits them to flourish.

Nor does the sense of life as a gift mean that parents must shrink from shaping and directing the development of their child. Just as athletes and artists have an obligation to cultivate their talents, so parents have an obligation to cultivate their children, to help them discover and develop their talents and gifts. As May points out, parents give their children two kinds of love: accepting love and transforming love. Accepting love affirms the being of the child, whereas transforming love seeks the well-being of the child. Each aspect corrects the excesses of the other, he writes: "Attachment becomes too quietistic if it slackens into mere acceptance of the child as he is." Parents have a duty to promote their children's excellence.

These days, however, overly ambitious parents are prone to get carried away with transforming love – promoting and demanding all manner of accomplishments from their children, seeking perfection. "Parents find it difficult to maintain an equilibrium between the two sides of love," May observes. "Accepting love, without transforming love, slides into indulgence and finally neglect. Transforming love, without accepting love, badgers and finally rejects." May finds in these competing impulses a parallel with modern science: it, too, engages us in beholding the given world, studying and savoring it, and also in molding the world, transforming and perfecting it.

The mandate to mold our children, to cultivate and improve them, complicates the case against enhancement. We usually admire parents who seek the best for their children, who spare no effort to help them achieve happiness and success. Some parents confer advantages on their children by enrolling them in expensive schools, hiring private tutors, sending them to tennis camp, providing them with piano lessons, ballet lessons, swimming lessons, SAT-prep courses, and so on. If it is permissible and even admirable for parents to help their children in these ways, why isn't it equally admirable for parents to use whatever genetic technologies may emerge (provided they are safe) to enhance their children's intelligence, musical ability, or athletic prowess?

The defenders of enhancement are right to this extent: improving children through genetic engineering is similar in spirit to the heavily managed, high-pressure child-rearing that is now common. But this similarity does not vindicate genetic enhancement. On the contrary, it highlights a problem with the trend toward hyperparenting. One conspicuous example of this trend is sports-crazed parents bent on making champions of their children. Another is the frenzied drive of overbearing parents to mold and manage their children's academic careers.

As the pressure for performance increases, so does the need to help distractible children concentrate on the task at hand. This may be why diagnoses of attention deficit and hyperactivity disorder have increased so sharply. Lawrence Diller, a pediatrician and the author of *Running on Ritalin*, estimates that five to six percent of American children under eighteen (a total of four to five million kids) are currently prescribed Ritalin, Adderall, and other stimulants, the treatment of choice for ADHD. (Stimulants counteract hyperactivity by making it easier to focus and sustain attention.) The number of Ritalin prescriptions for children and adolescents has tripled over the past decade, but not all users suffer from attention disorders or hyperactivity. High school and college students have learned that prescription stimulants improve concentration for those with normal attention spans, and some buy or borrow their classmates' drugs to enhance their performance on the SAT or other exams. Since stimulants work for both medical and nonmedical purposes, they raise the same moral questions posed by other technologies of enhancement.

However those questions are resolved, the debate reveals the cultural distance we have traveled since the debate over marijuana, LSD, and other drugs a generation ago. Unlike the drugs of the 1960s and 1970s, Ritalin and Adderall are not for checking out but for buckling down, not for beholding the world and taking it in but for molding the world and fitting in. We used to speak of nonmedical drug use as "recreational." That term no longer applies. The steroids and stimulants that figure in the enhancement debate are not a source of recreation but a bid for compliance – a way of answering a competitive society's demand to improve our performance and perfect our nature. This demand for performance and perfection animates the impulse to rail against the given. It is the deepest source of the moral trouble with enhancement.

Some see a clear line between genetic enhancement and other ways that people seek improvement in their children and themselves. Genetic manipulation seems somehow worse – more intrusive, more sinister – than other ways of enhancing performance and seeking success. But morally speaking, the difference is less significant than it seems. Bioengineering gives us reason to question the low-tech, high-pressure child-rearing practices we commonly accept. The hyperparenting familiar in our time represents an anxious excess of mastery and dominion that misses the sense of life as a gift. This draws it disturbingly close to eugenics.

The shadow of eugenics hangs over today's debates about genetic engineering and enhancement. Critics of genetic engineering argue that human cloning, enhancement, and the quest for designer children are nothing more than "privatized" or "free-market" eugenics. Defenders of enhancement reply that genetic choices freely made are not really eugenic – at least not in the pejorative sense. To remove the coercion, they argue, is to remove the very thing that makes eugenic policies repugnant.

Sorting out the lesson of eugenics is another way of wrestling with the ethics of enhancement. The Nazis gave eugenics a bad name. But what, precisely, was wrong with it? Was the old eugenics objectionable only insofar as it was coercive? Or is there something inherently

wrong with the resolve to deliberately design our progeny's traits?

James Watson, the biologist who, with Francis Crick, discovered the structure of DNA, sees nothing wrong with genetic engineering and enhancement, provided they are freely chosen rather than state-imposed. And yet Watson's language contains more than a whiff of the old eugenic sensibility. "If you really are stupid, I would call that a disease," he recently told The Times of London. "The lower 10 percent who really have difficulty, even in elementary school, what's the cause of it? A lot of people would like to say, 'Well, poverty, things like that.' It probably isn't. So I'd like to get rid of that, to help the lower 10 percent." A few years ago Watson stirred controversy by saying that if a gene for homosexuality were discovered, a woman should be free to abort a fetus that carried it. When his remark provoked an uproar, he replied that he was not singling out gays but asserting a principle: women should be free to abort fetuses for any reason of genetic preference – for example, if the child would be dyslexic, or lacking musical talent, or too short to play basketball.

Watson's scenarios are clearly objectionable to those for whom all abortion is an unspeakable crime. But for those who do not subscribe to the pro-life position, these scenarios raise a hard question: If it is morally troubling to contemplate abortion to avoid a gay child or a dyslexic one, doesn't this suggest that something is wrong with acting on any eugenic preference, even when no state coercion is involved?

Consider the market in eggs and sperm. The advent of artificial insemination allows prospective parents to shop for gametes with the genetic traits they desire in their offspring. It is a less predictable way to design children than cloning or pre-implantation genetic screening, but it offers a good example of a procreative practice in which the old eugenics meets the new consumerism. A few years ago some Ivy League newspapers ran an ad seeking an egg from a woman who was at least five feet ten inches tall and athletic, had no major family medical problems, and had a combined SAT score of 1400 or above. The ad offered $50,000 for an egg from a donor with these traits. More recently a Web site was launched claiming to auction eggs from fashion models whose photos appeared on the site, at starting bids of $15,000 to $150,000.

On what grounds, if any, is the egg market morally objectionable? Since no one is forced to buy or sell, it cannot be wrong for reasons of coercion. Some might worry that hefty prices would exploit poor women by presenting them with an offer they couldn't refuse. But the designer eggs that fetch the highest prices are likely to be sought from the privileged, not the poor. If the market for premium eggs gives us moral qualms, this, too, shows that concerns about eugenics are not put to rest by freedom of choice.

A tale of two sperm banks helps explain why. The Repository for Germinal Choice, one of America's first sperm banks, was not a commercial enterprise. It was opened in 1980 by Robert Graham, a philanthropist dedicated to improving the world's "germ plasm" and counteracting the rise of "retrograde humans." His plan was to collect the sperm of Nobel Prize-winning scientists and make it available to women of high intelligence, in hopes of breeding supersmart babies. But Graham had trouble persuading Nobel laureates to donate their sperm for his bizarre scheme, and so settled for sperm from young scientists of high promise. His sperm bank closed in 1999.

In contrast, California Cryobank, one of the world's leading sperm banks, is a for-profit company with no overt eugenic mission. Cappy Rothman, M.D., a co-founder of the firm, has nothing but disdain for Graham's eugenics, although the standards Cryobank imposes on the sperm it recruits are exacting. Cryobank has offices in Cambridge, Massachusetts, between Harvard and MIT, and in Palo Alto, California, near Stanford. It advertises for donors in campus newspapers (compensation up to $900 a month), and accepts less than five percent of the men who apply. Cryobank's marketing materials play up the prestigious source of its sperm. Its catalogue provides detailed information about the physical characteristics of each donor, along with his ethnic origin and college major. For an extra fee prospective customers can buy the results of a test that assesses the donor's temperament and character type. Rothman reports that Cryobank's ideal sperm donor is six feet tall, with brown eyes, blond hair, and dimples, and has a college degree – not because the company wants to propagate those traits, but because those are the traits his customers want: "If our customers wanted high school dropouts, we would give them high school dropouts."

Not everyone objects to marketing sperm. But anyone who is troubled by the eugenic aspect of the Nobel Prize sperm bank should be equally troubled by Cryobank, consumer-driven though it be. What, after all, is the moral difference between designing children according to an explicit eugenic purpose and designing children according to the dictates of the market? Whether the aim is to improve humanity's "germ plasm" or to cater to consumer preferences, both practices are eugenic insofar as both make children into products of deliberate design.

A number of political philosophers call for a new "liberal eugenics." They argue that a moral distinction can be drawn between the old eugenic policies and genetic enhancements that do not restrict the autonomy of the child. "While old-fashioned authoritarian eugenicists sought to produce citizens out of a single centrally designed mould," writes Nicholas Agar, "the distinguishing mark of the new liberal eugenics is state neutrality." Government may not tell parents what sort of children to design, and parents may engineer in their children only those traits that improve their capacities without biasing their choice of life plans. A recent text on genetics and justice, written by the bioethicists Allen Buchanan, Dan W. Brock, Norman Daniels, and Daniel Wikler, offers a similar view. The "bad reputation of eugenics," they write, is due to practices that "might be avoidable in a future eugenic program." The problem with the old eugenics was that its burdens fell disproportionately on the weak and the poor, who were unjustly sterilized and segregated. But provided that the benefits and burdens of genetic improvement are fairly distributed, these bioethicists argue, eugenic measures are unobjectionable and may even be morally required.

The libertarian philosopher Robert Nozick proposed a "genetic supermarket" that would enable parents to order children by design without imposing a single design on the society as a whole: "This supermarket system has the great virtue that it involves no centralized decision fixing the future human type(s)."

Even the leading philosopher of American liberalism, John Rawls, in his classic *A Theory of Justice* (1971), offered a brief endorsement of noncoercive eugenics. Even in a society that agrees to share the benefits and burdens of the genetic lottery, it is "in the interest of each to have greater natural assets," Rawls wrote. "This enables him to pursue a preferred plan of life." The parties to the social contract "want to insure for their descendants the best genetic endowment (assuming their own to be fixed)." Eugenic policies are therefore not only permissible but required as a matter of justice. "Thus over time a society is to take steps at least to preserve the general level of natural abilities and to prevent the diffusion of serious defects."

But removing the coercion does not vindicate eugenics. The problem with eugenics and genetic engineering is that they represent the one-sided triumph of willfulness over giftedness, of dominion over reverence, of molding over beholding. Why, we may wonder, should we worry about this triumph? Why not shake off our unease about genetic enhancement as so much superstition? What would be lost if biotechnology dissolved our sense of giftedness?

From a religious standpoint the answer is clear: To believe that our talents and powers are wholly our own doing is to misunderstand our place in creation, to confuse our role with God's. Religion is not the only source of reasons to care about giftedness, however. The moral stakes can also be described in secular terms. If bioengineering made the myth of the "self-made man" come true, it would be difficult to view our talents as gifts for which we are indebted, rather than as achievements for which we are responsible. This would transform three key features of our moral landscape: humility, responsibility, and solidarity.

In a social world that prizes mastery and control, parenthood is a school for humility. That we care deeply about our children and yet cannot choose the kind we want teaches parents to be open to the unbidden. Such openness is a disposition worth affirming, not only within families but in the wider world as well. It invites us to abide the unexpected, to live with dissonance, to rein in the impulse to control. A *Gattaca*-like world in which parents became accustomed to specifying the sex and genetic traits of their children would be a world inhospitable to the unbidden, a gated community writ large. The awareness that our talents and abilities are not wholly our own doing restrains our tendency toward hubris.

Though some maintain that genetic enhancement erodes human agency by overriding effort, the real problem is the explosion, not the erosion, of responsibility. As humility gives way, responsibility expands to daunting proportions. We attribute less to chance and more to choice. Parents become responsible for choosing, or failing to choose, the right traits for their children. Athletes become responsible for acquiring, or failing to acquire, the talents that will help their teams win.

One of the blessings of seeing ourselves as creatures of nature, God, or fortune is that we are not wholly responsible for the way we are. The more we become masters of our genetic endowments, the greater the burden we bear for the talents we have and the way we perform. Today when a basketball player misses a rebound, his coach can blame him for being out of position. Tomorrow the coach may blame him for being too short. Even now the use of performance-enhancing drugs in professional sports is subtly transforming the expectations players have for one another; on some teams players who take the field free from amphetamines or other stimulants are criticized for "playing naked."

The more alive we are to the chanced nature of our lot, the more reason we have to share our fate with others. Consider insurance. Since people do not know whether or when various ills will befall them, they pool their risk by buying health insurance and life insurance. As life plays itself out, the healthy wind up subsidizing the unhealthy, and those who live to a ripe old age wind up subsidizing the families of those who die before their time. Even without a sense of mutual obligation, people pool their risks and resources and share one another's fate.

But insurance markets mimic solidarity only insofar as people do not know or control their own risk factors. Suppose genetic testing advanced to the point where it could reliably predict each person's medical future and life expectancy. Those confident of good health and long life would opt out of the pool, causing other people's premiums to skyrocket. The solidarity of insurance would disappear as those with good genes fled the actuarial company of those with bad ones.

The fear that insurance companies would use genetic data to assess risks and set premiums recently led the Senate to vote to prohibit genetic discrimination in health insurance. But the bigger danger, admittedly more speculative, is that genetic enhancement, if routinely practiced, would make it harder to foster the moral sentiments that social solidarity requires.

Why, after all, do the successful owe anything to the least-advantaged members of society? The best answer to this question leans heavily on the notion of giftedness. The natural talents that enable the successful to flourish are not their own doing but, rather, their good fortune – a result of the genetic lottery. If our genetic endowments are gifts, rather than achievements for which we can claim credit, it is a mistake and a conceit to assume that we are entitled to the full measure of the bounty they reap in a market economy. We therefore have an obligation to share this bounty with those who, through no fault of their own, lack comparable gifts.

A lively sense of the contingency of our gifts – a consciousness that none of us is wholly responsible for his or her success – saves a meritocratic society from sliding into the smug assumption that the rich are rich because they are more deserving than the poor. Without this, the successful would become even more likely than they are now to view themselves as self-made and self-sufficient, and hence wholly responsible for their success. Those at the bottom of society would be viewed not as disadvantaged, and thus worthy of a measure of compensation, but as simply unfit, and thus worthy of eugenic repair. The meritocracy, less chastened by chance, would become harder, less forgiving. As perfect genetic knowledge would end the simulacrum of solidarity in insurance markets, so perfect genetic control would erode the actual solidarity that arises when men and women reflect on the contingency of their talents and fortunes.

Thirty-five years ago Robert L. Sinsheimer, a molecular biologist at the California Institute of Technology, glimpsed the shape of things to come. In an article titled "The Prospect of Designed Genetic Change" he argued that freedom of choice would vindicate the new genetics, and set it apart from the discredited eugenics of old.

> To implement the older eugenics ... would have required a massive social programme carried out over many generations. Such a programme could not have been initiated without the consent and cooperation of a major fraction of the population, and would have been continuously subject to social control. In contrast, the new eugenics could, at least in principle, be implemented on a quite individual basis, in one generation, and subject to no existing restrictions.

According to Sinsheimer, the new eugenics would be voluntary rather than coerced, and also more humane. Rather than segregating and eliminating the unfit, it would improve them. "The old eugenics would have required a continual selection for breeding of the fit, and a culling of the unfit," he wrote. "The new eugenics would permit in principle the conversion of all the unfit to the highest genetic level."

Sinsheimer's paean to genetic engineering caught the heady, Promethean self-image of the age. He wrote hopefully of rescuing "the losers in that chromosomal lottery that so firmly channels our human destinies," including not only those born with genetic defects but also "the 50,000,000 'normal' Americans with an IQ of less than 90." But he also saw that something bigger than improving on nature's "mindless, age-old throw of dice" was at stake. Implicit in technologies of genetic intervention was a more exalted place for human beings in the cosmos. "As we enlarge man's freedom, we diminish his constraints and that which he must accept as given," he wrote. Copernicus and Darwin had "demoted man from his bright glory at the focal point of the universe," but the new biology would restore his central role. In the mirror of our genetic knowledge we would see ourselves as more than a link in the chain of evolution: "We can be the agent of transition to a whole new pitch of evolution. This is a cosmic event."

There is something appealing, even intoxicating, about a vision of human freedom unfettered by the given. It may even be the case that the allure of that vision played a part in summoning the genomic age into being. It is often assumed that the powers of enhancement we now possess arose as an inadvertent by-product of biomedical progress — the genetic revolution came, so to speak, to cure disease, and stayed to tempt us with the prospect of enhancing our performance, designing our children, and perfecting our nature. That may have the story backwards. It is more plausible to view genetic engineering as the ultimate expression of our resolve to see ourselves astride the world, the masters of our nature. But that promise of mastery is flawed. It threatens to banish our appreciation of life as a gift, and to leave us with nothing to affirm or behold outside our own will.

Chapter 9

Nick Bostrom

HUMAN GENETIC ENHANCEMENTS: A TRANSHUMANIST PERSPECTIVE

1 What is transhumanism?

Transhumanism is a loosely defined movement that has developed gradually over the past two decades. It promotes an interdisciplinary approach to understanding and evaluating the opportunities for enhancing the human condition and the human organism opened up by the advancement of technology. Attention is given to both present technologies, like genetic engineering and information technology, and anticipated future ones, such as molecular nanotechnology and artificial intelligence.[1]

The enhancement options being discussed include radical extension of human health-span, eradication of disease, elimination of unnecessary suffering, and augmentation of human intellectual, physical, and emotional capacities.[2] Other transhumanist themes include space colonization and the possibility of creating superintelligent machines, along with other potential developments that could profoundly alter the human condition. The ambit is not limited to gadgets and medicine, but encompasses also economic, social, institutional designs, cultural development, and psychological skills and techniques.

Transhumanists view human nature as a work-in-progress, a half-baked beginning that we can learn to remold in desirable ways. Current humanity need not be the endpoint of evolution. Transhumanists hope that by responsible use of science, technology, and other rational means we shall eventually manage to become post-human, beings with vastly greater capacities than present human beings have.

Some transhumanists take active steps to increase the probability that they personally will survive long enough to become post-human, for example by choosing a healthy lifestyle or by making provisions for having themselves cryonically suspended in case of de-animation.[3] In contrast to many other ethical outlooks, which in practice often reflect a reactionary attitude to new technologies, the transhumanist view is guided by an evolving vision to take a more active approach to technology policy. This vision, in broad strokes, is to create the opportunity to live much longer and healthier lives, to enhance our memory and other intellectual faculties, to refine our emotional experiences and increase our subjective sense of well-being, and generally to achieve a greater degree of control over our own lives. This affirmation of human potential is offered as an alternative to customary injunctions against playing God, messing with nature, tampering with our human essence, or displaying punishable hubris.

Transhumanism does not entail technological optimism. While future technological capabilities carry immense potential for beneficial deployments, they also could be misused to cause enormous harm, ranging all the way to the extreme possibility of intelligent life becoming extinct. Other potential negative outcomes include widening social inequalities or a gradual erosion of the hard-to-quantify assets that we care deeply about but tend to neglect in our daily struggle for material gain, such as meaningful human relationships and ecological diversity. Such risks must be taken very seriously, as thoughtful transhumanists fully acknowledge.[4]

Transhumanism has roots in secular humanist thinking, yet is more radical in that it promotes not only traditional means of improving human nature, such as education and cultural refinement, but also direct application of medicine and technology to overcome some of our basic biological limits.

2 A core transhumanist value: exploring the post-human realm

The range of thoughts, feelings, experiences, and activities that are accessible to human organisms presumably constitute only a tiny part of what is possible. There is no reason to think that the human mode of being is any more free of limitations imposed by our biological nature than are the modes of being of other animals. Just as chimpanzees lack the brainpower to understand what it is like to be human, so too do we lack the practical ability to form a realistic intuitive understanding of what it would be like to be post-human.

This point is distinct from any principled claims about impossibility. We need not assert that post-human beings would not be Turing computable or that their concepts could not be expressed by any finite sentences in human language. The impossibility is more like the impossibility for us to visualize a twenty-dimensional hypersphere or to read, with

perfect recollection and understanding, every book in the Library of Congress. Our own current mode of being, therefore, spans but a minute subspace of what is possible or permitted by the physical constraints of the universe. It is not farfetched to suppose that there are parts of this larger space that represent extremely valuable ways of living, feeling, and thinking.

We can conceive of aesthetic and contemplative pleasures whose blissfulness vastly exceeds what any human being has yet experienced. We can imagine beings that reach a much greater level of personal development and maturity than current human beings do, because they have the opportunity to live for hundreds or thousands of years with full bodily and psychic vigor. We can conceive of beings that are much smarter than us, that can read books in seconds, that are much more brilliant philosophers than we are, that can create artworks, which, even if we could understand them only on the most superficial level, would strike us as wonderful masterpieces. We can imagine love that is stronger, purer, and more secure than any human being has yet harbored. Our everyday intuitions about values are constrained by the narrowness of our experience and the limitations of our powers of imagination. We should leave room in our thinking for the possibility that as we develop greater capacities, we shall come to discover values that will strike us as being of a far higher order than those we can realize as un-enhanced biological humans beings.

The conjecture that there are greater values than we can currently fathom does not imply that values are not defined in terms of our current dispositions. Take, for example, a dispositional theory of value such as the one described by David Lewis.[5] According to Lewis's theory, something is a value for you if and only if you would want to want it if you were perfectly acquainted with it and you were thinking and deliberating as clearly as possible about it. On this view, there may be values that we do not currently want, and that we do not even currently want to want, because we may not be perfectly acquainted with them or because we are not ideal deliberators. Some values pertaining to certain forms of post-human existence may well be of this sort; they may be values for us now, and they may be so in virtue of our current dispositions, and yet we may not be able to fully appreciate them with our current limited deliberative capacities and our lack of the receptive faculties required for full acquaintance with them. This point is important because it shows that the transhumanist view that we ought to explore the realm of post-human values does not entail that we should forego our current values. The post-human values can be our current values, albeit ones that we have not yet clearly comprehended. Transhumanism does not require us to say that we should favor post-human beings over human beings, but that the right way of favoring human beings is by enabling us to realize our ideals better and that some of our ideals may well be located outside the space of modes of being that are accessible to us with our current biological constitution.

We can overcome many of our biological limitations. It is possible that there are some limitations that are impossible for us to transcend, not only because of technological difficulties but on metaphysical grounds. Depending on what our views are about what constitutes personal identity, it could be that certain modes of being, while possible, are not possible for us, because any being of such a kind would be so different from us that they could not be us. Concerns of this kind are familiar from theological discussions of the afterlife. In Christian theology, some souls will be allowed by God to go to heaven after their time as corporal creatures is over. Before being admitted to heaven, the souls would undergo a purification process in which they would lose many of their previous bodily attributes. Skeptics may doubt that the resulting minds would be sufficiently similar to our current minds for it to be possible for them to be the same person. A similar predicament arises within transhumanism: if the mode of being of a post-human being is radically different from that of a human being, then we may doubt whether a post-human being could be the same person as a human being, even if the post-human being originated from a human being.

We can, however, envision many enhancements that would not make it impossible for the post-transformation someone to be the same person as the pre-transformation person. A person could obtain considerable increased life expectancy, intelligence, health, memory, and emotional sensitivity, without ceasing to exist in the process. A person's intellectual life can be transformed radically by getting an education. A person's life expectancy can be extended substantially by being unexpectedly cured from a lethal disease. Yet these developments are not viewed as spelling the end of the original person. In particular, it seems that modifications that add to a person's capacities can be more substantial than modifications that subtract, such as brain damage. If most of what someone currently is, including her most important memories, activities, and feelings, is preserved, then adding extra capacities on top of that would not easily cause the person to cease to exist.

Preservation of personal identity, especially if this notion is given a narrow construal, is not everything. We can value other things than ourselves, or we might regard it as satisfactory if some parts or aspects of ourselves survive and flourish, even if that entails giving up some parts of ourselves such that we no longer count as being the same person. Which parts of ourselves we might be willing to sacrifice may not become clear until we are more fully acquainted with the full meaning of the options. A careful, incremental exploration of the post-human realm may be indispensable for acquiring such an understanding, although we may also be able to learn from each other's experiences and from works of the imagination. Additionally, we may favor future people being posthuman rather than human, if the posthuman beings would lead lives more worthwhile than the alternative humans would lead. Any reasons stemming from such considerations would not depend on the assumption that we ourselves could become posthuman beings.

Transhumanism promotes the quest to develop further so that we can explore hitherto inaccessible realms of value. Technological enhancement of human organisms is a means that we ought to pursue to this end. There are limits to how much can be achieved by low-tech means such as education, philosophical contemplation, moral self-scrutiny and other such methods proposed by classical philosophers with perfectionist leanings, including Plato, Aristotle, and Nietzsche, or by means of creating a fairer and better society, as envisioned by social reformists such as Marx or Martin Luther King. This is not to denigrate what we can do with the tools we have today. Yet ultimately, transhumanists hope to go further.

3 The morality of human germ-line genetic engineering

Most potential human enhancement technologies have so far received scant attention in the ethics literature. One exception is genetic engineering, the morality of which has been extensively debated in recent years. To illustrate how the transhumanist approach can be applied to particular technologies, we shall therefore now turn to consider the case of human germ-line genetic enhancements.

Certain types of objection against germ-line modifications are not accorded much weight by a transhumanist interlocutor. For instance, objections that are based on the idea that there is something inherently wrong or morally suspect in using science to manipulate human nature are regarded by transhumanists as wrongheaded. Moreover, transhumanists emphasize that particular concerns about negative aspects of genetic enhancements, even when such concerns are legitimate, must be judged against the potentially enormous benefits that could come from genetic technology successfully employed.[6] For example, many commentators worry about the psychological effects of the use of germ-line engineering. The ability to select the genes of our children and to create so-called designer babies will, it is claimed, corrupt parents, who will come to view their children as mere products.[7] We will then begin to evaluate our offspring according to standards of quality control, and this will undermine the ethical ideal of unconditional acceptance of children, no matter what their abilities and traits. Are we really prepared to sacrifice on the altar of consumerism even those deep values that are embodied in traditional relationships between child and parents? Is the quest for perfection worth this cultural and moral cost? A transhumanist should not dismiss such concerns as irrelevant. Transhumanists recognize that the depicted outcome would be bad. We do not want parents to love and respect their children less. We do not want social prejudice against people with disabilities to get worse. The psychological and cultural effects of commodifying human nature are potentially important.

But such dystopian scenarios are speculations. There is no firm ground for believing that the alleged consequences would actually happen. What relevant evidence we have, for instance regarding the treatment of children who have been conceived through the use of in vitro fertilization or embryo screening, suggests that the pessimistic prognosis is alarmist. Parents will in fact love and respect their children even when artificial means and conscious choice play a part in procreation.

We might speculate, instead, that germ-line enhancements will lead to more love and parental dedication. Some mothers and fathers might find it easier to love a child who, thanks to enhancements, is bright, beautiful, healthy, and happy. The practice of germ-line enhancement might lead to better treatment of people with disabilities, because a general demystification of the genetic contributions to human traits could make it clearer that people with disabilities are not to blame for their disabilities and a decreased incidence of some disabilities could lead to more assistance being available for the remaining affected people to enable them to live full, unrestricted lives through various technological and social supports. Speculating about possible psychological or cultural effects of germ-line engineering can therefore cut both ways. Good consequences no less than bad ones are possible. In the absence of sound arguments for the view that the negative consequences would predominate, such speculations provide no reason against moving forward with the technology.

Ruminations over hypothetical side-effects may serve to make us aware of things that could go wrong so that we can be on the lookout for untoward developments. By being aware of the perils in advance, we will be in a better position to take preventive countermeasures. For instance, if we think that some people would fail to realize that a human clone would be a unique person deserving just as much respect and dignity as any other human being, we could work harder to educate the public on the inadequacy of genetic determinism. The theoretical contributions of well-informed and reasonable critics of germ-line enhancement could indirectly add to our justification for proceeding with germ-line engineering. To the extent that the critics have done their job, they can alert us to many of the potential untoward consequences of germ-line

engineering and contribute to our ability to take precautions, thus improving the odds that the balance of effects will be positive. There may well be some negative consequences of human germ-line engineering that we will not forestall, though of course the mere existence of negative effects is not a decisive reason not to proceed. Every major technology has some negative consequences. Only after a fair comparison of the risks with the likely positive consequences can any conclusion based on a cost-benefit analysis be reached.

In the case of germ-line enhancements, the potential gains are enormous. Only rarely, however, are the potential gains discussed, perhaps because they are too obvious to be of much theoretical interest. By contrast, uncovering subtle and non-trivial ways in which manipulating our genome could undermine deep values is philosophically a lot more challenging. But if we think about it, we recognize that the promise of genetic enhancements is anything but insignificant. Being free from severe genetic diseases would be good, as would having a mind that can learn more quickly, or having a more robust immune system. Healthier, wittier, happier people may be able to reach new levels culturally. To achieve a significant enhancement of human capacities would be to embark on the transhuman journey of exploration of some of the modes of being that are not accessible to us as we are currently constituted, possibly to discover and to instantiate important new values. On an even more basic level, genetic engineering holds great potential for alleviating unnecessary human suffering. Every day that the introduction of effective human genetic enhancement is delayed is a day of lost individual and cultural potential, and a day of torment for many unfortunate sufferers of diseases that could have been prevented. Seen in this light, proponents of a ban or a moratorium on human genetic modification must take on a heavy burden of proof in order to have the balance of reason tilt in their favor. Transhumanists conclude that the challenge has not been met.

4 Should human reproduction be regulated?

One way of going forward with genetic engineering is to permit everything, leaving all choices to parents. While this attitude may be consistent with transhumanism, it is not the best transhumanist approach. One thing that can be said for adopting a libertarian stance in regard to human reproduction is the sorry track record of socially planned attempts to improve the human gene pool. The list of historical examples of state intervention in this domain ranges from the genocidal horrors of the Nazi regime, to the incomparably milder but still disgraceful semi-coercive sterilization programs of mentally impaired individuals favored by many well-meaning socialists in the past century, to the controversial but perhaps understandable program of the current Chinese government to limit population growth. In each case, state policies interfered with the reproductive choices of individuals. If parents had been left to make the choices for themselves, the worst transgressions of the eugenics movement would not have occurred. Bearing this in mind, we ought to think twice before giving our support to any proposal that would have the state regulate what sort of children people are allowed to have and the methods that may be used to conceive them.[8]

We currently permit governments to have a role in reproduction and child-rearing and we may reason by extension that there would likewise be a role in regulating the application of genetic reproductive technology. State agencies and regulators play a supportive and supervisory role, attempting to promote the interests of the child. Courts intervene in cases of child abuse or neglect. Some social policies are in place to support children from disadvantaged backgrounds and to ameliorate some of the worst inequities suffered by children from poor homes, such as through the provision of free schooling. These measures have analogues that apply to genetic enhancement technologies. For example, we ought to outlaw genetic modifications that are intended to damage the child or limit its opportunities in life, or that are judged to be too risky. If there are basic enhancements that would be beneficial for a child but that some parents cannot afford, then we should consider subsidizing those enhancements, just as we do with basic education. There are grounds for thinking that the libertarian approach is less appropriate in the realm of reproduction than it is in other areas. In reproduction, the most important interests at stake are those of the child-to-be, who cannot give his or her advance consent or freely enter into any form of contract. As it is, we currently approve of many measures that limit parental freedoms. We have laws against child abuse and child neglect. We have obligatory schooling. In some cases, we can force needed medical treatment on a child, even against the wishes of its parents.

There is a difference between these social interventions with regard to children and interventions aimed at genetic enhancements. While there is a consensus that nobody should be subjected to child abuse and that all children should have at least a basic education and should receive necessary medical care, it is unlikely that we will reach an agreement on proposals for genetic enhancements any time soon. Many parents will resist such proposals on principled grounds, including deep-seated religious or moral convictions. The best policy for the foreseeable future may therefore be to not legally require any genetic enhancements, except perhaps in extreme cases for which there is no alternative treatment. Even in such cases, it is dubious that the social climate in many countries is ready for mandatory genetic interventions.

The scope for ethics and public policy, however, extends far beyond the passing of laws requiring or banning specific interventions. Even if a given enhancement option is neither outlawed nor legally required, we may still seek to discourage or encourage its use in a variety of ways. Through subsidies and taxes, research-funding policies, genetic counseling practices and guidelines, laws regulating genetic information and genetic discrimination, provision of health care services, regulation of the insurance industry, patent law, education, and through the allocation of social approbation and disapproval, we may influence the direction in which particular technologies are applied. We may appropriately ask, with regard to genetic enhancement technologies, which types of applications we ought to promote or discourage.

5 Which modifications should be promoted and which discouraged?

An externality, as understood by economists, is a cost or a benefit of an action that is not carried by a decision-maker. An example of a negative externality might be found in a firm that lowers its production costs by polluting the environment. The firm enjoys most of the benefits while escaping the costs, such as environmental degradation, which may instead be paid by people living nearby. Externalities can also be positive, as when people put time and effort into creating a beautiful garden outside their house. The effects are enjoyed not exclusively by the gardeners but spill over to passersby. As a rule of thumb, sound social policy and social norms would have us internalize many externalities so that the incentives of producers more closely match the social value of production. We may levy a pollution tax on the polluting firm, for instance, and give our praise to the home gardeners who beautify the neighborhood.

Genetic enhancements aimed at the obtainment of goods that are desirable only in so far as they provide a competitive advantage tend to have negative externalities. An example of such a positional good, as economists call them, is stature. There is evidence that being tall is statistically advantageous, at least for men in Western societies. Taller men earn more money, wield greater social influence, and are viewed as more sexually attractive. Parents wanting to give their child the best possible start in life may rationally choose a genetic enhancement that adds an inch or two to the expected length of their offspring. Yet for society as a whole, there seems to be no advantage whatsoever in people being taller. If everybody grew two inches, nobody would be better off than they were before. Money spent on a positional good like length has little or no net effect on social welfare and is therefore, from society's point of view, wasted.

Health is a very different type of good. It has intrinsic benefits. If we become healthier, we are personally better off and others are not any worse off. There may even be a positive externality of enhancing our own health. If we are less likely to contract a contagious disease, others benefit by being less likely to get infected by us. Being healthier, we may also contribute more to society and consume less of publicly funded healthcare.

If we were living in a simple world where people were perfectly rational self-interested economic agents and where social policies had no costs or unintended effects, then the basic policy prescription regarding genetic enhancements would be relatively straightforward. We should internalize the externalities of genetic enhancements by taxing enhancements that have negative externalities and subsidizing enhancements that have positive externalities. Unfortunately, crafting policies that work well in practice is considerably more difficult. Even determining the net size of the externalities of a particular genetic enhancement can be difficult. There is clearly an intrinsic value to enhancing memory or intelligence in as much as most of us would like to be a bit smarter, even if that did not have the slightest effect on our standing in relation to others. But there would also be important externalities, both positive and negative. On the negative side, others would suffer some disadvantage from our increased brainpower in that their own competitive situation would be worsened. Being more intelligent, we would be more likely to attain high-status positions in society, positions that would otherwise have been enjoyed by a competitor. On the positive side, others might benefit from enjoying witty conversations with us and from our increased taxes.

If in the case of intelligence enhancement the positive externalities outweigh the negative ones, then a *prima facie* case exists not only for permitting genetic enhancements aimed at increasing intellectual ability, but for encouraging and subsidizing them too. Whether such policies remain a good idea when all practicalities of implementation and political realities are taken into account is another matter. But at least we can conclude that an enhancement that has both significant intrinsic benefits for an enhanced individual and net positive externalities for the rest of society should be encouraged. By contrast, enhancements that confer only positional advantages, such as augmentation of stature or physical attractiveness, should not be socially encouraged, and we might even attempt to make a case for social policies aimed at reducing expenditure on such goods, for instance through a progressive tax on consumption.[9]

6 The issue of equality

One important kind of externality in germ-line enhancements is their effects on social equality. This has been a focus for many opponents of germ-line genetic engineering who worry that it will widen the gap between haves and have-nots. Today, children from wealthy homes enjoy many environmental privileges, including access to better schools and social networks. Arguably, this constitutes an inequity against children from poor homes. We can imagine scenarios where such inequities grow much larger thanks to genetic interventions that only the rich can afford, adding genetic advantages to the environmental advantages already benefiting privileged children. We could even speculate about the members of the privileged stratum of society eventually enhancing themselves and their offspring to a point where the human species, for many practical purposes, splits into two or more species that have little in common except a shared evolutionary history.[10] The genetically privileged might become ageless, healthy, supergeniuses of flawless physical beauty, who are graced with a sparkling wit and a disarmingly self-deprecating sense of humor, radiating warmth, empathetic charm, and relaxed confidence. The non-privileged would remain as people are today but perhaps deprived of some their self-respect and suffering occasional bouts of envy. The mobility between the lower and the upper classes might disappear, and a child born to poor parents, lacking genetic enhancements, might find it impossible to successfully compete against the super-children of the rich. Even if no discrimination or exploitation of the lower class occurred, there is still something disturbing about the prospect of a society with such extreme inequalities.

While we have vast inequalities today and regard many of these as unfair, we also accept a wide range of inequalities because we think that they are deserved, have social benefits, or are unavoidable concomitants to free individuals making their own and sometimes foolish choices about how to live their lives. Some of these justifications can also be used to exonerate some inequalities that could result from germ-line engineering. Moreover, the increase in unjust inequalities due to technology is not a sufficient reason for discouraging the development and use of the technology. We must also consider its benefits, which include not only positive externalities but also intrinsic values that reside in such goods as the enjoyment of health, a soaring mind, and emotional well-being.

We can also try to counteract some of the inequality-increasing tendencies of enhancement technology with social policies. One way of doing so would be by widening access to the technology by subsidizing it or providing it for free to children of poor parents. In cases where the enhancement has considerable positive externalities, such a policy may actually benefit everybody, not just the recipients of the subsidy. In other cases, we could support the policy on the basis of social justice and solidarity.

Even if all genetic enhancements were made available to everybody for free, however, this might still not completely allay the concern about inequity. Some parents might choose not to give their children any enhancements. The children would then have diminished opportunities through no fault of their own. It would be peculiar, however, to argue that governments should respond to this problem by limiting the reproductive freedom of the parents who wish to use genetic enhancements. If we are willing to limit reproductive freedom through legislation for the sake of reducing inequities, then we might as well make some enhancements obligatory for all children. By requiring genetic enhancements for everybody to the same degree, we would not only prevent an increase in inequalities but also reap the intrinsic benefits and the positive externalities that would come from the universal application of enhancement technology. If reproductive freedom is regarded

as too precious to be curtailed, then neither requiring nor banning the use of reproductive enhancement technology is an available option. In that case, we would either have to tolerate inequities as a price worth paying for reproductive freedom or seek to remedy the inequities in ways that do not infringe on reproductive freedom.

All of this is based on the hypothesis that germ-line engineering would in fact increase inequalities if left unregulated and no countermeasures were taken. That hypothesis might be false. In particular, it might turn out to be technologically easier to cure gross genetic defects than to enhance an already healthy genetic constitution. We currently know much more about many specific inheritable diseases, some of which are due to single gene defects, than we do about the genetic basis of talents and desirable qualities such as intelligence and longevity, which in all likelihood are encoded in complex constellations of multiple genes. If this turns out to be the case, then the trajectory of human genetic enhancement may be one in which the first thing to happen is that the lot of the genetically worst-off is radically improved, through the elimination of diseases such as Tay Sachs, Lesch-Nyhan, Downs Syndrome, and early-onset Alzheimer's disease. This would have a major leveling effect on inequalities, not primarily in the monetary sense, but with respect to the even more fundamental parameters of basic opportunities and quality of life.

7 Are germ-line interventions wrong because they are irreversible?

Another frequently heard objection against germ-line genetic engineering is that it would be uniquely hazardous because the changes it would bring are irreversible and would affect all generations to come. It would be highly irresponsible and arrogant of us to presume that we have the wisdom to make decisions about what should be the genetic constitutions of people living many generations hence. Human fallibility, on this objection, gives us good reason not to embark on germ-line interventions. For our present purposes, we can set aside the issue of the safety of the procedure, understood narrowly, and stipulate that the risk of medical side-effects has been reduced to an acceptable level. The objection under consideration concerns the irreversibility of germ-line interventions and the lack of predictability of its long-term consequences; it forces us to ask if we possess the requisite wisdom for making genetic choices on behalf of future generations.

Human fallibility is not a conclusive ground for resisting germ-line genetic enhancements. The claim that such interventions would be irreversible is incorrect. Germ-line interventions can be reversed by other germ-line interventions. Moreover, considering that technological progress in genetics is unlikely to grind to an abrupt halt any time soon, we can count on future generations being able to reverse our current germ-line interventions even more easily than we can currently implement them. With advanced genetic technology, it might even be possible to reverse many germ-line modifications with somatic gene therapy, or with medical nanotechnology.[11] Technologically, germ-line changes are perfectly reversible by future generations.

It is possible that future generations might choose to retain the modifications that we make. If that turns out to be the case, then the modifications, while not irreversible, would nevertheless not actually be reversed. This might be a good thing. The possibility of permanent consequences is not an objection against germ-line interventions any more than it is against social reforms. The abolition of slavery and the introduction of general suffrage might never be reversed; indeed, we hope they will not be. Yet this is no reason for people to have resisted the reforms. Likewise, the potential for everlasting consequences, including ones we cannot currently reliably forecast, in itself constitutes no reason to oppose genetic intervention. If immunity against horrible diseases and enhancements that expand the opportunities for human growth are passed on to subsequent generations in perpetuo, it would be a cause for celebration, not regret.

There are some kinds of changes that we need be particularly careful about. They include modifications of drives and motivations of our descendants. For example, there are obvious reasons why we might think it worthwhile to seek to reduce propensity of our children to violence and aggression. We would have to take care, however, that we do not do this in a way that would make future people overly submissive or complacent. We can conceive of a dystopian scenario along the lines of Brave New World, in which people are leading shallow lives but have been manipulated to be perfectly content with their sub-optimal existence. If the people transferred their shallow values to their children, humanity could get permanently stuck in a not-very-good state, having foolishly changed itself to lack any desire to strive for something better. This outcome would be dystopian because a permanent cap on human development would destroy the transhumanist hope of exploring the post-human realm. Transhumanists therefore place an emphasis on modifications which, in addition to promoting human well-being, also open more possibilities than they close and which increase our ability to make subsequent choices wisely. Longer active lifespans, better memory, and greater intellectual capacities are plausible candidates for enhancements that would improve our ability to figure out what we ought to do next. They would be a good place to start.[12]

Notes

1 See Eric K. Drexler, *Nanosystems: Molecular Machinery, Manufacturing, and Computation* (New York: John Wiley & Sons, Inc., 1992); Ray Kurzweil, *The Age of Spiritual Machines: When Computers Exceed Human Intelligence* (New York: Viking, 1999); Hans Moravec, *Robot: Mere Machine to Transcendent Mind.* (New York: Oxford University Press, 1999).

2 See Robert A. Freitas Jr., *Nanomedicine, Volume 1: Basic Capabilities* (Georgetown, Tex.: Landes Bioscience, 1999).

3 See Robert Ettinger, *The Prospect of Immortality* (New York: Doubleday, 1964); James Hughes, "The Future of Death: Cryonics and the Telos of Liberal Individualism," *Journal of Evolution and Technology* 6 (2001).

4 See Eric K. Drexler, *Engines of Creation: The Coming Era of Nanotechnology* (London: Fourth Estate, 1985).

5 See David Lewis, "Dispositional Theories of Value," *Proceedings of the Aristotelian Society Supp.* 63, pp. 113–37 (1989).

6 See Erik Parens, ed., *Enhancing Human Traits: Ethical and Social Implications.* (Washington, D. C: Georgetown University Press, 1998).

7 See Leon Kass, *Life, Liberty, and Defense of Dignity: The Challenge for Bioethics* (San Francisco: Encounter Books, 2002).

8 See Jonathan Glover, *What Sort of People Should There Be?* (New York: Penguin, 1984); Gregory Stock, *Redesigning Humans: Our Inevitable Genetic Future* (New York, Houghton Mifflin, 2002); and Allen Buchanan et al., *From Chance to Choice: Genetics & Justice* (Cambridge, England: Cambridge University Press, 2002).

9 See Robert H. Frank, *Luxury Fever: Why Money Fails to Satisfy in an Era of Excess* (New York: Free Press, 1999).

10 Cf. Lee M. Silver, *Remaking Eden: How Genetic Engineering and Cloning will Transform the American Family* (New York: Avon Books, 1997); and Nancy Kress, *Beggars in Spain* (Avon Books, 1993).

11 See Freitas, op. cit.

12 For their helpful comments I am grateful to Heather Bradshaw, Robert A. Freitas Jr., James Hughes, Gerald Lang, Matthew Liao, Thomas Magnell, David Rodin, Jeffrey Soreff, Mike Treder, Mark Walker, Michael Weingarten, and an anonymous referee of the *Journal of Value Inquiry*.

Is it wrong to clone human beings?

INTRODUCTION TO PART THREE

S OMATIC CELL NUCLEAR transfer (SCNT) involves replacing the nucleus of an unfertilised egg cell with the material from the nucleus of a somatic cell of the individual to be cloned, and stimulating the resulting cell to divide. This technique was famously employed to create Dolly the Sheep. Mass cloning of humans by this method is unlikely in the foreseeable future; nonetheless, cloning is a major bioethical issue. To see why, the distinction and relationship between two sorts of cloning are important.

'Therapeutic cloning' is undertaken to produce cloned human embryos for medical research and therapy; 'human reproductive cloning' would be undertaken in order to bring about a mature human being. Therapeutic cloning is permitted in some countries, whereas there is a global ban on human reproductive cloning. But an argument connecting the two is central to the topic: there is a 'slippery slope' from therapeutic cloning to human reproductive cloning; human reproductive cloning is wrong; therefore, therapeutic cloning is unethical. Two sorts of responses to slippery slope arguments are, first, to question whether the slope between two practices really is slippery and, second, to question whether the practice at the bottom of the slope really is impermissible.

Chadwick represents the latter sort of response, concluding from her survey of a number of arguments that cloning as an alternative method of reproduction 'is not necessarily undesirable in itself'. Kass is less sanguine. A useful way to approach Kass' article is to distinguish two ways of doing bioethics. The first, informed by analytic philosophy, consists in clarifying and evaluating specific considerations for or against permitting biomedical procedures. Kass illustrates this by 'three kinds of concerns and objections': cloning threatens confusion of identity and individuality, represents a step towards transforming procreation into manufacture, and would be a form of despotism. But he eloquently expresses a second way of doing bioethics, based less on analytic philosophy than his worldview. Human reproductive cloning appears repugnant on Kass' understanding of human nature, the natural world, and the nature of human sexuality. Kass suggests that it is wise to recognise repugnance as indicative of something seriously wrong with the biomedical procedure in question.

The problem with basing bioethical arguments on worldviews is that they tend not to be universally shared. So the next two pieces retreat from Kass' expansive style to the narrower analytic approach. Elliott takes up two of the objections to human reproductive cloning that Kass discusses. The first, the 'manufacturing objection', can be put in terms of a category error: cloning puts people and their relationships in the wrong category by treating babies as something to be made, not begotten. This recalls Kass, who states that, whilst 'mass-scale cloning ... makes the point vividly', the manufacturing objection arises even in the case of 'a single planned clone' because 'procreation dehumanized into manufacture is further degraded by commodification, a virtually inescapable result of allowing baby-making to proceed under the banner of commerce'. In his rejoinder, Elliott points out that parents ordinarily do much to select the traits of their offspring. He considers whether the worry is not with 'manufacturing' *per se*, but that parents who make too many manufacturing decisions are treating their child as an object to perfect rather than 'a person in his or her own right'; nonetheless, since the decision to clone is not inherently a manufacturing decision, cloning can be laudably undertaken.

The second objection to cloning with which both Elliott and Kass engage is that the cloned offspring would not be biologically unique because it would share a genome with the cloned parent, which compromises its identity and individuality. One response is that this smacks of genetic determinism: since environmental factors are important in determining who we are, clones would not be identical despite sharing a genome. Another response is that monozygotic twins do not evince deep problems about identity and individuality despite being biologically non-unique. Aside from these arguments, Elliott connects the non-uniqueness objection to repeatability, being replaceable and, therefore, lacking value. He responds that, on any plausible account of the value of persons, clones would be as valuable as non-clones. For example, according to Kant, what confers dignity on a person is their ability to conform their will to the moral law; since clones would have this ability, they would have dignity.

Sparrow considers 'the best case for reproductive human cloning' to be that it is the only way for some parents to have a child to whom they are genetically related. But a hidden premise here is that the genetic relatedness between clones – they 'share the same genes' – is of the right sort to establish parenthood. Sparrow suggests that, with regard to cloning, 'ordinary intuitions about genetic relatedness are unreliable here because the normal connection between genes and the history whereby they are transmitted has been severed'. A dilemma emerges. On the one hand, the natural parents of the cloned adult have the right sort of genetic relatedness to the cloned child to be the latter's parents. On the other hand, to deny this is to admit that it is the intention to parent that singles out the cloned adult as the cloned child's parent, and this conflicts with the main premise of the argument for reproductive cloning under discussion, namely, that genetic relatedness is what matters in parenthood. Sparrow concludes that, since cloning fails unambiguously to satisfy infertile couples' desires to have a child to whom they are genetically related, there are better uses of scarce resources than funding research into cloning technologies.

The final selection refocuses on the slippery slope argument against therapeutic cloning. Blackford calls this a 'horrible-result argument' that works only if reproductive cloning is 'sufficiently horrible', and decriminalising therapeutic cloning is really likely to lead to it. He suggests that the anti-cloning argument requires a multiplication of slippery slopes to various horrible results – for example, reproductive cloning puts us on a slippery slope to grotesque forms of genetic enhancement or Kass' 'brave new world' – that are increasingly unconvincing.

Selections

Chadwick, R. 1982. 'Cloning', *Philosophy*, 57 (220), 201–9.

Kass, L. 1997. 'The Wisdom of Repugnance', *New Republic*, 216 (22), 17–26.

Elliott, D. 1998. 'Uniqueness, Individuality, and Human Cloning', *Journal of Applied Philosophy*, 15 (3), 217–30.

Sparrow, R. 2006. 'Cloning, Parenthood, and Genetic Relatedness', *Bioethics*, 20 (6), 308–18.

Blackford, R. 2007. 'Slippery Slopes to Slippery Slopes: Therapeutic Cloning and the Criminal Law', *The American Journal of Bioethics*, 7 (2), 63–4.

Chapter 10

Ruth F. Chadwick

CLONING

Every body cell of an animal or human being contains the same complete set of genes. In theory any of these cells can be used to start a new embryo. The technique has been employed in the case of frogs. The nucleus is taken out of a body cell of a frog and implanted in an enucleated frog's egg. The resulting egg cell is stimulated to develop into a normal frog, and will be an exact copy of that frog which provided the nucleus with all the genetic information. In normal sexual reproduction, two parents each contribute half their genes, but in the case of cloning, one parent passes on all his or her genes.

The possibility of the cloning of human beings has had a great effect upon the popular imagination. It has caught the attention of science fiction writers, and evoked fears of the multiplication of undesirable types on a large scale. The claim that one child has already been produced by cloning[1] has been met with scepticism,[2] but if it is a possibility, we need to consider the moral problems that may be raised by it.

Here I shall confine my attention to cloning when seen as an alternative method of reproduction of living human individuals. The cloning of dead, possibly long-dead, individuals, may raise different problems.

As R. M. Hare has pointed out,[3] in order for a moral philosopher to apply moral philosophy to a practical problem, he must first of all have a theory to apply. The view put forward (but not defended) here is a form of preference utilitarianism, in which an attempt is made to maximize preference satisfaction. Such a position can provide reasons for holding that cloning may be justified in some circumstances. First, however, I shall consider alternative ways of looking at the question, which might be thought to offer arguments for taking the view that human beings should not practice clonal reproduction.

Arguments about the unnaturalness of cloning

It might be thought that the question of the relevance to moral philosophy of what is 'natural' is a dead issue. However, the notion still carries considerable intuitive appeal. What is claimed when anything is objected to on the ground that it is 'unnatural' is far from clear, and it is no clearer in the case of cloning than anything else. It is not that cloning does not occur 'in nature'.[4] Asexual reproduction took place in evolutionary history long before sexual reproduction did, and many species still reproduce by parthenogenesis. The point is rather that the human species has evolved as one that employs the practice of sexual reproduction, and thus cloning is unnatural for the species. The claim must be understood as making this restricted point.

What is its force? It is intended to give a reason to support the view that cloning is wrong.[5] If this enterprise is to have any chance of success, however, the statement that 'cloning is unnatural for the species' cannot be interpreted as saying simply that it has not been used as a means of reproduction before. This could not be taken seriously as a reason for adopting the view that cloning is wrong.

It seems that if the claim that cloning is unnatural for the species is to carry any weight it must be interpreted in one of two ways.

Argument from function

To say that a certain practice is unnatural for the species may be understood as claiming that the members of the species cannot function properly if it is carried out.

For some, such an argument may carry the unwelcome suggestion of design and thus of a designer, who has determined some function or purpose for members of the human species. But perhaps the argument can be stripped of such connotations. This would be done by an attempt to show that there are certain very basic features with which we associate being human, which are taken away by the practice. Thus, to take an analogy, it is often claimed that it is unnatural for hens to live in batteries because they cannot perform such basic hen-like activities as spreading their wings. This may provide a reason for supporting the view that it is wrong to keep hens in batteries.

A similar line of thought may be applied to human beings, with the intention of showing that certain ways of treating them should be ruled out because they are unnatural in this way. Such an attempt encounters two difficulties. First, there is the problem of trying to give clear criteria of what would be 'unnatural' in this sense for members of the human species. There is widespread disagreement over what

basic human functions are. Secondly, it is difficult to see how such an argument, in terms of what is unnatural for the species, could have an advantage over an approach to the problem in terms of what people want or what their preferences are. This way of looking at the question, to be considered in a later section, has at least the merit that it is easier to state what individuals want than it is to give criteria for what would be unnatural for them in the sense outlined.

Even if this could be done, however, it is not obvious that cloning would satisfy the criteria. Depriving human beings of opportunities for sex and reproduction might, but it is difficult to see why opportunities for asexual reproduction should.

Playing God

Those who try to gain some control over life and death are sometimes accused of 'playing God', with the implication that this is undesirable.[6] Man is seen as overreaching himself. The Greek hubris from which this idea derives was inevitably followed by nemesis, in the shape of very unpleasant consequences.

The 'playing God' argument is a version of the unnatural argument, because it is in situations where man interferes with the course of nature that the term is commonly used. Cloning could be seen as a playing God type of interference in the natural reproductive process, inasmuch as it may be a bid for a kind of immortality, with potentially disastrous consequences.

Should we take this 'playing God' argument seriously? Is it sufficient to label something as an instance of 'playing God' and therefore rule it out? One might object that since the reason why hubris is discouraged is that nemesis follows it, we should do better simply to assess actions in terms of their consequences, and discard the concept of 'playing God'.

However, while we may reject it as a reason for automatically ruling out an interference with nature, it has some usefulness as a *description* of one class of cases in which we should be concerned to consider whether interference is wise. The reasons for this are twofold: firstly, actions describable as 'playing God' have the tendency to arouse anxiety. This results from fear of the unknown, as playing God actions are usually ones not attempted before.

Secondly, this type of action, having consequences that are unforeseeable, may *in fact* lead to the terrible consequences that have been feared, or worse.

Rather than ruling out the action with no more ado, however, it may be preferable to consider the possible consequences, and adopt some kind of risk-assessment.

The two versions of the unnatural argument both seem to have considerable drawbacks, and cannot on their own provide reasons for the view that cloning is wrong. We must consider another approach.

Arguments about rights

It could be argued by those who speak in terms of rights that cloning is wrong because it violates certain rights, the right to be genetically unique and the right to privacy, for example.

Right to genetic uniqueness

Any such argument fails to gain a foothold in the cloning debate. Lederberg[7] points out that the Thirteenth Amendment to the American Constitution which prohibits slavery and involuntary servitude has been proposed to cover cloning since a 'genetic bondage' which diminishes autonomy is deliberately designed to reduce the option of choices which create individuality.

There is something rather odd about this. In what sense does someone who has been cloned have less autonomy than someone who has been produced by normal sexual reproduction? If our genes restrict our choices, as it is plausible to think they do, then everyone is faced with some restrictions determined by the genes they have.

What is really found objectionable is not the fact of 'genetic bondage' but the having of a genome identical to someone else's. But has someone's *right* been transgressed here? Is that the idea? It is difficult to see how those moral philosophers who speak in terms of rights could make out a case for this. If everyone has a right to be genetically unique, then identical twins are cheated of that right.

Further, in the cloning case, *who* is the possessor of the right? Not the person who is born as a result of cloning, because *he* would not have existed if he had not been cloned. There is no person who can be said to have the right to be genetically unique.

Right to privacy

It might be argued that cloning would violate a right to privacy. There seems to be some confusion, among those who hold that there is a right to privacy, as to what constitutes it.[8] Privacy seems to be an element of liberty: it involves the ability to control who has access to information about us, access to our person, and to intimacy with us. What force would the right to privacy have in the cloning debate? The argument would be that cloning constitutes a transgression of the right, because the child's genetic make-up is known and foretold. But again, he could not exist if he were not cloned. The right could carry force only after he was born, and could then generate conclusions about access to information about his genotype, as it might do in the case of anyone else's genetic make-up. The argument cannot therefore show that cloning is wrong.

Arguments about maximizing worthwhile life

It is interesting to note that anyone who, like Jonathan Glover,[9] holds that there is an impersonal principle urging us to maximize the *amount* of worthwhile life, and that there is no difference in principle between contraception and abortion, ought also to hold that there is no difference in principle between contraception and abstaining from using cells of one's own body in order to produce new individuals. The refusal to clone, contraception and abortion would be on a par.

A difference would of course be drawn in terms of side-effects, but a discussion of these will be postponed until we look at the application of preference utilitarianism to the problem.

Arguments about preferences

It may be claimed that while there is no *right* to be genetically unique, nevertheless most people would *prefer* to be genetically unique, and thus that cloning is wrong because it will produce people who are not.

One objection to this might be a factual one, that individuals do not, as a matter of fact, prefer to be genetically unique. Identical twins do not necessarily wish they were not twins. Also, the very fact that people want copies of themselves, as is claimed in David Rorvik's book *In His Image*, surely disproves the suggestion.

It seems, however, that there is a very important point to be made about priority. If X at 50 is intending to have a child Y by cloning, the way X and Y view the situation may be very different. X may like the idea of having a copy of *himself*, while Y may hate the fact that he *is a copy* of X. In the identical twin case neither is a copy of the other.

If we can say that there is a general preference for being unique rather than a copy, then that may give us a reason for opposing cloning, but perhaps it can be overridden by other preferences, both of the cloned individuals and of the people of whom they are copies.

As far as the former is concerned, it may be claimed that life as a cloned individual is better than no life at all. This individual would not exist were he not a product of the cloning technique.

Secondly, we must consider the desire to *be* the donor of a cell for cloning. It may be that it would be desired as a way of ensuring some kind of survival. By this method there is one real sense in which I survive. When I have a child by sexual reproduction, only half my genes will be passed on, but they will all be passed on in cloning.

It is not clear whether this would be regarded as a better way of survival than e.g. the brain transplant and bisection cases considered by Derek Parfit.[10] In the latter type of situation it is envisaged that the resulting person may have certain memories of the earlier person, so one may feel that he would be a future self to a greater degree than the genetic copy.

But cloning would bring to birth an individual who may not have had one's own experiences but who would have the potential for developing similar abilities and character.

It might be felt that this desire for survival is one that should not be given too much weight. But great efforts are made to help otherwise infertile people to pass on half their genes, because they so desire. Why should it not be possible for people to pass on *all* their genes, if they so desire? What is the significant difference here? If it is a desire that people have, it should be given some weight.

One possibility is that it could be justified in the case of couples in which the husband is sterile, as a preferable alternative to AID. The couple could have a copy of the man and one of the woman, thus producing a family with an ideally balanced sex ratio. But if it can be justified in the case of such a couple it is very hard to see why it should be wrong for other people who do not have problems of infertility but simply desire it.

On the basis of preference-satisfaction, then, it may be the case that the satisfaction of such desires, and of those of the person produced in this way, could outweigh the desire not to be a copy, and thus cloning might be justified in some cases.

To lessen the problem of the desire for uniqueness, it might be advisable to take the cells from young people, to minimize the age gap. Identical twins do not have the problem of thinking they are predetermined to grow up in a certain way. On the other hand, decreasing the age gap will also increase the physical similarity, and so such a policy would probably not be successful.

If many copies of one individual are produced the psychological effects may be even worse. One is suspicious of the desire to produce large numbers of copies of an individual. If it is the individual himself who desires it, then it seems unlikely that the utility arising out of the satisfaction of his desire will outweigh the disutility experienced by the members of the clone and the undesirable side-effects on society, which must now be discussed.

Side-effects on society

One point is that cloning might lead to a lessening of respect for individual persons, because of the feeling that they can easily be replaced. If we have clones with large memberships, the loss of one member may be insignificant. It is already argued that foetuses are replaceable. Cloning could lead to the extension of this argument to other ages.

Bernard Williams suggests (in another context) that such replaceability could have a far-reaching effect on our relationships:

If someone loved a token-person just as a Mary Smith, then it might well be unclear that the token-person was really what he loved. What he loves is Mary Smith, and that is to love the type-person. We can dimly see what this would be like. It would be like loving a work of art in some reproducible medium. One might start comparing, as it were, performances of the type; and wanting to be near the person one loved would be like wanting very much to hear some performance, even an indifferent one, of Figaro – just as one will go to the scratch provincial performance of Figaro rather than hear no Figaro at all, so one would see the very run-down Mary Smith who was in the locality, rather than see no Mary Smith at all.[11]

It appears, however, that this situation is very unlikely, because of the fact that love seems very much bound up with the phenotype.

Secondly, cloning could be employed to produce new forms of discrimination. Members of an élite clone may discriminate against members of a downtrodden one. This is the kind of situation described in *Brave New World*.

A third suggestion is that cloning might lead to the practice of 'wombhiring'. This need not occur, however, either when wives agree to bear copies of their husbands (or indeed of themselves), or when artificial wombs are available.

Less likely is the use of cloning to breed a class specifically for the purpose of military objectives. As Motulsky[12] points out, there are ways of producing the required mentality which are easier and more reliable than clonal reproduction.

One side-effect that is suggested as *good* is that members of a clone could easily exchange organs if the need arose. They would not experience the normal rejection problems

128   Ruth F. Chadwick

associated with transplants. However, this, though an advantage, could also produce further undesirable side-effects in itself. If one member of a clone had achieved a position of great importance, and then became seriously ill, needing a transplant, the others might, with good reason, feel that their lives were in danger. The general acceptance of the idea that the replacement value of individuals had been lessened would only make the individuals concerned less secure.

Side-effects on the gene pool

Would cloning have any eugenic effect?

If we had a policy of cloning genetically desirable types (assuming that we had adequate criteria of what these were) we should run the risk of having any benefit this might produce outweighed by the undesirability of the lack of genetic variation that would result. Cloning is conservative, preserving types that we already have rather than producing new genotypes that might be advantageous.

On the other hand it may be thought that if we had just a few copies of some extraordinarily desirable types, these could be used to good effect in interbreeding with other types. In this case, however, the eugenic effect on the species as a whole would be almost unnoticeable.

It is unlikely, then, to be of use in any plan to breed a super-race. It could, however, be used as a tool for dealing with genetic disease. If one partner has a history of e.g. Huntington's chorea in the family, then the couple could have a child by cloning the other parent.

Conclusion

It appears that cloning is not necessarily undesirable in itself, and can be used to good effect in e.g. treatment of genetic disease, as outlined above. The possible unwelcome side-effects,

such as the lessening of the replacement value of individuals and the lack of variation in the gene pool, would be likely to occur only when large numbers of copies of an individual were produced.

If cloning is employed to bring into existence a single copy of a person, then the worst of the possible side-effects would be avoided. Then the problem would be whether the individual so produced would wish he had never been born. But as we have seen it is not inconceivable that the satisfaction of any other preferences he may have when alive may outweigh a backward-looking preference not to have been a copy, especially when added to the desires of the parent to produce a copy of himself.

Notes

1   David M. Rorvik, In His Image (London: Sphere, 1978).
2   Derek Bromhall, 'The Great Cloning Hoax', New Statesman (2 June 1978) 734–36.
3   R. M. Hare, 'Abortion and the Golden Rule', Philosophy and Public Affairs 4 (1975) 201–22.
4   Cf. J. S. Mill, 'Nature', in Three Essays on Religion (London, 1874) 8, where 'nature' is defined as 'all the powers existing in either the outer or the inner world and everything which takes place by means of these powers', or as 'what takes place without the agency, or without the voluntary and intentional agency, of man'.
5   The naturalistic fallacy would be committed only if it were held that to say that cloning is 'unnatural' in some descriptive sense entailed that cloning is wrong. That is not the view under discussion.
6   See e.g. John Harris, 'The Survival Lottery', Philosophy 50 (1975) 84.
7   Seymour Lederberg, 'Law and Cloning: The State as Regulator of Gene Function', in A. Milunsky and G. J. Annas (eds), Genetics and the Law (New York: Plenum, 1976) 377–86.
8   See e.g. Judith Jarvis Thomson, 'The Right to Privacy', Philosophy and Public Affairs 4 (1975)

295–314; James Rachels, 'Why Privacy is Important', ibid. 323–33.

9 Jonathan Glover, *Causing Death and Saving Lives* (Harmondsworth: Penguin, 1977).

10 Derek Parfit, 'Personal Identity', in Jonathan Glover (ed.), *The Philosophy of Mind* (Oxford University Press, 1976) 142–62.

11 Bernard Williams, 'Are Persons Bodies?' in *Problems of the Self* (London: Cambridge University Press, 1973) 81.

12 Arno G. Motulsky, 'Brave New World?' in Thomas R. Mertens (ed.), *Human Genetics: Readings on the Implications of Genetic Engineering* (New York: John Wiley and Sons, 1975) 280–308.

Chapter 11

Leon R. Kass

WHY WE SHOULD BAN THE CLONING OF HUMANS: THE WISDOM OF REPUGNANCE

Our habit of delighting in news of scientific and technological breakthroughs has been sorely challenged by the birth announcement of a sheep named Dolly. Though Dolly shares with previous sheep the "softest clothing, woolly, bright," William Blake's question, "Little Lamb, who made thee?" has for her a radically different answer: Dolly was, quite literally, made. She is the work not of nature or nature's God but of man, an Englishman, Ian Wilmut, and his fellow scientists. What's more, Dolly came into being not only asexually – ironically, just like "He [who] calls Himself a Lamb" – but also as the genetically identical copy (and the perfect incarnation of the form or blueprint) of a mature ewe, of whom she is a clone. This long-awaited yet not quite expected success in cloning a mammal raised immediately the prospect – and the specter – of cloning human beings: "I a child and Thou a lamb," despite our differences, have always been equal candidates for creative making, only now, by means of cloning, we may both spring from the hand of man playing at being God.

After an initial flurry of expert comment and public consternation, with opinion polls showing overwhelming opposition to cloning human beings, President Clinton ordered a ban on all federal support for human cloning research (even though none was being

supported) and charged the National Bioethics Advisory Commission to report in ninety days on the ethics of human cloning research. The commission (an eighteen-member panel, evenly balanced between scientists and non-scientists, appointed by the president and reporting to the national Science and Technology Council) invited testimony from scientists, religious thinkers and bioethicists, as well as from the general public. It is now deliberating about what it should recommend, both as a matter of ethics and as a matter of public policy.

Congress is awaiting the commission's report, and is poised to act. Bills to prohibit the use of federal funds for human cloning research have been introduced in the House of Representatives and the Senate; and another bill, in the House, would make it illegal "For any person to use a human somatic cell for the process of producing a human clone." A fateful decision is at hand. To clone or not to clone a human being is no longer an academic question.

Taking cloning seriously, then and now

Cloning first came to public attention roughly thirty years ago, following the successful asexual production, in England, of a clutch of tadpole clones by the technique of nuclear transplantation. The individual largely responsible for bringing the prospect and promise of human cloning to public notice was Joshua Lederberg, a Nobel Laureate geneticist and a man of large vision. In 1966, Lederberg wrote a remarkable article in *The American Naturalist* detailing the eugenic advantages of human cloning and other forms of genetic engineering, and the following year he devoted a column in *The Washington Post*, where he wrote regularly on science and society, to the prospect of human cloning. He suggested that cloning could help us overcome the unpredictable variety that still rules human reproduction, and allow us to benefit from perpetuating superior genetic endowments. These writings sparked a small public debate in which I became a participant. At the time a young researcher in molecular biology at the National Institutes of Health (NIH), I wrote a reply to the *Post*, arguing against Lederberg's amoral treatment of this morally weighty subject and insisting on the urgency of confronting a series of questions and objections, culminating in the suggestion that "the programmed reproduction of man will, in fact, dehumanize him."

Much has happened in the intervening years. It has become harder, not easier, to discern the true meaning of human cloning. We have in some sense been softened up to the idea—through movies, cartoons, jokes and intermittent commentary in the mass media, some serious, most lighthearted. We have become accustomed to new practices in human reproduction: not just in vitro fertilization, but also embryo manipulation, embryo donation, and surrogate pregnancy. Animal biotechnology has yielded transgenic animals and a burgeoning science of genetic engineering, easily and soon to be transferable to humans.

Even more important, changes in the broader culture make it now vastly more difficult to express a common and respectful understanding of sexuality, procreation, nascent life, family, and the meaning of motherhood, fatherhood and the links between the generations. Twenty-five years ago, abortion was still largely illegal and thought to be immoral, the sexual revolution (made possible by the extramarital use of the pill) was still in its infancy, and few had yet heard about the reproductive rights of single women, homosexual men, and lesbians. (Never mind shameless memoirs about one's own incest!) Then one could argue, without embarrassment, that the new technologies of human reproduction—babies without sex—and their confounding of normal kin relations – who's the mother: the egg donor, the surrogate who carries and delivers, or the one who rears? – would "undermine the justification and support that biological parenthood gives to the monogamous marriage." Today, defenders of stable, monogamous marriage risk charges of giving offense to those adults who are living in "new family forms" or to those children who, even without the benefit of assisted reproduction, have acquired either three or four parents or one or none at all. Today, one must even apologize for voicing opinions that twenty-five years ago were nearly universally regarded as the core of our culture's wisdom on these matters. In a world whose once-given natural boundaries are blurred by technological change and whose moral boundaries are seemingly up for grabs, it is much more difficult to make persuasive the still compelling case against cloning human beings. As Raskolnikov put it, "man gets used to everything – the beast!"

Indeed, perhaps the most depressing feature of the discussions that immediately followed the news about Dolly was their ironical tone, their genial cynicism, their moral fatigue: "AN UDDER WAY OF MAKING LAMBS" (*Nature*), "WHO WILL CASH IN ON BREAKTHROUGH IN CLONING?" (*The Wall Street Journal*), "IS CLONING BAAAAAAAAD?" (*The Chicago Tribune*). Gone from the scene are the wise and courageous voices of Theodosius Dobzhansky (genetics), Hans Jonas (philosophy) and Paul Ramsey (theology) who, only twenty-five years ago, all made powerful moral arguments against ever cloning a human being. We are now too sophisticated for such argumentation; we wouldn't be caught in public with a strong moral stance, never mind an absolutist one. We are all, or almost all, postmodernists now.

Cloning turns out to be the perfect embodiment of the ruling opinions of our new age. Thanks to the sexual revolution, we are able to deny in practice, and increasingly in thought, the inherent procreative teleology of sexuality itself. But, if sex has no intrinsic connection to generating babies, babies need have no necessary connection to sex. Thanks to feminism and the gay rights movement, we are increasingly encouraged to treat the natural heterosexual difference and its preeminence as a matter of "cultural construction." But if male and female are not normatively complementary and generatively significant, babies need not come from male and female complementarity. Thanks to the prominence and the acceptability of divorce and out-of-wedlock births, stable, monogamous marriage as the ideal home for procreation is no longer the agreed-upon cultural norm. For this new dispensation, the clone is the ideal emblem: the ultimate "single-parent child."

Thanks to our belief that all children should be *wanted* children (the more high-minded principle we use to justify contraception and abortion), sooner or later only those children who fulfill our wants will be fully acceptable. Through cloning, we can work our wants and wills on the very identity of our children, exercising control as never before. Thanks to modern notions of individualism and the rate of cultural change, we see ourselves not as linked to ancestors and defined by traditions, but as projects for our own self-creation, not only as self-made men but also man-made selves; and self-cloning is simply an extension of such rootless and narcissistic self-re-creation.

Unwilling to acknowledge our debt to the past and unwilling to embrace the uncertainties and the limitations of the future, we have a false relation to both: cloning personifies our desire fully to control the future, while being subject to no controls ourselves. Enchanted and enslaved by the glamour of technology, we have lost our awe and wonder before the deep mysteries of nature and of life. We cheerfully take our own beginnings in our hands and, like the last man, we blink.

Part of the blame for our complacency lies, sadly, with the field of bioethics itself, and its claim to expertise in these moral matters. Bioethics was founded by people who understood that the new biology touched and threatened the deepest matters of our humanity: bodily integrity, identity and individuality, lineage and kinship, freedom and self-command, eros and aspiration, and the relations and strivings of body and soul. With its capture by analytic philosophy, however, and its inevitable routinization and professionalization, the field has by and large come to content itself with analyzing moral arguments, reacting to new technological developments and taking on emerging issues of public policy, all performed with a naïve faith that the evils we fear can all be avoided by compassion, regulation and a respect for autonomy. Bioethics has made some major contributions in the protection of human subjects and in other areas where personal freedom is threatened; but its practitioners, with few exceptions, have turned the big human questions into pretty thin gruel.

One reason for this is that the piecemeal formation of public policy tends to grind down large questions of morals into small questions of procedure. Many of the country's leading bioethicists have served on national commissions or state task forces and advisory boards, where, understandably, they have found utilitarianism to be the only ethical vocabulary acceptable to all participants in discussing issues of law, regulation and public policy. As many of these commissions have been either officially under the aegis of NIH or the Health and Human Services Department, or otherwise dominated by powerful voices for scientific progress, the ethicists have for the most part been content, after some "values clarification" and wringing of hands, to pronounce their blessings upon the inevitable. Indeed, it is the bioethicists, not the scientists, who are now the most articulate defenders of human cloning: the two witnesses testifying before the National Bioethics Advisory Commission in favor of cloning human beings were bioethicists, eager to rebut what they regard as the irrational concerns of those of us in opposition. One wonders whether this commission, constituted like the previous commissions, can tear itself sufficiently free from the accommodationist pattern of rubber-stamping all technical innovation, in the mistaken belief that all other goods must bow down before the gods of better health and scientific advance.

If it is to do so, the commission must first persuade itself, as we all should persuade ourselves, not to be complacent about what is at issue here. Human cloning, though it is in some respects continuous with previous reproductive technologies, also represents something radically new, in itself and in its easily foreseeable consequences. The stakes are very high indeed. I exaggerate, but in the direction of the truth, when I insist that we are faced with having to decide nothing less than whether human procreation is going to remain human, whether children are going to be made rather than begotten, whether it is a good thing, humanly speaking, to say yes in principle to the road which leads (at best) to the dehumanized rationality of *Brave New World*. This is not business as usual, to be fretted about for a while but finally to be given our seal of approval. We must rise to the occasion and make our judgments as if the future of our humanity hangs in the balance. For so it does.

The State of the Art

If we should not underestimate the significance of human cloning, neither should we exaggerate its imminence or misunderstand just what is involved. The procedure is conceptually simple. The nucleus of a mature but unfertilized egg is removed and replaced with a nucleus obtained from a specialized cell of an adult (or fetal) organism (in Dolly's case, the donor nucleus came from mammary gland epithelium). Since almost all the hereditary material of a cell is contained within its nucleus, the renucleated egg and the individual into which this egg develops are genetically identical to the organism that was the source of the transferred nucleus. An unlimited number of genetically identical individuals – clones – could be produced by nuclear transfer. In principle, any person, male or female, newborn or adult, could be cloned, and in any quantity. With laboratory cultivation and storage of tissues, cells outliving their sources make it possible even to clone the dead.

The technical stumbling block, overcome by Wilmut and his colleagues, was to find a means of reprogramming the state of the DNA in the donor cells, reversing its differentiated expression and restoring its full totipotency, so that it could again direct the entire process of producing a mature organism. Now that this problem has been solved, we should expect a rush to develop cloning for other animals, especially livestock, in order to propagate in perpetuity the champion meat or milk producers. Though exactly how soon someone will succeed in cloning a human being is anybody's guess, Wilmut's technique, almost certainly applicable to humans, makes *attempting* the feat an imminent possibility.

Yet some cautions are in order and some possible misconceptions need correcting. For a start, cloning is not Xeroxing. As has been reassuringly reiterated, the clone of Mel Gibson, though his genetic double, would enter the world hairless, toothless and peeing in his diapers, just like any other human infant. Moreover, the success rate, at least at first, will probably not be very high: the British transferred 277 adult nuclei into enucleated sheep eggs, and implanted twenty-nine clonal embryos, but they achieved the birth of only one live lamb clone. For this reason, among others, it is unlikely that, at least for now, the practice would be very popular, and there is no immediate worry of mass-scale production of multicopies. The need of repeated surgery to obtain eggs and more crucially, of numerous borrowed wombs for implantation will surely limit use, as will the expense; besides, almost everyone who is able will doubtless prefer nature's sexier way of conceiving.

Still, for the tens of thousands of people already sustaining over 200 assisted-reproduction clinics in the United States and already availing themselves of in vitro fertilization, intracytoplasmic sperm injection, and other techniques of assisted reproduction, cloning would be an option with virtually no added fuss (especially when the success rate improves). Should commercial interests develop in "nucleus-banking," as they have in sperm-banking; should famous athletes or other celebrities decide to market their DNA the way they now market their autographs and just about everything else; should techniques of embryo and germline genetic testing and manipulation arrive as anticipated, increasing the use of laboratory assistance in order to obtain "better" babies – should all this come to pass, then cloning, if it is permitted, could become more than a marginal practice simply on the basis of free reproductive choice, even without any social encouragement to upgrade the gene pool or to replicate superior

types. Moreover, if laboratory research on human cloning proceeds, even without any intention to produce cloned humans, the existence of cloned human embryos in the laboratory, created to begin with only for research purposes, would surely pave the way for later baby-making implantations.

In anticipation of human cloning, apologists and proponents have already made clear possible uses of the perfected technology, ranging from the sentimental and compassionate to the grandiose. They include: providing a child for an infertile couple; "replacing" a beloved spouse or child who is dying or has died; avoiding the risk of genetic disease; permitting reproduction for homosexual men and lesbians who want nothing sexual to do with the opposite sex; securing a genetically identical source of organs or tissues perfectly suitable for transplantation; getting a child with a genotype of one's own choosing, not excluding oneself; replicating individuals of great genius, talent or beauty – having a child who really could "be like Mike"; and creating large sets of genetically identical humans suitable for research on, for instance, the question of nature versus nurture, or for special missions in peace and war (not excluding espionage), in which using identical humans would be an advantage. Most people who envision the cloning of human beings, of course, want none of these scenarios. That they cannot say why is not surprising. What is surprising, and welcome, is that, in our cynical age, they are saying anything at all.

The wisdom of repugnance

"Offensive." "Grotesque." "Revolting." "Repugnant." "Repulsive." These are the words most commonly heard regarding the prospect of human cloning. Such reactions come both from the man or woman in the street and from the intellectuals, from believers and atheists, from humanists and scientists. Even Dolly's creator has said he "would find it offensive" to clone a human being.

People are repelled by many aspects of human cloning. They recoil from the prospect of mass production of human beings, with large clones of look-alikes, compromised in their individuality; the idea of father-son or mother-daughter twins; the bizarre prospects of a woman giving birth to and rearing a genetic copy of herself, her spouse or even her deceased father or mother; the grotesqueness of conceiving a child as an exact replacement for another who had died; the utilitarian creation of embryonic genetic duplicates of oneself, to be frozen away or created when necessary, in case of need for homologous tissues or organs for transplantation; the narcissism of those who would clone themselves and the arrogance of others who think they know who deserves to be cloned or which genotype any child-to-be should be thrilled to receive; the Frankensteinian hubris to create human life and increasingly to control its destiny; man playing God. Almost no one finds any of the suggested reasons for human cloning compelling; almost everyone anticipates its possible misuses and abuses. Moreover, many people feel oppressed by the sense that there is probably nothing we can do to prevent it from happening. This makes the prospect all the more revolting.

Revulsion is not an argument; and some of yesterday's repugnances are today calmly accepted – though, one must add, not always for the better. In crucial cases, however, repugnance is the emotional expression of deep wisdom, beyond reason's power fully to articulate it. Can anyone really give an argument fully adequate to the horror which is father-daughter incest (even with consent), or having sex with animals, or mutilating a corpse, or eating human flesh, or even just (just!) raping or murdering another human being? Would anybody's failure to give full rational justification for his or her revulsion

at these practices make that revulsion ethically suspect? Not at all. On the contrary, we are suspicious of those who think that they can rationalize away our horror, say, by trying to explain the enormity of incest with arguments only about the genetic risks of inbreeding.

The repugnance at human cloning belongs in this category. We are repelled by the prospect of cloning human beings not because of the strangeness or novelty of the undertaking, but because we intuit and feel, immediately and without argument, the violation of things that we rightfully hold dear. Repugnance, here as elsewhere, revolts against the excesses of human willfulness, warning us not to transgress what is unspeakably profound. Indeed, in this age in which everything is held to be permissible so long as it is freely done, in which our given human nature no longer commands respect, in which our bodies are regarded as mere instruments of our autonomous rational wills, repugnance may be the only voice left that speaks up to defend the central core of our humanity. Shallow are the souls that have forgotten how to shudder.

The goods protected by repugnance are generally overlooked by our customary ways of approaching all new biomedical technologies. The way we evaluate cloning ethically will in fact be shaped by how we characterize it descriptively, by the context into which we place it, and by the perspective from which we view it. The first task for ethics is proper description. And here is where our failure begins.

Typically, cloning is discussed in one or more of three familiar contexts, which one might call the technological, the liberal and the meliorist. Under the first, cloning will be seen as an extension of existing techniques for assisting reproduction and determining the genetic makeup of children. Like them, cloning is to be regarded as a neutral technique, with no inherent meaning or goodness, but subject to multiple uses, some good, some bad. The morality of cloning thus depends absolutely on the goodness or badness of the motives and intentions of the cloners: as one bioethicist defender of cloning puts it, "the ethics must be judged [only] by the way the parents nurture and rear their resulting child and whether they bestow the same love and affection on a child brought into existence by a technique of assisted reproduction as they would on a child born in the usual way."

The liberal (or libertarian or liberationist) perspective sets cloning in the context of rights, freedoms and personal empowerment. Cloning is just a new option for exercising an individual's right to reproduce or to have the kind of child that he or she wants. Alternatively, cloning enhances our liberation (especially women's liberation) from the confines of nature, the vagaries of chance, or the necessity for sexual mating. Indeed, it liberates women from the need for men altogether, for the process requires only eggs, nuclei and (for the time being) uteri – plus, of course, a healthy dose of our (allegedly "masculine") manipulative science that likes to do all these things to mother nature and nature's mothers. For those who hold this outlook, the only moral restraints on cloning are adequately informed consent and the avoidance of bodily harm. If no one is cloned without her consent, and if the clonant is not physically damaged, then the liberal conditions for licit, hence moral, conduct are met. Worries that go beyond violating the will or maiming the body are dismissed as "symbolic" – which is to say, unreal.

The meliorist perspective embraces valetudinarians and also eugenicists. The latter were formerly more vocal in these discussions, but they are now generally happy to see their goals advanced under the less threatening banners of freedom and technological growth. These people see in cloning a new prospect for improving human beings – minimally, by ensuring the perpetuation of healthy individuals by avoiding the risks of genetic disease inherent in the lottery of sex, and maximally, by producing "optimum babies," preserving outstanding genetic material, and (with the help of soon-to-come techniques for precise genetic engineering) enhancing inborn human capacities on many fronts. Here the morality of cloning as a means is justified solely by the excellence of the end, that is, by the outstanding traits or individuals cloned – beauty, or brawn, or brains.

These three approaches, all quintessentially American and all perfectly fine in their places, are sorely wanting as approaches to human procreation. It is, to say the least, grossly distorting to view the wondrous mysteries of birth, renewal and individuality, and the deep meaning of parent-child relations, largely through the lens of our reductive science and its potent technologies. Similarly, considering reproduction (and the intimate relations of family life!) primarily under the political-legal, adversarial, and individualistic notion of rights can only undermine the private yet fundamentally social, cooperative, and duty-laden character of child-bearing, child-rearing, and their bond to the covenant of marriage. Seeking to escape entirely from nature (in order to satisfy a natural desire or a natural right to reproduce!) is self-contradictory in theory and self-alienating in practice. For we are erotic beings only because we are embodied beings, and not merely intellects and wills unfortunately imprisoned in our bodies. And, though health and fitness are clearly great goods, there is something deeply disquieting in looking on our prospective children as artful products perfectible by genetic engineering, increasingly held to our willfully imposed designs, specifications, and margins of tolerable error.

The technical, liberal and meliorist approaches all ignore the deeper anthropological, social and, indeed, ontological meanings of bringing forth new life. To this more fitting and profound point of view, cloning shows itself to be a major alteration, indeed, a major violation, of our given nature as embodied, gendered and engendering beings – and of the social relations built on this natural ground. Once this perspective is recognized, the ethical judgment on cloning can no longer be reduced to a matter of motives and intentions, rights and freedoms, benefits and harms, or even means and ends. It must be regarded primarily as a matter of meaning: Is cloning a fulfillment of human begetting and belonging? Or is cloning rather, as I contend, their pollution and perversion? To pollution and perversion, the fitting response can only be horror and revulsion; and conversely, generalized horror and revulsion are prima facie evidence of foulness and violation. The burden of moral argument must fall entirely on those who want to declare the widespread repugnances of humankind to be mere timidity or superstition.

Yet repugnance need not stand naked before the bar of reason. The wisdom of our horror at human cloning can be partially articulated, even if this is finally one of those instances about which the heart has its reasons that reason cannot entirely know.

The profundity of sex

To see cloning in its proper context, we must begin not, as I did before, with laboratory technique, but with the anthropology – natural and social – of sexual reproduction.

Sexual reproduction – by which I mean the generation of new life from (exactly) two complementary elements, one female, one male, (usually) through coitus – is established (if that is the right term) not by human decision, culture or tradition, but by nature; it is the natural way of all mammalian reproduction. By nature, each child has two complementary biological progenitors. Each child thus stems from and unites exactly two lineages. In natural

generation, moreover, the precise genetic constitution of the resulting offspring is determined by a combination of nature and chance, not by human design: each human child shares the common natural human species genotype, each child is genetically (equally) kin to each (both) parent(s), yet each child is also genetically unique.

These biological truths about our origins foretell deep truths about our identity and about our human condition altogether. Every one of us is at once equally human, equally enmeshed in a particular familial nexus of origin, and equally individuated in our trajectory from birth to death – and, if all goes well, equally capable (despite our mortality) of participating, with a complementary other, in the very same renewal of such human possibility through procreation. Though less momentous than our common humanity, our genetic individuality is not humanly trivial. It shows itself forth in our distinctive appearance through which we are everywhere recognized; it is revealed in our "signature" marks of fingerprints and our self-recognizing immune system; it symbolizes and foreshadows exactly the unique, never-to-be-repeated character of each human life.

Human societies virtually everywhere have structured child-rearing responsibilities and systems of identity and relationship on the bases of these deep natural facts of begetting. The mysterious yet ubiquitous "love of one's own" is everywhere culturally exploited, to make sure that children are not just produced but well cared for and to create for everyone clear ties of meaning, belonging and obligation. But it is wrong to treat such naturally rooted social practices as mere cultural constructs (like left- or right-driving, or like burying or cremating the dead) that we can alter with little human cost. What would kinship be without its clear natural grounding? And what would identity be without kinship? We must resist those who have begun to refer to sexual reproduction as the "traditional method of reproduction," who would have us regard as merely traditional, and by implication arbitrary, what is in truth not only natural but most certainly profound.

Asexual reproduction, which produces "single-parent" offspring, is a radical departure from the natural human way, confounding all normal understandings of father, mother, sibling, grandparent, etc., and all moral relations tied thereto. It becomes even more of a radical departure when the resulting offspring is a clone derived not from an embryo, but from a mature adult to whom the clone would be an identical twin; and when the process occurs not by natural accident (as in natural twinning), but by deliberate human design and manipulation; and when the child's (or children's) genetic constitution is preselected by the parent(s) (or scientists). Accordingly, as we will see, cloning is vulnerable to three kinds of concerns and objections, related to these three points: cloning threatens confusion of identity and individuality, even in small-scale cloning; cloning represents a giant step (though not the first one) toward transforming procreation into manufacture, that is, toward the increasing depersonalization of the process of generation and, increasingly, toward the "production" of human children as artifacts, products of human will and design (what others have called the problem of "commodification" of new life); and cloning – like other forms of eugenic engineering of the next generation – represents a form of despotism of the cloners over the cloned, and thus (even in benevolent cases) represents a blatant violation of the inner meaning of parent-child relations, of what it means to have a child, of what it means to say "yes" to our own demise and "replacement."

Before turning to these specific ethical objections, let me test my claim of the profundity of the natural way by taking up a challenge recently posed by a friend. What if the given natural human way of reproduction were asexual, and we now had to deal with a new technological innovation – artificially induced sexual dimorphism and the fusing of complementary gametes – whose inventors argued that sexual reproduction promised all sorts of advantages, including hybrid vigor and the creation of greatly increased individuality? Would one then be forced to defend natural asexuality because it was natural? Could one claim that it carried deep human meaning?

The response to this challenge broaches the ontological meaning of sexual reproduction. For it is impossible, I submit, for there to have been human life – or even higher forms of animal life – in the absence of sexuality and sexual reproduction. We find asexual reproduction only in the lowest forms of life: bacteria, algae, fungi, some lower invertebrates. Sexuality brings with it a new and enriched relationship to the world. Only sexual animals can seek and find complementary others with whom to pursue a goal that transcends their own existence. For a sexual being, the world is no longer an indifferent and largely homogeneous *otherness*, in part edible, in part dangerous. It also contains some very special and related and complementary beings, of the same kind but of opposite sex, toward whom one reaches out with special interest and intensity. In higher birds and mammals, the outward gaze keeps a lookout not only for food and predators, but also for prospective mates; the beholding of the many splendored world is suffused with desire for union, the animal antecedent of human eros and the germ of sociality. Not by accident is the human animal both the sexiest animal – whose females do not go into heat but are receptive throughout the estrous cycle and whose males must therefore have greater sexual appetite and energy in order to reproduce successfully – and also the most aspiring, the most social, the most open, and the most intelligent animal.

The soul-elevating power of sexuality is, at bottom, rooted in its strange connection to mortality, which it simultaneously accepts and tries to overcome. Asexual reproduction may be seen as a continuation of the activity of self-preservation. When one organism buds or divides to become two, the original being is (doubly) preserved, and nothing dies. Sexuality, by contrast, means perishability and serves replacement; the two that come together to generate one soon will die. Sexual desire, in human beings as in animals, thus serves an end that is partly hidden from, and finally at odds with, the self-serving individual. Whether we know it or not, when we are sexually active we are voting with our genitalia for our own demise. The salmon swimming upstream to spawn and die tell the universal story: sex is bound up with death, to which it holds a partial answer in procreation.

The salmon and the other animals evince this truth blindly. Only the human being can understand what it means. As we learn so powerfully from the story of the Garden of Eden, our humanization is coincident with sexual self-consciousness, with the recognition of our sexual nakedness and all that it implies: shame at our needy incompleteness, unruly self-division and finitude; awe before the eternal; hope in the self-transcending possibilities of children and a relationship to the divine. In the sexually self-conscious animal, sexual desire can become eros, lust can become love. Sexual desire humanly regarded is thus sublimated into erotic longing for wholeness, completion and immortality, which drives us knowingly into the embrace and its generative fruit — as well as into all the higher human possibilities of deed, speech, and song.

Through children, a good common to both husband and wife, male and female achieve some genuine unification (beyond the mere sexual "union," which fails to do so). The two become one through sharing generous (not needy) love for this third being as good. Flesh of their flesh, the child is the parents' own commingled being externalized, and given a separate and persisting existence. Unification is enhanced also by their commingled work of rearing. Providing an opening to the future beyond the grave, carrying not only our seed but also our names, our ways and our hopes that they will surpass us in goodness and happiness, children are a testament to the possibility of transcendence. Gender duality and sexual desire, which first draws our love upward and outside of ourselves, finally provide for the partial overcoming of the confinement and limitation of perishable embodiment altogether.

Human procreation, in sum, is not simply an activity of our rational wills. It is a more complete activity precisely because it engages us bodily, erotically and spiritually, as well as rationally. There is wisdom in the mystery of nature that has joined the pleasure of sex, the inarticulate longing for union, the communication of the loving embrace and the deep-seated and only partly articulate desire for children in the very activity by which we continue the chain of human existence and participate in the renewal of human possibility. Whether or not we know it, the severing of procreation from sex, love, and intimacy is inherently dehumanizing, no matter how good the product.

We are now ready for the more specific objections to cloning.

The perversities of cloning

First, an important if formal objection: any attempt to clone a human being would constitute an unethical experiment upon the resulting child-to-be. As the animal experiments (frog and sheep) indicate, there are grave risks of mishaps and deformities. Moreover, because of what cloning means, one cannot presume a future cloned child's consent to be a clone, even a healthy one. Thus, ethically speaking, we cannot even get to know whether or not human cloning is feasible.

I understand, of course, the philosophical difficulty of trying to compare a life with defects against nonexistence. Several bioethicists, proud of their philosophical cleverness, use this conundrum to embarrass claims that one can injure a child in its conception, precisely because it is only thanks to that complained-of conception that the child is alive to complain. But common sense tells us that we have no reason to fear such philosophisms. For we surely know that people can harm and even maim children in the very act of conceiving them, say, by paternal transmission of the AIDS virus, maternal transmission of heroin dependence or, arguably, even by bringing them into being as bastards or with no capacity or willingness to look after them properly. And we believe that to do this intentionally, or even negligently, is inexcusable and clearly unethical.

The objection about the impossibility of presuming consent may even go beyond the obvious and sufficient point that a clonant, were he subsequently to be asked, could rightly resent having been made a clone. At issue are not just benefits and harms, but doubts about the very independence needed to give proper (even retroactive) consent, that is, not just the capacity to choose but the disposition and ability to choose freely and well. It is not at all clear to what extent a clone will truly be a moral agent. For, as we shall see, in the very fact of cloning, and of rearing him as a clone, his makers subvert the cloned child's independence, beginning with that aspect that comes from knowing that one was an unbidden surprise, a gift, to the world, rather than the designed result of someone's artful project.

Cloning creates serious issues of identity and individuality. The cloned person may experience concerns about his distinctive identity not only because he will be in genotype and appearance identical to another human being, but, in this case, because he may also be twin to the person who is his "father" or "mother" – if one can still call them that. What would be the psychic burdens of being the "child" or "parent" of your twin? The cloned individual, moreover, will be saddled with a genotype that has already lived. He will not be fully a surprise to the world. People are likely always to compare his performances in life with that of his alter ego. True, his nurture and his circumstance in life will be different: genotype is not exactly destiny. Still, one must also expect parental and other efforts to shape this new life after the original – or at least to view the child with the original version always firmly in mind. Why else did they clone from the star basketball player, mathematician, and beauty queen – or even dear old dad – in the first place?

Since the birth of Dolly, there has been a fair amount of doublespeak on this matter of genetic identity. Experts have rushed in to reassure the public that the clone would in no way be the same person, or have any confusions about his or her identity: as previously noted, they are pleased to point out that the clone of Mel Gibson would not be Mel Gibson. Fair enough. But one is shortchanging the truth by emphasizing the additional importance of the intrauterine environment, rearing and social setting: genotype obviously matters plenty. That, after all, is the only reason to clone, whether human beings or sheep. The odds that clones of Wilt Chamberlain will play in the NBA are, I submit, infinitely greater than they are for clones of Robert Reich.

Curiously, this conclusion is supported, inadvertently, by the one ethical sticking point insisted on by friends of cloning: no cloning without the donor's consent. Though an orthodox liberal objection, it is in fact quite puzzling when it comes from people (such as Ruth Macklin) who also insist that genotype is not identity or individuality, and who deny that a child could reasonably complain about being made a genetic copy. If the clone of Mel Gibson would not be Mel Gibson, why should Mel Gibson have grounds to object that someone had been made his clone? We already allow researchers to use blood and tissue samples for research purposes of no benefit to their sources: my falling hair, my expectorations, my urine, and even my biopsied tissues are "not me" and not mine. Courts have held that the profit gained from uses to which scientists put my discarded tissues do not legally belong to me. Why, then, no cloning without consent – including, I assume, no cloning from the body of someone who just died? What harm is done the donor, if genotype is "not me"? Truth to tell, the only powerful justification for objecting is that genotype really does have something to do with identity, and everybody knows it. If not, on what basis could Michael Jordan object that someone cloned "him," say, from cells taken from a "lost" scraped-off piece of his skin? The insistence on donor consent unwittingly reveals the problem of identity in all cloning.

Genetic distinctiveness not only symbolizes the uniqueness of each human life and the independence of its parents that each human child rightfully attains. It can also be an important support for living a worthy and dignified life. Such arguments apply with great force to any large-scale replication of human individuals. But they are sufficient, in my view, to rebut even the first attempts to clone a human being. One must never forget that these are human beings upon whom our eugenic or merely playful fantasies are to be enacted.

Troubled psychic identity (distinctiveness), based on all-too-evident genetic identity (sameness), will be made much worse by the utter confusion of social identity and kinship ties. For, as already noted, cloning radically confounds lineage and social relations, for "offspring" as

for "parents." As bioethicist James Nelson has pointed out, a female child cloned from her "mother" might develop a desire for a relationship to her "father," and might understandably seek out the father of her "mother," who is after all also her biological twin sister. Would "grandpa," who thought his paternal duties concluded, be pleased to discover that the clonant looked to him for paternal attention and support?

Social identity and social ties of relationship and responsibility are widely connected to, and supported by, biological kinship. Social taboos on incest (and adultery) everywhere serve to keep clear who is related to whom (and especially which child belongs to which parents), as well as to avoid confounding the social identity of parent-and-child (or brother-and-sister) with the social identity of lovers, spouses and co-parents. True, social identity is altered by adoption (but as a matter of the best interest of already living children: we do not deliberately produce children for adoption). True, artificial insemination and in vitro fertilization with donor sperm, or whole embryo donation, are in some way forms of "prenatal adoption" – a not altogether unproblematic practice. Even here, though, there is in each case (as in all sexual reproduction) a known male source of sperm and a known single female source of egg – a genetic father and a genetic mother – should anyone care to know (as adopted children often do) who is genetically related to whom.

In the case of cloning, however, there is but one "parent." The usually sad situation of the "single-parent child" is here deliberately planned, and with a vengeance. In the case of self-cloning, the "offspring" is, in addition, one's twin; and so the dreaded result of incest – to be parent to one's sibling – is here brought about deliberately, albeit without any act of coitus. Moreover, all other relationships will be confounded. What will father, grandfather, aunt, cousin, sister mean? Who will bear what ties and what burdens? What sort of social identity will someone have with one whole side – "father's" or "mother's" – necessarily excluded? It is no answer to say that our society, with its high incidence of divorce, remarriage, adoption, extramarital child-bearing and the rest, already confounds lineage and confuses kinship and responsibility for children (and everyone else), unless one also wants to argue that this is, for children, a preferable state of affairs.

Human cloning would also represent a giant step toward turning begetting into making, procreation into manufacture (literally, something "handmade"), a process already begun with in vitro fertilization and genetic testing of embryos. With cloning, not only is the process in hand, but the total genetic blueprint of the cloned individual is selected and determined by the human artisans. To be sure, subsequent development will take place according to natural processes; and the resulting children will still be recognizably human. But we here would be taking a major step into making man himself simply another one of the man-made things. Human nature becomes merely the last part of nature to succumb to the technological project, which turns all of nature into raw material at human disposal, to be homogenized by our rationalized technique according to the subjective prejudices of the day.

How does begetting differ from making? In natural procreation, human beings come together, complementarily male and female, to give existence to another being who is formed, exactly as we were, *by what we are*: living, hence perishable, hence aspiringly erotic, human beings. In clonal reproduction, by contrast, and in the more advanced forms of manufacture to which it leads, we give existence to a being not by what we are but by what we intend and design. As with any product of our making, no matter how excellent, the artificer stands above

it, not as an equal but as a superior, transcending it by his will and creative prowess. Scientists who clone animals make it perfectly clear that they are engaged in instrumental making; the animals are, from the start, designed as means to serve rational human purposes. In human cloning, scientists and prospective "parents" would be adopting the same technocratic mentality to human children: human children would be their artifacts.

Such an arrangement is profoundly dehumanizing, no matter how good the product. Mass-scale cloning of the same individual makes the point vividly; but the violation of human equality, freedom and dignity are present even in a single planned clone. And procreation dehumanized into manufacture is further degraded by commodification, a virtually inescapable result of allowing baby-making to proceed under the banner of commerce. Genetic and reproductive biotechnology companies are already growth industries, but they will go into commercial orbit once the Human Genome Project nears completion. Supply will create enormous demand. Even before the capacity for human cloning arrives, established companies will have invested in the harvesting of eggs from ovaries obtained at autopsy or through ovarian surgery, practiced embryonic genetic alteration, and initiated the stockpiling of prospective donor tissues. Through the rental of surrogate-womb services, and through the buying and selling of tissues and embryos, priced according to the merit of the donor, the commodification of nascent human life will be unstoppable.

Finally, and perhaps most important, the practice of human cloning by nuclear transfer – like other anticipated forms of genetic engineering of the next generation – would enshrine and aggravate a profound and mischievous misunderstanding of the meaning of having children and of the parent-child relationship. When a couple now chooses to procreate, the partners are saying yes to the emergence of new life in its novelty, saying yes not only to having a child but also, tacitly, to having whatever child this child turns out to be. In accepting our finitude and opening ourselves to our replacement, we are tacitly confessing the limits of our control. In this ubiquitous way of nature, embracing the future by procreating means precisely that we are relinquishing our grip, in the very activity of taking up our own share in what we hope will be the immortality of human life and the human species. This means that our children are not our children: they are not our property, not our possessions. Neither are they supposed to live our lives for us, or anyone else's life but their own. To be sure, we seek to guide them on their way, imparting to them not just life but nurturing, love, and a way of life; to be sure, they bear our hopes that they will live fine and flourishing lives, enabling us in small measure to transcend our own limitations. Still, their genetic distinctiveness and independence are the natural foreshadowing of the deep truth that they have their own and never-before-enacted life to live. They are sprung from a past, but they take an uncharted course into the future.

Much harm is already done by parents who try to live vicariously through their children. Children are sometimes compelled to fulfill the broken dreams of unhappy parents; John Doe Jr. or the III is under the burden of having to live up to his forebear's name. Still, if most parents have hopes for their children, cloning parents will have expectations. In cloning, such overbearing parents take at the start a decisive step which contradicts the entire meaning of the open and forward-looking nature of parent-child relations. The child is given a genotype that has already lived, with full expectation that this blueprint of a past life ought to be controlling of the life that is to come. Cloning is inherently despotic, for it seeks to make one's children (or someone else's children) after one's own image (or an image of

one's choosing) and their future according to one's will. In some cases, the despotism may be mild and benevolent. In other cases, it will be mischievous and downright tyrannical. But despotism – the control of another through one's will – it inevitably will be.

Meeting some objections

The defenders of cloning, of course, are not wittingly friends of despotism, Indeed, they regard themselves mainly as friends of freedom: the freedom of individuals to reproduce, the freedom of scientists and inventors to discover and devise and to foster "progress" in genetic knowledge and technique. They want large-scale cloning only for animals, but they wish to preserve cloning as a human option for exercising our "right to reproduce" – our right to have children, and children with "desirable genes." As law professor John Robertson points out, under our "right to reproduce" we already practice early forms of unnatural, artificial and extramarital reproduction, and we already practice early forms of eugenic choice. For this reason, he argues, cloning is no big deal.

We have here a perfect example of the logic of the slippery slope, and the slippery way in which it already works in this area. Only a few years ago, slippery slope arguments were used to oppose artificial insemination and in vitro fertilization using unrelated sperm donors. Principles used to justify these practices, it was said, will be used to justify more artificial and more eugenic practices, including cloning. Not so, the defenders retorted, since we can make the necessary distinctions. And now, without even a gesture at making the necessary distinctions, the continuity of practice is held by itself to be justificatory.

The principle of reproductive freedom as currently enunciated by the proponents of cloning logically embraces the ethical acceptability of sliding down the entire rest of the slope – to producing children ectogenetically from sperm to term (should it become feasible) and to producing children whose entire genetic makeup will be the product of parental eugenic planning and choice. If reproductive freedom means the right to have a child of one's own choosing, by whatever means, it knows and accepts no limits.

But, far from being legitimated by a "right to reproduce," the emergence of techniques of assisted reproduction and genetic engineering should compel us to reconsider the meaning and limits of such a putative right. In truth, a "right to reproduce" has always been a peculiar and problematic notion. Rights generally belong to individuals, but this is a right which (before cloning) no one can exercise alone. Does the right then inhere only in couples? Only in married couples? Is it a (woman's) right to carry or deliver or a right (of one or more parents) to nurture and rear? Is it a right to have your own biological child? Is it a right only to attempt reproduction, or a right also to succeed? Is it a right to acquire the baby of one's choice?

The assertion of a negative "right to reproduce" certainly makes sense when it claims protection against state interference with procreative liberty, say, through a program of compulsory sterilization. But surely it cannot be the basis of a tort claim against nature, to be made good by technology, should free efforts at natural procreation fail. Some insist that the right to reproduce embraces also the right against state interference with the free use of all technological means to obtain a child. Yet such a position cannot be sustained: for reasons having to do with the means employed, any community may rightfully prohibit surrogate pregnancy, or polygamy, or the sale of babies to infertile couples, without violating anyone's basic human "right to reproduce." When the exercise of a previously innocuous freedom now

involves or impinges on troublesome practices that the original freedom never was intended to reach, the general presumption of liberty needs to be reconsidered.

We do indeed already practice negative eugenic selection, through genetic screening and prenatal diagnosis. Yet our practices are governed by a norm of health. We seek to prevent the birth of children who suffer from known (serious) genetic diseases. When and if gene therapy becomes possible, such diseases could then be treated, in utero or even before implantation – I have no ethical objection in principle to such a practice (though I have some practical worries), precisely because it serves the medical goal of healing existing individuals. But therapy, to be therapy, implies not only an existing "patient." It also implies a norm of health. In this respect, even germline gene "therapy," though practiced not on a human being but on egg and sperm, is less radical than cloning, which is in no way therapeutic. But once one blurs the distinction between health promotion and genetic enhancement, between so-called negative and positive eugenics, one opens the door to all future eugenic designs. "To make sure that a child will be healthy and have good chances in life": this is Robertson's principle, and owing to its latter clause it is an utterly elastic principle, with no boundaries. Being over eight feet tall will likely produce some very good chances in life, and so will having the looks of Marilyn Monroe, and so will a genius-level intelligence.

Proponents want us to believe that there are legitimate uses of cloning that can be distinguished from illegitimate uses, but by their own principles no such limits can be found. (Nor could any such limits be enforced in practice.) Reproductive freedom, as they understand it, is governed solely by the subjective wishes of the parents-to-be (plus the avoidance of bodily harm to the child). The sentimentally appealing case of the childless married couple is, on these grounds, indistinguishable from the case of an individual (married or not) who would like to clone someone famous or talented, living or dead. Further, the principle here endorsed justifies not only cloning but, indeed, all future artificial attempts to create (manufacture) "perfect" babies.

A concrete example will show how, in practice no less than in principle, the so-called innocent case will merge with, or even turn into, the more troubling ones. In practice, the eager parents-to-be will necessarily be subject to the tyranny of expertise. Consider an infertile married couple, she lacking eggs or he lacking sperm, that wants a child of their (genetic) own, and propose to clone either husband or wife. The scientist-physician (who is also co-owner of the cloning company) points out the likely difficulties – a cloned child is not really their (genetic) child, but the child of only *one* of them; this imbalance may produce strains on the marriage: the child might suffer identity confusion; there is a risk of perpetuating the cause of sterility; and so on – and he also points out the advantages of choosing a donor nucleus. Far better than a child of their own would be a child of their own choosing. Touting his own expertise in selecting healthy and talented donors, the doctor presents the couple with his latest catalog containing the pictures, the health records and the accomplishments of his stable of cloning donors, samples of whose tissues are in his deep freeze. Why not, dearly beloved, a more perfect baby?

The "perfect baby," of course, is the project not of the infertility doctors, but of the eugenic scientists and their supporters. For them, the paramount right is not the so-called right to reproduce but what biologist Bentley Glass called, a quarter of a century ago, "the right of every child to be born with a sound physical and mental constitution, based on a sound genotype . . . the inalienable right to a sound heritage."

But to secure this right, and to achieve the requisite quality control over new human life, human conception and gestation will need to be brought fully into the bright light of the laboratory, beneath which it can be fertilized, nourished, pruned, weeded, watched, inspected, prodded, pinched, cajoled, injected, tested, rated, graded, approved, stamped, wrapped, sealed, and delivered. There is no other way to produce the perfect baby.

Yet we are urged by proponents of cloning to forget about the science fiction scenarios of laboratory manufacture and multiple-copied clones, and to focus only on the homely cases of infertile couples exercising their reproductive rights. But why, if the single cases are so innocent, should multiplying their performance be so off-putting? (Similarly, why do others object to people making money off this practice, if the practice itself is perfectly acceptable?) When we follow the sound ethical principle of universalizing our choice – "would it be right if everyone cloned a Wilt Chamberlain (with his consent, of course)? Would it be right if everyone decided to practice asexual reproduction?"– we discover what is wrong with these seemingly innocent cases. The so-called science fiction cases make vivid the meaning of what looks to us, mistakenly, to be benign.

Though I recognize certain continuities between cloning and, say, in vitro fertilization, I believe that cloning differs in essential and important ways. Yet those who disagree should be reminded that the "continuity" argument cuts both ways. Sometimes we establish bad precedents, and discover that they were bad only when we follow their inexorable logic to places we never meant to go. Can the defenders of cloning show us today how, on their principles, we will be able to see producing babies ("perfect babies") entirely in the laboratory or exercising full control over their genotypes (including so-called enhancement) as ethically different, in any essential way, from present forms of assisted reproduction? Or are they willing to admit, despite their attachment to the principle of continuity, that the complete obliteration of "mother" or "father," the complete depersonalization of procreation, the complete manufacture of human beings and the complete genetic control of one generation over the next would be ethically problematic and essentially different from current forms of assisted reproduction? If so, where and how will they draw the line, and why? I draw it at cloning, for all the reasons given.

Ban the cloning of humans

What, then, should we do? We should declare that human cloning is unethical in itself and dangerous in its likely consequences. In so doing, we shall have the backing of the overwhelming majority of our fellow Americans, and of the human race, and (I believe) of most practicing scientists. Next, we should do all that we can to prevent the cloning of human beings. We should do this by means of an international legal ban if possible, and by a unilateral national ban, at a minimum. Scientists may secretly undertake to violate such a law, but they will be deterred by not being able to stand up proudly to claim the credit for their technological bravado and success. Such a ban on clonal baby-making, moreover, will not harm the progress of basic genetic science and technology. On the contrary, it will reassure the public that scientists are happy to proceed without violating the deep ethical norms and intuitions of the human community.

This still leaves the vexed question about laboratory research using early embryonic human clones, specially created only for such research purposes, with no intention to implant them into a uterus. There is no question that such research holds great promise for gaining fundamental knowledge about normal (and abnormal) differentiation, and for developing tissue lines for transplantation that might be used, say, in treating leukemia or in repairing brain or spinal cord injuries – to mention just a few of the conceivable benefits. Still, unrestricted clonal embryo research will surely make the production of living human clones much more likely. Once the genies put the cloned embryos into the bottles, who can strictly control where they go (especially in the absence of legal prohibitions against implanting them to produce a child)?

I appreciate the potentially great gains in scientific knowledge and medical treatment available from embryo research, especially with cloned embryos. At the same time, I have serious reservations about creating human embryos for the sole purpose of experimentation. There is something deeply repugnant and fundamentally transgressive about such a utilitarian treatment of prospective human life. This total, shameless exploitation is worse, in my opinion, than the "mere" destruction of nascent life. But I see no added objections, as a matter of principle, to creating and using *cloned* early embryos for research purposes, beyond the objections that I might raise to doing so with embryos produced sexually.

And yet, as a matter of policy and prudence, any opponent of the manufacture of cloned humans must, I think, in the end oppose also the creating of cloned human embryos. Frozen embryonic clones (belonging to whom?) can be shuttled around without detection. Commercial ventures in human cloning will be developed without adequate oversight. In order to build a fence around the law, prudence dictates that one oppose – for this reason alone – all production of cloned human embryos, even for research purposes. We should allow all cloning research on animals to go forward, but the only safe trench that we can dig across the slippery slope,

I suspect, is to insist on the inviolable distinction between animal and human cloning.

Some readers, and certainly most scientists, will not accept such prudent restraints, since they desire the benefits of research. They will prefer, even in fear and trembling, to allow human embryo cloning research to go forward.

Very well. Let us test them. If the scientists want to be taken seriously on ethical grounds, they must at the very least agree that embryonic research may proceed if and only if it is preceded by an absolute and effective ban on all attempts to implant into a uterus a cloned human embryo (cloned from an adult) to produce a living child. Absolutely no permission for the former without the latter.

The National Bioethics Advisory Commission's recommendations regarding this matter should be watched with the greatest care. Yielding to the wishes of the scientists, the commission will almost surely recommend that cloning human embryos for research be permitted. To allay public concern, it will likely also call for a temporary moratorium – not a legislative ban – on implanting cloned embryos to make a child, at least until such time as cloning techniques will have been perfected and rendered "safe" (precisely through the permitted research with cloned embryos). But the call for a moratorium rather than a legal ban would be a moral and a practical failure. Morally, this ethics commission would (at best) be waffling on the main ethical question, by refusing to declare the production of human clones unethical (or ethical). Practically, a moratorium on implantation cannot provide even the minimum protection needed to prevent the production of cloned humans.

Opponents of cloning need therefore to be vigilant. Indeed, no one should be willing even to consider a recommendation to allow the embryo research to proceed unless it is accompanied by a call for *prohibiting* implantation and until steps are taken to make such a prohibition effective.

Technically, the National Bioethics Advisory Commission can advise the president only on federal policy, especially federal funding policy. But given the seriousness of the matter at hand, and the grave public concern that goes beyond federal funding, the commission should take a broader view. (If it doesn't, Congress surely will.) Given that most assisted reproduction occurs in the private sector, it would be cowardly and insufficient for the commission to say, simply, "no federal funding" for such practices. It would be disingenuous to argue that we should allow federal funding so that we would then be able to regulate the practice; the private sector will not be bound by such regulations. Far better, for virtually everyone concerned, would be to distinguish between research on embryos and baby-making, and to call for a complete national and international ban (effected by legislation and treaty) of the latter, while allowing the former to proceed (at least in private laboratories).

The proposal for such a legislative ban is without American precedent, at least in techno-logical matters, though the British and others have banned cloning of human beings, and we ourselves ban incest, polygamy and other forms of "reproductive freedom." Needless to say, working out the details of such a ban, especially a global one, would be tricky, what with the need to develop appropriate sanctions for violators. Perhaps such a ban will prove ineffective; perhaps it will eventually be shown to have been a mistake. But it would at least place the burden of practical proof where it belongs: on the propo-nents of this horror, requiring them to show very clearly what great social or medical good can be had only by the cloning of human beings.

We Americans have lived by, and prospered under, a rosy optimism about scientific and technological progress. The technological imperative – if it can be done, it must be done— has probably served us well, though we should admit that there is no accurate method for weighing benefits and harms. Even when, as in the cases of environmental pollution, urban decay or the lingering deaths that are the unintended by-products of medical success, we recognize the unwelcome outcomes of technological advance, we remain confident in our ability to fix all the "bad" consequences – usually by means of still newer and better technologies. How successful we can continue to be in such post hoc repairing is at least an open question. But there is very good reason for shifting the paradigm around, at least regarding those technological interventions into the human body and mind that will surely effect fundamental (and likely irreversible) changes in human nature, basic human relationships, and what it means to be a human being. Here we surely should not be willing to risk everything in the naïve hope that, should things go wrong, we can later set them right.

The president's call for a moratorium on human cloning has given us an important opportunity. In a truly unprecedented way, we can strike a blow for the human control of the technological project, for wisdom, prudence, and human dignity. The prospect of human cloning, so repulsive to contemplate, is the occasion for deciding whether we shall be slaves of unregulated progress, and ultimately its artifacts, or whether we shall remain free human beings who guide our technique toward the enhancement of human dignity. If we are to seize the occasion, we must, as the late Paul Ramsey wrote:

> raise the ethical questions with a serious and not a frivolous conscience. A man of frivolous conscience announces that there are ethical quandaries ahead that we must urgently consider before the future catches up with us. By this he often means that we need to devise a new ethics that will provide the rationalization for doing in the future what men are bound to do because of new actions and interventions science will have made possible. In contrast a man of serious conscience means to say in raising urgent ethical questions that there may be some things that men should never do. The good things that men do can be made complete only by the things they refuse to do.

Chapter 12

David Elliott

UNIQUENESS, INDIVIDUALITY, AND HUMAN CLONING

The successful cloning of a sheep at the Roslin Institute in Edinburgh has raised the prospect of human cloning[1] to a level that many people – including a U.S. President – find alarming.[2] Even prior to the Roslin announcement in February 1997, Canadian legislators had been working on legislation that would ban almost any attempt to develop human asexual reproduction techniques.[3] Many other European countries also seem to have taken the same approach.[4] Many of these efforts seem motivated by the idea that, even apart from the possible human and social consequences of such technology (popularly believed to be extremely dangerous), there just is something deeply morally objectionable with the entire business of 'duplicating' individual human genomes.

My purpose here is to challenge this negative, deontologically motivated assumption. To do so, I want to focus on the most broad, and hence the most minimal, description of what human cloning involves and then try to determine what might be morally objectionable about such an activity. I will maintain that two standard counter-arguments which purport to show that human cloning is morally objectionable *per se* are generally unsound. The two arguments are that cloning manufactures (rather than reproduces) persons and that it violates human uniqueness. I conclude that if cloning is morally wrong, it could only be because of present or potential harm imposed on the person cloned, on women who might participate in the procedure or on society generally.

The definition of 'cloning'

Cloning involves, at the very least, the reproduction of another entity which is, in some sense, identical to an original. Cloning can be achieved, not only with entire organisms, but at the cellular and molecular levels as well. I am, of course, only interested in the first form. The qualification 'identical in some sense' acknowledges straightforwardly that it simply would be a mistake to suggest or imply that cloning a human being (or any other animal, for that matter) allows the complete 'duplication' or 'xeroxing' of that individual. An organism is the expression, not merely of its genome, but of its interactive development within some particular environment. Thus the only thing that is 'duplicated' by cloning an organism is that organism's genome, along with those features that are unequivocally tied to its phenotypic expression. In fact, this qualified understanding of what is copied requires even further qualification.[5] Some cloning techniques – e.g., the one used in the Roslin sheep experiment – do not always result in the exact duplication of the cloned individual's genome. The reason for this is that such techniques rely on transferring DNA from a cell nucleus to a donor cell or an (empty) egg. But not all of the operative DNA in an organism is found in a cell's nucleus. The mitochondria are part of a cell that passes its DNA along only in the mother's egg. This mitochondrial DNA apparently does not control very many genes – perhaps only one percent of the entire genome – but it does control genes that regulate an organism's metabolism.[6] What this means is that the only ways to get a truly identical clone would be either to have nuclear DNA from a woman put into her own egg or to engage in embryo splitting. (Embryo splitting, which I will explain a bit further below, has the same result because each blastomere that is separated from an early embryo has the same mitochondrial DNA.) In all other cases of nuclear transfer, where another individual's cells or eggs are used, the 'copy' is, strictly speaking, genetically non-exact.[7]

There is some resistance in the scientific literature to the straightforward definition that I have just presented.[8] Scientists often tie the definition of 'cloning' more carefully to specific techniques that can be used to reproduce organisms asexually that otherwise reproduce sexually. When this is done, the moral issues surrounding cloning seem to change. In view of my general purposes here, however, these more careful understandings do not add up to any important moral differences. Let me briefly explain why.

There are two standard techniques that have been used to clone organisms: nuclear transfer (or transplantation) and embryo splitting (often referred to as blastomere separation). Embryo splitting involves separating cells from very early embryos and then growing them into separate embryos. Because these cells were derived from a single embryo, each separate embryo has the same genome as the original. In 1993, Jerry Hall

and other researchers at George Washington University caused great media uproar when they performed this procedure on human cells, but it has been done repeatedly on other animal species since the early 1930s.[9] Nuclear transfer, on the other hand, involves taking a host cell, removing its nucleus, and then replacing it with the nucleus of another donor cell that is to be cloned. Nuclear transfer can be performed using either a cell taken from an embryo (a totipotent cell, i.e., an undifferentiated cell capable of becoming any other cell) or a cell taken from some other part of the body (a non-totipotent cell, i.e., a cell which has differentiated itself into a particular kind of cell, say of skin or muscle). The nuclear transfer of donor DNA from totipotent cells into host embryo cells (blastomeres) has been performed regularly and successfully on domestic animals since the mid-1980's. These techniques were developed from successful cloning experiments done on amphibians in the early 1950's. The breakthrough in the recent Roslin experiment was that, apparently for the first time, an adult body cell (a cell from the udder of a pregnant ewe) was used instead of an embryonic cell. Adult body cells are, of course, non-totipotent, so the breakthrough here was the demonstration that such cells could be returned to a totipotent state and re-start the process of embryonic development.

The point here is that these different techniques could have very different moral implications.[10] First, embryo splitting requires two progenitors; nuclear transfer (in principle) requires only one, Second, embryo splitting produces only limited numbers of the original; nuclear transfer allows large numbers of duplicates. Third, embryo splitting does not entail direct manipulation or selection of genetic material, whereas nuclear transfer does.[11] Since embryo splitting is standardly done on a genetically unique, undeveloped individual, the decision to clone is usually made in the absence of detailed information about what features that individual will come to have. Nuclear transfer, on the other hand, can be done on an adult (as was the case in the Roslin experiment) where selection could be made on the basis of traits that the adult might possess. Hence cloning by nuclear transfer raises more acutely the moral spectre of eugenics, allowing many more opportunities for 'positive' or 'negative' selection. Finally, embryo splitting is not nearly as technically difficult as nuclear transfer. As a result, it is a much less risky procedure offering fewer chances for things to go wrong. This alone might make a moral difference, since many important concerns about cloning are based on issues surrounding the safety of the procedure. Taking a cell from an adult as opposed to a newly fertilised embryo, for example, means starting a new life from a cell that has acquired genetic mutations over an individual's lifetime. These mutations are often believed to be the basis of both aging and cancer. It is therefore unclear at the present time whether undue health risks or diminished life expectancy will be one of the serious risks involved in the procedure. Again, embryo splitting, since it proceeds from sexual reproduction, and hence its development is based on the event of a new and unique genetic code, is much more like the natural procedure of twinning, and hence seems to present us with fewer unknown risks.

The point is that if we were to accept cloning and then pursue the issue of *how* we should clone an individual, the differences between these procedures might become morally important. But since any of these procedures effectively allow the biological duplication of another individual, they raise the same general issue of whether it is right to do this sort of thing. Hence, for my purposes, it is not necessary to attend to the otherwise important moral differences that might hold between cloning techniques.

Why is cloning morally wrong?

Cloning is an instance of assisted reproduction. As such, it could be morally condemned for reasons that could be raised against any form of assisted reproduction. I will refer to arguments that employ such a strategy as global anti-assisted reproduction arguments. These arguments, again, appeal to a *general* position about the immorality of assisted reproduction and then apply this general claim to particular cases of it. Recent philosophical and theological literature on the ethics of assisted reproduction presents many examples of such arguments. Paul Ramsey, for example, a pioneering bioethicist, reasons that since God's creative activity was performed and motivated by love, '. . . neither should there be among men and women, whose man-womanhood . . . is in the image of God, any lovemaking set out of the context of responsibility for procreation or any begetting apart from the sphere of human love and responsiveness'.[12] The idea here is that since most forms of assisted reproduction are not reasonably seen as expressions of inter-personal love, they become morally suspect. Some feminists, too, seem to defend global arguments against assisted reproductive technologies. A standard argument here is that most forms of assisted reproduction are morally wrong or should be banned because they play into existing patterns of patriarchal power and sexist oppression. The main work of such an argument then becomes showing how this general principle applies to each particular case of assisted reproduction. Whatever the value of this type of argument, I mention it only by way of contrast with the approach that I want to adopt here. My approach begins by looking much more specifically at the particular case of assisted reproduction and then trying to determine whether this would present moral concerns, even a positive social context where sophisticated social or legal regulations might surround the practice, and where there are no serious, known harms associated with it. And, again, what I want to defend now is the claim that human cloning passes this methodological threshold.

If cloning is to be singled out among the forms of assisted reproduction as being wrong *per se*, its wrong-making property must be connected to the fact that cloning is minimally the attempt to intentionally duplicate, as closely as possible, some individual's genome. Why would this intrinsically be wrong or evil? Two main responses are standardly given to this question: Cloning is wrong 1) because it involves the intention to produce a person with a certain genome, and 2) because it violates some right or value that humans have in being biologically unique. I will refer to the former as the manufacturing argument, and the latter as the uniqueness argument (or the non-uniqueness objection). I will consider each of these proposals in turn.

The manufacturing objection

When you think about it, cloning a human would be a very simple, and yet enormously efficient way to select for an individual with certain biological features. The other way to do this would be to engage in genetic alchemy – gene therapy as it is now (perhaps euphemistically) called – and try to change that individual's genome. But this technology is nascent, uncertain, risky, complicated, enormously expensive, and to date, not terribly successful. This fact alone seems to be why certain forms of mammalian cloning in domestic animal husbandry have been largely favoured over biotechnology that involves manipulating genomes. You simply find the cow that you like, and then go about producing 'copies' of it. Human cloning would seem directly to involve these same 'manufacturing' or selection opportunities. Indeed,

someone might suggest that these capacities are built right into the decision to clone. It is inherently a decision to produce an individual of a certain type, with certain features that we hope are based in his or her genome. This is how Jeremy Rifkin, a popular biotechnology critic, sees the matter. As he puts it: 'It's a horrendous crime to make a Xerox of someone . . . [because] you're putting a human into a genetic straitjacket. For the first time, we've taken the principles of industrial design – quality control and predictability – and applied them to a human being'.[8] The moral idea here seems to be that in manufacturing people, we devalue them; we treat them as objects to be designed rather than as potential subjects or agents capable of their own making.

There are several familiar problems with this familiar line of argument. First, let us assume that the decision to clone is inherently a manufacturing choice (I will question this assumption in a moment). The problem is that these sorts of choices seem to typify so many choices that people make which, even if they are not conscious choices of this sort, at least have the effect of shaping or selecting the traits of their child. The process, for example, of selecting a partner where children could arrive in the future, while it surely is not (and I would hope it never should be) *merely* a decision to select the traits of one's children, it does in part present an opportunity for just this sort of selection. Additionally, if we really do believe that 'manufacturing' choices are morally objectionable, then it becomes difficult to imagine why people seeking to adopt a child should have to be consulted about the adoption of any particular child once they have expressed a general interest in adoption. The same might hold for a woman seeking to have a child through artificial insemination by donor. It seems rather strong to hold that her moral qualities as a future parent are seriously diminished by any interest in the

adoption, choosing father, choosing sperm donor. You choose in this process too.

general features of the donor, or that she would be an ideal parent if she were to accept sperm only if she could know nothing at all about the physical appearance, family medical history, etc., of its donor.

Furthermore, even if it were just false that people really do make many choices that have the effect of trait selection prior to the birth of their children, they certainly do go to considerable trouble to see that their children develop certain traits after they are born. All of this 'manufacturing' seems appropriate, however, given certain standard moral assumptions – e.g., when it is at least not harmful to the child, when it does not severely restrict her opportunities to become an autonomous, self-affirming individual or when it is in her interest to have (or avoid) certain characteristics (say, a debilitating disease). The same could be said about pre-conception selection decisions. Many of these decisions could indeed be frivolous, selfish, and so on. But many of them might be capable of moral defence by showing how they might be important for cultivating the capacities and opportunities for a person's self-development.

What tends to upset many of us about manufacturing decisions, I suspect, is not really that *some* of them might occur in ordinary choices about having a child, but rather that *too many* of them might be present. What might be objectionable, then, is the total quantity of the same sort of choice. Too many manufacturing choices, it might be suggested, push us to the point where we would be treating a (potential) child as an object of his or her parent's desires and goals, rather than a person in his or her own right. Furthermore, being able to determine a person's traits to a very considerable extent might raise concerns about whether that parent is capable of valuing or loving a child unconditionally, or loving him for the person that he might become through self-development. It could also raise questions about whether the potential person would be able adequately to develop her own sense of self and personal agency. In this regard, Joseph Fletcher, another early bioethicist, surely overstated his response to the manufacturing argument when he enthusiastically *celebrates* our potential to manufacture people. 'Man,' he writes, 'is a maker and a selecter [sic] and a designer, and the more rationally contrived and deliberate anything is, the more human it is.' Fletcher even goes so far as to claim that laboratory reproduction is 'radically human compared to conception by ordinary heterosexual intercourse' because the manufacturing is willed, chosen, purposed, controlled; it is a matter of 'choice, and not chance'.[14]

Whatever merit there might be in responding to the manuacturing objection by arguing either that manufacturing is a standard decision most parents make or that it is not itself morally objectionable, let us for the moment set both of these suggestions aside. There is another consideration that is, I would suggest, even more decisive. The decision to clone is not *inherently* a choice to manufacture a particular individual in a certain way, even though this consequence may be foreseeable. It can simply be a choice to have a child of one's own in the only way possible. The familiar examples standardly offered in the literature as morally defensible reasons to clone illustrate this point. These examples can usually be classified under two main categories: 1) the prevention (by bypassing) of infertility and 2) the avoidance of genetic disease. Given either one or both of these situations, some cloning technique may be the only way that some people might be able to have children of their own. In these cases, however, the decision to have a child could be only that of having a child of one's own; it need not be a detailed manufacturing decision – a point which seems particularly true with respect to bypassing infertility. This is just the sort of outlook that we might arguably suggest is the ideal outlook that couples having children through sexual reproduction should have. Cloning, of course, does come with biological foreknowledge; one would have a fairly good idea of what the child's genome would be, and all that this entails. But, again, simply because there is foreknowledge that a duplication of one's genome will be the outcome of one's decision to have a child, it simply does not follow that the decision to have a child involves or entails a determination to have a child for these reasons. Imagine an analogy with a couple where infertility and genetic disease is not a known consideration, but who may know in advance (say due to some established medical condition) that all of their offspring will be female. It is not obvious that their choice to have a child should be regarded as an instance of sex-selection. If this is right, then there is simply no reason to regard all instances of cloning as instances of manufacturing humans. Other more general, recognizable, and morally defensible motivations, I would suggest, can be present.

The non-uniqueness objection

The other main reason that many bioethicists give as a principled objection to cloning is that it creates a biologically non-unique individual. Daniel Callahan, along with other prominent bioethicists, has strenuously voiced this particular concern. He writes:

> For all of its haphazard qualities, there is one enormous advantage in the current lottery: save for the occasional natural twinning, it gives each of us our own unique identity. There is no one else in the world like us. This is a precious gift of nature. It allows us to become our own person, to have some of our parents' genetic traits, but to have even more of our own. Nature does not make us in our parents' image; it makes us in our own unrepeatable image. Cloning would deprive the products of an engineered conception of that gift.[15]

There are two ways of reading Callahan's statement here. He could be claiming either that human cloning would result in 1) an *objective* loss of uniqueness/individuality for the clone or 2) a *subjective* (i.e., a perceived or believed) loss of uniqueness or individuality by the clone. In either case, of course, Callahan seems to be suggesting that the respective loss of uniqueness or individuality is morally significant. The most natural interpretation of Callahan's argument seems to be 1), but as we shall see, this really is an implausible line of argument. I will also argue that even if 1) were true, and human cloning could result in duplicating persons and their characters or temperaments, cloning would still not be morally objectionable (at least not solely on the grounds that it creates similar persons). I will also hold that while it seems reasonable to think that 2) is empirically unlikely, it seems more plausible to expect that at least some clones would come to feel this way about their situation. I will maintain, however, that this feeling, even if it occurs, is still not a sufficiently morally serious reason not to clone – something more serious would need to be presented, and this probably cannot be found solely in an argument from the moral value of uniqueness.

I have already suggested above why it is a mistake to believe that human cloning would result in an objective loss of uniqueness/individuality for the clone. The reason is that if this claim were true, some very strong kind of biological determinism would also have to be true. For the only thing that we are really considering in cloning humans is the duplication of an individual's genome (and perhaps this too is only partial). So the only way that we could duplicate a person by duplicating that person's genome would be if most, if not all, of what makes up a person comes from the genes. Although there are exceptions, most biologically inclined psychologists seem to resist this idea, and posit that environment plays some role in shaping personality. Furthermore, however similar two persons might be genetically, they clearly would have different experiences, and it seems reasonable to believe that these different experiences would have some differential effect on a person's character and personality.

My focus in this paper, of course, is not on psychology or philosophy of mind. So I do not want to defend rigorously the claims that I have just suggested. Let us, then, for the sake of argument assume that in duplicating a person's genome, we really would be duplicating a person in some interesting sense. Is there something morally objectionable about this? I do not believe so, for three main reasons. First, Callahan's suggestion that non-uniqueness implies repeatability does not lead in any interesting philosophical direction. Secondly, if we accept that non-uniqueness is a moral wrong or evil for persons, then we must entertain rather bizarre beliefs about the moral status or situation of natural clones – i.e., twins, triplets, etc. Underlining both of these arguments will be a third reason – viz., that individuality or uniqueness yields no interesting account of what makes persons or their lives morally valuable.

As we have seen, Callahan suggests that there is a connection between the idea of an individual's uniqueness and the idea of that individual's non-repeatability. If the possibility of repeating another person is to be morally troublesome, it must imply that in being repeated (or in being rendered non-unique) the original and the repeated individuals are somehow devalued as persons. This suggestion, however, is going to need some further explanation, since it is not obvious why non-repeatability should be considered a morally significant feature.

One way that it might take on some importance would be to suggest that a person's repeatability implies that he or she is replaceable. Viewing a being as replaceable does not seem to denote a very strong notion of that being's moral worth or value. It seems to imply that we would not be doing anything wrong if we 'removed' a particular individual provided that we 'replaced' that individual with a new one. Intuitively, at least, many people (I include myself) are unwilling to consider persons in these moral terms, and hence do not want to think of humans as beings whose value as individuals is replaceable. So one way of expanding on Callahan's notion of human value being linked to an individual's non-repeatability is to claim that individuals have value or status when they are viewed as being non-replaceable.

This is the way that Immanuel Kant has conceived of the value of persons. In the *Groundwork for the Metaphysics of Morals*, Kant notes a difference between the concepts of the value and dignity of persons. He claims that the idea of value involves the notion of replaceability. Something that is of one value can always replace some other thing of the same value. Value, then, is something that admits of equivalence and hence of exchange. But exchange value is only relative value – i.e., value relative to the interests that some other being takes in that thing. When something possesses dignity, however, it is not merely valuable, it is beyond all value. Such a being is, therefore, not replaceable. It seems appropriate, then, to think of persons as non-replaceable, and hence to think of them as possessing a dignity rather than intrinsic value.

The argument from uniqueness suggested by Callahan implies that we must think of persons as unique – i.e., that we should find their dignity in their uniqueness. It is worth noting that this is a considerable distance from Kant's general position, since Kant insists that the idea of dignity is formal and abstract. The idea is not that every person is valuable in virtue of being empirically different from everyone else. Rather, dignity is grounded in the rational capacity to conform one's will to the moral law, and there is no reason to think that this capacity would be diminished by duplication. So for Kant, dignity certainly comes from being non-replaceable, but this is simply not the same thing as claiming that every human being finds their dignity in being non-identical.

Defenders of the argument from uniqueness (like Callahan) seem to extend Kant's notion of dignity much further than what he had in mind. They maintain that humans are properly seen as possessing dignity rather than merely possessing value, that this dignity comes from being non-replaceable, and that the only way we can properly understand non-replaceability is in terms of genetic uniqueness. That is, the argument from uniqueness contends that humans are rightly seen as non-replaceable only when they are valued as unique, individual, persons. If they were the same, then they would not really be non-replaceable.

This, however, is not a promising line of argument. It is simply implausible to conceive of non-replaceability solely or even mainly in terms of uniqueness. This, again, is not the way that Kant thinks of the matter, and since he defends the idea of human worth as linked to the idea of non-replaceability, his position is worth considering as an alternative. Kant claims that our dignity comes from our capacity for rational agency, for legislating our lives by universal moral laws. In fact, he insists that it is morality – or moral action – which has dignity. Humans acquire dignity by being rational beings who can act morally. As he puts it:

> Nothing can have any worth other than what the law determines. But the legislation itself which determines all worth must for that very reason have dignity, i.e., unconditional and incomparable worth; and the word

'respect' alone provides a suitable expression for the esteem which a rational being must have for it. Hence autonomy is the ground of the dignity of human nature and of every rational nature.[16]

The point here (quite apart from Kant's more controversial claim that only the moral law has dignity) is that psychologically normal clones would surely have the capacity for moral agency which Kant refers to here and hence, under his moral schema, would possess dignity. The problem with the uniqueness argument, then, is that it is far too narrow in its implications for an account of the moral status or worth of persons.

What would be helpful now would be a thought experiment imagining the situation that clones might find themselves in, and how we might regard their moral value. The fact is, however, that we really do not need to speculate about such matters. Cloning, even in human reproduction, is a naturally occurring event. Identical twins or triplets regularly appear in the human population. And it is important to stress just what a threat natural clones (monozygotic twins, triplets, etc.) present for concerns about the moral value of individual uniqueness. They are clearly the most serious threat to human uniqueness that we can find, much more so than would be the case with technologically assisted clones. Monozygotic or 'identical' twins not only share the same nuclear DNA, but have the same mitochondrial DNA as well.[17] They are usually gestated in the same woman at the same time and circumstances in her body, and are standardly raised in the same cultural and familial contexts. So with natural clones (identical twins) both genetic and environmental forces standardly conspire to make them relatively non-unique individuals.

In contrast, clones resulting from technically assisted reproduction would probably not face this convergence of standardizing forces. They might not, and probably would not, be strictly genetically identical (e.g., they could easily not share the same mitochondrial DNA); they probably would not have the same gestational mother at the same time and circumstances in her body, and they could be raised temporally and spatially apart from their 'siblings.' The point here is that if cloning is morally objectionable because it would produce non-unique human beings, then it must be the case that identical twins are either 1) less morally valuable or 2) somehow worse off than the rest of us.

Before we consider the defensibility of these two claims, we should pause and reconsider the analogy that I am drawing between technologically assisted clones and natural ones. My claim is that if there is a serious threat to non-uniqueness worthy of our moral attention, it seems to point more to natural rather than technically assisted clones. Even if someone accepts this argument, however, it might still be pointed out that, setting aside differences in the technical methods involved in cloning (for reasons that I have set out above), a significant moral difference between these two cases still remains. Whereas the production of non-unique genomes in technically assisted cloning would be intentional and deliberate, in most cases of natural twinning it is not. Moreover, it seems that in many contexts we do right by accepting circumstances as we find them, but it would often be clearly wrong to set out to bring about such circumstances. Thus the analogy that I want to draw between technically assisted and natural clones (twins) fails because there is a relevant moral difference between these two cases: intentionality.

This objection, however, really does not point to a significant disanalogy. The problem is that intentionality and deliberateness do not, *on their own*, add any moral significance. The only way that they could would be if natural twins either possessed a different moral status or are somehow worse off than genetically unique

individuals – i.e., only if either option 1) or 2) above is true. If the two cases are otherwise morally similar, then intentionality points to no important moral difference between them.

To clarify this argument further, consider a similar analogy. Imagine two situations, one of a couple who, through no avoidable fault, awareness or intention of their own, have a child with a severely disabling genetic disease, and another of a couple who, by clear intentional means (say by direct genetic manipulation of an embryo), set out to have a child with the same disease. We probably would want to say that the former couple has not done anything wrong, but the latter arguably has. But the mere fact that in the latter case the action is intentional adds no moral weight unless it is already true that the disease makes either child somehow worse off than healthy individuals. If all else is equal between two cases like this, it is simply unclear what makes intentionality morally significant. Intentionality is morally significant only if there is an intention to do something that is *otherwise* morally good or bad, right or wrong. The analogy, therefore, between natural and technically assisted clones stands up.

We are left, then, with suggestions 1) and 2). If twins are somehow morally worse off than the rest of us, we need to be shown exactly what it is that twins are deprived of and its moral nature needs to be explained. It is very difficult to accept that twins are deprived of something essentially related to their inherent moral value as persons. And it does not seem any more obvious that in not being biologically unique twins are made worse off.

It might be suggested that they are psychologically damaged in some way because of their biological non-uniqueness. But evaluations of the 'similarity' of twins are often psychologically and morally ambivalent. In many cases this similarity can be an *advantage* in that they can share an enviable sense of empathy and human relatedness. Given this ambivalence, it is not clear how one could strongly maintain that non-uniqueness is objectionable, something to be avoided as a matter of moral principle.

It could be simply denied, of course, that twins or triplets are morally worse off than the rest of us. But if the defender of the argument from uniqueness were to do this, she would have to claim something like the following: Twins, although biologically identical, have not been wronged or are not worse off than people who are unique *and* being non-unique is morally worse than being unique. This is, of course, inconsistent. To address this inconsistency she might claim that what is wrong here has nothing to do with the twins, their state of mind, or even their state of being, but (again) with someone else's *intention* to create biologically non-unique individuals. But now it is just not clear what work the appeal to non-uniqueness is doing in the overall argument. If there is nothing wrong with (the fact of) non-uniqueness itself, it is not clear why intending to do it would be wrong. That is, it is not obvious how some action or state of affairs, X, could *never* be bad (either instrumentally or intrinsically), whereas intending to do X or to bring about X, would *always* be wrong or bad. As a result of this problem, I see no sound way for someone to defend both a) the claim that twins are not worse off than the rest of us in being non-unique and b) the claim that non-uniqueness for clones is morally bad.

Cloning may diminish the cloned person's subjective sense of uniqueness, but no more, perhaps, than an adopted child's sense of alienation when she is totally genetically unrelated to her (social) parents. The threat of this sense of alienation and lack of self-identity – and I admit it can often be very real and painful – do not seem to be any more sufficient reasons for arguing against adoption than is the risk of a lack of sense of uniqueness which might arise in

a person who is totally unrelated (genetically) to her mother.

Adoption, of course, is not strictly analogous to the decision to clone. Adoption is the best solution to an apparently much worse situation – i.e., being raised without a deep, lasting relationship with at least one adult guardian. As such, the risk of self-alienation may be worth taking because the alternative is much worse. Cloning, on the other hand, is not the solution to a graver outcome. The alternatives are either not being born or parents not having genetically related children – neither of which seems nearly as bad as a child going through her childhood without a relationship to a loving and committed guardian (or guardians). But this difference does not seem important. At least it does not take away from my claim that the risk of feeling non-unique is no more morally decisive against cloning than is the risk of feeling alienated in adoption.

It is also worth stressing that, as the practice of adoption shows, the relation of father or mother and child is also a profoundly *social* relation; it cannot be reduced merely to biological relations. There seems to be no compulsion for a child who is a clone of the person who has raised her to understand or interpret that relation as a sibling one. And it is not clear that the cloned individual, like any other natural twin, needs to see her individuality as importantly connected to her genome, or even to phenotypic similarities with others. Furthermore, there is some evidence that identical twins separated from birth tend to show comparatively more divergent personalities and behaviours than twins who are not separated.[18] It seems reasonable, then, to believe that identical twins separated by temporal distance will show even more differences. Hence a psychological sense of uniqueness may not be seriously threatened. All of the differences implied in temporal distance would probably give rise to sufficient grounds for a feeling of difference. Thus it seems that the claim about felt identity being psychologically damaging is not very plausible.

The argument that I have just made, however, is an empirically based one, and I could simply be wrong about the relevant facts. Even if this kind of negative feeling with regard to one's individuality could arise, it still does not seem that it *must* arise, nor that if it does arise, it raises serious moral concerns in relation to the basic value of persons. This lack of necessity is morally relevant because in the case of most adults it is the individual him/herself who is probably best seen as the person who takes some final sense of responsibility for self-understanding and self-development. Where this is not an appropriate allocation of responsibility, this is usually due to the way that the individuals have been raised by their parents or guardians. Parents can, and often do, raise genetically non-identical children to be and appear like other persons. There is surely no need to clone someone in order to threaten his feeling of uniqueness and individuality. A parent, then, who is acting as I would suggest a good parent should, will address this issue, and will encourage difference and individuality.

Suppose, however, that a cloned child has all of these environmental counterbalances in place and still comes to feel non-unique. It is again not clear that this consequence finally renders cloning immoral. The problem now is that whatever the foundation of an individual's inherent moral worth is, it surely must not rest importantly on being genetically different from other people. If it did, then the disparate value of twins problem that I have just been discussing resurfaces. Since there seems to be every reason, intuitively at least, to regard twins as possessing equal moral status or value as the rest of us, we should reject the idea that there is anything of any great moral import to be found in possessing different or unique genomes (or perhaps even unique phenotypes). So if I am right about the ground of inherent moral worth, duplicating a person's genome (or phenotype) cannot morally violate or affront that worth.

We have seen that even a philosopher with Kantian convictions, who values the role of moral principle to a high degree, might not object to cloning, since he or she would find inherent moral worth in the possession of rational agency. Surely clones would have this in spite of their biological identity. Someone defending the uniqueness objection, however, could put forward a non-Kantian notion of rationality, something much less formal or abstract. Perhaps, to connect back to the manufacturing objection, the claim could be made that a non-unique genome imposes limitations on an individual's capacity for self-development, leading to a sense of disempowerment, and personal inadequacy. It would be odd, however, to suggest that inherent human worth is somehow connected to being born without these limitations. Whether I have this sense of empowerment or not, it seems that it should never diminish my claim for others to respect my value as a person. Forms of social oppression usually have the effect of diminishing self-respect or self-empowerment in their victims which seriously impedes their own self- and social development. But, again, this fact should never detract from their abstract moral worth.[19] Thus it would provide no good reasons for treating a person lacking this self-empowerment badly, nor for holding that such persons should not be born simply because they will come to lack this quality.

Conclusion

It is understandable why individual uniqueness might be seen as morally significant. It certainly seems *biologically* important. Evolution by natural selection benefits considerably from individual genetic diversity, and sexual reproduction by and large ensures that this diversity will continue. Hence, given this biological norm, and the role that it has played in our own evolutionary history, it seems understandable why we might see moral significance in it. This may equally explain why many people are disturbed by the prospect of human clones, and for that matter, why we are in awe of twins, triplets, and so on. But claiming that individual uniqueness is of paramount moral importance is another matter entirely, and we have seen that it does not present much philosophical promise as a way of soundly articulating moral concerns about human cloning.

The argument that I have defended here should never provide a defence for further research and experimentation into the prospects of cloning. It is important to stress this point since it might seem like a very easy step from the argument that some particular activity is not wrong in itself to the argument that it is at least not impermissible to try to realise it. This is a misguided assumption, however, for two main reasons. First, it should be obvious that merely showing that something is not intrinsically wrong goes nowhere to show that it would not have morally bad consequences, and if it were readily available, that it would not provide some people with what Ruth Macklin has described as a tremendous 'opportunity for mischief'.

The second reason is that the research or clinical procedures necessary to provide any opportunities to clone human genomes could simply be immoral. This seems to be the case with human cloning at the present time – especially if the procedure under consideration is nuclear transfer using somatic cell nuclei. The Roslin experiment was so dreadfully inefficient that an attempt to try the same procedure on humans would almost certainly raise very serious moral questions. Ian Wilmut and his colleagues began with 434 eggs – many times the number of eggs in a ewe's lifetime ovulation. Fusion with the DNA from somatic cells was successful only 277 times. Worse still, only 29 of these transfers divided sufficiently for implantation. And of these 29, only 13 were implanted in ewes. Only *one* of these ewes became pregnant and then gave birth to the only successful clone.

Now since domestic animals are much more fertile than humans – perhaps three or four times – it would follow that about 1,200 to 1,500 human eggs would be needed to conduct a similar experiment in humans – several times more than a woman's entire lifetime supply. Furthermore, the procedure would probably require that about 50 women should be willing to be impregnated and possibly give birth at a very high rate of implantation failure and miscarriage.[20] These estimates assume, of course, that there would be no species-specific obstacles in using the Roslin procedure in humans. But there have been such problems. In 1981, Karl Illmensee and Peter Hoppe cloned mice by nuclear transfer.[21] For reasons that are not really understood, other scientists have had considerable difficulties reproducing this experiment. It is widely believed, however, that there is something specific in the reproductive cycle in mice that has prevented the widespread cloning of mice by nuclear transfer. Similar problems might arise if the procedure were tried in humans, resulting in a much more severe rate of pregnancy failure, and of increased health risks for either the mothers involved or the individuals who were cloned.

My point is that these, and many other issues, pose questions about risk assessments which, it seems to me, would never be outweighed by the argument that I have presented above. Thus it would simply be a misuse of my argument to suggest that it in any way implies that research on humans would be justified. Rather, the argument assumes a counterfactual context in which there are no serious risks attending to the procedure of nuclear transfer itself. The counterfactuality of this procedure, however, may not always be present. Sadly, in my view, scientific research proceeds without anywhere near the same moral reservations in its use of non-human animals as with human experimentation. It could well be that the gap between the applicability of this procedure in animals and humans could narrow severely through research conducted on non-human animals, even primates. There is currently a considerable interest in promoting this sort of research in non-human animals in order to produce genetically engineered pharmaceuticals and even organs for human benefit. Indeed, this, and the commercial benefits likely to follow from such discoveries, are the main reasons why the Roslin Institute conducted its experiment in the first place. If so, the possibility of developing this sort of technique through experiments in humans may not present as many difficulties as it now seems to. If I am right about the argument presented above, however, only harm or other consequentialist considerations should give us reasons for moral pause on this issue.

Acknowledgements

I am grateful to Phil Gosselin, Stephen Clark, and one of the reviewers for this journal for helpful comments on earlier versions of this paper. I am particularly indebted to Eldon Soifer for detailed written comments and lengthy discussions about the issues raised in this paper.

Notes

1 See Ian Wilmut et al. (1997) 'Viable offspring derived from fetal and adult mammalian cells,' *Nature*, 385, 27 February, pp. 810–13, and K.H.S. Campbell et al. (1996) 'Sheep cloned by nuclear transfer from a cultured cell line,' *Nature*, 380, 7 March, pp. 64–66. I will use the term 'cloning' in this paper to refer exclusively to human cloning.

2 See Meredith Wadman (1997) 'White House bill would ban human cloning,' *Nature*, 387, 12 June, p. 644.

3 Bill C-47 'An act respecting human reproductive technologies and commercial transactions relating to human reproduction.' The Bill, which did not pass, proposed to make it illegal, on a maximal fine of $500,000 or 10 years (or both), to 'manipulate an ovum, zygote or embryo for the purpose of producing a zygote or embryo that contains the same genetic information as a living or deceased human being or a zygote or embryo or foetus, or implant in a woman a zygote or embryo so produced' (section 4.1a).

4 See Declan Butler (1997) 'European ethics advisers back cloning ban,' *Nature*, 387, 5 June, p. 536, and Robin Herman (1997) 'European Bioethics Panel Denounces Human Cloning,' *The Washington Post*, 10 June, p. Z19. An interesting exception to this is Australia, where the Infertility (Medical Procedures) Act (Victoria) passed restrictive legislation much earlier (1984) than many other countries. This Australian legislation, it seems, was the product of a similar flurry of concern about technological developments in the late 1970s surrounding the success of in vitro fertilisation techniques. For more about this legislation see Margaret Brumby and Pascal Kasimba (1970) 'When Is Cloning Lawful?' *Journal of In Vitro Fertilization and Embryo Transfer*, 4, pp. 198–204. See also Peter Singer and Deane Wells (1985) *Making Babies: The New Science and Ethics of Conception* (New York, Scribner) pp. 146–49 for discussion of an important background report on which this legislation was based. In 1990, Britain introduced legislation (the Human Fertilization and Embryology Act) which ostensibly banned human cloning. But there have been concerns raised about whether this law forbids cloning only human embryos, and if so, whether it would thereby allow cloning adults by nuclear transfer as per the Roslin experiment, since such a procedure does not initially involve manipulating embryos. For more on this see Ehsan Masood (1997) 'Cloning technique "reveals legal loophole",' *Nature*, 385, 27 February, p. 757.

5 I will use terms like 'duplicate' and 'copy' throughout this paper in terms of the two qualifications that I have just noted in the previous sentence.

6   See *A Dictionary of Biology* (New York, Oxford University Press, 1996, 3rd ed.), p. 325.

7   For more about such matters, see Richard Lewontin (1997) 'The confusion over cloning,' *New York Review of Books*, 23 October.

8   For more about this dispute see Rebecca Voelker (1994) 'A clone by any other name is still an ethical concern,' *Journal of the American Medical Association*, 271, p. 331, and J. Cohen and Giles Tompkin (1994) 'The science, fiction, and reality of embryo cloning,' *Kennedy Institute of Ethics Journal*, 4, p. 194.

9   See J. L. Hall et al. (1993) 'Experimental cloning of human polypoid embryos using an artificial zona pellucida,' *The American Fertility Society* conjointly with the Canadian Fertility and Andrology Society, Program Supplement, Abstracts of the Scientific Oral and Poster Sessions, Abstract 0–001, S1.

10   For more on this see National Advisory Board on Ethics in Reproduction [NABER] (1994) 'Report on human cloning through embryo splitting: an amber light,' *Kennedy Institute of Ethics Journal*, 4, p. 252.

11   An exception to this might be where several blastomeres are produced by separation and then cryopreserved for possible implantation after one of them has developed to a stage where distinct features can be observed and expected. But even here the initial decision to clone by blastomere separation must still be done in the absence of detailed phenotypic information. And this decision would have to be aligned with the decision whether to discontinue cryopreservation or to implant the embryo.

12   (1970) *Fabricated Man: The ethics of genetic control* (New Haven, Yale University Press) p. 88.

13   Quoted in Jeffrey Kluger (1997) 'Will we follow the sheep?' *Time*, 10 March, p. 40.

14   (1971) "Ethical aspects of genetic control," *New England Journal of Medicine*, 285, pp. 780–81.

15   (1993) 'Perspective on cloning: a threat to individual uniqueness; an attempt to aid childless couples by engineered conceptions could transform the idea of human identity,' *Los Angeles Times*, 12 November, p. B7.

16   (1993) *Groundwork for the Metaphysics of Morals*, trans. James W. Ellington, (Indianapolis, Hackett), p. 41 (436, Academy pagination).

17   This point is stressed by Stephen Jay Gould (1997) 'Individuality: cloning and the discomfiting case of Siamese twins,' *The Sciences*, 37, September/October, p. 16.

18   For more about this see Richard C. Lewontin (1982) *Human Diversity* (New York, Scientific American Library).

19   I do not intend to suggest here that moral worth should be seen as an exclusively abstract, objective property of persons, and never a subjectively felt or valued empowerment. I only intend to commit myself to the view that an account of moral worth or value should be at least in part objective and abstract.

20   The numbers in this, and the previous, sentence are from Stephen Strauss (1997) 'Hello Dolly, it's so good to see you,' *Globe and Mail*, 1 March, p. A5.

21   See Jean L. Marx (1981) 'Three mice "cloned" in Switzerland,' *Science*, 211, pp. 375–76.

Chapter 13

Robert Sparrow

CLONING, PARENTHOOD, AND GENETIC RELATEDNESS

Introduction

Numerous authors have tried to give substance to the intuition that human cloning would violate something fundamental to our sense of the dignity of persons; others have argued that cloning would risk, or even necessarily involve, unjustifiable harms to the clone, or the person cloned.[1] My approach to the issue will be slightly different. Instead of arguing directly against cloning, I will examine what I take to be the best case for reproductive human cloning, as a medical procedure designed to overcome infertility, and argue that it founders on an irresolvable tension in the attitude towards the importance of being 'genetically related' to our children implied in the desire to clone.

Some initial distinctions

Discussions of the possible applications and ethics of human cloning typically distinguish between two different sorts of cloning, with different motivations: 'therapeutic' and 'reproductive' cloning. 'Therapeutic' cloning is hypothetical cloning of an individual for the purpose of procuring tissues from the clone which will serve some therapeutic purpose in relation to the person cloned. 'Reproductive' cloning aims at the creation of a whole person in order to satisfy the reproductive desires of some couple or individual. Strictly speaking, 'embryo splitting', a process already available in some IVF clinics, whereby a fertilised ovum is allowed to divide and then is split so as to produce a number of viable zygotes, is a form of reproductive human cloning, because it may result in the birth of a number of individuals with identical genetic make-up. However, this is not the technology that springs to the minds of most people when they hear the phrase 'human cloning'.[2] While it raises a number of important issues, this technology is arguably continuous with existing IVF technologies and may be defended as the generation of identical twins by artificial means.[3] What most people think of when they think of human cloning is the cloning of an existing or past individual using DNA extracted from cells taken from their body. That is, human cloning via somatic cell nuclear transfer (SCNT).

In this paper I will be concerned to examine only arguments for reproductive cloning via SCNT. Therapeutic cloning arguably raises more, and more difficult, issues than reproductive cloning because it typically involves the creation of a human embryo with the intention of later destroying it, but also because the potential benefits it offers, in terms of life-saving medical procedures, are so great. In comparison, the issues raised by reproductive cloning are perhaps more straightforward. Reproductive cloning aims at the birth of a child and need not involve the intentional destruction of human embryos – so this major source of objections, at least, does not apply.[4] On the other hand, the 'benefits' of human cloning for reproductive purposes are not so obvious that it is likely that they will provide reasons to allow it if it turns out that there are significant moral dangers involved. One would expect, therefore, that the question of the ethics of reproductive human cloning would be more easily resolved. Despite this, argument about the matter continues to rage. My contribution to the debate will be oblique. I shall argue that the circumstances in which there is a strong argument for cloning as a reproductive technology are much, much, narrower than is currently recognised.

In order to understand how this conclusion may bear on the debate about human cloning, we need to distinguish the ethics of *research* into human cloning from the ethics of the act of creating a clone itself. There are questions about the ethics of researching a technology that are distinct from those involved in the decision of whether to employ it once it exists. Most obviously in this case, given the large number of urgent social and medical challenges facing humanity, especially in the Third World, which would benefit from research into their amelioration, is it ethical to devote scientific energies to researching human cloning?

Recognising that the cases in which human cloning might be useful as a reproductive technology are very rare – even rarer than has been suggested in the literature – will have few, if any implications for the ethics of human cloning itself. However, it has obvious implications for the question of the ethics of funding – especially public funding – for cloning research. The ethics of funding research depends importantly on the extent to which it contributes, or might contribute, to meeting important human needs.

If it turns out that cloning could only do this in a tiny fraction of cases, that will mitigate against funding for cloning research.

The case for human reproductive cloning

Much of the public's interest in, as well as hostility towards, cloning derives from misconceptions about what the technology involves. Contrary to popular belief, cloning technology will not allow the copying or replication of persons. Those people who hope to generate identical copies of themselves or of some human archetype through cloning will inevitably be disappointed, as environmental and cultural factors will result in every clone becoming a different and unique individual.[5] The main reason for popular interest in cloning as a reproductive technology turns out to be misplaced; the technology simply cannot do what is required of it. It can only produce children with the same genotype, not the same character. Furthermore, the desire to produce a child with our own genotype, presumably for reasons of curiosity or vanity, seems insufficient to justify the use of scarce medical resources to this purpose.[6]

In fact it is quite difficult to imagine a case for reproductive human cloning where the technology is being used to meet an important human need. Arguments for cloning as a reproductive technology must rely on cases where couples are unable to become parents by any other means. Given that couples can always become parents by adopting a child or, more controversially, by arranging for a child to be conceived for them to adopt, arguments for cloning must rely on the importance we place on parents being able to have children that are genetically related to them.[7]

There are three scenarios wherein reproductive human cloning might be thought to have a useful role to play.

The first is where one or both members of a heterosexual couple are unable to make a genetic contribution to the genotype of a child because of their failure to produce or possess viable gametes (i.e. sperm or ova).[8] In this case, cloning via somatic cell nuclear transfer would allow the couple to bring into being a child that is genetically related, indeed genetically identical, to *one* of them.[9]

If the woman is capable of producing gametes, while the man is not, then the couple could use the man's somatic DNA and the woman's ovum in the cloning process, which would mean that she would have made a material contribution to the birth of the child and also have contributed a small amount of mitochondrial DNA. If she is capable of bringing the child to term then she could also be the gestational mother of the child. This arrangement would arguably allow both parents to feel that they had both played an important part in the creation of a child that was genetically related to at least one of them.[10]

However, if the woman is unable to bring the child to term, the couple will need to make use of a surrogate mother. If she is also unable to produce gametes, then cloning will only make it possible for a couple to raise a child genetically related to one of its parents, without the other partner playing a material role in this process.

It is important to note in relation to this scenario, that, except in the case where neither partner is capable of providing viable gametes, other techniques, such as the use of donor sperm or of donor ova – and if necessary the aid of a surrogate mother – could just as well achieve the same result: a child that was genetically related to one of its (social) parents.

The second scenario where cloning might be thought appropriate is perhaps a version of this first scenario, but where the inability to provide viable gametes results from a same sex couple, or perhaps even a single person, being unwilling to allow another person to make a genetic contribution to the process of reproduction.[11] Perhaps one of the strongest cases for cloning is that it might allow lesbian couples an opportunity to bear and raise children that were genetically related to at least one parent, without the need for a genetic contribution from a man.[12] One member of a couple might provide the genetic material for the nuclear transfer, the other provide the ovum and carry the child. In this way both women would make a substantial contribution to the health and character of the child and no men need be involved.[13] The couple might have personal, psychological, or perhaps political reasons, for wishing not to involve a man in the process of conception. Of course, in a male dominated research and medical environment, it is likely that the assistance of men will be essential in other ways. However, this might be more acceptable to some women than employing donor sperm.[14] Similarly, a single woman might wish to have a child without a genetic contribution from any other person.[15] Cloning herself using SCNT and then gestating the resulting embryo would allow her to do this.[16]

Again, note that unless both members of a lesbian couple, or a single person who wishes to clone themselves, are unable to provide viable gametes, they will equally well be able to have a child that is related to one of its parents through the use of donor gametes. Any defence of the value of cloning in these circumstances must therefore rest on the moral weight of their desire not to involve another person – or perhaps more specifically, a man – in the process of conception. I will argue below that this desire is of negligible moral weight. Furthermore, unfortunately, the political reality is that there is unlikely to be much support for cloning on the grounds of its purported utility for homosexual and single prospective parents. Conservative and religious objections to homosexual and single parenting will most likely ensure that the public justification of cloning refers to the needs of heterosexual couples in the other scenarios presented here.

A third scenario involves a couple who have already conceived a child and who are unable to conceive another by any means. By cloning their existing child they could provide him/her with an identical sibling, that would be related to both his/her parents.[17] This option would be available even after the death of the original child, as long as they could source DNA from a cell recovered after death or stored prior to death.

The justification of human cloning as an assisted reproductive technology therefore appeals to the desires of a small number of persons in unusual circumstances to rear children that are genetically related to (usually, only one of) them. If either partner is capable of providing viable gametes then reproduction involving the use of donor gametes (and perhaps a surrogate) would equally well allow a couple to bring into being a child that was genetically related to one of them. Only if neither partner is capable of providing viable gametes will cloning be the only way to satisfy this desire.

Genes, relatedness, and genetic parenthood

In order for our concern for parents' desires to have children who are genetically related to them to ground an argument to justify funding for cloning research, two things must be true; first, the desire of parents to have children who are genetically related to them must be important enough to justify the use of the resources required to satisfy it *and*, second, cloning must produce a child that is genetically related to the couple in the appropriate way.

For the purposes of this paper I am going to assume that there is something important about the relation of genetic parenthood. That is, that

we are right to care for our children because they are *our* children, in the sense that we are their genetic parents.

In fact I believe this assumption is (very) questionable and that the social relation of parenting, marked by the provision of love and care, is more important to the well-being of both parents and children than any genetic relation and should carry most of the weight in establishing a parental relationship. There are many individual cases where parents love and care for their adopted children, who have no genetic relation to them, as much and as well as any natural parent. Furthermore, the children do not seem to suffer by virtue of being cared for by parents other than their genetic parents. Similarly there have been many societies where a genetic relationship has not been the major factor determining the parents of a child, where children have been cared for by other relatives, or adopted out as a matter of course, or cared for communally. All of which suggests that the genetic relation – if any – between parent and child is much less important in the establishing of meaningful relationships between adults and children than is commonly believed.

Nevertheless, it is clear that many people do feel that there is something special about having children that are related to them 'by blood'. Widespread support for, and public funding of, IVF programs reflects this concern.

Further, I am going to assume that this concern is sufficiently well grounded to justify spending public funds on research into technologies that might allow parents, who otherwise might be unable to do so, to have their 'own' children. Again, I am personally inclined to doubt this, for much the same reasons. If our current concern for genetic ties is excessive then so too is the amount of effort we, as a society, put into trying to make them possible. Moreover, the level of scientific research and medical effort dedicated to overcoming infertility through IVF and other medical technologies may contribute to harmful cultural preconceptions: that adoption can never be as satisfying as rearing one's 'own' child, that childlessness is the worst thing that can happen to a couple, and that women are essentially defined by their reproductive role. It also risks contributing to and reinforcing the pernicious genetic determinism that grounds public support for these technologies, by publicly affirming the superiority of genetic ties.[18] Yet, regardless of my own reservations, IVF programs are not just popular, but publicly funded. Clearly most people do think that parents' desires to have children who are genetically related to them justifies devoting scientific and medical resources to this project.

It is the relevance of this belief to the case for reproductive cloning that I wish to challenge here. Are parents in the situations described above 'genetically related' to their child in such a way as should engage the intuitions we have about the importance of genetic parenthood in normal circumstances?

In assessing the nature and significance of the genetic relation between persons and their clones it is, I think, difficult to escape the effects of an influential 'informational' metaphor in contemporary understanding of the role and significance of genes. Talk of genetic make-up as a 'blueprint', 'code' or 'program' for an individual encourages us to measure the relation between two individuals through comparison between their genetic 'blueprints', with the result that two individuals with the same genes, i.e. clones, are maximally related. This would suggest that cloning is the ideal way to produce a child that is genetically related to (one of) its parents. Indeed, from the perspective of someone in the grip of this informational metaphor, it seems to be a *better* way to have a child than through normal procreation.

However, as a number of writers have observed, there are deep problems with understanding DNA as a code, blueprint or program for anything.

To begin with, these metaphors do little justice to the specificity of the biological and chemical processes underlying phylogeny and evolution. If DNA is the 'code', what is the language? If it is a 'program', what is the 'machine' upon which it runs? If it is a 'blueprint' then who or what does the building? Answers have been suggested to all of these questions, but in providing them there is a tendency to sacrifice the detail and distinctiveness of the actual processes involved in the service of the metaphor. Overreliance on the informational metaphor may lead us to neglect the many ways in which genes are *not* like a blueprint, code or program.[19]

Describing DNA as a code, etc., also tends to exaggerate the importance of genes and to disguise the role of the environment in the development of the organism. It encourages us to see individuals' genetic make-up as determining their character (phenotype), which is then merely modified by the environment in which they grow up. In fact the phenotype of an individual is always the result of an interaction between his/her genetics and the environment, in which neither of these should be thought of as prior to the other. Genes 'code' for a phenotype in *an environment*. In a different environment, the 'same' genes code for a different result. Thus without specifying an environment we can say nothing about the future of an individual, regardless of our knowledge of his/her genetic 'code'.[20] Furthermore, even where genetics does play a large role in determining a character trait, often the best way to ensure a particular phenotypical outcome, such as, for instance, a child with a 'high IQ', is to modify the environment in which he or she is raised. The informational metaphor can blind us to the essential role played by the environment in which an organism develops.

The informational metaphor also leads to paradoxical results when we consider the question of who are – or perhaps should be – the parents of clones. The genetic relation between DNA donors and their clones is *not* the relation that parents have with their children. Normally a genetic parent shares only approximately half their genes with their child, but clones share *all* of their DNA with the person from whom they were cloned.[21] As we noted earlier, the genetic relation between the DNA donor and their clone is the same as the genetic relation between identical twins. Yet we do not think that the relation identical twins have is such as to ground the claim that they are each other's genetic *parents*. A relationship of genetic identity is not the right sort of relationship to ground a claim to parenthood. Clones share *too much* genetic material with their donors to be their children.

Indeed, we do not even need to be concerned about the case of cloning to perceive problems with the informational model of genetic relatedness. As Barbara Katz Rothman has observed, the 'genetic relation' between siblings, who share roughly half their DNA with each other, is the same as the relation between parents and their children.[22] Yet this relation is not one of parenthood. 'Genetically', the offspring of my identical twin with my partner will have the same relation to me as my own children do, yet I do not consider them to be my children. A concern with genetic similarity will even allow total strangers to be 'more related' to us than our own children. While, statistically, my child is likely to have a phenotype (partially) determined by the expression of 50 percent of my genes, their character may in fact reflect the expression of many more, or less, of my genes. Moreover, while the particular set of genes that my cells contain – my total genetic makeup – is almost certainly unique, the individual genes within that set are not. Except for the small number of mutations that occur with each conception, I share all my individual genes with thousands, and probably millions, of other people. Any of these people may in fact have more of 'my' genes than my own children.[23] If this should chance to happen, then according to the informational metaphor I am more related to such a person than I am to my own children.[24]

The problem with the informational metaphor is that it ignores the role played by history in determining even our sense of *genetic* relatedness. What is missing in these cases is the appropriate *causal* connection between the genetic make-up of the parties involved. Our genetic relation to others is not merely a question of the genes we happen to share, but also a question of the history of how we came to share those genes.[25] Until relatively recently these histories involved a series of couplings. They took the form of family trees wherein the points of branching necessarily involved coitus and conception.

However, the relatively primitive technology of artificial insemination by donor separated coitus from conception.[26] But until the advent of cloning, conception at least was an essential part of any reproductive endeavour. Our genetic relation to others was a question of who had conceived whom. The invention of cloning establishes the possibility of a radical break in these histories – a birth without a conception.

The existence of this break poses a genuine question as to who are, or should be, the parents of a cloned child.[27] Our ordinary intuitions about genetic relatedness are unreliable here because the normal connection between genes and the history whereby they are transmitted has been severed.[28] As Alpern puts it, ordinarily, 'the meaning and significance of having (one's own) child essentially involves reference to the child's *genesis*; that is, reference not only to patterns of genes, but to the processes and activities through which a child comes to be.'[29] The 'processes and activities' through which a child cloned by SCNT is brought into existence are so far removed from those ordinarily involved in becoming a parent that it is difficult to know if they sustain a genetic relation at all, let alone a parental relation. The open nature of this question is disguised by a reliance on the informational metaphor, which implies that sharing the 'same' genes is enough to make two individuals genetically related.

Cloning and parental intentions

However, even if sharing the 'same genes' was enough to establish a genetic relation between a DNA donor and their clone, it would not be enough to secure a claim to parenthood by the DNA donor of a clone – let alone their partner. Any argument about genetic relatedness which purports to establish that I am the parent of my clone will ground a stronger claim by my parents to be the parents of my clone, given that they have the paradigmatic version of this relation to the clone. Even if I am related to my clone, my parents are *more* related in the appropriate way. That is, 'genetically', the parents of my clone are my parents.[30]

We might therefore imagine a scenario where the natural parents of the person who had themselves cloned sued for custody of the clone. Perhaps they do not feel that their (natural) child is a suitable parent for any child, let alone a clone. Or perhaps, on seeing a child who looks just like their own daughter or son, they feel a desire to exercise their 'parental rights'. If it is genetic relatedness that grounds a claim to parenthood then they would surely have a better claim than the DNA donor. Any suggestion that the DNA donor should be granted custody of the clone against their own parents, on the grounds that they (the donor) are 'more related' to the clone, risks the ludicrous conclusion that I am more the parent of my identical twin than are my own parents.

In order to resist these deliriously implausible conclusions, defenders of cloning may wish to insist that in the case of a cloned child it is the *intention* to bring the child into the world that makes the donor the parent (rather than the donor's parents). What makes my clone, or my partner's clone, our child is that we have made the decision to bring it into being with the intention of raising it as our child. Couples might emphasise the effort to which they have gone to create a child, and the psychological and emotional significance of the history of this project, to lend weight to their claim to stand in a more parental relation to the child than the genetic parents of the DNA donor.

There is clearly something to be said for this move. If the parents of a clone were the parents of the DNA source, or even the person who is cloned, then there is a real risk that any of us could become parents unintentionally when someone clones our children, or perhaps us, without our knowledge by using DNA obtained surreptitiously. Most of us would, I suspect, object if told that we had become a parent without doing anything at all because we, or our children, had been cloned without our knowledge, let alone consent. We would object even more strenuously if we were further expected to fulfil our 'parental duties' and look after and support the child. What seems to be missing in cases like this is any intention on the part of the donor to become a parent, or even to participate in any activity that might reasonably be held to involve a foreseeable risk of becoming a parent.

Insisting that intention plays a crucial role in determining who are, or should be considered, the parents of a clone is all very well. However, we should note that, legally speaking, becoming a genetic parent through coitus is typically *not* a question of intention. In many jurisdictions, claims for child support payments can be made against a father who went to substantial effort in relation to contraception to *avoid* fathering a child. The genetic relation of being a natural parent is sufficient to ground a responsibility for the child, regardless of intention.

Paradoxically then, in order to resist the claims of the parents of the donor to the cloned child, this argument for human reproductive cloning must place *more* weight on the intention to parent a child, than we do in cases of ordinary reproduction. It must insist that the parental relation is established by the intentions of the labours of the couple who bring a clone into the world and not by their genetic relation to the child.

The emphasis placed on intention as establishing the parental relationship therefore works to undermine the justification for cloning in the first place. For cloning to play a useful role as a reproductive technology, it must allow couples to become parents who could do so no other way. However, to the extent that intention is sufficient to establish parenthood, then adoption or surrogacy, which are existing alternatives to cloning, will allow couples to become parents equally as well. These projects also involve an extended effort from prospective parents and generate a history linking them to the particular child that results. Indeed, the nature of the project of becoming a parent through surrogacy or adoption is much closer to that embarked upon by ordinary parents because, unlike the case of cloning, the prospective parents experience the uncertainty that results from an act of conception.[31] In the case of cloning the genotype of the child (with the exception of the possible contribution of mitochondrial DNA from the donor ovum) is fixed from the beginning of this project. This is a substantial disanalogy with the ordinary parental relationship which actually weakens the claim of the source of DNA for a clone to be its parent.

Might cloning still serve some useful purpose?

Defenders of the utility of cloning may at this point concede that it is intention that plays the primary role in explaining why the donor and her/his partner are the parents of a clone, but insist that couples nevertheless do wish to have *a* genetic relation to their child if possible, and that a clone is *more related* to them than any other child they could have.

If this defence of cloning works at all, it only works in the cases where neither member of a couple has any viable gametes, or where a couple is cloning a child they have previously conceived using gametes from each of them. In any case where one member of a couple has viable gametes and can produce a child through conception using donor sperm or ova, then this child is more related to them in the relevant sense required to establish parenthood, than their clone. This child is only related to one of its 'parents', but then so too is a cloned child. Moreover, the nature of the relation between the parent and a child conceived using donor gametes is more clearly that between parent and child than is the relation between the DNA donor and their clone.

In cases where neither member of a couple possesses viable gametes, cloning will allow them to have a child that arguably has some genetic relation to one of them, which an adopted child will not.[32] But, as I argued above, they will only be this child's *parents* on the proviso that their intention to be so is sufficient to make them so. The genetic relation which they might achieve by cloning – and which will exist only between one member of the couple and the child – will be that of (identical) sibling rather than parent. Given that cloning can only achieve this unorthodox genetic relation between one member of the couple and the child, and that they can only assume a parental relation if the alternative of adoption would also allow them to become a parent, the argument for cloning on the basis of this relationship must be said to be weak.

The possibility of cloning establishing a genetic relationship of the sort that would normally justify any claim to be a parent is confined to the case where a couple can clone a child they have previously conceived.[33] Even here, as we have seen, we may have reason to doubt that the causal/historical relation between them and their child is sufficient to establish that such a child was related to them in the appropriate fashion.[34] Again, a lot of weight must be placed on the question of intention, lest couples be at risk of acquiring parental duties when others clone their children.

Defenders of cloning may still wish to insist that cloning is a valuable reproductive option that should be made available as an alternative to artificial insemination by donor or use of donor ova. What is perhaps important to many parents is not the particular nature of their genetic connection to their child but that no-one else should have one. They do not want to rely on a genetic contribution from a third party. They cannot bear the thought that their child is someone else's child.[35]

However, as I have argued, their cloned child is at the very least the child of the natural parents of the donor. That is, 'genetically speaking', a clone is already someone else's child.[36] This may be acceptable to some couples, as they know and presumably like their parents. However, note that the genes that the parents of the DNA donor contribute came to them from their parents, and their four grandparents, and their eight grandparents, etc. No matter how precious we are about tracing our family tree, there will always come a point where we can no longer track the origins of genes that we possess. Ultimately, therefore, all of our genes come to us from strangers – and this will of course be true too of clones. Indeed the whole project of trying to 'own' our children's genetic make-up is misguided. Our total genetic make-up is likely to be unique but 'our' genes are not and these are all that we can hope to pass on to our children. We all share all these genes with the entire

human gene pool (and with most other living things besides). Thus while it may avoid the need to use donor gametes, cloning will not allow couples to escape their child having genes that come to them from strangers. The argument for the potential value of cloning on these grounds is correspondingly weak.

Alternatives to cloning

In any case other than that where couples are cloning a child they have already conceived, then, aspiring parents are equally well able to become parents through the use of donor gametes, or adoption, or perhaps (and more controversially) by arranging to have a child conceived for them for the purposes of adoption.

Our assessment of the worth of cloning as a reproductive technology will therefore turn partially on our attitudes towards conception using donor gametes, the availability of children for adoption in a society, and on our feelings about the ethics of conceiving children for the purposes of adoption. If we object to the use of donor gametes in reproductive procedures, or if there are few children available for adoption and we believe it unethical to conceive a child for the purposes of adoption, then the case for reproductive cloning is stronger than I allow here.

The availability of children for adoption is a contingent matter that will depend on the society in which the question is asked. In societies where few children are available for adoption then childless couples' desires to have children must be served by other means. However, note that it may be hard for advocates of reproductive human cloning to object to the use of donor gametes, and even to the conceiving of children for the purposes of adoption, and still defend cloning. For cloning to be an option in several of the scenarios described above, the use of donor gametes (ova to be de-nucleated) and/or a surrogate mother is necessary. If these procedures are unacceptable then so too will be cloning in these cases. It is also hard to see how there is much difference between a scenario in which a child is brought into being, through the use of donor gametes (although not the genetic material therein) and a surrogate mother, in order to be turned over to another couple, and the situation where a child is conceived and gestated for the purposes of adoption. Indeed, except in the case where a couple clone their own child, cloning already in a sense always involves conception for the purposes of adoption, in that the cloned child is deliberately brought into being with the intention that it will then be taken from its 'genetic parents'. Those who would defend the utility of reproductive human cloning may therefore be hard pressed to object to the conceiving of children for the purposes of adoption using more traditional methods.

Thus while there are profound ethical and policy issues associated with the existing technologies, to which cloning might serve as an alternative, it is highly unlikely that cloning will not raise these issues to the same extent. This mitigates against the idea that we have strong reasons to pursue human cloning as a reproductive technology.

Conclusion

The only situation where cloning can unambiguously serve couples' desires to have children, where existing reproductive technologies cannot, is where parents who are unable to conceive wish to clone a child they have previously conceived naturally. The case for cloning as a reproductive technology is therefore far far weaker than is generally recognised.

This conclusion will have few, if any, implications for the question as to whether or not it will be ethical to clone human beings once it becomes possible and safe to do so. Establishing

that there are few, if any, good reasons for wishing to have a child through cloning is not the same as establishing that there are reasons not to do it. Once cloning becomes possible many people may want to use cloning to reproduce themselves. Some of them may be motivated by a desire to be genetically related to their children. Once the technology exists, it may be wrong to prevent them from attempting to realise this desire, even if we firmly believe it to be misguided, or that cloning is ill suited to this purpose. Nothing I have said here has been addressed to this question.

However, the scarcity of good arguments for cloning does have significant implications for the ethics of funding for research into human cloning as a reproductive technology. Given the relative scarcity of medical and scientific resources and the many other important projects to which they could be directed it is arguably unethical to devote funding to researching a technology which will serve important human needs in only a very small number of cases.

This is not as strong a conclusion as many opponents of human cloning would hope for. The argument I have explored here can ground only the argument that funding for research into cloning is unethical in the same way as research into (purely) cosmetic surgery is unethical. In a world of relative scarcity of resources for medical research we simply cannot justify it.[37] But it is cloning's lack of utility rather than its ethics that motivates this conclusion.

This is not to say that that there may not be other reasons why cloning might be unethical. Indeed, achieving a proper understanding of what cloning can and cannot achieve as a reproductive technology may also have implications for our overall assessment of the ethics of cloning itself. Any such assessment would require consideration of a multitude of arguments which are beyond the scope of this paper, but are explored in the large and polarised literature that exists around human cloning. Meanwhile, in the absence of a public consensus on the ethics of human cloning, a realistic appraisal of how little the technology has to offer may serve to avoid or at least defer any ethical questions it may raise.

Notes

1 An early survey of potential issues is J.A. Robertson. "The Question of Human Cloning," *Hastings Cent Rep* 1994; 24(2): 6–15. This paper is notable for denying the likelihood of cloning mammals by SCNT in 'even the mid-range future' (p. 6); a vivid reminder of just how rapid technological development in the area is. Useful collections of post-'Dolly' discussions of cloning include, B. MacKinnon, ed. 2000. *Human Cloning: Science, Ethics, and Public Policy.* Urbana and Chicago: University of Illinois Press; G. McGee, ed. 2000. *The Human Cloning Debate.* Berkeley, California: Berkeley Hills Books; M.C. Nussbaum & C. Sunstein, eds. 1998. *Clones and Clones: Facts and Fantasies About Human Cloning.* New York and London: W. W. Norton & Company.

2 Robertson, *op. cit.* note 1, p. 6.

3 Ibid.

4 I am presuming of course that cloning technology will improve to such an extent that prohibitively large numbers of embryos need not be created to produce one clone. Note also here that we currently tolerate the creation and eventual destruction of a certain number of human embryos in the course of existing IVF procedures. For that matter, even a 'natural' pregnancy may have involved the creation and destruction of any number of embryos before a pregnancy comes to term.

5 B. Rollin. "Send in the Clones ... Don't Bother, They're Here," *J Agric Environ Ethics* 1997; 10: 25–40; J. Harris, 1999. " 'Goodbye Dolly?' The Ethics of Human Cloning," in *Bioethics: An Anthology.* H. Kuhse & P. Singer, eds. Oxford: Blackwell: 143–152.

6 It is easy to imagine medical or therapeutic reasons to desire the birth of a child with a certain genotype, for instance, to serve as a source of tissue for transplant or other therapy, but these are arguments for *therapeutic* cloning, which I am not discussing here.

7 See C. Strong. 2000. "Cloning and Infertility," in *The Human Cloning Debate.* G. McGee, ed. Berkeley, California: Berkeley Hills Books: 184–211; Harris, *op. cit.* note 5, pp. 148–49; D.W. Brock. 1998. "Cloning Human Beings: An Assessment of the Ethical Issues Pro and Con," in *Clones and Clones: Facts and Fantasies About Human Cloning.* M.C. Nussbaum & C. Sunstein, eds. New York and London: W. W. Norton & Company: 141–64. The strength of the desire for children who are 'genetically related' is likely to vary amongst individuals and across cultures. Some cultures, especially in Asia and the Middle East, may place a very high value on the continuation of the family line through the birth of children who are genetically related to their parents.

8 One important case where parents are unable to provide 'viable' gametes is where both members of a couple carry a recessive gene for a lethal or severely debilitating genetic condition, such that they are unwilling to risk conceiving a child by 'normal' means. In this case, however, conception using existing IVF technology and pre-implantation diagnosis would allow them to conceive a child with each other and avoid the risk of a child inheriting the lethal genes. For cloning to serve a useful role, this option must be ruled out for some reason.

9 Strong, *op. cit.* note 7, p. 185.

10 Ibid: 190.

11 T.F. Murphy. 2000. "Entitlement to Cloning: A Response to Strong" in *The Human Cloning Debate.* G. McGee, ed. Berkeley, California: Berkeley Hills Books: 212–20.

12 The possibility that lesbian couples might have good grounds to use cloning technology is mentioned in P. Kitcher. 2000. "There Will Never Be Another You," In *Human Cloning: Science, Ethics and Public Policy.* Barbara McKinnon, ed. Urbana and Chicago: University of Illinois Press: 53–67. However, in his treatment of this scenario, Kitcher neglects the possibility that lesbian couples might prefer to use cloning rather than donor sperm because of a desire to avoid a male contribution to the pregnancy. Similarly, Timothy Murphy's

defence of the rights of same sex couples to use cloning to produce children neglects the particular (political) benefits for lesbian couples. See Murphy, *op. cit.* note 11, pp. 212–20.

13 In fact, because there will be a small contribution from the mitochondrial DNA of the (other?) mother's ovum both women might also be said to play a role in determining the genetic make-up of the child.

14 Cloning seems less likely to be of value to male homosexual couples. While cloning would allow a gay male couple the option of having a child that was genetically related to one of them, without a direct genetic contribution of a woman, creating the clone will still require a woman to provide an ovum and also, for the foreseeable future, to bring the child to term. The social/political motivation for the attempt to create a child without involving a member of the opposite sex is therefore missing.

15 My thanks to Patricia Peterson for drawing this possibility to my attention.

16 Cloning would allow a single man to reproduce himself without a genetic contribution from another person (with the exception of mitochondrial DNA) but would require use of a donated denucleated ovum and the assistance of a surrogate mother. Given this, it is difficult to see that this process has many advantages over reproduction involving conception with a donor ovum and the assistance of a surrogate mother, or indeed natural reproduction.

17 In this case, unlike those above, *both* parents are genetically related to the cloned child.

18 Jean Bethke Elshtain. 1998. "To Clone or Not to Clone," in *Clones and Clones: Facts and Fantasies About Human Cloning.* M.C. Nussbaum & C. Sunstein, eds. New York and London: W.W. Norton & Company: 181–89.

19 Barbara Katz Rothman. 1998. *Genetic maps and human imaginations: the limits of science in understanding who we are.* New York: Norton & Co: 21–25.

20 Ibid.

21 In fact the description of the genetic relation between parents and children given here and below is not strictly speaking accurate. Given that all human beings share the vast majority of their DNA with each other (as well as with rabbits, fish and bacterium), 'half their genes' here can only refer to the genetic variation within the human population. Furthermore, this Mendelian assumption about the genetic relation between parents and their children presumes that choice of mates is random. If people tend to choose partners who share genetic similarities with them, then each partner will tend to share more than 50 percent of their genes, within the range of human variation, with their offspring. There is some evidence that this is the case. See L. Dicks. Like Father, Like Husband. *New Sci* 2002; 2 February: 26–29. (My thanks to Ashley Sparrow, of the University of Canterbury, for drawing my attention to this paper.) Nevertheless, the basic point – that the relation between parents and children is not one of genetic identity – stands.

22 Barbara Katz Rothman. 1989. *Recreating Motherhood: Ideology and Technology in a Patriarchal Society.* New York and London: W.W. Norton & Company: 37–39.

23 Barbara Katz Rothman, *op. cit.* note 22, pp. 66–71.

24 K.D. Alpern. 1992. "Genetic Puzzles and Stork Stories," in *The Ethics of Reproductive Technology.* K.D. Alpern, ed. Oxford: Oxford University Press: 147–69. Although it does not mention cloning, and was written before human cloning was thought to be a serious possibility, Alpern's discussion of the significance of an engineered relation of genetic identity is eerily prescient and extremely relevant today.

25 Ibid: 160–164. See also Avery Kolers. Cloning and Genetic Parenthood. *Camb Q Healthc Ethics* 2003; 12: 401–10. Unfortunately, I only became aware of Kolers' excellent paper, which also deals with the complexities of our concept of genetic relatedness, after submitting this paper for publication. Revising this paper to take account of Kolers' arguments would have required extending it beyond the length appropriate for this journal and I have therefore chosen to publish it in its original form except for the inclusion of references to Kolers where they illuminate the argument.

26 This development itself has stripped away some of the context of reproduction that explains why it is important to couples to have 'their own' children.

For instance, children conceived using artificial insemination are no longer a direct expression and result of sexual intimacy between their parents. Nor, if conception involves the use of donor gametes, need it affirm the love and mutual regard of the (genetic) parents. See Strong, *op. cit.* note 7, pp. 87–88.

27 For a discussion of the plausibility of various bases for parenthood, see, A. Kolers & T. Bayne. " 'Are you my mommy?' On the Genetic Basis of Parenthood." *J Appl Philos* 2001; 18(3): 273–85.

28 Alpern, *op. cit.* note 24, pp. 160–64. See also Kolers, *op. cit.* note 25.

29 Ibid: 163.

30 Harris, *op. cit.* note 5, p. 148. See also Kolers, *op. cit.* note 25.

31 In the case of adoption this uncertainty may be reduced if prospective parents can select the child they wish to adopt. Even in this case the genetic inheritance of the child will to a certain extent represent an unknown. In many cases of adoption, parents will also confront the uncertainty of not knowing which child may become available for adoption.

32 But note the reservations expressed above about whether donors and their clones have any genetic relation at all – which explain why this matter remains arguable.

33 Note that if the argument for couples' rights to have access to a technology to allow them to become parents is grounded in the importance we place on participation in child rearing, or – more plausibly – in participation in conception, their rights to access in this case will be weak, because they have already conceived, and perhaps raised, a child. It is implausible to think that the 'right' to have children extends to the right to bring them up successfully.

34 Alpern, *op. cit.* note 24, pp. 160–64. An important difference between a child produced through cloning and the original child that is being cloned,

for instance, is that the cloned child does not represent a *mixing* of the genetic character of its parents, so much as the reproduction of the result of a previous such mixing. This itself might be thought to constitute a significant difference in the process whereby the child is brought into being.

35 Strong, *op. cit.* note 7, pp. 202–04.

36 In passing, we are now in a position to note that the case that I identified earlier as perhaps the most promising justification for reproductive cloning, that of a lesbian couple who wanted to have a child without requiring a genetic contribution from a man, is much weaker than first appeared. Because the genetic make-up of the clone will be identical to the donor, who presumably did have a male parent, clones will always have 'male' genes. Indeed the whole idea of distinguishing between 'male' and 'female' DNA is misguided. Except for those genes that can be linked to one of the sex chromosomes there is no way to distinguish 'male' from female human inheritance.

37 In particular it may be unethical to devote *public* funding to researching cloning. The ethical standards for the use of public funds seem to be higher than those for the use of private resources because we feel that decisions about the use of public monies should be appropriately responsive to public opinion, and that they express a society's aspirations and priorities in a way that private funding perhaps does not. Moreover, given the tightly inter-woven nature of privately – and publicly – funded research science, and the extent to which private research often 'piggybacks' on public research, through making use of techniques and results developed in publicly funded institutions and by employing researchers educated and trained in publicly funded institutions, a decision about the ethics of public funding may also have substantial implications for the future of *any* research into human cloning.

Chapter 14

Russell Blackford

SLIPPERY SLOPES TO SLIPPERY SLOPES: THERAPEUTIC CLONING AND THE CRIMINAL LAW

In principle, somatic cell nuclear transfer can be used to create human embryos for research or, more speculatively, to provide cells or tissues for medical therapy – these uses are often referred to collectively as *therapeutic cloning*. Somatic cell nuclear transfer also could be used for (human) reproductive cloning, i.e., to create a human embryo that would be gestated and brought to birth as a genetic near-twin of the somatic cell donor who contributed the embryo's nuclear DNA. One argument for the criminal prohibition of therapeutic cloning is that it will lead to reproductive cloning. The analysis by Caulfield and Bubela (2007) shows that this argument was one of the important secondary arguments used to support the prohibition of therapeutic cloning in Canada, although the main arguments used in the public debate were predicated on the supposed moral status of human embryos.

If we set aside contested claims about the moral status of embryos, the secondary argument has little force as a reason to enact criminal legislation. Once the argument is required to stand alone, it proceeds on the basis that there is nothing so harmful, or otherwise horrible, about therapeutic cloning itself as to merit the intervention of the criminal law. However, so it is argued, it will put Canadian society on a slippery slope to something else – reproductive

cloning – that is, indeed, sufficiently horrible. Therapeutic cloning must, then, be prohibited in order to avert a significant evil some time in the future. This kind of argument will be successful only if it surmounts seemingly impossible hurdles.

In this case, the appeal to a slippery slope takes the more specific form of what Williams refers to as a "horrible-result argument" (1985, 126–27). Its logic can be represented as follows:

Premise 1. It would be a horrible outcome if reproductive cloning became a practice in Canadian society.

Premise 2. Such an outcome is likely if the Canadian parliament does not prohibit therapeutic cloning.

Conclusion. The Canadian parliament should prohibit therapeutic cloning.

For this to compel assent, two requirements must be met. First, the outcome described in premise 1 must be sufficiently horrible. Second, the causal connection between this outcome and permitting (i.e., not prohibiting) therapeutic cloning must be strong enough to support premise 2. Moreover, the term *likely* needs to be interpreted in a fairly strong sense: although the term may refer to something less than inevitability, the term *likely* must mean something more than a remote chance or even a significant risk that could be averted by prudent social policy.

Advocates of horrible-result arguments face the problem of simultaneously meeting both the first and the second requirement. In one regard, if reproductive cloning is horrible, whereas therapeutic cloning itself is not, then it should be relatively straightforward for the law to make a logical distinction between them. So why not allow therapeutic cloning while, if necessary, prohibiting reproductive cloning? In another regard, if reproductive cloning is not horrible (after all), then its future prospect does not (after all) support prohibition of therapeutic cloning.

To satisfy the requirements needed to make the argument stick, some strong empirical claim will need to be defended. For example, it might be claimed that reproductive cloning really is horrible, but if we go ahead and allow therapeutic cloning we will somehow be corrupted or seduced into overlooking the horror: though the logical distinction is there to be drawn, we will be psychologically incapable of drawing it. But an empirical claim such as this cannot just be asserted without extensive and convincing support, and it is most unclear what support could be forthcoming to make it believable to anyone not already predisposed to accept it.

In any event, what *makes* reproductive cloning so horrible? Perhaps it is the risk that congenitally malformed children will be born. If that is the mischief for the law to address, reproductive cloning is clearly distinguishable from therapeutic cloning, where the value of preventing birth deformities and similar undesirable effects is simply not at stake. Note that it is currently difficult to imagine how we could refine the practice of reproductive cloning to avoid the risk without conducting long-term trials that involve the very same risk – which seems obviously

unethical. Unless there is some kind of unforeseeable breakthrough, we appear to be at an impasse in developing what might be called a "safe" technology for reproductive cloning in human beings. All this, however, suggests that there is no realistic prospect, let alone a likelihood, of therapeutic cloning leading Canadians or anyone else to practice reproductive cloning.

However, what if some radical breakthrough – perhaps even something spinning off from therapeutic cloning research – actually did lead us to a safe technology of reproductive cloning? At that point, the question would become, "What exactly, is *now* wrong with (safe) reproductive cloning?" In such circumstances, therapeutic cloning might have led to reproductive cloning, but it would not be at all obvious that this outcome was horrible. Indeed, in the absence of concerns that it would produce congenitally malformed children, the case against reproductive cloning appears rather weak (Blackford 2005, 12–14).

At this stage, it might be suggested that the real problem with reproductive cloning is neither something intrinsic to it nor any direct consequence of particular acts of cloning. Rather, it would put us on a second slippery slope to yet another practice. This, in turn, might be something sufficiently horrible that we must avert it. We must do so even at the cost of enacting criminal legislation to ban reproductive cloning and anything else that might foreseeably lead us that far. For example, it might be argued that the availability of a form of safe reproductive cloning would put us on a slippery slope to allowing certain kinds of genetic enhancement of children. More vaguely, it might be said to put us on a slippery slope to a "brave new world" as Kass has tried to show (2001, 33–35). The revised argument for prohibiting therapeutic cloning, then, is that allowing it would put us on a slippery slope to a further slippery slope that, in turn, would eventually lead us to a horrible result.

Leaving aside how implausibly indirect and tenuous this is starting to appear, the allegation of a new slippery slope from *reproductive* cloning to something worse must meet requirements similar to those already discussed above (Blackford 2005, 15–16). First, what is so bad about the social practices that are contemplated – whatever, exactly, they are? Second, will reproductive cloning really lead us to whichever practices we choose to specify? If they are truly horrible, why will we not be able to distinguish them from reproductive cloning itself and adopt appropriate policy responses, including the enactment of any necessary criminal prohibitions? Once again, if the claim is that our psychological ability to make logical distinctions will somehow be corrupted – or overcome by a sort of technological seduction – where is the evidence?

Perhaps there are certain kinds of genetic engineering that are not harmful or horrible in themselves, and which we might be tempted to practice if we could, but which would lead to some even more distant prospect that must be averted. For example, differential access to genetic enhancement of intelligence might lead to greater stratification of society. Why, however, can *that* outcome not be prevented by appropriate policies? By this point, the connection to therapeutic cloning has become absurdly remote, and, worse, there is the danger of an infinite regress: slippery slope arguments cannot be allowed to go on forever.

A point must come when the proponents of horrible-result-style slippery slope arguments either abandon their approach or actually produce the goods: they must specify a result that is adequately horrible, while also demonstrating that it is likely to eventuate if we take even the first step onto the slope, or onto the first of however many slopes there are supposed to be. Once a horrible result is clearly and uncontroversially specified, however, much more will need to be said as to why we cannot avoid *that* result while also obtaining the benefits of new biomedical technologies.

References

Blackford, R. 2005. "Human cloning and 'posthuman' society." *Monash Bioethics Review* 24:10–26.

Caulfield, T. and T. Bubela. 2007. "Why a criminal ban? Analyzing the arguments against somatic cell nuclear transfer in the Canadian parliamentary debate." *American Journal of Bioethics* 7(2): 51–61.

Kass, L. R. 2001. "Preventing a brave new world: Why we should ban human cloning now." *The New Republic* 21 May: 30–39.

Williams, B. 1985. "Which slopes are slippery?" in *Moral Dilemmas in Modern Medicine*, ed. M. Lockwood, pp. 126–37. Oxford, UK: Oxford University Press.

What uses of animals for biomedical purposes are permissible?

INTRODUCTION TO PART FOUR

TWO ISSUES ABOUT the ethical treatment of non-human animals that have been prominent in bioethics are medical research involving animal experimentation, and xenografting (or xenotransplantation), i.e., using animals as sources of transplant materials for human patients. Such biomedical practices raise the question as to whether medical benefits to human beings justify inflicting suffering or death on animals.

Singer argues from the 'basic idea that the interests of all beings should be given equal consideration irrespective of their species'. This deceptively simple premise has three main components. First, a capacity to suffer entails having interests; for example, a sentient creature capable of feeling pain has an interest in being pain-free. Second, 'equal consideration' does not mean that all specific interests have equal moral weight, but that equal consideration should be given to creatures' interests; for example, we should choose to inflict a small amount of pain on an animal rather than a large amount of pain on a human because, having given the interests of both sorts of creatures equal consideration, we thereby minimise suffering. Third, the phrase, 'irrespective of their species' warns against preferring to meet the interests of creatures solely because they are members of one's own species; Singer calls this 'speciesism' to imply an analogy with racism and sexism because, like race and sex, species membership is a morally irrelevant characteristic.

Harrison queries Singer's assumption that non-human animals have the capacity to suffer. He counters three arguments for the view that animals consciously experience pain, based on animal behaviour, animal nervous systems, and evolutionary theory, respectively. He also presents two arguments for believing that only human beings feel pain. The first of these starts with the claim that ascribing mental states to animals is conducive to our survival by enabling us to predict their behaviour. This argument seems weak because the fact that ascribing mental states to animals has this evolutionary pay-off does not count against the veracity of those ascriptions. But the point of the argument is in the analogy between animals and chess-playing computers; in both cases, 'there will always be instances where this intentional model will break down' and it makes more sense to think of them as 'designed' as opposed to thinking things. Harrison's second argument for believing that only human beings feel pain is that animal pain would be superfluous. Pain provides practical reasons that can be weighed against others; to cite an example of Harrison's, that touching it will hurt is a practical reason for avoiding hot metal, one that can be outweighed by the fact that the sleepy village is in danger. Since this sort of thinking is not characteristic of animals, their consciously experiencing pain would be superfluous. Harrison concludes with 'other considerations [that] can give us a more solid foundation for an "animal ethic" '.

Cohen disputes the two main arguments for opposing the use of animals in medical research, based on rights and sentience, respectively. Concerning the latter, he is at odds with Harrison in assuming that '[a]nimals can certainly suffer'. He is also at odds with Singer: first, by defending speciesism; second, by claiming that, even if we ought to consider the pains of all animate beings equally, 'a cogent utilitarian calculation' is bound to favour the use of animals in medical research. Although these seem like clear battle lines, neither Cohen nor Harrison advocates doing whatever we please with animals; Cohen states that '[w]e surely do have obligations to animals', which recalls Harrison's list of considerations other than the capacity for suffering that can ground an 'animal ethic'. Conversely, Singer's principle of equal consideration of interests does not rule out all uses of animals by humans. So, despite their theoretical differences, there is scope for agreement between these bioethicists on practical matters.

The last two selections debate the ethics of xenografting, i.e. using animals as sources of transplant organs for humans. Caplan sets out the rationale for developing animal-to-human transplants and usefully distinguishes ethical issues associated with the experimentation required to develop xenografting from ethical issues associated with its widespread therapeutic employment. Asking whether humans and apes are 'moral equivalents', he alludes to humans' greater capacities, and acknowledges the rejoinder that 'some individual animals ... have more capacities and abilities than certain individual human beings'. Caplan's response to this is interestingly different to Cohen's. Cohen insists that the 'issue is one of kind'; i.e., the point about capacities applies at the species level so it is a category error to apply it at the level of individual human beings and animals. By contrast, Caplan presents two reasons for using a chimp rather than a human being with reduced capacities: 'respect for their former existence [and] the impact using them would have upon other human beings'.

Nelson's critique focuses on the latter of these reasons. He sets out the marginal cases argument, clarifying that it can support two quite opposing conclusions: that it is permissible to use both animals and marginal humans, and that it is impermissible to use either animals or marginal humans, as organ sources. Nelson subtly reconstructs Caplan's thinking in terms of 'a general system of sensibilities that runs through our species'. Nonetheless, he notices a fact-value gap: how do we get from the fact that there are 'specific patterns of feeling' about using marginal humans as organ sources to a 'morally relevant difference between the humans and the nonhumans in question'? After all, not all partialities are morally defensible. He considers whether we might as well abide by our feelings and kill animals rather than marginal humans, since death is not a harm to any of the beings in question. But he goes on to suggest that this solves one problem only at the expense of creating another: if killing animals for organs is justified by the fact that their deaths do not harm them, then killing marginal humans for organs is justified on the same grounds.

Selections

Singer, P. 2003. 'Animal Liberation at 30', *New York Review of Books*, 50 (8), May 15th, 23–6.

Harrison, P. 1991. 'Do Animals Feel Pain?' *Philosophy*, 66 (255), 25–40.

Cohen, C. 1986. 'The Case for the Use of Animals in Biomedical Research', *New England Journal of Medicine*, 315 (14), 865–70.

Caplan, A. 1992. 'Is Xenografting Morally Wrong?' *Transplantation Proceedings*, 24 (2), 722–7.

Nelson, J. 1993. 'Moral Sensibilities and Moral Standing: Caplan on Xenograft "Donors"', *Bioethics*, 7 (4), 315–22.

Chapter 15

Peter Singer

ANIMAL LIBERATION AT 30

1

The phrase "Animal Liberation" appeared in the press for the first time on the April 5, 1973, cover of *The New York Review of Books*. Under that heading, I discussed *Animals, Men and Morals*, a collection of essays on our treatment of animals, which was edited by Stanley and Roslind Godlovitch and John Harris.[1] The article began with these words:

> We are familiar with Black Liberation, Gay Liberation, and a variety of other movements. With Women's Liberation some thought we had come to the end of the road. Discrimination on the basis of sex, it has been said, is the last form of discrimination that is universally accepted and practiced without pretense, even in those liberal circles which have long prided themselves on their freedom from racial discrimination. But one should always be wary of talking of "the last remaining form of discrimination."

In the text that followed, I urged that despite obvious differences between humans and nonhuman animals, we share with them a capacity to suffer, and this means that they, like us, have interests. If we ignore or discount their interests, simply on the grounds that they are not members of our species, the logic of our position is similar to that of the most blatant racists or sexists who think that those who belong to their race or sex have superior moral status, simply in virtue of their race or sex, and irrespective of other characteristics or qualities. Although most humans may be superior in reasoning or in other intellectual capacities to nonhuman animals, that is not enough to justify the line we draw between humans and animals. Some humans – infants and those with severe intellectual disabilities – have intellectual capacities inferior to some animals, but we would, rightly, be shocked by anyone who proposed that we inflict slow, painful deaths on these intellectually inferior humans in order to test the safety of household products. Nor, of course, would we tolerate confining them in small cages and then slaughtering them in order to eat them. The fact that we are prepared to do these things to nonhuman animals is therefore a sign of "speciesism" – a prejudice that survives because it is convenient for the dominant group – in this case not whites or males, but all humans.

That essay and the book that grew out of it, also published by *The New York Review*,[2] are often credited with starting off what has become known as the "animal rights movement" – although the ethical position on which the movement rests needs no reference to rights. Hence the essay's thirtieth anniversary provides a convenient opportunity to take stock both of the current state of the debate over the moral status of animals and of how effective the movement has been in bringing about the practical changes it seeks in the way we treat animals.

2

The most obvious difference between the current debate over the moral status of animals and that of thirty years ago is that in the early 1970s, to an extent barely credible today, scarcely anyone thought that the treatment of individual animals raised an ethical issue worth taking seriously. There were no animal rights or animal liberation organizations. Animal welfare was an issue for cat and dog lovers, best ignored by people with more important things to write about. (That's why I wrote to the editors of The New York Review with the suggestion that they might review Animals, Men and Morals, whose publication the British press had greeted a year earlier with total silence.)

Today the situation is very different. Issues about our treatment of animals are often in the news. Animal rights organizations are active in all the industrialized nations. The US animal rights group called People for the Ethical Treatment of Animals has 750,000 members and supporters. A lively intellectual debate has sprung up. (The most comprehensive bibliography of writings on the moral status of animals lists only ninety-four works in the first 1970 years of the Christian era, and 240 works between 1970 and 1988, when the bibliography was completed.[3] The tally now would probably be in the thousands.) Nor is this debate simply a Western phenomenon – leading works on animals and ethics have been translated into most of the world's major languages, including Japanese, Chinese, and Korean.

To assess the debate, it helps to distinguish two questions. First, can speciesism itself – the idea that it is justifiable to give preference to beings simply on the grounds that they are members of the species Homo sapiens – be defended? And secondly, if speciesism cannot be defended, are there other characteristics about human beings that justify them in placing far greater moral significance on what happens to them than on what happens to nonhuman animals?

The view that species is in itself a reason for treating some beings as morally more significant than others is often assumed but rarely defended. Some who write as if they are defending speciesism are in fact defending an affirmative answer to the second question, arguing that there are morally relevant differences between human beings and other animals that entitle us to give more weight to the interests of humans.[4] The only argument I've come across that looks like a defense of speciesism itself is the claim that just as parents have a special obligation to care for their own children in preference to the children of strangers, so we have a special obligation to other members of our species in preference to members of other species.[5]

Advocates of this position usually pass in silence over the obvious case that lies between the family and the species. Lewis Petrinovich, professor emeritus at the University of California, Riverside, and an authority on ornithology and evolution, says that our biology turns certain boundaries into moral imperatives – and then lists "children, kin, neighbors, and species."[6] If the argument works for both the narrower circle of family and friends and the wider sphere of the species, it should also work for the middle case: race. But an argument that supported our preferring the interests of members of our own race over those of members of other races would be less persuasive than one that allowed priority only for kin, neighbors, and members of our species. Conversely, if the argument doesn't show race to be a morally relevant boundary, how can it show that species is?

The late Harvard philosopher Robert Nozick argued that we can't infer much from the fact that we do not yet have a theory of the moral importance of species membership. "No one," he wrote, "has spent much time trying to formulate" such a theory, "because the issue hasn't seemed pressing."[7] But now that nearly twenty years have passed since Nozick wrote those words, and many people have, during those years, spent quite a lot of time trying to defend the importance of species membership, Nozick's comment takes on a different weight. The continuing failure of philosophers to produce a plausible theory of the moral importance of species membership indicates, with increasing probability, that there can be no such thing.

That takes us to the second question. If species is not morally important in itself, is there something else that happens to coincide with the human species, on the basis of which we can justify the inferior consideration we give to nonhuman animals?

Peter Carruthers argues that it is the lack of a capacity to reciprocate. Ethics, he says, arises out of an agreement that if I do not harm you, you will not harm me. Since animals cannot take part in this social contract we have no direct duties to them.[8] The difficulty with this approach to ethics is that it also means we have no direct duties to small children, or to future generations yet unborn. If we produce radioactive waste that will be deadly for thousands of years, is it unethical to put it into a container that will last 150 years and drop it into a convenient lake? If it is, ethics cannot be based on reciprocity.

Many other ways of marking the special moral significance of human beings have been suggested: the ability to reason, self-awareness, possession of a sense of justice, language, autonomy, and so on. But the problem with all of these allegedly distinguishing marks is, as noted above, that some humans are entirely lacking in these characteristics and few want to consign them to the same moral category as nonhuman animals.

This argument has become known by the tactless label of "the argument from marginal cases," and has spawned an extensive literature of its own.[9] The attempt by the English philosopher and conservative columnist Roger Scruton to respond to it in *Animal Rights and Wrongs* illustrates both the strengths and weaknesses of the argument. Scruton is aware that if we accept the prevailing moral rhetoric that asserts that all human beings have the same set of basic rights, irrespective of their intellectual level, the fact that some nonhuman animals are at least as rational, self-aware, and autonomous as some human beings looks like a firm basis for asserting that all animals have these basic rights. He points out, however, that this prevailing moral rhetoric is not in accord with our real attitudes, because we often regard "the killing of a human vegetable" as excusable. If human beings with profound intellectual disabilities do not have the same right to life as normal human beings, then there is no inconsistency in denying that right to nonhuman animals as well.

In referring to a "human vegetable," however, Scruton makes things too easy for himself, for that expression suggests a being that is not even conscious, and thus has no interests at all that need to be protected. He might be less comfortable making his point with respect to a human being who has as much awareness and ability to learn as the foxes he wants to continue being permitted to hunt. In any case, the argument from marginal cases is not limited to the question of what beings we can justifiably kill. In addition to killing animals, we inflict suffering on them, in a wide variety of ways. So the defenders of common practices involving animals owe us an explanation for their willingness to make animals suffer when they would not be willing to do the same to humans with similar intellectual capacities. (Scruton, to his credit, is opposed to the close confinement of modern animal raising, saying that "a true morality of animal welfare ought to begin from the premise that this way of treating animals is wrong.")

Scruton is in fact only half-willing to acknowledge that a "human vegetable" may be treated differently from other human beings. He muddies the waters by claiming that it is "part of human virtue to acknowledge human life as sacrosanct." In addition, he argues that because in normal conditions human beings are members of a moral community protected by rights, even deeply serious abnormality does not cancel membership of this community. Thus even though humans with profound intellectual disability do not really have the same claims on us as normal humans, we would do well, Scruton says, to treat them as if they did. But is this defensible? Certainly if any sentient being, human or nonhuman, can feel pain or distress, or conversely can enjoy life, we ought to give the interests of that being the same consideration as we give to the similar interests of normal human beings with unimpaired capacities. To say, however, that species alone is both necessary and sufficient for being a member of our moral community, and for having the basic rights granted to all members of that community, requires further justification. We return to the core question: Should all and only human beings be protected by rights, when some nonhuman animals are superior in their intellectual capacities, and have richer emotional lives, than some human beings?

One well-known argument for an affirmative answer to this question asserts that unless we can draw a clear boundary around the moral community, we will find ourselves on a slippery slope.[10] We may start by denying rights to Scruton's "human vegetable," that is, to those

who can be shown to be irreversibly unconscious, but then we may gradually extend the category of those without rights to others, perhaps to the intellectually disabled, or to the demented, or just to those whose care is a burden on their family and the community, until in the end we have reached a situation that none of us would have accepted if we had known we were heading there when we denied the irreversibly unconscious a right to life. This is one of several arguments critically examined by the Italian animal activist Paola Cavalieri in *The Animal Question: Why Nonhuman Animals Deserve Human Rights*, a rare contribution to the English-language debate by a writer from continental Europe. Cavalieri points to the ease with which slave-owning societies were able to draw lines between humans with rights and humans without rights.

That slaves were human beings was acknowledged both in ancient Greece and in the slave-holding states of the US – Aristotle explicitly says that barbarians are human beings who exist to serve the good of the more rational Greeks,[11] and Southern whites sought to save the souls of the Africans they enslaved by making them Christians. Yet the line between slaves and free people did not slip significantly, even when some barbarians and some Africans became free, or when slaves produced children of mixed race. So, Cavalieri suggests, there is no reason to doubt our ability to deny that some humans have rights, while keeping the rights of other humans as secure as ever. But she is certainly not advocating that we do this. Her concern is rather to undermine the argument for drawing the boundaries of the sphere of rights so as to include all and only humans.

Cavalieri also responds to the argument that all humans, including the irreversibly unconscious, are to be elevated above other animals because of the characteristics they "normally" possess, rather than those they actually have. This argument seems to appeal to a kind of unfairness in excluding those who "fortuitously" fail to have the required characteristics. Cavalieri replies that if the "fortuitousness" is merely statistical, it carries no moral relevance, and if it is intended to suggest that the lack of the required characteristics is not the fault of those with profound intellectual disability, then that is not a basis for separating such humans from nonhuman animals.

Cavalieri states her own position in terms of rights, and in particular the basic rights that constitute what, following Ronald Dworkin, she calls the "egalitarian plateau." We want, Cavalieri insists, to secure a basic form of equality for all human beings, including the "non-paradigmatic" ones (her term for "marginal cases"). If the egalitarian plateau is to have a defensible, nonarbitrary boundary that safeguards all humans from being pushed off the edge, we must select as a criterion for that boundary a standard that allows a large number of nonhuman animals inside the boundary as well. Hence we must allow onto the egalitarian plateau beings whose intellect and emotions are at a level that is shared by, at least, all birds and mammals.

Cavalieri does not argue that the rights of birds and mammals can be derived from self-evidently true moral premises. Her starting point, rather, is our prevailing belief in human rights. She seeks to show that all who accept this belief must also accept that similar rights apply to other animals. Following Dworkin, she sees human rights as part of the basic political framework of a decent society. They set limits to what the state may justifiably do to others. In particular, institutions like slavery or other invidious forms of racial discrimination that are based on violating the human rights of some of those over whom the state rules are, for that reason alone, illegitimate. Our acceptance of the idea of human rights therefore requires the abolition of all practices that routinely overlook the basic interests of rights-holders. Hence, if

Cavalieri's argument is sound, our belief in rights commits us to an extension of rights beyond humans, and that in turn requires us to abolish all practices, like factory farming and the use of animals as subjects of painful and lethal research, that routinely overlook the basic interests of nonhuman rights-holders.

On the other hand, the rights for which Cavalieri argues are not supposed to resolve every situation in which there is a conflict of interests or of rights. Her notion of rights as part of the basic political framework of a decent society is compatible with specific restrictions of rights, as occurred for example when "Typhoid Mary" was compulsorily quarantined because she carried a lethal disease. A government may be entitled to restrict the movements of humans or animals who are a danger to the public, but it must still show them the concern and respect due to them as possessors of basic rights.[12]

My own opposition to speciesism is based, as I have already mentioned, not on rights, but on the thought that a difference of species is not an ethically defensible ground for giving less consideration to the interests of a sentient being than we give to similar interests of a member of our own species. David DeGrazia skillfully defends equal consideration for all sentient beings in *Taking Animals Seriously*. Such a position need not rely on prior acceptance of our current view of human rights – a view that, though widespread, can be rejected, especially once its implications in regard to animals are drawn out as Cavalieri draws them out. While the principle of equal consideration of interests is therefore more solidly based than Cavalieri's argument, however, it must face the difficulties that follow from the fact that interests, not rights, are now the focus of attention. That requires us to estimate what the interests are in an endless variety of different circumstances.

To take one case of particular ethical significance: the interest a being has in continued life

– and hence, on the interests view, the wrong-ness of taking that being's life – will depend in part on whether the being is aware of itself as existing over time, and is capable of forming future-directed desires that give it a particular kind of interest in continuing to live. To that extent Roger Scruton is right about our attitudes to the deaths of members of our own species who lack these characteristics. We see it as less of a tragedy than the death of a being who is future-oriented, and whose desires to do things in the medium- and long-term future will there-fore be thwarted if he or she dies.[13] But this is not a defense of speciesism, for it implies that killing a self-aware being like a chimpanzee causes a greater loss to the being killed than does killing a human being with an intellectual dis-ability so severe as to preclude the capacity to form desires for the future.

We then need to ask what other beings may have this kind of interest in living into the future. DeGrazia combines philosophical insights and scientific research to help us answer such ques-tions about specific species of animals, but there is often room for doubt, and the calculations required for applying the principle of equal consideration of interests can only be rough approximations, if they can be done at all. Perhaps, though, that is just the nature of our ethical situation, and rights-based views avoid such calculations at the cost of leaving out some-thing relevant to what we ought to do.

The most recent addition to the literature of the animal movement has come from a surprising quarter, one deeply hostile to any discussion of the possibility of justifying the killing of human beings, no matter how severely disabled they may be. In *Dominion: The Power of Man, the Suffering of Animals, and the Call to Mercy* Matthew Scully, a conservative Christian, past literary editor of *National Review* and now speechwriter to President George W. Bush, has written an eloquent polemic against human abuse of animals, culminating with a devastating descrip-tion of factory farming.

Since the animal movement has, for the past thirty years, generally been associated with the left, it is curious now to see Scully make a case for many of the same goals within the perspec-tive of the Christian right, replete with refer-ences to God, interpretation of the scripture, and attacks on "moral relativism, self-centered mate-rialism, license passing itself off as freedom, and the culture of death"[14] – but this time aimed at condemning not victimless crimes like homo-sexuality or physician-assisted suicide, but the needless suffering inflicted by factory farming and the modern slaughterhouse. Scully calls on all of us to show mercy toward animals and abandon ways of treating them that fail to respect their nature as animals. The result is a work that, although not philosophically rigorous, has had a remarkable amount of sympathetic publicity in the conservative press, which usually sneers at animal advocates.

3

The history of the modern animal movement makes a nice counterexample to skepticism about the impact of moral argument on real life.[15] As James Jasper and Dorothy Nelkin observed in *The Animal Rights Crusade: The Growth of a Moral Protest*, "Philosophers served as midwives of the animal rights movement in the late 1970s."[16] The first successful protest against animal exper-iments in the United States was the 1976–1977 campaign against experiments conducted at the American Museum of Natural History on the sexual behavior of mutilated cats. Henry Spira, who conceived and ran the campaign, had a background of working in the union and civil rights movements, and had not considered, until he read the 1973 *New York Review* article, that animals are also worth the attention of those concerned about the exploitation of the weak.

Spira went on to take on larger targets, such as the testing of cosmetics on animals. His technique was to target a prominent corporation that used animals – in the cosmetics campaign, he started with Revlon – and ask them to take reasonable steps to find alternatives to the use of animals. Always willing to engage in dialogue, and never one to paint the abusers of animals as evil sadists, he was remarkably successful in stimulating interest in developing ways of testing products without using animals, or with using fewer animals in less painful ways.[17]

Partly as a result of his work, there has also been a sizable drop in the number of animals used in research. In Britain official statistics show that roughly half as many animals are now experimented upon as were used in 1970. Estimates for the United States – where no official statistics are kept – suggest a similar story. From the standpoint of a nonspeciesist ethic there is still a long way to go for animals used in research, but the changes the animal movement has brought about mean that every year millions fewer animals are forced to undergo painful procedures and slow deaths.

The animal movement has had other successes too. Despite "fur is back" claims by the industry, fur sales have still not recovered to their level in the 1980s, when the animal movement began to target it. Since 1973, while the number of dogs and cats owned has nearly doubled, the number of stray and unwanted animals killed in pounds and shelters has been cut by more than half.[18]

These modest gains are dwarfed, however, by the huge increase in animals kept confined, some so tightly that they are unable to stretch their limbs or walk even a step or two, on America's factory farms. This is by far the greatest source of human-inflicted suffering on animals, simply because the numbers are so great. Animals used in experiments are numbered in the tens of millions annually, but last year ten billion birds and mammals were raised and killed for food in the United States alone. The increase over the previous year is, at around 400 million animals, more than the total number of animals killed in the US by pounds and shelters, for research, and for fur combined. The overwhelming majority of these factory-reared animals now live their lives entirely indoors, never knowing fresh air, sunshine, or grass until they are trucked away to be slaughtered.

Against the confinement and slaughter of farm animals in America, the animal movement has, until quite recently, been impotent. Gail Eisnitz's 1997 book *Slaughterhouse* contains shocking, well-authenticated accounts of animals in major American slaughterhouses being skinned and dismembered while still conscious.[19] If such incidents had been documented in Britain they would have led to major news stories and the national government would have been forced to do something about it. Here the book passed virtually unnoticed outside the animal movement.

The situation is very different in Europe. Americans have often looked down on some European nations, especially the Mediterranean countries, for tolerating cruelty to animals. Now the accusing glance goes in the opposite direction. Even in Spain, with its culture of bullfighting, most animals are better cared for than in America. By 2012, European egg producers will be required to give their hens access to a perch and a nesting box to lay their eggs in, and to allow at least 750 square centimeters, or 120 square inches, per bird – dramatic changes that will transform the living conditions of more than two hundred million hens. United States egg producers haven't even started thinking about perches or nesting boxes, and typically give their fully grown hens just forty-eight square inches, or about half the area of a sheet of 8 1/2-X-11-inch letter paper per bird.[20]

In the US veal calves are deliberately kept anemic, deprived of straw for bedding, and confined in individual crates so narrow that they cannot even turn around. That system of keeping calves has been illegal in Britain for many years, and will become illegal throughout the European Union by 2007. Keeping pregnant sows in individual crates for their entire pregnancy, also the standard American practice, was banned in Britain in 1998, and is being phased out in Europe. These changes have wide support throughout the European Union, and the backing of leading European experts on the welfare of farm animals. They are a vindication of much that animal advocates have been saying for the past thirty years.

Are Americans simply less concerned with animal suffering than their European counterparts? Perhaps, but in *Political Animals: Animal Protection Policies in Britain and the United States*, Robert Garner explores several other possible explanations for the widening gap in animal welfare standards between the two nations.[21] By comparison with Britain, the US political process is more corrupt. Elections are many times more costly — the entire 2001 British general election cost less than John Corzine spent to win a single Senate seat in 2000. With money playing a greater role, American candidates are more beholden to their donors. Moreover, fund raising in Europe is largely done by the political parties, not by individual candidates, which makes it more open to public scrutiny and more likely to produce an electoral backlash for the entire party if it is seen to be in the pocket of a particular industry. These differences allow the agribusiness industry far greater control over Congress than it can hope to have over the political processes in Europe.

Consistent with that explanation, the most successful American campaigns — like Spira's campaign against the use of animals to test cosmetics — have concentrated on corporations rather than on the legislature or the government. Recently a ray of hope has come from an unlikely vehicle for change. After protracted discussions with animal advocates, started by Henry Spira before his death and then taken up by People for the Ethical Treatment of Animals, McDonald's agreed to set and enforce higher standards for the slaughterhouses that supply it with meat, and then announced that it would require its egg suppliers to provide each hen with a minimum of seventy-two square inches of living space — a 50 percent improvement for most American hens, but still only enough to bring these producers up to a level that is already on its way out in Europe. Burger King and Wendy's followed suit. These steps were the first hopeful signs for American farm animals since the modern animal movement began.

An even greater triumph was achieved last November by using another route around the legislative roadblock: the citizen-initiated referendum. With support from a number of national animal organizations, a group of animal activists in Florida succeeded in gathering 690,000 signatures to put on the ballot a proposal to change the constitution of Florida so as to ban the keeping of pregnant sows in crates so narrow that they cannot even turn around. Changing the constitution is the only way citizens can get a direct vote on a measure in Florida. Opponents of the measure, obviously unwilling to argue that pigs don't need to be able to turn around or walk, instead tried to persuade Florida voters that the confinement of pigs was not an appropriate subject for the state constitution. But by a margin of 55 to 45 percent, voters said no to sow crates, thus making Florida the first jurisdiction in the United States to ban a major form of farm-animal confinement. Though Florida has only a small number of intensive piggeries, the vote supports the idea that it is not hard hearts or lack of sympathy for animals but a failure of democracy that causes America to lag so far behind Europe in abolishing the worst features of factory farming.

4

My original article in *The New York Review* ended with a paragraph that saw the challenge of the animal movement as a test of human nature:

> Can a purely moral demand of this kind succeed? The odds are certainly against it. The book [*Animals, Men and Morals*] holds out no inducements. It does not tell us that we will become healthier, or enjoy life more, if we cease exploiting animals. Animal Liberation will require greater altruism on the part of mankind than any other liberation movement, since animals are incapable of demanding it for themselves, or of protesting against their exploitation by votes, demonstrations, or bombs. Is man capable of such genuine altruism? Who knows? If this book does have a significant effect, however, it will be a vindication of all those who have believed that man has within himself the potential for more than cruelty and selfishness.

So how have we done? Both the optimists and the cynics about human nature could see the results as confirming their views. Significant changes have occurred, in animal testing and other forms of animal abuse. In Europe, entire industries are being transformed because of the concern of the public for the welfare of farm animals. Perhaps most encouraging for the optimists is the fact that millions of activists have freely given up their time and money to support the animal movement, many of them changing their diet and lifestyle to avoid supporting the abuse of animals. Vegetarianism and even veganism (avoiding all animal products) are far more widespread in North America and Europe than they were thirty years ago, and although it is difficult to know how much of this relates to concern for animals, undoubtedly some of it does.

On the other hand, despite the generally favorable course of the philosophical debate about the moral status of animals, popular views on that topic are still very far from adopting the basic idea that the interests of all beings should be given equal consideration irrespective of their species. Most people still eat meat, and buy what is cheapest, oblivious to the suffering of the animal from which the meat comes. The number of animals being consumed is much greater today than it was thirty years ago, and increasing prosperity in East Asia is creating a demand for meat that threatens to boost that number far higher still. Meanwhile the rules of the World Trade Organization threaten advances in animal welfare by making it doubtful that Europe will be able to keep out imports from countries with lower standards. In short, the outcome so far indicates that as a species we are capable of altruistic concern for other beings; but imperfect information, powerful interests, and a desire not to know disturbing facts have limited the gains made by the animal movement.

Notes

1 Taplinger, 1972.
2 Peter Singer, *Animal Liberation* (New York Review/ Random House, 1975; revised edition, New York Review/Random House, 1990; reissued with a new preface, Ecco, 2001).
3 Charles Magel, *Keyguide to Information Sources in Animal Rights* (McFarland, 1989).
4 See, for example, Carl Cohen, "The Case for the Use of Animals in Biomedical Research," *New England Journal of Medicine*, Vol. 315 (1986), pp. 865–70; and Michael Leahy, *Against Liberation: Putting Animals in Perspective* (London: Routledge, 1991).
5 See Mary Midgley, *Animals and Why They Matter* (University of Georgia Press, 1984); Jeffrey Gray, "On the Morality of Speciesism," *Psychologist*, Vol. 4, No. 5 (May 1991), pp. 196–98, and "On Speciesism and Racism: Reply to Singer and Ryder,"

Psychologist, Vol. 4, No. 5 (May 1991), pp. 202–03; and Lewis Petrinovich, Darwinian Dominion: Animal Welfare and Human Interests (MIT Press, 1999).

6 Petrinovich, Darwinian Dominion, p. 29.

7 Robert Nozick, "About Mammals and People," The New York Times Book Review, November 27, 1983, p. 11; I draw here on Richard I. Arneson, "What, If Anything, Renders All Humans Morally Equal?" in Singer and His Critics, edited by Dale Jamieson (Blackwell, 1999), p. 123.

8 Peter Carruthers, The Animals Issue: Moral Theory in Practice (Cambridge University Press, 1992).

9 Daniel Dombrowski, Babies and Beasts: The Argument from Marginal Cases (University of Illinois Press, 1997).

10 See, for example, Peter Carruthers, The Animals Issue.

11 Aristotle, Politics (London: J.M. Dent and Sons, 1916), p. 16.

12 As Dworkin himself argued in regard to the detention of suspected terrorists; see "The Threat to Patriotism," The New York Review, February 28, 2002.

13 See my Practical Ethics (Cambridge University Press, 1993), especially Chapter 4.

14 Quoted from Kathryn Jean Lopez, "Exploring 'Dominion': Matthew Scully on Animals," National Review Online, December 3, 2002.

15 See, for example, Richard A. Posner, The Problematics of Moral and Legal Theory (Belnap Press/Harvard University Press, 1999).

16 Free Press, 1992, p. 90.

17 See Peter Singer, Ethics into Action: Henry Spira and the Animal Rights Movement (Rowman and Littlefield, 1998).

18 The State of the Animals 2001, edited by Deborah Salem and Andrew Rowan (Humane Society Press, 2001).

19 Prometheus, 1997.

20 See Karen Davis, Prisoned Chickens, Poisoned Eggs: An Inside Look at the Modern Poultry Industry (Book Publishing Company, 1996).

21 St. Martin's, 1998.

Chapter 16

Peter Harrison

DO ANIMALS FEEL PAIN?

In an oft-quoted passage from *The Principles of Morals and Legislation* (1789), Jeremy Bentham addresses the issue of our treatment of animals with the following words: 'the question is not, can they *reason*? nor, can they *talk*? but, can they *suffer*?'[1] The point is well taken, for surely if animals suffer, they are legitimate objects of our moral concern. It is curious therefore, given the current interest in the moral status of animals, that Bentham's question has been assumed to be merely rhetorical. No-one has seriously examined the claim, central to arguments for animal liberation and animal rights, that animals actually feel pain. Peter Singer's *Animal Liberation* is perhaps typical in this regard. His treatment of the issue covers a scant seven pages, after which he summarily announces that 'there are no good reasons, scientific or philosophical, for denying that animals feel pain'.[2] In this paper I shall suggest that the issue of animal pain is not so easily dispensed with, and that the evidence brought forward to demonstrate that animals feel pain is far from conclusive.

Three kinds of argument are commonly advanced to support the contention that animals feel pain. The first involves the claim that animal behaviours give us clues to alleged mental states, about what animals are feeling. Thus animals confronted with noxious stimuli which would cause human beings pain, react in similar ways. They attempt to avoid the stimulus, they show facial contortions, they may even cry out. From these 'pain behaviours' it is inferred that the animals must be experiencing pain.

A second argument asserts that by virtue of a similarity in structure and function of nervous systems it is likely that human beings and animals closely related to the human species will experience the external environment in much the same way. It is assumed, for example, that primates have visual experiences similar to our own, feel hunger and thirst as we do, and so on. Presumably when they encounter noxious stimuli, they, like us, feel pain.

A third line of argument is derived from evolutionary theory. Organic evolution implies that there is no radical discontinuity between human and other species. It is likely, on this view, that human minds evolved from animal minds, and that closely related species would experience similar mental events. The evolutionary model would also suggest that pain is an essential adaptation for organisms in that it helps them avoid those things which would reduce their chances of survival and reproduction.

Let us consider these arguments in turn.

I

The argument based on 'pain behaviours' is the most intuitive. Considered in isolation, however, it is the least compelling. Even the simplest representatives of the animal kingdom exhibit rudimentary 'pain behaviours'. Single-celled organisms, for example, will withdraw from harmful stimuli. Insects struggle feebly after they have been inadvertently crushed underfoot. Yet few would want to argue that these behaviours resulted from the experience of pain. Certainly we show little sympathy for those unfortunate ants which are innocent casualties of an afternoon stroll, or the countless billions of micro-organisms destroyed by the chlorination of our water supplies. For all practical purposes we discount the possibility that such simple forms of life feel pain, despite their behaviours.

In more elevated levels of the animal kingdom there are also instances of 'pain behaviours' which undoubtedly occur in the absence of pain. Some parent birds, for instance, will feign injury to lure predators away from their young. The converse is also true. Animals might have sustained considerable tissue damage, but display none of the signs which we imagine would usually attend such trauma. This is because immobility is the best response to certain kinds of injury.[3] Pain behaviours, in any case, can be ably performed by non-living entities. If we were to construct a robot which was devoid of speech, yet was to have an active and independent existence, it would be necessary to programme it with mechanisms of self-preservation. Of the many objects it might encounter, it would need to be able to detect and respond to those likely to cause it most harm. Properly programmed, such a machine would manifest its own 'pain behaviour'. If we lit a fire under it, it would struggle to escape. If it found itself in a dangerous situation from which it could not extricate itself (say it fell into an acid bath) it would attempt to summon aid with shrill cries. If it were immobilized after a fall, it might, by facial contortions, indicate that it was damaged. But this 'pain behaviour' would

convey nothing about what it was feeling, for robots, on most accounts, can feel nothing. All that could be learned from such behaviour was how well the robot had been programmed for self-preservation. Mutatis mutandis, the 'pain behaviours' of animals demonstrate, in the first instance, how well natural selection has fitted them for encounters with unfriendly aspects of their environment. For neither animals, nor our imaginary robot, is 'pain behaviour' primarily an expression of some internal state.

I think these examples are sufficient to show that the argument from behaviours alone is fairly weak. But the reason we are inclined to deny that simple animals and computers feel pain is that despite their competent performance of 'pain behaviours', their internal structure is sufficiently dissimilar to our own to warrant the conclusion that they do not have a mental life which is in any way comparable. Animals closely related to the human species, however, possess at least some of the neural hardware which in human beings is thought to be involved in the experience of pain. It might be that the behavioural argument is stronger when considered together with the second argument – that based on the affinity of nervous systems.

II

Pain is a mental state. It might be caused by, or correlated with, brain states. It might have behavioural or psychological indicators. Yet it remains intractably mental. Herein lies the stumbling block of the second argument, for the closest scrutiny of the nervous systems of human beings and animals has never progressed beyond, and arguably never will progress beyond, the description of brain states to arrive at mental states. Thus the introduction of the structure and function of nervous systems into this discussion brings with it that whole constellation of difficulties which revolve around the problem of psycho-physical reductionism. Can mental states be reduced to physical states, and is it possible to project mental states from appropriate anatomical and physiological data? To be successful, the second argument for animal pain must answer both of these questions in the affirmative.

Descartes, in his *Meditations* (1641), quite correctly pointed out that there is no necessary *logical* relation between propositions about mental states and propositions about physical states. We may doubt the existence of our bodies, but not our minds. A disembodied mind is a logical possibility. Conversely, there is no logical impropriety in imagining bodies behaving in quite complex ways, without those behaviours being necessarily accompanied by relevant mental processes. Our robot, for example, would fit the bill, and indeed for Descartes, animals too were merely automatons, albeit organic ones.

Of course from the fact that there is no *logical* connection between mental states and physical states it cannot be inferred that no *contingent* connection is possible. Descriptions of mental and physical states may be linked in a number of ways, and it is upon such linkages that the second argument for animal pain depends. The most compelling evidence of connection between the physical state of the brain and the mental life of the individual comes from instances of brain pathology or brain surgery. The fact that damage to the cerebral cortex can reduce individuals to a 'mindless' state would suggest that observable brain states cause mind states, or at the very least are a necessary condition of mind states. More specifically, neurologists have had some success in identifying those parts of the brain which seem to be responsible for particular conscious states.

Our experience of pain, for example, seems to be mediated through a complicated physical network involving the neospinothalamic projection system (sensory aspects of pain), reticular and limbic structures (motivational aspects of pain), and the neocortex (overall control of sensory and motivational systems).[4] (It may be significant that this latter structure we share only with the primates. An argument could be made on this basis alone that the experience which we designate 'pain' is peculiar to us and a few primate species.) But despite such well-established connections between observable brain structures and more elusive mental states, it would be rash to attempt to predict the mental states of individuals on the basis of the presence or absence of certain structures, or even on the basis of the physiological status of those structures.[5] The well-known literature on the psychology of pain illustrates that the same stimulus may prove intensely painful to one individual, and be of little concern to another. The use of placebos to control pain, the influence of hypnosis or suggestion to influence pain perception, national differences in pain thresholds, all such aspects of the psychology of pain illustrate that the presence of certain brain structures and requisite sensory inputs are not sufficient conditions for the prediction of mental states.

Not only does the psychology of pain afford instances in which the same neural hardware might give rise to a variety of different conscious states, but the human brain itself exhibits an amazing ability to generate certain mental states in the absence of the relevant physical structures. Phantom pain is perhaps the most obvious example. Amputees frequently report awareness of a limb which has been recently amputated. In a minority of cases a phantom limb may become an ongoing source of severe pain. Often the pain is located in a quite specific part of the missing appendage.

An even more compelling illustration of the generation of certain mental states in the absence of appropriate structures comes from John Lorber's engaging paper 'Is Your Brain Really Necessary?'[6] Paediatric neurologist Lorber reports on a number of individuals with hydro-cephalus – a condition which resulted in their having virtually no cerebral cortex. The most intriguing case cited by Lorber is that of a mathematician with IQ of 126. A brain scan revealed that this young man had, in Lorber's words, 'virtually no brain'. The supratentorial part of the intracranial cavity contained only a thin layer of brain tissue, between one and two millimetres thick, attached to the skull wall. No 'visual cortex' was evident, yet the individual, who by all accounts should have been blind, had above average visual perception. It is likely that the functions which would normally have taken place in the missing cerebral cortex had been taken over by other structures. Cases such as this show that certain aspects of human consciousness have a tenacity which confounds our understanding of the link between brain structure and consciousness.

Lorber's discoveries are a striking example of the fact that an advancing neuroscience, far from establishing concrete links between brain states and mental states, is actually deepening the mystery of how the brain is causally related to human consciousness. It need hardly be said that when we cross the species boundary and attempt to make projections about animals' putative mental lives based on the structures of their nervous systems we are in murky waters indeed. Two further examples illustrate this.

The brains of birds, such as they are, do not contain a 'visual cortex'. Thus if we are to argue that similar brain structures give rise to similar experiences, then it is unlikely that the visual experiences of birds will be qualitatively similar to our own. On the other hand, the behaviour of birds would seem to indicate that they can 'see'. While we assume from the behaviour of birds that their visual experience of the world is much the same as ours, if we are committed to the view that like mental states are generated by like brain stuctures, we are bound to admit that this assumption is unfounded. We might of course

be tempted to revert to the first argument – that behaviour, not structure, gives the correct cues to mental states. But this seems to commit us to the view that computers, flies, and amoebas have states of consciousness like our own.

Another illustration which concerns visual experiences is the much-discussed phenomenon of 'blind-sight'.[7] As we have already mentioned, the 'visual' or striate cortex is thought to be necessary for human vision. Individuals suffering from damage to the striate cortex may lose sight in part of their visual field. Larry Weisenkrantz and his colleagues have carried out a number of experiments on one such individual who claimed to be blind in his left field of view. Simple shapes were presented to this subject in his blind field of view. Though he denied being able to see anything, the subject could, with reasonable consistency, describe the shape of the object and point to it. In each instance he insisted that his correct response was merely a guess.[8] Examples of blindsight indicate, amongst other things, that it is possible to have visual experiences of which we are unaware. The blindsight phenomenon thus opens up the possibility that there might be *non-conscious* experiences to which we can nonetheless respond with the appropriate behaviour.[9] Blindsighted individuals can learn to respond *as if* they see, even though they have no conscious awareness of seeing anything. The significance of this for a discussion of animal behaviours is that animals might respond to stimuli as if they were conscious of them, while in fact they are not. Thus birds which lack the human apparatus of conscious vision (as do blindsighted subjects) might not simply have qualitatively different visual experiences as suggested above, they might not have conscious visual experiences at all. It may be concluded that an animal's experience of stimuli which we would find painful might be qualitatively different (that is, not painful) or may even be non-conscious. Animals might react to such stimuli by exhibiting 'pain behaviour' and yet not have that mental experience which we call 'pain', or perhaps not have any conscious experience at all.[10]

So far our discussion of neural circuitry and how it relates to putative mental states has focused upon the inability of contemporary neuroscience to bridge the gap between brain and mind. There are those, of course, who have asserted that it is impossible in principle to bridge that gap. It is significant that Thomas Nagel, one of the chief spokesmen for this group, has alluded to animal consciousness to make his point.

In the seminal paper 'What is it Like to be a Bat?',[11] Nagel leads us into the subjective world of the bat. These curious mammals, he reminds us, perceive the external world using a kind of sonar. By emitting high-pitched squeals and detecting the reflections, they are able to create an accurate enough image of their environment to enable them to ensnare small flying insects, while they themselves are airborne. Nagel points out that we might observe and describe in detail the neurophysiology which makes all this possible, but that it is unlikely that any amount of such observation would ever give us an insight into the bat's subjective experience of the world – into what it is like to be a bat. As Nagel himself puts it:

> For if the facts of experience – facts about what it is like for the experiencing organism – are available only from one point of view, then it is a mystery how the true character of experience could be revealed in the physical operation of that organism.[12]

Nagel thus asserts that the construction of subjective experiences from the observation of brain states is in *principle* impossible.[13]

For our present purposes it is not necessary to enter into the argument about whether mind states are reducible to brain states. Suffice it to say that there is sufficient confusion about how brain structure and function relate to mental states to rule out any simple assertion that animal nervous systems which resemble our own will give rise to mental states like ours.

It seems then, that pain, a mental state, can be neither perceived nor inferred by directing the senses on to behaviours or on to the brain itself. But what of the third argument for animal pain – that based on evolutionary theory?

III

Evolutionary theory provides the most convincing case for animal pain. Because evolution stresses continuities in the biological sphere, it breaks down the distinction between human and animal. Thus any special claims made on behalf of the human race – that they alone experience pain, for example – require justification. Before examining how, in evolutionary terms, we might justify treating *Homo sapiens* as a unique case, we ought to consider first how animal pain might conceivably fit into the evolutionary scheme of things.

Natural selection 'designs' animals to survive and reproduce. An important sort of adaptation for organisms to acquire would be the ability to avoid aspects of the environment which would reduce their chances of survival and reproduction. Pain, we might suppose, plays this adaptive rôle by compelling organisms to avoid situations in their world which might harm them. This view of the matter receives some measure of support from cases of individuals born with a congenital insensitivity to pain. Such unfortunate people frequently injure themselves quite severely in their early childhood, and must be taught how to avoid inflicting damage upon themselves. That such a condition can lead eventually to permanent disability or death would suggest that pain has considerable adaptive value for human beings at least.[14] Animals which were similarly insensitive to damaging stimuli, we might reasonably infer, would have little chance of survival. Yet there are difficulties with this interpretation.

Strictly, it is not pain (real or imputed) which is the adaptation, but the *behaviour* which is elicited when the damaging stimulus is applied. Those who are insensitive to pain are not disadvantaged by the absence of unpleasant mental states, but by a lack of those behavioural responses which in others are prompted by pain. We tend to lose sight of the primacy of behaviour because we get caught up in the connotations of 'expression'. That is to say, we consider some animal behaviours to be expressions of a particular mental state. Even Darwin, who should have known better, was guilty of this infelicity when he spoke of the 'expression of the emotions in man and animals'. Such locutions are misleading because they suggest that certain aspects of animal behaviour are arbitrary outward signs which signify some conscious state. But the simplest application of the theory of natural selection would only allow that such behaviours as violent struggling, grimacing and crying out, serve some more direct purpose in enhancing an animal's chances of survival and reproduction. (Darwin admittedly stressed the communicative aspects of these signs.) To exploit another example which I have drawn upon in another context, a wildebeest which is being torn apart by dogs will die in silence, while a chimpanzee will screech out in response to some trivial hurt like a thorn puncturing its foot.[15] It seems that the chimp gives expression to its pain, whereas the wildebeest does not. Yet neither expresses its pain. Rather, each behaves in a way likely to enhance the survival of the species. The chimpanzee communicates either to warn its conspecifics, or to summon aid. The wildebeest remains silent so that others will not be lured to their deaths. It is the behaviour, rather than some hypothetical mental state, which adapts the organism.

Another linguistic usage which holds us in thrall is the language of 'detection'. We assume that 'detection' entails 'conscious awareness of'. This leads us to believe that an animal cannot respond to a stimulus unless in some sense it consciously 'knows' what it has encountered. The reason such insectivorous plants as the venus fly trap capture our imagination is that they behave as if they are aware. How, we ponder, do they 'know' that the fly is there? Again we need to remind ourselves that the simplest of organisms are able to detect and respond to stimuli, yet we are not thereby committed to the view that they have knowledge or beliefs. The same is true of more neurologically complex organisms. There is an important truth in that litany of behaviourists: animals acquire behaviours, not beliefs.

If it is granted that the behaviour rather than some postulated mental state is what adapts an organism, we are next led to inquire whether organisms might exhibit 'pain behaviours' without that attendant mental state which we call 'pain'. As we noted at the outset, many invertebrates to which we do not generally attribute feelings of pain exhibit 'pain behaviour'. In higher animals too, as we have already seen, it is possible that relevant behaviours might be performed in the absence of any conscious experience. But is it probable? Must pain be introduced to cause the behaviours, or might these be caused more directly by the stimulus, or perhaps by indifferent conscious states? We might at this point simply opt for the most parsimonious explanation. This is in fact the upshot of Lloyd Morgan's famous dictum: 'In no case may we interpret an action as the outcome of the exercise of a higher psychical faculty, if it can be interpreted as the outcome of the exercise of one which stands lower in the psychological scale.'[16] We must ask, in other words, if we can explain all animals' reactions to noxious stimuli without recourse to particular mental states. Our blindsight examples show that it is possible for organisms to respond appropriately to stimuli in the complete absence of mental states. If the general case is true, then the same might be said for the specific performance of 'pain behaviours' in the absence of pain.

The thrust of Morgan's canon can be reinforced epistemologically with the arguments of Descartes. As we know, Descartes' radical doubt led him to propose that all we can know for certain are the truths of logic and the existence of our own mental states.[17] Fortunately one of the truths of logic was the existence of a God who could guarantee, to some extent, the veracity of perceptions of the world. Yet strict application of the criterion of doubt permits us to ascribe minds to other creatures only if they demonstrate (verbally, by signs, or by rational behaviour) evidence of mental activity. From the lack of such indications from animals, Descartes concluded that we have no evidence which would enable us legitimately to infer that animals have minds.[18] Not having minds, they cannot feel pain. Descartes thus provides *epistemological* grounds for denying that animals feel pain.[19]

If we adopt the conservative stance of Morgan or Descartes, then it seems that we have no grounds, *scientific* or *philosophical*, for asserting that animals feel pain. Yet this is a much weaker claim than the positive assertion that we have good reasons for believing that animals do not feel pain, or, to put it another way, that only human beings feel pain.[20] Certainly a reasonable case could be advanced that given our admitted ignorance, we have *moral* grounds for giving animals the benefit of the doubt. We shall return to this point later. For the moment, let us consider the positive statement of the case. Do we have reasons for believing that only human beings feel pain? Or, recasting the question in evolutionary terms, why should pain have adaptive value for the human species, if it would serve no purpose in other species?

IV

Pain is a mental state, and mental states require minds. Our inquiry, then, is in part an investigation of the selective advantage conferred by the possession of a mind. A mind's reflection on its own activities, amongst other things, enables us to predict the behaviour of other human beings, and to a lesser extent, animals. By reflecting upon our reasons for behaving in certain ways, and by assuming that our fellow human beings are similarly motivated, we can make predictions about how they are likely to behave in certain situations. But more than this, by ascribing consciousness and intelligence to other organisms we can also make predictions about how they will behave. Such ascriptions, whether they have any basis in fact or not, can thus help the human species survive. As H. S. Jennings remarked almost ninety years ago, if an amoeba 'were as large as a whale, it is quite conceivable that occasions might arise when the attribution to it of the elemental states of consciousness might save the unsophisticated human from destruction that would result from lack of such attribution.'[21] Along with human self-awareness then, came a tendency to attribute a similar awareness to other creatures. That animals might have beliefs, mental images, intentions and pains like our own could be nothing more than a useful fiction which gives us a shorthand method of predicting their behaviour.

There is, then, some value in the belief that animals suffer pain, for it provides a reasonably reliable guide to how they will behave. But it is not an infallible guide. If, for example, we were to pit ourselves against a chess-playing computer, the best strategy to adopt would be to act as if the machine were a skilled human opponent, possessed of certain intentional states – a desire to win, particular beliefs about the rules, and so on. However, there might be occasions when it would be better to adopt another attitude towards the computer. Let us imagine that the computer was programmed to play at three levels — beginner, intermediate, and advanced. Set at the 'beginner' level, the computer might show itself to be vulnerable to a basic 'fool's mate', so that whenever this simple gambit was used, it inevitably lost. A human opponent could thus be confident of beating the computer whenever he or she wished. Now this exploitation of the computer's weakness would result from the adoption of quite a different stance. No longer would the computer be treated as if it had desires and beliefs (or more importantly as if it had the ability to acquire new beliefs), for a human opponent in the same situation would quickly learn to counter the 'fool's mate'. Instead, predictions of the computer's behaviour would be based on the way it had been designed to operate. Thus, our wildebeest, on an intentional account, should exhibit 'pain behaviour'. Only when we adopt a 'design stance' (the animal was 'designed' by natural selection to behave in ways which would enhance the survival of the species) do we get a reasonable explanation of why it dies in silence.[22] The general point is this. The ascription to animals of certain mental states usually enables us to predict their behaviour with some accuracy (such ascription increasing our own chances of survival). But there will always be instances where this intentional model will break down and explanations which refer to selective advantages will be preferred.

Another reason for attributing pain experiences only to human beings is to do with free-will and moral responsibility. While there has been some dispute about whether animals ought to be the object of our moral concern, we do not usually consider animals to be moral agents. Animals are not generally held to be morally responsible for their own acts, and notwithstanding some rather odd medieval judicial practices, animals do not stand trial for antisocial acts which they might have committed.

What is absent in animals which is thought to be crucial to the committing of some wrong is the mens rea — the evil intent. Animals are not morally responsible for the acts they commit because while they may have behavioural dispositions, they do not have thoughts and beliefs about what is right and wrong, nor can they, whatever their behavioural disposition, form a conscious intent. Or at least, so we generally believe. Animals, in short, are not 'free agents', and this is why they are not regarded as being morally responsible. But what does the determined nature of animal behaviour have to do with pain? Simply this, that if animals' behaviours are causally determined, it makes no sense to speak of pain as an additional causal factor.

One way of seeing the force of this is to explore some of the contexts in which we use the term 'pain'. There are many ways we have of talking about pain which exclude animals. Consider the following: (1) 'For the long-distance runner, it is a matter of mind over matter. He must break through the pain barrier'. (2) 'The hunger striker finally succumbed and died'. (3) 'Even though she knew it would mean a horrible death at the stake, she refused to recant'. (4) 'The pain became unbearable. He cried out'. If we attempt to substitute animals for the human agents in these statements, the result becomes complete nonsense. Our inability to fit animals into the logic of these expressions is not merely because animals are not (contingently) long-distance runners, or hunger strikers, or religious martyrs. The key lies in statement (4). We must ask: Do animals ever find pain unbearable?, and, What reasons could they have for bearing it?

Consider this sentence in which a suitable substitution might be made. 'The man's hand reached into the flames, and was immediately withdrawn with a cry'. We could easily substitute 'ape' for 'man' here and the statement will retain its sense. But what about this: 'The man plunged his hand into the flames again, knowing that only he could reach the valve and stem the flow of petrol which threatened to turn the sleepy village into an inferno.' Now the substitution becomes impossible, for what could conceivably cause the ape to plunge its hand back into the flames? Nothing, I suspect, for apes do not have reasons for bearing pain.

Now it may seem unsatisfactory to proceed on the basis of certain linguistic practices to make some claim about how things really are. (This, I suspect, is why Anselm's ontological argument always leaves one feeling a little uneasy.) But the exclusive nature of the grammar of 'pain', or more correctly of 'bearing pain', reveals the unique province of pain. Pain operates as one kind of reason which free agents are bound to take into consideration when they decide on a particular course of action. Pain can be borne if there are *reasons*. But an animal never has reasons either to bear pain, or to succumb to pain. And if pain never need be brought into the sphere of reasons – the mind – then there is no need for it, *qua* unpleasant mental event, at all. Thus, while it is undeniable that animals sense noxious stimuli and react to them, these stimuli only need be represented as unpleasant mental states if they are to become the body's reasons in the context of other reasons. Only as various degrees of unpleasantness can they be taken seriously amongst reasons, and this is only necessary in the mind of a rational agent.

Another way of thinking about this is to consider the attributes of the long-distance runner, the hunger striker, the martyr, the hero of the sleepy village. We could say that they had mental strength, great courage, or moral character. But we would never predicate these of animals. The wildebeest dies silently and does not endanger the herd. But does it die courageously? Does it bear the pain to the end? Does it have a reason for remaining silent? No, because it does not have a choice. All wildebeest behave in this fashion. And if it does not have a choice,

there is no requirement for the dismemberment of its body to be represented mentally as pain.

Pain is the body's representative in the mind's decision-making process. Without pain, the mind would imperil the body (as cases of insensitivity to pain clearly show). But without the rational, decision-making mind, pain is superfluous. Animals have no rational or moral considerations which might overrule the needs of the body. It is for this reason that Descartes referred to pain, hunger and thirst as 'confused modes of *thought*', which can only be predicated of creatures which can think.[23]

V

We may now return to the original issue which prompted this examination of the reasons for ascribing pains to animals – the moral question of how we should treat animals. The arguments set out above do not constitute a conclusive disproof of animal pain. Indeed if the mind-body problem is as intractable as I have suggested, then the best we can manage is to arrive at some degree of probability. This much should be clear, however: First, there are reasons for claiming that only human beings feel pain; second, our treatment of animals cannot be based on dubious speculations about their mental lives. It follows, at the very least, that Bentham's question cannot provide a sound basis for an ethic which is to extend to animals. How then do we proceed from here?

It will seem to some that while there remains even a small possibility that animals (or certain kinds of animals) feel pain, these creatures ought to be given the benefit of the doubt. This is true to a point. Animal liberationists and animal rights activists have performed a valuable service in exposing many frivolous and mischievous practices which resulted in the unnecessary mutilation and deaths of animals. Such practices should cease, and many have. On the other hand, there are many animal experiments which improve, or might lead to the improvement of, the human lot. Even if a utilitarian equation which balances net pleasures over net pains can provide a rational basis for making moral choices in these matters (and this is doubtful), the balance should be tipped in favour of human beings, given our uncertainty about animal pain. Further, it virtually goes without saying that if it is doubtful that animals experience physical pain, even more groundless are claims that animals have other kinds of mental states – anxiety, the desire for freedom, and so on. Concerns for the psychological well-being of battery hens, veal calves, penned dolphins, and the like, would seem to be fundamentally misplaced. Our moral sensibilities have gone sadly awry when we expend effort on determining 'what animals prefer' before inquiring into whether 'preference' can be sensibly applied to animals. This is especially so when we are in little doubt as to what human beings prefer, and yet so many of them exist in conditions little different from those of battery hens.

None of this means, however, that there are no strictures on how we ought to behave towards animals. Other considerations – aesthetic, ecological, sentimental, psychological, and pedagogical – can give us a more solid foundation for an 'animal ethic'. Briefly, it would be morally wrong to attack Michelangelo's 'Pieta' with a hammer, despite the fact that this beautifully crafted piece of marble cannot feel pain. If animals are mere machines, they are, for all that, intricate and beautiful machines (most of them), which like old buildings, trees and works of art, can greatly enrich our lives. Accordingly, rational arguments can be mounted against acts which would damage or destroy them.

There is also a growing awareness in the Western world that human beings and animals form part of a global biological community. While at times this awareness expresses itself in rather silly ways, it is still true that if we carelessly alter the balance of that community by the slaughter of certain animals for pleasure or short-term economic gain, we place at risk the quality of life of ourselves and that of future generations.

At a more personal level, many people form strong emotional attachments to animals. Domestic animals traditionally have served as playmates for children and as company for the elderly. If mistreating these animals causes human beings to suffer, then such mistreatment is clearly wrong. Moreover, as the notorious Milgram experiments have shown, the belief that one is causing pain to another, even if false, can do great psychological harm.[24] When we believe we are being cruel to animals we do ourselves damage, even though our belief might be mistaken.

Finally, there is surely some value in the observation of Thomas Aquinas that kindness to animals might help to teach kindness to human beings.[25] Considerations of these kinds, though they require further development, can provide a far more certain guide to how we should treat animals.

Notes

1   (Oxford: Clarendon Press, 1907), 310f. n. 1 (XVII, 1, iv).

2   Peter Singer, *Animal Liberation* (London: Cape, 1976), 16.

3   Thus Dennis and Melzack: 'The appropriate behavioural response to overt damage may be inactivity; pain arising from trauma should presumably promote such behaviour. However, the appropriate behavioural response to threat may be vigorous activity; pain arising from threat should therefore promote this sort of activity. Thus the overt expression of pain sensation may actually be a combination of inherently contradictory processes and behavioural tendencies.' S. Dennis and R. Melzack, 'Perspectives on Phylogenetic Evolution of Pain Expression', *Animal Pain: Perception and Alleviation*, R. L.

Kitchell and H. H. Erickson (eds), (Bethesda: American Physiological Society, 1983), p. 155.

4 See, e.g., Ronald Melzack, *The Puzzle of Pain* (Ringwood: Penguin, 1973) pp. 93–103, 162f.

5 Thus Theodore Barber reports of individuals chronically insensitive to pain that for most, if not all, 'no distinct localized damage exists in the central nervous system'. 'Toward a Theory of Pain', *Psychological Bulletin* 56 (1959), 443. It is true that Barber cites no evidence from autopsies, and that more sophisticated scanning apparatus has been developed since this publication, but the fact that this insensitivity to pain can be reversed without surgical intervention would support Barber's observation.

6 See David Paterson's article of the same name in *World Medicine* 3 May 1980, 21–24. Also see Norton Nelkin, 'Pains and Pain Sensations', *The Journal of Philosophy* 83 (1986), pp. 129–48.

7 On 'blindsight' see Larry Weisenkrantz, 'Varieties of Residual Experience', *Quarterly Journal of Experimental Psychology* 32 (1980), pp. 365–386; Thomas Natsoulas, 'Conscious Perception and the Paradox of "Blindsight" ', in *Aspects of Consciousness*, III, Geoffrey Underwood (ed.), (London: Academic Press, 1982), pp. 79–109.

8 Larry Weisenkrantz, 'Trying to Bridge some Neurophysiological Gaps between Monkey and Man', *British Journal of Psychology* 68 (1977), pp. 431–35.

9 On the possibility of 'non-conscious experience', see Peter Carruthers, 'Brute Experience', *The Journal of Philosophy* 86 (1989), pp. 258–69.

10 This is also suggested by Carruthers, ibid., pp. 266–269.

11 *The Philosophical Review* 83 (1974), pp. 435–50.

12 Ibid., p. 442.

13 Colin McGinn has made a similar point from a different perspective. He argues that the mystery of our mental life arises out of the fact that we simply do not possess the cognitive faculties necessary to solve the mind-body problem. 'Cognitive closure' prevents our ever having access to that vital natural link which presumably exists

between brain states and conscious states. See Colin McGinn, 'Can We Solve the Mind-Body Problem?', *Mind* 98 (1989), pp. 349–66.

14 On congenital insensitivity to pain see Melzack, *The Puzzle of Pain*, 15f.

15 David McFarland, 'Pain', *The Oxford Companion to Animal Behaviour*, David McFarland (ed.), (Oxford University Press, 1981), p. 439.

16 Quoted in Robert Boakes, *From Darwin to Behaviourism* (Cambridge University Press, 1984), 40. This dictum is actually a version of the Aristotelian principle, 'Nature does nothing in vain', couched in evolutionary terms.

17 *Meditations* II.

18 Descartes' clearest explanation of the matter comes in a letter to the English Platonist, Henry More. See Descartes, *Philosophical Letters*, Anthony Kenny (ed.), (Oxford: Clarendon Press, 1970), pp. 243–45.

19 It may seem that Morgan and Descartes are making the same point, but they are not. Morgan's canon was virtually a biological application of the second law of thermodynamics, asserting that a complex biological system would not evolve if a simpler one could perform the same function. Of course, in applying this canon to 'psychical' functions, Morgan seems to have committed himself to the view that more complex mental states require a more complex physical apparatus.

20 Thus Descartes admitted in his letter to More that his thesis about animals was only probable. *Philosophical Letters*, p. 244.

21 Quoted in Larry Weisenkrantz, 'Neurophysiology and the Nature of Consciousness', *Mindwaves*, C. Blakemore and S. Greenfield (eds), (Oxford: Blackwell, 1987), p. 309.

22 The terms 'intentional stance' and 'design stance' are D. C. Dennett's. See his *Brainstorms* (Hassocks: Harvester Press, 1978), pp. 3–22.

23 *Meditation* IV (HR I, 192) my emphasis. Cf. Norton Nelkin, who states that pain is an attitude not a sensation. 'Pains and Pain Sensations', p. 148.

24 See Stanley Milgram, *Obedience to Authority: An Experimental View* (London: Tavistock, 1974).

25 *Summa theologiae*, 1a, 2ae. p. 102, 6.

Chapter 17

Carl Cohen

THE CASE FOR THE USE OF ANIMALS IN BIOMEDICAL RESEARCH

Using animals as research subjects in medical investigations is widely condemned on two grounds: first, because it wrongly violates the *rights* of animals,[1] and second, because it wrongly imposes on sentient creatures much avoidable *suffering.*[2] Neither of these arguments is sound. The first relies on a mistaken understanding of rights; the second relies on a mistaken calculation of consequences. Both deserve definitive dismissal.

Why animals have no rights

A right, properly understood, is a claim, or potential claim, that one party may exercise against another. The target against whom such a claim may be registered can be a single person, a group, a community, or (perhaps) all humankind. The content of rights claims also varies greatly: repayment of loans, nondiscrimination by employers, noninterference by the state, and so on. To comprehend any genuine right fully, therefore, we must know *who* holds the right, *against whom* it is held, and to *what* it is a right.

Alternative sources of rights add complexity. Some rights are grounded in constitution and law (e.g., the right of an accused to trial by jury); some rights are moral but give no legal claims (e.g., my right to your keeping the promise you gave me); and some rights (e.g., against theft or assault) are rooted both in morals and in law.

The differing targets, contents, and sources of rights, and their inevitable conflict, together weave a tangled web. Notwithstanding all such complications, this much is clear about rights in general: they are in every case claims, or potential claims, within a community of moral agents. Rights arise, and can be intelligibly defended, only among beings who actually do, or can, make moral claims against one another. Whatever else rights may be, therefore, they are necessarily human; their possessors are persons, human beings.

The attributes of human beings from which this moral capability arises have been described variously by philosophers, both ancient and modern: the inner consciousness of a free will (Saint Augustine[3]); the grasp, by human reason, of the binding character of moral law (Saint Thomas[4]); the self-conscious participation of human beings in an objective ethical order (Hegel[5]); human membership in an organic moral community (Bradley[6]); the development of the human self through the consciousness of other moral selves (Mead[7]); and the underivative, intuitive cognition of the rightness of an action (Prichard[8]). Most influential has been Immanuel Kant's emphasis on the universal human possession of a uniquely moral will and the autonomy its use entails.[9] Humans confront choices that are purely moral; humans – but certainly not dogs or mice – lay down moral laws, for others and for themselves. Human beings are self-legislative, morally *auto-nomous*.

Animals (that is, nonhuman animals, the ordinary sense of that word) lack this capacity for free moral judgment. They are not beings of a kind capable of exercising or responding to moral claims. *Animals therefore have no rights, and they can have none.*

This is the core of the argument about the alleged rights of animals. The holders of rights must have the capacity to comprehend rules of duty, governing all including themselves. In applying such rules, the holders of rights must recognize possible conflicts between what is in their own interest and what is just. Only in a community of beings capable of self-restricting moral judgments can the concept of a right be correctly invoked.

Humans have such moral capacities. They are in this sense self-legislative, are members of communities governed by moral rules, and do possess rights. Animals do not have such moral capacities: They are not morally self-legislative, cannot possibly be members of a truly moral community, and therefore cannot possess rights. In conducting research on animal subjects, therefore, we do not violate their rights, because they have none to violate.

To animate life, even in its simplest forms, we give a certain natural reverence. But the possession of rights presupposes a moral status not attained by the vast majority of living things. We must not infer, therefore, that a live being has, simply in being alive, a "right" to its life. The assertion that all animals, only because they are alive and have interests, also possess the "right to life"[10] is an abuse of that phrase, and wholly without warrant.

It does not follow from this, however, that we are morally free to do anything we please to animals. Certainly not. In our dealings with animals, as in our dealings with other human beings, we have obligations that do not arise from claims against us based on rights. Rights entail obligations, but many of the things one ought to do are in no way tied to another's entitlement. Rights and obligations are not reciprocals of one another, and it is a serious mistake to suppose that they are.

Illustrations are helpful. Obligations may arise from internal commitments made: physicians have obligations to their patients not grounded merely in their patients' rights. Teachers have such obligations to their students, shepherds to their dogs, and cowboys to their horses. Obligations may arise from differences of status: adults owe special care when playing with young children, and children owe special care when playing with young pets. Obligations may arise from special relationships: the payment of my son's college tuition is something to which he may have no right, although it may be my obligation to bear the burden if I reasonably can; my dog has no right to daily exercise and veterinary care, but I do have the obligation to provide these things for her. Obligations may arise from particular acts or circumstances: one may be obliged to another for a special kindness done, or obliged to put an animal out of its misery in view of its condition – although neither the human benefactor nor the dying animal may have had a claim of right.

Plainly, the grounds of our obligations to humans and to animals are manifold and cannot be formulated simply. Some hold that there is a general obligation to do no gratuitous harm to sentient creatures (the principle of nonmaleficence); some hold that there is a general obligation to do good to sentient creatures when that is reasonably within one's power (the principle of beneficence). In our dealings with animals, few will deny that we are at least obliged to act humanely – that is, to treat them with the decency and concern that we owe, as sensitive human beings, to other sentient creatures. To treat animals humanely, however, is not to treat them as humans or as the holders of rights.

A common objection, which deserves a response, may be paraphrased as follows:

> If having rights requires being able to make moral claims, to grasp and apply moral laws, then many humans – the brain-damaged, the comatose, the senile – who plainly lack those capacities must be without rights. But that is absurd. This proves [the critic concludes] that rights do not depend on the presence of moral capacities.[1,10]

This objection fails; it mistakenly treats an essential feature of humanity as though it were a screen for sorting humans. The capacity for moral judgment that distinguishes humans from animals is not a test to be administered to human beings one by one. Persons who are unable, because of some disability, to perform the full moral functions natural to human beings are certainly not for that reason ejected from the moral community. The issue is one of kind. Humans are of such a kind that they may be the subject of experiments only with their voluntary consent. The choices they make freely must be respected. Animals are of such a kind that it is impossible for them, in principle, to give or withhold voluntary consent or to make a moral choice. What humans retain when disabled, animals have never had.

A second objection, also often made, may be paraphrased as follows:

> Capacities will not succeed in distinguishing humans from the other animals. Animals also reason; animals also communicate with one another, animals also care passionately for their young; animals also exhibit desires and preferences.[11,12] Features of moral relevance – rationality, interdependence, and

love – are not exhibited uniquely by human beings. Therefore [this critic concludes], there can be no solid moral distinction between humans and other animals.[10]

This criticism misses the central point. It is not the ability to communicate or to reason, or dependence on one another, or care for the young, or the exhibition of preference, or any such behavior that marks the critical divide. Analogies between human families and those of monkeys, or between human communities and those of wolves, and the like, are entirely beside the point. Patterns of conduct are not at issue. Animals do indeed exhibit remarkable behavior at times. Conditioning, fear, instinct, and intelligence all contribute to species survival. Membership in a community of moral agents nevertheless remains impossible for them. Actors subject to moral judgment must be capable of grasping the generality of an ethical premise in a practical syllogism. Humans act immorally often enough, but only they – never wolves or monkeys – can discern, by applying some moral rule to the facts of a case, that a given act ought or ought not to be performed. The moral restraints imposed by humans on themselves are thus highly abstract and are often in conflict with the self-interest of the agent. Communal behavior among animals, even when most intelligent and most endearing, does not approach autonomous morality in this fundamental sense. Genuinely moral acts have an internal as well as an external dimension. Thus, in law, an act can be criminal only when the guilty deed, the *actus reus*, is done with a guilty mind, *mens rea*. No animal can ever commit a crime; bringing animals to criminal trial is the mark of primitive ignorance. The claims of moral right are similarly inapplicable to them. Does a lion have a right to eat a baby zebra? Does a baby zebra have a right not to be eaten? Such questions, mistakenly invoking the concept of right where it does not belong, do not make good sense. Those who condemn biomedical research because it violates "animal rights" commit the same blunder.

In defense of "speciesism"

Abandoning reliance on animal rights, some critics resort instead to animal sentience – their feelings of pain and distress. We ought to desist from the imposition of pain insofar as we can. Since all or nearly all experimentation on animals does impose pain and could be readily forgone, say these critics, it should be stopped. The ends sought may be worthy, but those ends do not justify imposing agonies on humans, and by animals the agonies are felt no less. The laboratory use of animals (these critics conclude) must therefore be ended – or at least very sharply curtailed.

Argument of this variety is essentially utilitarian, often expressly so[13]; it is based on the calculation of the net product, in pains and pleasures, resulting from experiments on animals. Jeremy Bentham, comparing horses and dogs with other sentient creatures, is thus commonly quoted: "The question is not, Can they reason? nor Can they talk? but, Can they suffer?"[14]

Animals certainly can suffer and surely ought not to be made to suffer needlessly. But in inferring, from these uncontroversial premises, that biomedical research causing animal distress is largely (or wholly) wrong, the critic commits two serious errors.

The first error is the assumption, often explicitly defended, that all sentient animals have equal moral standing. Between a dog and a human being, according to this view, there is no moral difference; hence the pains suffered by dogs must be weighed no differently from the pains suffered by humans. To deny such equality, according to this critic, is to give unjust preference to one species over another; it is "speciesism." The most influential statement of this moral equality of species was made by Peter Singer:

> The racist violates the principle of equality by giving greater weight to the interests of members of his own race when there is a clash between their interests and the interests of those of another race. The sexist violates the principle of equality by favoring the interests of his own sex. Similarly the speciesist allows the interests of his own species to override the greater interests of members of other species. The pattern is identical in each case.[2]

This argument is worse than unsound; it is atrocious. It draws an offensive moral conclusion from a deliberately devised verbal parallelism that is utterly specious. Racism has no rational ground whatever. Differing degrees of respect or concern for humans for no other reason than that they are members of different races is an injustice totally without foundation in the nature of the races themselves. Racists, even if acting on the basis of mistaken factual beliefs, do grave moral wrong precisely because there is no morally relevant distinction among the races. The supposition of such differences has led to outright horror. The same is true of the sexes, neither sex being entitled by right to greater respect or concern than the other. No dispute here.

Between species of animate life, however – between (for example) humans on the one hand and cats or rats on the other – the morally relevant differences are enormous, and almost universally appreciated. Humans engage in moral reflection; humans are morally autonomous; humans are members of moral communities, recognizing just claims against their own interest. Human beings do have rights; theirs is a moral status very different from that of cats or rats.

I am a speciesist. Speciesism is not merely plausible; it is essential for right conduct, because those who will not make the morally relevant distinctions among species are almost certain, in consequence, to misapprehend their true obligations. The analogy between speciesism and racism is insidious. Every sensitive moral judgment requires that the differing natures of the beings to whom obligations are owed be considered. If all forms of animate life – or vertebrate animal life? – must be treated equally, and if therefore in evaluating a research program the pains of a rodent count equally with the pains of a human, we are forced to conclude (1) that neither humans nor rodents possess rights, or (2) that rodents possess all the rights that humans possess. Both alternatives are absurd. Yet one or the other must be swallowed if the moral equality of all species is to be defended.

Humans owe to other humans a degree of moral regard that cannot be owed to animals. Some humans take on the obligation to support and heal others, both humans and animals, as a principal duty in their lives; the fulfillment of that duty may require the sacrifice of many animals. If biomedical investigators abandon the effective pursuit of their professional objectives because they are convinced that they may not do to animals what the service of humans requires, they will fail, objectively, to do their duty. Refusing to recognize the moral differences among species is a sure path to calamity. (The largest animal rights group in the country is People for the Ethical Treatment of Animals; its codirector, Ingrid Newkirk, calls research using animal subjects "fascism" and "supremacism." "Animal liberationists do not separate out the human animal," she says, "so there is no rational basis for saying that a human being has special rights. A rat is a pig is a dog is a boy: They're all mammals."[15])

Those who claim to base their objection to the use of animals in biomedical research on their reckoning of the net pleasures and pains produced make a second error, equally grave. Even if it were true – as it is surely not – that the pains of all animate beings must be counted equally, a cogent utilitarian calculation requires that we weigh all the consequences of the use, and of the nonuse, of animals in laboratory research. Critics relying (however mistakenly) on animal rights may claim to ignore the beneficial results of such research, rights being trump cards to which interest and advantage must give way. But an argument that is explicitly framed in terms of interest and benefit for all over the long run must attend also to the disadvantageous consequences of not using animals in research, and to all the achievements attained and attainable only through their use. The sum of the benefits of their use is utterly beyond quantification. The elimination of horrible disease, the increase of longevity, the avoidance of great pain, the saving of lives, and the improvement of the quality of lives (for humans and for animals) achieved through research using animals is so incalculably great that the argument of these critics, systematically pursued, establishes not their conclusion but its reverse: to refrain from using animals in biomedical research is, on utilitarian grounds, morally wrong.

When balancing the pleasures and pains resulting from the use of animals in research, we must not fail to place on the scales the terrible pains that would have resulted, would be suffered now, and would long continue had animals not been used. Every disease eliminated, every vaccine developed, every method of pain relief devised, every surgical procedure invented, every prosthetic device implanted – indeed, virtually every modern medical therapy is due, in part or in whole, to experimentation using animals. Nor may we ignore, in the balancing process, the predictable gains in human (and animal) well-being that are probably achievable in the future but that will not be achieved if the decision is made now to desist from such research or to curtail it.

Medical investigators are seldom insensitive to the distress their work may cause animal subjects. Opponents of research using animals are frequently insensitive to the cruelty of the results of the restrictions they would impose.[2] Untold numbers of human beings – real persons, although not now identifiable – would suffer grievously as the consequence of this well-meaning but shortsighted tenderness. If the morally relevant differences between humans and animals are borne in mind, and if all relevant considerations are weighed, the calculation of long-term consequences must give overwhelming support for biomedical research using animals.

Concluding remarks

Substitution

The humane treatment of animals requires that we desist from experimenting on them if we can accomplish the same result using alternative methods – in vitro experimentation, computer simulation, or others. Critics of some experiments using animals rightly make this point.

It would be a serious error to suppose, however, that alternative techniques could soon be used in most research now using live animal subjects. No other methods now on the horizon – or perhaps ever to be available – can fully replace the testing of a drug, a procedure, or a vaccine, in live organisms. The flood of new medical possibilities being opened by the successes of recombinant DNA technology will turn to a trickle if testing on live animals is forbidden. When initial trials entail great risks, there may be no forward movement whatever without the use of live animal subjects. In seeking knowledge that may prove ethical in later clinical applications, the unavailability of

animals for inquiry may spell complete stymie. In the United States, federal regulations require the testing of new drugs and other products on animals, for efficacy and safety, before human beings are exposed to them.[16,17] We would not want it otherwise.

Every advance in medicine – every new drug, new operation, new therapy of any kind – must sooner or later be tried on a living being for the first time. That trial, controlled or uncontrolled, will be an experiment. The subject of that experiment, if it is not an animal, will be a human being. Prohibiting the use of live animals in biomedical research, therefore, or sharply restricting it, must result either in the blockage of much valuable research or in the replacement of animal subjects with human subjects. These are the consequences – unacceptable to most reasonable persons – of not using animals in research.

Reduction

Should we not at least reduce the use of animals in biomedical research? No, we should increase it, to avoid when feasible the use of humans as experimental subjects. Medical investigations putting human subjects at some risk are numerous and greatly varied. The risks run in such experiments are usually unavoidable, and (thanks to earlier experiments on animals) most such risks are minimal or moderate. But some experimental risks are substantial.

When an experimental protocol that entails substantial risk to humans comes before an institutional review board, what response is appropriate? The investigation, we may suppose, is promising and deserves support, so long as its human subjects are protected against unnecessary dangers. May not the investigators be fairly asked, Have you done all that you can to eliminate risk to humans by the extensive testing of that drug, that procedure, or that device on animals? To achieve maximal safety for humans we are right to require thorough experimentation on animal subjects before humans are involved.

Opportunities to increase human safety in this way are commonly missed; trials in which risks may be shifted from humans to animals are often not devised, sometimes not even considered. Why? For the investigator, the use of animals as subjects is often more expensive, in money and time, than the use of human subjects. Access to suitable human subjects is often quick and convenient, whereas access to appropriate animal subjects may be awkward, costly, and burdened with red tape. Physician-investigators have often had more experience working with human beings and know precisely where the needed pool of subjects is to be found and how they may be enlisted. Animals, and the procedures for their use, are often less familiar to these investigators. Moreover, the use of animals in place of humans is now more likely to be the target of zealous protests from without. The upshot is that humans are sometimes subjected to risks that animals could have borne, and should have borne, in their place. To maximize the protection of human subjects, I conclude, the wide and imaginative use of live animal subjects should be encouraged rather than discouraged. This enlargement in the use of animals is our obligation.

Consistency

Finally, inconsistency between the profession and the practice of many who oppose research using animals deserves comment. This frankly ad hominem observation aims chiefly to show that a coherent position rejecting the use of animals in medical research imposes costs so high as to be intolerable even to the critics themselves.

One cannot coherently object to the killing of animals in biomedical investigations while continuing to eat them. Anesthetics and thoughtful animal husbandry render the level of actual animal distress in the laboratory generally lower than that in the abattoir. So long as death and discomfort do not substantially differ in the two contexts, the consistent objector must not only refrain from all eating of animals but also protest as vehemently against others eating them as against others experimenting on them. No less vigorously must the critic object to the wearing of animal hides in coats and shoes, to employment in any industrial enterprise that uses animal parts, and to any commercial development that will cause death or distress to animals.

Killing animals to meet human needs for food, clothing, and shelter is judged entirely reasonable by most persons. The ubiquity of these uses and the virtual universality of moral support for them confront the opponent of research using animals with an inescapable difficulty. How can the many common uses of animals be judged morally worthy, while their use in scientific investigation is judged unworthy?

The number of animals used in research is but the tiniest fraction of the total used to satisfy assorted human appetites. That these appetites, often base and satisfiable in other ways, morally justify the far larger consumption of animals, whereas the quest for improved human health and understanding cannot justify the far smaller, is wholly implausible. Aside from the numbers of animals involved, the distinction in terms of worthiness of use, drawn with regard to any single animal, is not defensible. A given sheep is surely not more justifiably used to put lamb chops on the supermarket counter than to serve in testing a new contraceptive or a new prosthetic device. The needless killing of animals is wrong; if the common killing of them for our food or convenience is right, the less common but more humane uses of animals in the service of medical science are certainly not less right.

Scrupulous vegetarianism, in matters of food, clothing, shelter, commerce, and recreation, and in all other spheres, is the only fully coherent position the critic may adopt. At great human cost, the lives of fish and crustaceans must also be protected, with equal vigor, if speciesism has been forsworn. A very few consistent critics adopt this position. It is the reductio ad absurdum of the rejection of moral distinctions between animals and human beings.

Opposition to the use of animals in research is based on arguments of two different kinds – those relying on the alleged rights of animals and those relying on the consequences for animals. I have argued that arguments of both kinds must fail. We surely do have obligations to animals, but they have, and can have, no rights against us on which research can infringe. In calculating the consequences of animal research, we must weigh all the long-term benefits of the results achieved – to animals and to humans – and in that calculation we must not assume the moral equality of all animate species.

Notes

1 Regan T. *The Case for Animal Rights.* Berkeley, CA: University of California Press, 1983.

2 Singer P. *Animal Liberation.* New York: Avon Books. 1977.

3 St. Augustine. *Confessions: Book Seven.* 397 A.D. New York: Pocketbooks. 1957: 104–26.

4 St. Thomas Aquinas. *Summa Theologica.* 1273 A.D. Philosophic texts. New York: Oxford University Press. 1960: 353–66.

5 Hegel G.W.F. *Philosophy of Right.* London: Oxford University Press, 1952: 105–10.

6 Bradley F.H. "Why Should I Be Moral?" 1876. In: Malden A.I., ed. *Ethical Theories.* New York: Prentice-Hall 1959: 345–59.

7 Mead G.H. "The Genesis of the Self and Social Control". 1925. In: Keck A.K., ed. *Selected Writings.* Indianapolis: Bobbs-Merrill, 1964: 264–93.

8 Prichard H.A. "Does Moral Philosophy Rest on a Mistake?" 1912. In: Cellars W., Hospers J., eds. *Readings in Ethical Theory.* New York: Appleton-Century-Crofts, 1952: 149–63.

9 Kant I. *Fundamental Principles of the Metaphysic of Morals.* 1785. New York: Liberal Arts Press, 1949.

10 Rollin B. *Animal Rights and Human Morality.* New York: Prometheus Books, 1981.

11 Hoff C. "Immoral and Moral Uses of Animals". *N Engl J Med* 1980; 302: 115–18.

12 Jamieson D. "Killing Persons and Other Beings." In: Miller H.B., Williams W.H., eds. *Ethics and Animals.* Clifton, N.J.: Humana Press, 1983: 135–46.

13 Singer P. "Ten Years of Animal Liberation." *New York Review of Books.* 1985: 31: 46–52.

14 Bentham J. *Introduction to the Principles of Morals and Legislation.* London: Athlone Press, 1970.

15 McCabe K. "Who Will Live, Who Will Die?" *Washingtonian Magazine.* August 1986: 115.

16 *U.S. Code of Federal Regulations,* Title 21, Sect. 505(i). Food, Drug, and Cosmetic Regulations.

17 *U.S. Code of Federal Regulations,* Title 16, Sect. 1500.40–2. Consumer Product Regulations.

Chapter 18

A.L. Caplan

IS XENOGRAFTING MORALLY WRONG?

It is tempting to think that a decision about whether or not it is immoral to use animals as sources for transplantable organs and tissues hinges only upon the question of whether or not it is ethical to kill them. But the ethics of xenografting involves more than an analysis of that question. And even the assessment of the morality of killing animals to obtain their parts to use in human beings is more complicated than it might at first glance appear to be.

To decide whether it is ethical to kill animals, a variety of subsidiary questions must be considered. Is it ethical to kill animals to obtain organs and tissues to save human lives or alleviate severe disability when it might not be ethical to kill them for food or sport?[1] Why are animals being considered as sources of organs and tissues – do alternative methods for obtaining replacement parts for human beings exist? What sorts of animals would have to be killed, how would they be killed, and how would they be stored, handled, and treated prior to their deaths?

If it is possible to defend the killing of animals for xenografting then questions as to the morality of subjecting human beings to the risks, both physical and psychological, associated with xenografting must also be weighed. In undertaking a xenograft on a human subject the focus of moral concern ought not to be solely on the animal that will be killed.

Even for those who eat meat or hunt, it might well seem immoral to kill animals for their parts if alternative sources of replacement parts were or might soon be available. The moral acceptability of xenografting will for many, including the prospective recipients of animal parts, be contingent on the presumption that there is no other plausible alternative source of transplantable organs and tissues. Unfortunately, the scarcity that is behind the current interest in xenografting is all too real.

Why pursue xenografting?

The supply of organs and tissues available from human cadaveric sources for transplantation in the United States and other nations is entirely inadequate. Many children and adults die or remain disabled due to the shortage of transplantable organs and tissues. Unless some solution is found to the problem of scarcity, the plight of those in need of organs and tissues will only grow worse and the numbers who die solely for want of an organ will continue to grow.

More than one-third of those now awaiting liver transplants die for want of a donor organ. Well over one-half of all children born with fatal, congenital deformities of the heart or liver die without a transplant due to the shortage of organs. The percentage of those who die while waiting would actually be higher if all potential candidates were on waiting lists.[2]

Some Americans are not referred for transplants because they cannot afford them. If those with organ failure from economically underdeveloped nations were wait-listed at North American and European transplant centers, the percentage of those who die while awaiting a transplant would be much larger.[3]

More than 150,000 Americans with kidney failure are kept alive by renal dialysis. The cost for this treatment exceeded five billion dollars in 1988. It would be far cheaper and, from the patient perspective as to the quality of life, far more desirable to treat kidney failure by means of transplants. But there are simply not enough cadaveric kidneys for all who desire and could benefit from a transplant.

The supply of cadaveric organs for pancreas, small intestine, lung, heart-lung transplants for those dying of a wide variety of diseases affecting these organs is not adequate. The same story holds for bone, ligament, dural matter, heart valves, and skin. Moreover, demand for the limited supply of organs and tissues is increasing as more and more medical centers become capable of offering this form of surgery and as techniques for managing rejection and infection improve.[4]

The shortage of organs and tissues for transplantation has led researchers to pursue a variety of alternatives in order to bridge the gap between supply and demand. Some suggest changing existing public policies governing cadaveric procurement. Others focus on locating alternatives to human cadaveric organs.

Efforts could be made to modify public policy to encourage more persons to serve as organ and tissue donors. Legislation mandating that the option of organ and tissue donation be presented whenever a person dies in a hospital setting has been enacted in the United States but, while leading to increases in both tissue

and organ availability, has not been adequately implemented.[5,6] Organs and tissues are still lost because families are not approached about donation. Hospital compliance is not what it should be and the training of those who must identify potential donors and approach families is woefully lacking. Refusal rates to requests to donate are high and there exists a significant degree of mistrust and misunderstanding about donation on the part of the public.[6] Public policies could and should be changed to rectify these problems.

Other proposals to change public policy involve changing laws to permit payment to those who agree to donate or whose families consent to cadaveric donation[7,8] or to move toward a presumed consent system in which the burden of proof is placed on those who do not wish to donate to carry cards or other evidence of their nondonor status. However, cultural and religious attitudes in large segments of American society and in other societies will not support either the creation of markets, bounties, or property status for body parts[9] or the extension of state authority to the seizure of cadaveric organs and tissues.

Allowing markets may lead to criminal activities.[10] Selling irreplaceable body parts is an especially repugnant way to ask a person to earn income or benefits.[11] There is much reluctance on the part of the general public to swap the presumption of individual control over the body, either in life or death, for a policy that might benefit the common good by risking the loss of personal autonomy.[4,11] Nor has the actual experience with presumed consent laws in European nations been such as to justify enthusiasm for the likely results of a shift away from individualism and personal autonomy with respect to the control of cadavers.[12]

Even if drastic changes were made in existing public policies, other factors are working against the prospects for large increases in the human cadaveric organ supply. Improvements in emergency room access and care, the onset of the acquired immunodeficiency syndrome (AIDS) epidemic, ambivalence about cadaveric donation on the part of the public, plus laudable advances in public health measures such as mandatory seatbelt use, raising the age for legally purchasing liquor, and tougher laws against drunk driving mean that a tremendous increase in the supply of cadaveric organs is unlikely to occur no matter what public policies are adopted.

Most importantly with respect to the moral defensibility of exploring xenografting as an alternative source of organs, even if all human cadaveric organs were somehow made available for transplant, the supply would still not meet the potential demand. The hidden pool of potential transplant recipients would quickly become visible were these organs and tissues to become available.[3,6] The search for alternatives to the use of human cadaveric organs rests on the recognition that scarcity is an insurmountable obstacle.

In recent years surgeons have tried to solve the problem of shortage by using kidneys, and segments of liver, lung, and pancreas from living donors. Transplant teams in a few nations have been testing the feasibility of transplanting lobes of livers between biologically related individuals.[13,14] Teams at Stanford University and the University of Minnesota have used parents as lung donors for their child. Minnesota and some other centers have been experimenting for many years with transplants of the kidney and pancreas from related and unrelated living donors.[15] Many centers around the world have performed bone marrow transplants between biologically unrelated persons.

There are serious problems with and limits to the use of living donors as alternative sources of organs and tissues. The most inviolate limit is that unmatched vital organs, such as the heart, cannot be used. Using living donors for other organs requires subjecting the donors to life-threatening risks, some pain and disfigurement, and some risk of disability. The legitimacy of consent on the part of living donors, especially among family members, is hard to assess. And, since it is not yet known whether transplanting lobes or segments of organs or unrelated bone marrow is efficacious, it is not certain that this strategy for getting more organs and tissue is a realistic option much less whether it will prove attractive to a sufficient number of actual donors.

Another alternative to human cadaveric organ transplantation is the development of mechanical or artificial organ and tissue substitutes. Kidney dialysis is one such substitute. The widely publicized efforts to create a total artificial heart, first at the University of Utah and then later at Humana Audubon Hospital in Louisville, Kentucky, represent another, albeit failed effort to create an alternative to transplantation using cadaveric hearts. New generations of mechanical hearts are in the research pipeline as are artificial insulin pumps, artificial lungs, and various types of artificial livers. But, safe, effective, and reliable artificial organs and tissues appear to still be decades away. Ironically, the immediate impact of the available forms of artificial organs is to increase the problem of allocation in the face of scarcity since these devices permit temporary "bridging" of children and adults in need of transplants, thereby increasing the pool of prospective recipients.

It is the plight of those dying of end-stage diseases for want of donor organs from both living and cadaveric human sources that has led a number of research groups to explore the option of using animals as the source of transplantable organs and tissues. Transplant researchers at Loma Linda, the University of Pittsburgh, Stanford, Columbia, and Minnesota as well as in England, China, Belgium, Sweden, Japan, and France, among other countries, are conducting research on xenografting organs and tissues. Some are exploring the feasibility of primate to human transplants. Others are pursuing lines of research that would allow them to utilize animals other than primates as sources.

It is scarcity that grounds the moral case for thinking about animals as sources of organs and tissues. In light of current and potential demand, no other options exist for alleviating the scarcity in the supply of replacement parts. If, however, society were to decide not to perform transplants or to perform only a limited number of them, then the case for xenografting would be considerably weakened. While transplant surgeons and those on waiting lists may find strategies for finding more organs or tissues attractive, another possible response to scarcity is to simply live with it. Those who advocate this response could do so on the grounds that it is not morally necessary to transplant all persons who are in need especially if doing so requires the systematic killing of animals for human purposes.[1,16,17]

Is the problem of scarcity resolvable by some other strategy?

Instead of pursuing the option of xenografting it is possible to argue that the answer to the problem of scarcity is that medicine should simply stop doing transplants entirely or only do as many as can be done with whatever human cadaveric organs and tissues are available. This moral stance might rest on the claim that transplants are simply too expensive and do not work well enough to justify a hunt for alternative sources to human organs. Or someone might view xenografting as unnecessary and immoral on the grounds that prevention makes far more sense than salvage and rescue in dealing with end-stage organ failure. Some forms of transplantation are notoriously expensive. A critic of the xenografting option might argue that it is

not wise to spend hundreds of thousands of dollars on heart, liver, or nonrelated bone marrow transplants when the number of people requiring these treatments could be drastically reduced by decreasing the incidence of smoking, alcohol consumption, and the exposure to toxic substances in the workplace and the environment.

The arguments in favor of the "live within your means" position are not persuasive. While prevention of organ and tissue failure is surely to be preferred to rescue by means of transplants, large numbers of persons suffer organ and tissue failure for reasons that are poorly understood and, thus, not amenable to prevention. While individuals should certainly be encouraged to adopt more healthful lifestyles (including the consumption of less animal fat!) there are no proven techniques available for ensuring that people will behave wisely or prudently. Moreover, those who will need transplants in the next few decades are persons for whom prevention is too late. So, while prevention is desirable, the demand for transplants will not diminish for a significant period of time regardless of the efforts undertaken to improve public health.

Transplants are expensive, but many types of transplantation, especially for children and young adults, are very effective in providing a good quality of life for many, many years. The wisdom of doing any medical procedure cannot simply be equated with its overall cost. A more reasonable measure would be to see what is purchased for the price that is charged. If the moral value of spending money for health services is not total price but cost per year of life saved or cost relative to the likelihood of saving a life, then there are many other areas of health care that ought to be restricted or abandoned long before accepted forms of transplantation such as heart, liver, and kidney are deemed too expensive or not cost worthy.

Ethical problems with the xenograft option

If it is true that the case for pursuing xenografting is a persuasive one then the question of whether xenografting is morally wrong shifts to an analysis of which animals will be used, how they will be kept and killed, and the risks and dangers involved for potential human recipients. Xenografting is still evolving so the ethical issues must be considered under two broad headings; issues associated with basic and clinical experimentation and, if this proves successful, those that would then be associated with the widespread use of xenografts as therapies.

Ethical issues raised by basic and clinical research

In thinking about the ethics of research on xenografting a couple of assumptions can safely be made. The number and type of animals used will be very much a function of cost, prior knowledge of the species, inbred characteristics, ease in handling, and availability. Gorillas are not going to be used in research simply because there are too few of them and they are unlikely to make compliant subjects. Rats and mice will dominate the early stages of research (as they already do) because they are relatively well understood, special-purpose bred, and cheap to acquire and maintain. Few primates will be used for basic or clinical research simply because they are too scarce, too expensive, and too complex to permit controlled study for most experimental purposes.

Even if the number of animals to be used is relatively small, the question still must be faced as to whether it is ethical to use animal models involving species such as rats, chickens, sheep, pigs, monkeys, baboons, and chimps in order to study the feasibility of cross-species xenografting. In part, the answer to this question pivots on

whether or not there are plausible alternative models to the use of animals for exploring the two critical steps required for successful xenografting – overcoming immunologic rejection and achieving long-term physiologic function in an organ or tissue.

To some extent, immunologic problems can be examined without killing primates or higher animals by using lower animals or cellular models. But there would not appear to be any viable alternatives or nonhuman substitutes available for understanding the processes involved in rejection. At best it may be possible to utilize animals that have fewer cognitive and intellectual capacities for most forms of basic research with respect to understanding the immunology of xenografting.

When research gets closer to the clinical stage, especially when it becomes possible to examine the extent to which xenografted organs and tissues can function post-transplant, it will be necessary to use some animals as donors and recipients that are closely related to human beings. If, in light of the scarcity of human organs, it is ethical to pursue the option of xenografting, then it would be unethical to subject human beings to any form of xenografting that has not undergone a prior demonstration of both immunologic and physiologic feasibility in animals closely analogous to humans.

The use of animals analogous to humans for basic research on xenografting means that some form of primates must be used both as donors and as recipients. Is it ethical to kill primates, even if only a small number, to demonstrate the feasibility of xenografting in human beings? If primates and humans have the same moral status then it is hard to see how the use of primates could be justified.[1,16,18] Are humans and ape moral equivalents?

At this point, those who want to deny the validity of moral equivalence begin to look for morally significant properties uniquely present in human beings but not in primates. Strong candidates for the property that might make a moral difference sufficient to allow the killing of primates to advance human interests are language, tool-use, rationality, intentionality, consciousness, conscience, and/or empathy. The debate about the morality of killing a primate of some sort to advance human interests by saving human lives then hinges on empirical facts about what particular species of primates can or cannot do in comparison to what humans are capable of doing.[19]

It is indisputable that there are some differences in the capacities and abilities of humans and primates. Chimps can sign but humans have much more to say. Gorillas seem to reason but humans have calculus, novels, and quantum theory. Humans are capable of a much broader range of behavior and intellectual functioning than are any specific primate species.

Many who would protest the use of primates in xenografting research are keen to illustrate that primates possess many of the properties and abilities that are found to contribute moral standing to humans. The fact that one species or another of primate is capable of some degree of intellectual or behavioral ability that seems worthy of moral respect when manifest by humans does not mean that human beings are the moral equivalents of primates. It is one thing to argue that primates ought to have moral standing. It is a very different matter to argue that humans and primates are morally equivalent. One can grant that primates deserve moral consideration without conceding that, on average, the death of a human being is of greater moral significance than is the death of a baboon, a green monkey, or a chimpanzee.

Xenografting involving primates can be morally justified on the grounds that, in general, human beings possess capacities and abilities that confer more moral value upon them than do primates. This is not "speciesism"[1,20]

but, rather, a claim of comparative worth that is based on important empirical differences between two classes or sets of creatures.

Perhaps there are empirical reasons to support the claim that it is worth killing primates for humans on the grounds that as groups humans have properties that confer greater moral worth and standing upon them than do other animal species. Human beings are after all moral agents while, at most, animals, even primates, are moral subjects.

Even conceding this point, there is still a problem when it comes time to kill a baboon and a chimp to see if xenografting between them is possible. Critics might ask whether scientists would be willing to kill a retarded child or an adult in a permanent vegetative state in the service of the same scientific goal. It is indisputable that human beings have on average more capacities and abilities than do animals. But there are some individual animals, many of them primates, that have more capacities and abilities than certain individual human beings who lack them due to congenital disorders or as the result of disease or injury. If it is argued that we ought to use animals instead of humans to assess the feasibility of xenografting because humans are more highly developed in terms of intellectual and emotional capacities, capacities that make a moral difference in that they are the basis for moral agency, then why should we not use a severely retarded child instead of a bright chimp or gorilla[20] as subjects in basic or clinical research? Unless those who are doing or wish to engage in basic research on xenografting can answer this question they will be open to the charge of immorality even if they kill primates or other "higher" animals in order to benefit humans who are as a species of more moral worth than are animals.

One line of response is to simply say that we are powerful and the primates are less, therefore they must yield to human purposes if we choose to experiment upon them rather than retarded children. This line of response sounds rather far removed from the kinds of arguments we expect to be mustered in the name of morality. A more promising line of attack on the view that humans and primates or other animals are morally equivalent is to examine a bit more closely why it is that we would not want to use a retarded child instead of a chimp in basic research.

Two reasons might be given for picking a chimp instead of a human being with limited or damaged capacities. We might decide not to use a human being who has lost his or her capacities and abilities out of respect for their former existence. If the person makes a conscious decision to allow his or her body to be used for scientific research prior to having become comatose or brain dead then perhaps those wishes should be honored. If no such advance notice has been given then we ought not to presume anything about what they would have wanted and should forego any involvement in medical research generally and xenografting in particular on the grounds that this is what is demanded out of respect for the persons they once were.

However, severely retarded children or those born with devastating conditions such as anencephaly have never had the capacities and abilities that confer a greater moral standing on humans as compared with animals. Should they be used as the first donors and recipients in xenografting research instead of primates?

The reason they should not has nothing to do with the properties, capacities, and abilities of children or infants who lack and have always lacked significant degrees of intellectual and cognitive function. The reason they should not be used is because of the impact using them would have upon other human beings, especially their parents and relatives. A severely retarded child can still be the object of much love, attention, and devotion from his or her parents. These feelings and the abilities and capacities that generate them are deserving of moral respect. I do not believe animals including primates are capable of such feelings.

If a human mother were to learn that her severely retarded son had been used in lethal xenografting research she would mourn this fact for the rest of her days. A baboon, monkey, or orangutan would not. The difference counts in terms of whether it is a monkey or a retarded human being who is selected as a subject in xenografting research.

It may be that parents would want to volunteer their child's organ or tissue for research or they might wish to have their baby with anencephaly serve as the first recipient of an animal organ or tissue. It may be necessary to honor such a choice. Whatever public policies are created to govern our actions toward severely retarded children or babies born with most of their brain missing, these are policies that are meant to be respectful of the sensibilities and interests of other human beings.[21] They do not find their source in some inherent property of the anencephalic infant. It is in the relationships with others, both family and strangers, that the moral worth and standing of these children are grounded.

The case for using animals, even primates, before using a human being with severely limited abilities and capacities is based on the relationships that exist among human beings, which do not have parallels in the animal kingdom. These relationships, such as love, loyalty, empathy, sympathy, family-feeling, protectiveness, shame, community-mindedness, a sense of history, and a sense of responsibility, which ground many moral duties and set the backdrop for distinguishing virtuous conduct and character, do not, despite the sociality of some species, appear to exist among animals.

If animals are to be used then what sorts of guidelines should animal care and use committees or other review bodies follow in reviewing basic research proposals? These committees must ensure that basic research is designed in such a fashion as to minimize the need for animal subjects while maximizing the opportunity to obtain generalizable knowledge. They must also ensure that the animals used are kept in optimal conditions and are handled humanely and killed without pain. These steps will be necessary in order to both respect the moral standing of the animals and to maximize the chance for generating useful knowledge from the use of these animals in research.

Perhaps the most difficult question arising when sufficient data have been obtained to make clinical trials plausible is who ought to be the first subjects in clinical xenografting trials. The Baby Fae xenograft case involved an infant because the researchers felt that the scarcity of organs for infants born with congenital defects was so great that morality demanded that infants be the initial subjects selected. Many argued that this choice was mistaken since at the level of initial clinical trials it is morally wrong to use infants, children, or other human beings incapable of giving informed consent to their participation.[17]

If it is true that clinical trials involving xenografting should avoid the initial use of infants, children, and adults who lack the capacity for consent then who ought to go first? Perhaps adults who would not be otherwise eligible for transplants under existing exclusion criteria, the imminently dying, terminally ill volunteers who agree to serve, those who are brain dead, or those needing a second or third transplant when scarcity and prognosis would make it most unlikely they would receive another human organ or tissue.

Similarly, for clinical trials the question arises as to which type of organ or tissue ought to be the subject of initial research efforts? Those for which the scarcity of human organs and tissues is greatest or those for which alternatives to

animal organs seem the least promising? A strong case can be made that the selection of an organ or tissue to xenograft should be guided by the scarcity of human organs and tissues as well as the results achieved in animal to animal xenografting during basic research.

When clinical trials are designed for human subjects, those undertaking the research must have the qualifications and the background to make it likely that they will generate reliable and replicable results that are quickly made available in the literature. While the use of primates and human beings in research may be morally justified their use can only be justified when research is designed and conducted in circumstances most likely to maximize the chances of creating knowledge.

Those who will be recruited for clinical trials have the right to know the risks and benefits of the research that can best be inferred from animal and any other relevant studies. They should also know how many subjects will be used before the end-point of this phase of research is reached. The measures taken for coping with any psychological issues raised by the use of animal organs for subjects must be presented. Subjects should also be told about the steps that have been taken to ensure their privacy and confidentiality to the extent they wish these preserved. It is of special importance given the high odds of failure in the initial phase of xenografting on human subjects that procedures be in place for ending the experiment if the subject wishes to withdraw from the research.

Ethical issues raised by therapy

Perhaps the most obvious moral problem that would arise if xenografting proved to be a viable source of organs and tissue for transplantation is whether prospective recipients would be able to accept animal parts when in need of transplants. Some might feel that it is unnatural to do so. But, naturalness is very much a function of familiarity. One hundred years ago surgery and anesthesia were viewed by many as unnatural. People may require support and counseling when faced with the option of xenografting but, facing death, most will probably accept a transplant and decide to deal with the naturalness issue later.

What about systematically breeding, raising, and killing animals for their parts on a large scale? Is it moral to systematically farm and kill animals for spare parts for humans? Would it be right to systematically farm and kill animals for spare parts for other animals, say companion animals?

One response to the issue of breeding and raising animals in order to have a regular supply of organs for xenografting is to argue that this practice does not raise any new or special moral issues since huge numbers of animals are currently bred, raised, and killed solely for consumption. However, animals raised for food are often raised under inhumane and brutal conditions. Nor is there much consideration given to the techniques involved in their slaughter.[1] Animals that are to be used to generate a constant source of organs and tissues for transplant must be raised under conditions that would ensure the healthiest possible animals. The moral obligation to potential recipients would seem to require that systematic farming of animals only be permitted under the most humane circumstances.

Interestingly the issue of whether it is morally permissible to systematically breed, raise, and kill animals to obtain their parts does not raise the same issues of moral equivalence between animals and humans that arise with respect to basic and clinical research. It would obviously be immoral to breed, raise, and kill humans for their parts.

The availability of animal organs for transplant on a therapeutic basis is likely to become a matter of economics. The economics of this sort of animal breeding must take into account the costs of creating the healthiest possible animals and the most painless modes of killing. However, should xenografting evolve to the point of therapy, those who perform xenografts must also strive to ensure that access to transplants is equitable.

Conclusion

The morality of using animals as sources of organs is contingent on the need to turn to animals as sources. The scarcity of organs and tissues from human sources is real, growing, and unlikely to be solved by any other alternative policies or approaches in the foreseeable future. If xenografting must be explored as an option then the moral justification for doing basic and clinical research involving animals is that it would not be possible to learn about the feasibility of overcoming immunologic and physiologic problems without using animals and that animals are to be used instead of human subjects whenever possible since human beings have more moral worth than do animals. If xenografting evolves into a therapy then provisions must be made for ensuring the welfare and health of the animals that would be bred, raised, and killed to supply organs and tissues to human beings.

Notes

1 Singer P: *Animal Liberation*. New York: Random House, 1975
2 Baum B, Bernstein D, Starnes VA, et al: *Pediatrics* 88:203, 1991
3 Caplan A: *Transplant Proc* 21:3381, 1989
4 Caplan A, Siminoff L, Arnold B, et al: *Surgeon General's Workshop on Organ Donation*. Washington, DC: HRSA, (in press)
5 Caplan A, Virnig B: *Crit Care Clin* 6:1007, 1990
6 Caplan A: *J Transplant Coordination* 1:78, 1991

7   Peters TG: *JAMA* 265:1302, 1991

8   Wight JP: *Br Med J* 303:110, 1991

9   John Paul II: Statement on Organ Transplantation. International Congress of the Society for Organ Sharing, Rome, Italy, June 20, 1991

10 *New Scientist*: July 13, p 1991

11 Tufts A: *Lancet* 337:1403, 1991

12 Eurotransplant Foundation: *Eurotransplant Newsletter* 87:1, 1991

13 Raia S, Nery JR, Mies S: *Lancet* 26:497, 1989

14 Strong RW, Lynch SV, Ong TH, et al: *N Engl J Med* 322:1505, 1990

15 Elick BA; Sutherland DER, Gillingham K, et al: *Transplant Proc* 22, 1990

16 Regan T, Singer P: *Animal Rights and Human Obligations*. Englewood Cliffs, NJ: Prentice-Hall, 1976

17 Caplan A: *JAMA* 254:3339, 1985

18 Regan T, Van De Veer D: *And Justice for All*. Totowa: Rowman and Littlefield, 1982

19 Jasper JM, Nelkin D: *The Animal Rights Crusade*. New York: The Free Press, 1992

20 Singer P: *Transplant Proc* 24:728, 1992

21 Caplan A: *Bioethics* 1:119, 1987

Chapter 19

James Lindemann Nelson

MORAL SENSIBILITIES AND MORAL STANDING: CAPLAN ON XENOGRAFT "DONORS"

On June 28th of 1992, a 35 year old man dying from hepatitis received a baboon's liver at the University of Pittsburgh; it was hoped that the nonhuman organ might resist the destructive effects of the virus inhabiting the man's body. Less than four months later, on October 11th, a 26 year old woman received a pig's liver at Los Angeles' Cedars-Sinai Medical Center; the hope in this case was that the animal organ might serve as a bridge to transplant with a suitable human liver. And on January 10th of 1993, surgeons at Pittsburgh performed another baboon-to-human liver transplant in a 62 year old man; he thus became the second subject in a proposed experimental series of four hepatic xenografts.

All these cases ended sadly: the first Pittsburgh gentleman, whose identity was never made public, died on September 6 of the same year; the second, whose name was also withheld, died 26 days after the transplant. Ms. Susan Fowler, the patient at Cedars-Sinai, died the day following her operation. But despite the tragic conclusion to these recent attempts at xenotransplantation, interest in animals as a source of organs and tissues for human beings remains strong.[1] New developments in immunosuppression technology promise to lower the technical barriers to a routine use of nonhumans as organ donors, and the image of colonies of animals kept at the ready for supplying the growing human need for new organs seems a much more plausible scenario now than it did when broached by transplantation specialists in the Sixties.[2] As Arthur Caplan has powerfully argued, the prospects that other sources of organs may resolve the supply problem are grim; easing the constriction on available organs by more effectively harnessing human altruism, or moving to routine retrieval or commodification of organs is unlikely fully to meet the need, which is severe; he reports that over a third of those currently waiting for liver transplantation in the U.S. die for the lack of an organ, and well over half of all children born with fatal, congenital deformities of the heart or liver die without a transplant due to organ shortage.[3] Even were supply-enhancing strategies morally acceptable and pragmatically successful, the expanded supply would simply tend to reveal the full scope of the need, by making visible the "hidden pool" of potential recipients (p. 723). Further, the advent of truly effective artificial organs seems remote (p. 723).

In the face of these "pro-xenograft" pressures, it becomes all the more significant to assess arguments against the practice that rest on considerations of the moral status of the nonhumans from whom the organs are taken. To be sure, xenograft faces other moral difficulties

– for example, concerns about the quality of informed consent obtained for recipients, worries about the possibility that xenografting will serve as a vector by which new and possibly virulent viruses become established in humans,[4] and problems about whether such spending is equitable in the light of other unresolved human needs. Yet whether we morally wrong animals in taking their organs and their lives remains a decidedly central issue here, one that cannot be finessed away by developing better informed consent procedures, better anti-viral strategies, or by situating transplantation medicine in a just health care system.

One of the most powerful arguments against the use of animals in xenograft (and against certain instrumental uses of many kinds of animals generally) is the so-called marginal cases argument.[5] As applied to xenograft, the reasoning runs as follows:

If it is morally wrong to use unconsenting humans as sources of organs, but not morally wrong to use unconsenting nonhumans, there must be something upon which this moral difference supervenes. But whatever criterion or set of criteria could plausibly be advanced to distinguish humans and nonhumans in this respect shall fail. For species membership, considered strictly on its own, is no more directly morally relevant than are other biological categories, such as race or gender; the moral weight must be born by properties which correlate with species. But the correlation between plausibly morally relevant properties and species is not exact. If the criteria are intended to range over all humans – e.g., sentience, or having a subjective welfare – they will include at least many nonhumans (including, interestingly, those most useful for xenograft); if, on the other hand, the criteria chosen are styled to exclude all nonhumans (e.g., high intelligence, complex

communicative and relational abilities) they will exclude some humans as well (roughly, those who, due to congenital dysfunction or trauma, are on a psychological par with the most developed nonhumans). Hence, there are no moral criteria which will include all and only humans inside a circle of moral considerability strong enough to preserve them from being unwilling sources of organs.

Many philosophers have accepted this kind of reasoning, using it in the service of the view that animals should be accorded the kind of moral respect we pay to humans handicapped in the relevant ways; others have drawn the conclusion that humans so handicapped ought to be accorded the same moral status as we now extend to animals.[6] Either approach makes things look rather dark for xenograft. On the former view, it will be flatly immoral; on the latter, it will not be the therapy of choice, as human organs would likely be preferable to nonhuman organs, since interspecific biological problems would not arise.

There have also been a number of attempts to refute this argument.[7] Caplan provides a thoughtful example; in his recent *Transplantation Proceedings* paper, he has argued that the moral disanalogy between severely retarded humans and primates is supplied not by intrinsic properties of either the human or the primate, but by relational properties enjoyed by the human, but not by the non-human. The reason that severely retarded or anencephalic children[8] should not be used as donors in xenograft research, Caplan writes,

has nothing to do with the properties, capacities, and abilities of children or infants who lack and have always lacked significant degrees of intellectual and cognitive function. The reason they should not be used is because of the impact using them would have upon other human beings, especially their parents and relatives (p. 726).

He makes this idea particularly vivid by pointing out that "if a human mother were to learn that her severely retarded son had been used in lethal xenografting research she would mourn this fact for the rest of her days" (p. 726). In contrast, we have no reason to believe that a primate mother would be similarly affected by the death of her children.

I find this an intuitively riveting example. This mother's pain is altogether compelling, and the special affection she feels for her child is worthy of respect. It is true that some philosophers, insisting on the impartiality of moral judgment, admittedly would challenge the legitimacy of such special affections as justification for treating otherwise morally indistinguishable beings differently,[9] but it is surely respectable, even among philosophers, to believe that such attitudes as love and loyalty are both inherently preferential, and morally defensible.[10]

But even if we reject the rigorous impartialist's challenge to the legitimacy of our special affections, on reflection Caplan's argument still seems to leave disturbing lacunae: what about severely mentally handicapped humans who lack special ties to particular people, those who are orphans? And what about nonhumans who are especially loved? He makes much of the fact that nonhumans don't mourn their dead as profoundly as do humans, but why should it be essential that the love come from a conspecific? If an animal can be rendered morally immune from xenograft by an intense kind of "emotional adoption", a new avenue of approach would seem opened to animal rights activists.

Given his stress on the significance of concrete relationships, it is a bit puzzling why Caplan doesn't see the phenomena of the love of a human for a nonhuman as worth discussing. A partial solution to this puzzle is suggested by his claim that "it is in the relationships with others, both family and strangers, that the moral worth and standing of these children are grounded" (p. 726); this hints that the patterns of feelings and relationships characteristic of humans have an importance which goes beyond that possessed by their individual instances, and therefore that the mother's response to the sacrifice of her child is not simply evidence of a morally distinct kind of relationship – parental love – but a particularly vivid example of a general system of sensibilities that runs through our species as well. Or so I take him to mean when he says

> The case for using animals, even primates, before using a human being with severely limited abilities and capacities is based on the relationships that exist among human beings, which do not have parallels in the animal kingdom. These relationships, such as love, loyalty, empathy, sympathy, family-feeling, protectiveness, shame, community-mindedness, a sense of history, and a sense of responsibility, which ground many moral duties and set the backdrop for distinguishing virtuous conduct and character, do not, despite the sociality of some species, appear to exist among animals.
>
> (p. 726)

These words are open to a number of readings, but I think the most plausible rendering is that there are human-specific relationships and concomitant emotions that are both the background conditions of moral agency, and that make people generally vulnerable to special harms were handicapped humans to be used as unconsenting organ sources. Xenograft, on the other hand, will not threaten, dishonor, or harm these relationships and feelings. These dispositions and patterns of affection and relationship are worthy of moral respect because of their involvement in making moral agency possible and because not respecting them puts moral agents at risk of serious harm.

But surely the general points about the significance of relationship and emotion may be allowed, without thereby granting that specific patterns of feeling make a morally relevant difference between the humans and the nonhumans in question. While the use of, say, handicapped children as organ sources would no doubt be greeted with general outrage among persons, that no more demonstrates or constitutes the wrongfulness of such behavior than the calm acceptance of medical experiments on handicapped Jewish children among Nazis would show the moral acceptability of such research. Feelings and relationships may well be essential to moral agency, and surely do make us liable to great harms, but not just any expression of feeling or form of relationship is legitimate. The question must be whether such expressions are apt or inappropriate. And this is not a matter that can be shown by relying on our intuitions about the significance of parental affections, for the existence of some kinds of praiseworthy partialities does not guarantee the acceptability of all partialities. My special love for my own children is, I think, morally required, and justifies the preferences I extend them. But if I had a special love for white children as such, matters would be much more dubious, particularly if I proposed to enact some sort of public policy of preference for white children on such a basis. Nor does the general moral significance of love, loyalty, family-feeling and the rest show the appropriateness of our preference for other human beings; such feelings may themselves be morally admirable, part of what makes life worth living, but they are as much at the disposal of bigots as paragons. How we entertain and express such feelings, for whom and in what ways, are all open to moral scrutiny.

An interesting reply to this criticism, one which is open to Caplan, runs along these lines:

the mistake is to think that death is some kind of harm to both the severely mentally handicapped and nonhumans, and that we must scrupulously look for just grounds to distribute that harm, if inflicting it is morally acceptable at all. If that were so, then preferential feelings as a basis for a biased distribution of the harms would be suspicious indeed. But suppose that death were not a harm at all to either severely mentally handicapped humans or to nonhumans; this could be seen as an amplification of Caplan's claim that what makes killing a handicapped person wrong has nothing to do with the human's own properties. If death is no harm to handicapped humans, considered on their own, and hence no harm to animals either, then no question of wronging them, no concern about injustice in the distribution of burdens, would seem to arise; we might as well kill the one as the other. If we have even a whimsical preference for one, there's no harm in indulging it – no one gets hurt. And the preferences here, on Caplan's account, are not whimsical at all, but deeply felt and significant to our whole moral system.

But if this is a defensible reading of Caplan's position, then, while he would no longer have a problem justifying removing animal organs, he would have a problem justifying why we don't regard the severely mentally handicapped as organ sources. Those so handicapped would, ex hypothesi, lose nothing of value if they lost their lives; the humans who die who would otherwise have lived have lost something of immense value. This seems rather a high price to pay for maintaining the integrity of our ethical sensibilities – if indeed that is what is at issue. For on this understanding of Caplan's analysis, it looks as though we're paying in a rather deflated kind of coin.

Surely, part of what grounds the general repugnance against the killing of the mentally handicapped is the belief that such killing takes advantage of their vulnerability to rob them of something of great value. If this belief is wrong, then that moral repugnance is left without one of its chief supports, and it is far from clear that it should be allowed to deprive those to whom death is a great harm, and who will die without organs, of what they need to live. Further, if killing the severely mentally handicapped truly doesn't harm them, it's hard to see how our using them as organ sources must damage our own moral discernment, or our moral system as a whole. Indeed, it would seem a positive good to educate ourselves into people whose moral sensibilities more carefully track what is really of value, and aren't beguiled by superficialities.

Caplan's argument in favor of the use of nonhumans as organ donors, even granted the significance of parental partialities and allowing this implicit analysis of what makes death a harm, is not ultimately persuasive; so far as "following the argument where it leads" goes, he really ought to make common cause with R.G. Frey, who has called for the use of handicapped children in medical research on terms similar to those currently experienced by nonhumans. Of course, as a matter of practical politics, xenograft is at hand, and the use of handicapped orphan children is not, and it might seem perfectly appropriate to accept it as a sort of compromise position. In that case, however, it becomes very important for anyone holding such a position to have a well-worked out explanation of the role of mental complexity of a certain order in transforming death from a matter of merest indifference to a profound harm.

But even if such an analysis were available, the resulting position is still intellectually unstable and emotionally unsatisfying. Caplan's stress on relationships and emotions over individual properties has the effect of cutting off, as it were, the emotion from the object to which it is directed. This neglects that emotions carry with them certain understandings, as well as feelings – the

mother's passion for her child includes (and perhaps partially constitutes) the belief that the child is "her own" – as Aristotle says, one of the great reasons for caring about anything. But the fact that the emotions of strangers are more engaged by the death of their handicapped conspecifics than by the death of those nonhumans who are their peers in vulnerabilities and capacities is not so innocent; such emotions carry with them beliefs and implications that are factually and morally doubtful.[11] Caplan's position has the advantage of leading us to focus on the significance of characteristically human sensibilities, marking in particular how deeply we mourn the death of handicapped children. But if we were to get clearer about what it is we mourn for in mourning the death of such children, and clearer about what nonhumans may lose and suffer as well, we might find that we had more to grieve about than we now tend to think.

Notes

1 For discussion see, for instance, Marjorie Shaffer, "Baboon Liver in Man: The Beginning of the Xenograft Era?" *Medical World News*, August 1992, pp. 27–28, B.D. Colen, "A Baboon Dies, A Man Lives", *New York Newsday*, Tuesday, July 7, 1992, p.51; 56–57, and "Pig Liver Transplant for Dying Patient is Defended", *New York Times*, Wednesday, October 14, 1992, p. A16. For an account of the second Pittsburgh transplant and its sequel, see Lawrence K. Altman, "Second Baboon-Liver Recipient Dies of Infection", *New York Times*, Saturday, February 6, 1993, p.6.

2 See the December, 1970 issue of *Transplantation Proceedings*; Discussion section, p.554.

3 Arthur Caplan, "Is Xenografting Morally Wrong?" *Transplantation Proceedings*, 24 (2), April, 1992: 722–27. All other references to this article will occur in the main text.

4 See Rachel Nowak, "Hope or Horror? Primate-to-Human Organ Transplants", *The Journal of NIH Research*, 4, September, 1922, 37–38.

5 So-called first by Jan Narveson, "Animal Rights", *Canadian Journal of Philosophy* 7 (2), March, 1977: 161–78.

6 Tom Regan is a good example of a philosopher whose conviction that the moral standing of animals should be substantially enhanced is substantially nourished by marginal case argument kind of concerns. See his *The Case for Animal Rights*, Berkeley and Los Angeles: The University of California Press, 1983. R.G. Frey also accepts the argument, but uses it to support a call for the use of handicapped children as research subjects. See his *Rights, Killing and Suffering*, Oxford: Basil Blackwell, 1983.

7 For surveys of the relevant literature, see David DeGrazia, "The Moral Status of Animals and Their use in Research: A Philosophical Review", *Kennedy Institute of Ethics Journal* 1 (1) (March, 1991): 48–70, who calls the argument "perhaps the most agonizing of the unresolved issues (in animal ethics)" (p.59), and James Nelson, "Recent Studies in Animal Ethics," *American Philosophical Quarterly* 22 (1), January, 1985: 13–24.

8 The focus is on severely retarded *children* to sidestep the possibility that those who have been full moral agents, and then have lost those abilities, have a claim to our respect based on their previous status.

9 See, for example, James Rachels, "Morality, Parents and Children", in George Graham and Hugh LaFollette, eds, *Person to Person*, Philadelphia: Temple University Press, 1989.

10 See, for example, Philip Pettit, "Social Holism and Moral Theory: A Defense of Bradley's Thesis", *Proceedings of the Aristotelian Society* 86 (New Series), 1986: 173–97.

11 There may indeed be some morally relevant disanalogies between the situation of severely handicapped humans and psychologically comparable nonhumans, and our feelings may, with careful reflection, give us a clue to what they might be. For example, our sense that the birth of a severely handicapped infant is a tragic occurrence, while the birth of a nonhuman whose species-normal capacities are no greater than the humans is not, may point toward some kind of morally salient distinction between the two. But simply pointing out differences in our emotional reactions, whether or not they are plausibly correlated with morally significant differences, is not sufficient. It has to be shown that any such differences justify the gulf-like dimensions of the gap we place between them. For a discussion of counter-factual differences between nonhumans and severely mentally handicapped humans, see James Lindemann Nelson, "Animals, Handicapped Children, and the Tragedy of Marginal Cases", *Journal of Medical Ethics* 14 (4), December, 1988: 191–93.

How should more human transplant organs be acquired?

INTRODUCTION TO PART FIVE

THE DISCUSSION OF xenotransplantation in the previous chapter introduced the 'transplant gap', i.e. the shortfall between the demand for, and supply of, transplantable materials. The present chapter focuses on ways of procuring more organs from human donors, as opposed to animals.

A feasible way of closing the transplant gap is to permit a market in organs. The dispute between Kass and Radcliffe-Richards *et al.* captures two styles of bioethics. According to Kass, organ donation and transplantation are improprieties we overcome for the sake of medical benefits by thinking of them in terms of gift-giving. He attempts to articulate the sources of our disquiet about transplantation which are 'not readily rationalizable or measurable but not for that reason unreasonable or irrational', and which can influence policy. To sell organs would be a step too far because it amounts to commodifying the body and denigrating the gift of organs. By contrast, Radcliffe-Richards *et al.* argue that the benefits to both recipients and vendors of permitting organ sales put the onus on opponents to a market in organs. At the end of their piece, Radcliffe-Richards *et al.* refer to 'deep feelings of repugnance'; whereas these are crucial for Kass, for Radcliffe-Richards *et al.* they merely explain the weakness of familiar arguments for prohibition.

A response to questionable proposals such as a market in organs is that intuitively less dubious policy changes are as yet untried. A deceptively simply suggestion is to switch from an opt-in policy – *de facto*, a policy of allowing decedents' relatives to decide whether to allow organ retrieval – to a policy of presumed consent to donate. Cohen's case for such a switch rests on two premises: the empirical claim that '[m]ost persons, when asked, express without qualification their willingness to donate', and the principle of respect for decedent autonomy. Putting the two together, Cohen concludes that decedent autonomy is best protected by presuming consent. Veatch endorses the principle of autonomy but disputes Cohen's empirical premise. In an instructive parallel, he points out that the presumption of consent to emergency treatment is surely valid because it would attract 'agreement . . . close to unanimous' whereas 'if we presume consent in the case of organ procurement, we will be wrong at least 30 per cent of the time'. But Veatch also thinks there is something more serious at stake here, namely 'the ethics of the relation of the individual to the society'. Veatch suggests that presumed consent and routine salvaging are underpinned by different assumptions about the relationship between the individual and the state; routine salvaging indicates the authority of the state over the individual, whereas presumed consent is in keeping with our liberal assumptions about the pre-eminence of the individual.

Autonomy is clearly central to the debate, the problem being how to procure more organs whilst respecting individual autonomy. Spital points out that the current practice of asking families to decide whether organs can be retrieved causes undue distress and undermines decedents' autonomy. He suggests a variant of the policy of presumed consent, namely, mandated choice: 'all competent adults would be required to decide and record whether or not they wish to become organ donors upon their deaths'. This would, he claims, both increase the supply of organs and promote autonomy. One response is to ask after the limits of autonomy; for example, should donors be allowed to decide who would receive their organs, since this might well increase supply? Wilkinson focuses on a case of donation conditional on the organs being transplanted into a white recipient. The response to the offer comprised recommendations in an NHS report, summarised by Beecham, which became official policy. The panel's first reason for disallowing conditional donations is that they would fail to be altruistic; Wilkinson retorts that conditional offers can be altruistic. The panel's second reason is that conditional donations would fail to meet the greatest need; Wilkinson retorts by clarifying that the fundamental needs principle is that it is important that needs are met; this does not rule out conditional donations that meet needs that would not otherwise be met, even though they fail to meet the greatest need.

The fact that most transplant organs are retrieved from the dead suggests two ways in which we could increase the number procured. First, we could redefine death so that more people are dead than we originally thought. For example, (re)defining death in terms of brain death meant that artificially maintained brain dead patients were identified as dead and their organs became available for transplantation. Second, we could retain our definition of death but relax the 'dead donor rule' that no one should be killed by organ retrieval, thereby permitting organ procurement from some living donors. Both suggestions are relevant to the policy of donation after cardiac or circulatory death (DCD), as discussed by Bernat, and Truog and Miller. These disputants agree that the organ procurement described in the case study is ethically acceptable but 'disagree sharply, however, over *why* it is acceptable'. A central issue is whether the donor is dead at the time of organ retrieval and this, in turn, depends on whether death occurs when circulatory function is irreversibly or permanently lost. As the debate makes clear, this is more than merely academic because changes to the determination of death or the dead donor rule would have profound effects on end-of-life practices.

Selections

Kass, L.R. 1992. 'Organs for Sale? Propriety, Property, and the Price of Progress', *Public Interest*, 107, 65–86.

Radcliffe-Richards, J., Daar, A.S., Guttmann, R.D., Hoffenberg, R., Kennedy, I., Lock, M., Sells, R.A. and Tilney, N. for the International Forum for Transplant Ethics 1998. 'The Case for Allowing Kidney Sales', *The Lancet*, 351 (9120), 1950–2.

Cohen, C. 1992. 'The Case for Presumed Consent to Transplant Human Organs after Death', *Transplantation Proceedings*, 24 (5), 2168–72.

Veatch, R.M. and Pitt, J.B. 1995. 'The Myth of Presumed Consent: Ethical Problems in New Organ Procurement Strategies', *Transplantation Proceedings*, 27 (2), 1888–92.

Spital, A. 1996. 'Mandated Choice for Organ Donation: Time to Give it a Try', *Annals of Internal Medicine*, 125 (1), 66–9.

Beecham, L. 2000. 'Donors and Relatives Must Place No Conditions on Organ Use', *British Medical Journal*, 320 (7234), 534.

Wilkinson, T.M. 2003. 'What's Not Wrong with Conditional Organ Donation?' *Journal of Medical Ethics*, 29 (3), 163–4.

Bernat, J.L. 2010. 'Point: Are Donors After Circulatory Death Really Dead, and Does it Matter? Yes and Yes', *Chest*, 138 (1), 13–16.

Truog, R.D. and Miller, F.G. 2010. 'Counterpoint: Are Donors After Circulatory Death Really Dead, and Does it Matter? No and Not Really', *Chest*, 138 (1), 16–18.

Bernat, J.L. 2010. 'Rebuttal', *Chest*, 138 (1), 18–19.

Chapter 20

Leon R. Kass

ORGANS FOR SALE? PROPRIETY, PROPERTY, AND THE PRICE OF PROGRESS

Just in case anyone is expecting to read about new markets for Wurlitzers, let me set you straight. I mean to discuss organ transplantation and, especially, what to think about recent proposals to meet the need for transplantable human organs by permitting or even encouraging their sale and purchase. If the reader will pardon the impropriety, I will not beat around the bush: the subject is human flesh, the goal is the saving of life, the question is, "To market or not to market?"

Such blunt words drive home a certain impropriety not only in my topic but also in choosing to discuss it in public. But such is the curse of living in interesting times. All sorts of shameful practices, once held not to be spoken of in civil society, are now enacted with full publicity, often to applause, both in life and in art. Not the least price of such "progress" is that critics of any impropriety have no choice but to participate in it, risking further blunting of sensibilities by plain overt speech. It's an old story: opponents of unsavory practices are compelled to put them in the spotlight. Yet if we do not wish to remain in the dark, we must not avert our gaze, however unseemly the sights, especially if others who do not share our sensibilities continue to project them – as they most certainly will. Besides, in the present matter, there is more than impropriety before us – there is the very obvious and unquestionable benefit of saving human lives.

About two years ago I was asked by a journal to review a manuscript that advocated overturning existing prohibitions on the sale of human organs, in order to take advantage of market incentives to increase their supply for transplantation. Repelled by the prospect, I declined to review the article, but was punished for my reluctance by finding it in print in the same journal. Reading the article made me wonder at my own attitude: What precisely was it that I found so offensive? Could it be the very idea of treating the human body as a heap of alienable spare parts? If so, is not the same idea implicit in organ *donation*? Why does payment make it seem worse? My perplexity was increased when a friend reminded me that, although we allow no commerce in organs, transplant surgeons and hospitals are making handsome profits from the organ-trading business, and even the not-for-profit transplant registries and procurement agencies glean for their employees a middleman's livelihood. Why, he asked, should everyone be making money from this business except the person whose organ makes it possible? Could it be that my real uneasiness lay with organ donation or with transplantation itself, for, if not, what would be objectionable about its turning a profit?

Profit from human tissue was centrally the issue in a related development two years ago, when the California Supreme Court ruled that a patient had no property rights in cells removed from his body during surgery, cells which, following commercial genetic manipulation, became a patented cell-line that now produces pharmaceutical products with a market potential estimated at several billion dollars, none of it going to the patient. Here we clearly allow commercial ownership of human tissue, but not to its original possessor. Is this fair and just? And quite apart from who reaps the profits, are we wise to allow patents for still-living human tissue? Is it really necessary, in order to encourage the beneficial exploitation of these precious resources, to allow the usual commercial and market arrangements to flourish?

With regard to obtaining organs for transplantation, voluntary donation rather than sale or routine salvage has been the norm until now, at least in the United States. The Uniform Anatomical Gift Act, passed in all fifty states some twenty years ago, altered common-law practices regarding treatment of dead bodies to allow any individual to donate all or any part of his body, the gift to take place upon his death. In 1984, Congress passed the National Organ Transplantation Act to encourage and facilitate organ donation and transplantation, by means of federal grants to organ-procurement agencies and by the creation of a national procurement and matching network; this same statute prohibited and criminalized the purchase or sale of all human organs for transplant (if the transfer affects interstate commerce). Yet in the past few years, a number of commentators have been arguing for change, largely because of the shortage in organs available through donation. Some have, once again, called for a system of routine salvage of cadaveric organs, with organs always removed unless there is prior objection from the deceased or, after death, from his family – this is the current practice in most European countries (but not in Britain). Others, believing that it is physician diffidence or neglect that is to blame for the low yield, are experimenting with a system of required request, in which physicians are legally obliged to ask next of kin for permission to donate. Still others, wishing not to intrude upon either individual rights or family feelings regarding the body of the deceased, argue instead for allowing financial incentives to induce donation, some by direct sale, others by more ingenious methods. For example, in a widely discussed article, Lloyd Cohen proposes and defends a futures market in organs, with individuals selling (say, to the government) future rights to their cadaveric organs for money that will accrue to their estate if an organ is taken upon their death and used for transplant.

In this business, America is not the leader of the free-market world. Elsewhere, there already exist markets in organs, indeed in live organs. In India, for example, there is widespread and open buying and selling of kidneys, skin, and even eyes from living donors – your kidney today would fetch about 25,000 rupees, or about $1,200, a lifetime savings among the Indian poor. Rich people come to India from all over the world to purchase. Last summer, the New York Times carried a front-page story reporting current Chinese marketing practices, inviting people from Hong Kong to come to China for fixed-price kidney-transplant surgery, organs – from donors unspecified – and air fare included in the price. A communist country, it seems, has finally found a commodity offering it a favorable balance of trade with the capitalist West.

What are we to think of all this? It is, for me, less simple than I first thought. For notwithstanding my evident revulsions and repugnances, I am prepared to believe that offering financial incentives to prospective donors could very well increase the supply – and perhaps even the quality – of organs. I cannot deny that the dead human body has become a valuable resource which, rationally regarded, is being allowed to go to waste – in burial or cremation. Because of our scruples against sales, potential beneficiaries of transplantation are probably dying; less troubling but also true, their benefactors, actual and potential – unlike the transplant surgeons – are not permitted to reap tangible rewards for their acts of service. Finally, and most troublesome to me, I suspect that regardless of all my arguments to the contrary, I would probably make every effort and spare no expense to obtain a suitable life-saving kidney for my own child – if my own were unusable. And though I favor the pre-modern principle, "One man, one liver," and am otherwise disinclined to be an organ donor, and though I can barely imagine it, I think I would readily sell one of my own kidneys, were the practice legal, if it were the only way to pay for a life-saving operation for my children or my wife. These powerful feelings of love for one's own are certainly widely shared; though it is far from clear that they should be universalized to dictate mores or policy in this matter, they cannot be left out of any honest consideration.

The question "Organs for sale?" is compelling and confusing also for philosophical reasons. For it joins together some of the most powerful ideas and principles that govern and enrich life in modern, liberal Western society: devotion to scientific and medical progress for the relief of man's estate; private property, commerce, and free enterprise; and the primacy of personal autonomy and choice, including freedom of contract. And yet, seen in the mirror of the present question, these principles seem to reach their natural limit or at least lose some of their momentum. For they painfully collide here with certain other notions of decency and propriety, pre-modern and quasi-religious, such as the sanctity of man's bodily integrity and

respect owed to his mortal remains. Can a balance be struck? If not, which side should give ground? The stakes would seem to be high – not only in terms of lives saved or lost but also in terms of how we think about and try to live the lives we save and have.

How to proceed? Alas, this, too, poses an interesting challenge – for in whose court should one conduct the inquiry? Shall we adopt the viewpoint of the economist or the transplant facilitator or the policy analyst, each playing largely by rational rules under some version of the utilitarian ethic: find the most efficient and economical way to save lives? Or shall we adopt the viewpoint of the strict libertarian, and place the burden of proof on those who would set limits to our autonomy to buy and sell or to treat our bodies in any way we wish? Or shall we adopt a moralist's position and defend the vulnerable, to argue that a great harm – say, the exploitation or degradation of even one person – cannot be overridden by providing greater goods to others, perhaps not even if the vulnerable person gives his less-than-fully-free consent?

Further, whichever outlook we choose, from which side shall we think about restrictions on buying and selling – what the experts call "inalienability"? Do we begin by assuming markets, and force opponents to defend non-sale as the exception? Or do we begin with some conception of human decency and human flourishing, and decide how best to pursue it, electing market mechanisms only where they are appropriate to enhancing human freedom and welfare, but remaining careful not to reduce the worth of everything to its market price? Or do we finesse such questions of principle altogether and try to muddle through, as we so often do, refining our policies on an *ad hoc* basis, in light of successes, costs, and public pressures? Whose principles and procedures shall we accept? And on whom shall we place the burden of what sort of proof?

Because of the special nature of this topic, I will not begin with markets and not even with rational calculations of benefits and harms. Indeed, I want to step back from policy questions altogether and consider more philosophically some aspects of the *meaning* of the idea of "organs for sale." I am especially eager to understand how this idea reflects and bears on our cultural and moral attitudes and sensibilities about our own humanity and, also, to discover the light it sheds on the principles of property, free contract, and medical progress. I wish, by this means, also to confront rational expertise and policy analysis with some notions outside of expertise, notions that are expressed and imbedded in our untutored repugnance at the thought of markets in human flesh. One would like to think that a proper understanding of these sentiments and notions – not readily rationalizable or measurable but not for that reason unreasonable or irrational – might even make a difference to policy.

I Propriety

The non-expert approaching the topic of organ transplantation will begin with questions of propriety, for it is through the trappings of propriety that we normally approach the human body; indeed, many of our evolved conventions of propriety – of manners and civility – are a response to the fact and problem of human embodiment. What, then, is the fitting or suitable or seemly or decent or proper way to think about and treat the human body, living and dead? This is, indeed, a vast topic, yet absolutely central to our present concern; for what is permissible to do to and with the body is partly determined by what we take the human body to be and how it is related to our own being.

I have explored these questions at some length elsewhere, in an essay entitled "Thinking About the Body,"[†] from which I transplant some conclusions without the argument. Against our

dominant philosophical outlooks of reductive corporealism (that knows not the soul) and person-body dualism (that deprecates the body), I advance the position of psychophysical unity, a position that holds that a human being is largely, if not wholly, self-identical with his enlivened body. Looking up to the body and meditating on its upright posture and on the human arm and hand, face and mouth, and the direction of our motion (with the help of Erwin Straus's famous essay on "The Upright Posture"), I argue for the body's intrinsic dignity:

> The dumb human body, rightly attended to, shows all the marks of, and creates all the conditions for, our rationality and our special way of being-in-the-world. Our bodies demonstrate, albeit silently, that we are more than just a complex version of our animal ancestors, and, conversely, that we are also more than an enlarged brain, a consciousness somehow grafted onto or trapped within a blind mechanism that knows only survival. The body-form *as a whole* impresses on us its inner powers of thought and action. Mind and hand, gait and gaze, breath and tongue, foot and mouth – all are part of a single package, suffused with the presence of intelligence. We are *rational* (i.e., *thinking*) animals, down to and up from the very tips of our toes. No wonder, then, that even a corpse still shows the marks of our humanity.

And, of course, it shows too the marks of our particular incarnation of humanity, with our individual and unique identity.

Yet this is only part of the story. We are *thinking* animals, to be sure, but we are simultaneously also and merely thinking *animals*. Looking down on the body, and meditating on the meaning of its nakedness (with the help of the story of man and woman in the Garden of Eden), we learn of human weakness and vulnerability, and especially of the incompleteness, insufficiency, needy dependence, perishability, self-division, and lack of self-command implicit in our sexuality. Yet while perhaps an affront to our personal dignity, these bodily marks of human abjection point also to special interpersonal relationships, which are as crucial to our humanity as is our rationality:

> For in the navel are one's forebears, in the genitalia our descendants. These reminders of perishability are also reminders of perpetuation; if we understand their meaning, we are even able to transform the necessary and shameful into the free and noble ... [The body, rightly considered,] reminds us of our debt and our duties to those who have gone before, [teaches us] that we are not our own source, neither in body nor in mind. Our dignity [finally] consists not in denying but in thoughtfully acknowledging and elevating the necessity of our embodiment, rightly regarding it as a gift to be cherished and respected. Through ceremonious treatment of mortal remains and through respectful attention to our living body and its inherent worth, we stand rightly when we stand reverently before the body, both living and dead.

This account of the meaning of the human body helps to make sense of numerous customs and taboos, some of them nearly universal. Cannibalism – the eating of human flesh, living and dead – is the preeminent defilement of the body; its humanity denied, the human body is treated as mere meat. Mutilation and dismemberment of corpses offend against bodily integrity; even surgery involves overcoming repugnance at violating wholeness and taboos against submitting to self-mutilation, overridden here only in order to defend the imperiled body against still greater threats to its integrity. Voyeurism, that cannibalism of the

eyes, and other offenses against sexual privacy invade another's bodily life, objectifying and publicizing what is, in truth, immediate and intimate, meaningful only within and through shared experience. Decent burial – or other ceremonial treatment – of the mortal remains of ancestors and kin pays honor to both personal identity and generational indebtedness, written, as it were, into the body itself. How these matters are carried out will vary from culture to culture, but no culture ignores them – and some cultures are more self-consciously sensitive to these things than others.

Culture and the body

The Homeric Greeks, who took embodiment especially to heart, regarded failure to obtain proper burial as perhaps the greatest affront to human dignity. The opposite of winning great glory is not cowardice or defeat, but becoming an unburied corpse. In his invocation to the Muse at the start of the *Iliad*, Homer deplores how the wrath of Achilles not only caused strong souls of heroes to be sent to Hades, but that *they themselves* were left to be the delicate feastings of birds and dogs; and the *Iliad* ends with the funeral of Hector, who is thus restored to his full humanity (above the animals) after Achilles's shameful treatment of his corpse: "So they buried Hector, breaker of horses." A similarly high regard for bodily integrity comes down to us through traditional Judaism and Christianity. Indeed, the Biblical tradition extends respect for bodily wholeness even to animals: while sanctioning the eating of meat, the Noachide code – widely regarded as enunciating natural rather than divine law – prohibits tearing a limb from a living animal.

Most of our attitudes regarding invasions of the body and treatment of corpses are carried less by maxims and arguments, more by sentiments and repugnances. They are transmitted inadvertently and indirectly, rarely through formal instruction. For this reason, they are held by some to be suspect, mere sentiments, atavisms tied to superstitions of a bygone age. Some even argue that these repugnances are based mainly on strangeness and unfamiliarity: the strange repels *because* it is unfamiliar. On this view, our squeamishness about dismemberment of corpses is akin to our horror at eating brains or mice. Time and exposure will cure us of these revulsions, especially when there are – as with organ transplantation – such enormous benefits to be won.

These views are, I believe, mistaken. To be sure, as an empirical matter, we can probably get used to many things that once repelled us – organ swapping among them. As Raskolnikov put it, and he should know, "Man gets used to everything – the beast." But I am certain that the repugnances that protect the dignity and integrity of the body are not based solely on strangeness. And they are certainly not irrational. On the contrary, they may just be – like the human body they seek to protect – the very embodiment of reason. Such was the view of Kant, whose title to rationality is second to none, writing in *The Metaphysical Principles of Virtue*:

> To deprive oneself of an integral part or organ (to mutilate oneself), e.g., to *give away* or *sell* a tooth so that it can be planted in the jawbone of another person, or to submit oneself to castration in order to gain an easier livelihood as a singer, and so on, belongs to partial self-murder. But this is not the case with the amputation of a dead organ, or one on the verge of mortification and thus harmful to life. Also, it cannot be reckoned a crime against one's own person to cut off something which is, to be sure, a part, but not an organ of the body, e.g., the hair, although selling one's hair for gain is not entirely free from blame.

Kant, rationalist though he was, understood the rational man's duty to himself as an animal body, precisely because this special animal body was the incarnation of reason:

> [T]o dispose of oneself as a mere means to some end of one's own liking is to degrade the humanity in one's person (*homo noumenon*), which, after all, was entrusted to man (*homo phenomenon*) to preserve.

Man contradicts his rational being by treating his body as a mere means.

Respect for the living and the dead

Beginning with notions of propriety, rooted in the meaning of our precarious yet dignified embodiment, we start with a series of presumptions and repugnances *against* treating the human body in the ways that are required for organ transplantation, which really is – once we strip away the trappings of the sterile operating rooms and their astonishing technologies – simply a noble form of cannibalism. Let me summarize these *prima facie* points of departure.

(1) Regarding *living donors*, there is a presumption against self-mutilation, even when good can come of it, a presumption, by the way, widely endorsed in the practice of medicine: Following venerable principles of medical ethics, surgeons are loath to cut into a healthy body not for its own benefit. As a result, most of them will not perform transplants using kidneys or livers from unrelated living donors.

(2) Regarding *cadaver donation*, there is a *beginning* presumption that mutilating a corpse defiles its integrity, that utilization of its parts violates its dignity, that ceremonial disposition of the total remains is the fitting way to honor and respect the life

that once this body lived. Further, because of our body's inherent connection with the embodied lives of parents, spouses, and children, the common law properly mandates the body of the deceased to next of kin, in order to perform last rites, to mourn together in the presence of the remains, to say ceremonial farewell, and to mark simultaneously the connection to and the final separation from familial flesh. The deep wisdom of these sentiments and ways explains why it is a strange and indeed upsetting departure to allow the will of the deceased to determine the disposition of his remains and to direct the donation of his organs after death: for these very bodily remains are proof of the limits of his will and the fragility of his life, after which they "belong" properly to the family for the reasons and purposes just indicated. These reflections also explain why doctors – who know better than philosophers and economists the embodied nature of all personal life – are, despite their interest in organ transplantation, so reluctant to press the next of kin for permission to remove organs. This, and not fear of lawsuit, is the reason why doctors will not harvest organs without the family's consent, even in cases in which the deceased was a known, card-carrying organ donor.

(3) Regarding the *recipients of transplantation*, there is some primordial revulsion over confusion of personal identity, implicit in the thought of walking around with someone else's liver or heart. To be sure, for most recipients, life with mixed identity is vastly preferable to the alternative, and the trade is easily accepted. Also, the alien additions are tucked safely inside, hidden from sight. Yet transplantation as such – especially of vital organs – troubles the

easygoing presumption of self-in-body, and ceases to do so only if one comes to accept a strict person-body dualism or adopts, against the testimony of one's own lived experience, the proposition that a person is or lives only in his brain-and-or-mind. Even the silent body speaks up to oppose transplantation, in the name of integrity, selfhood, and identity: its immune system, which protects the body against all foreign intruders, naturally rejects tissues and organs transplanted from another body.

(4) Finally, regarding *privacy and publicity*, though we may celebrate the life-saving potential of transplantation or even ordinary surgery, we are rightly repelled by the voyeurism of the media, and the ceaseless chatter about this person's donation and that person's new heart. We have good reason to deplore the coarsening of sensibilities that a generation ago thought it crude of Lyndon Johnson to show off his surgical scar, but that now is quite comfortable with television in the operating suite, requests for organ donation in the newspaper, talk-show confessions of conceiving children to donate bone marrow, and the generalized talk of spare parts and pressed flesh.

I have, I am aware, laid it on thick. But I believe it is necessary to do so. For we cannot begin in the middle, taking organ transplantation simply for granted. We must see that, from the point of view of decency and seemliness and propriety, there are scruples to be overcome and that organ transplantation must bear the burden of proof. I confess that, on balance, I believe the burden can be easily shouldered, for the saving of life is indeed a great good, acknowledged by all. Desiring the end, we will the means, and reason thus helps us overcome our repugnances – and, unfortunately, leads us to forget what this costs us, in coin of shame and propriety. We are able to overcome the restraints against violating the integrity of dead bodies; less easily, but easily enough for kin, we overcome our scruple against self-mutilation in allowing and endorsing living donation – though here we remain especially sensitive to the dangers of coercion and manipulation of family ties.

How have we been able to do so? Primarily by insisting on the principle not only of voluntary consent but also of *free donation*. We have avoided the simple utilitarian calculation and not pursued the policy that would get us the most organs. We have, in short, acknowledged the weight of the non-utilitarian considerations, of the concerns of propriety. Indeed, to legitimate the separation of organs from bodies, we have insisted on a principle which obscures or even, in a sense, denies the fact of ultimate separation. For in a *gift* of an organ – by its living "owner" – as with any gift, what is given is not merely the physical entity. Like any gift, a *donated* organ carries with it the donor's generous good will. It is accompanied, so to speak, by the generosity of soul of the donor. Symbolically, the "aliveness" of the organ requisite for successful transplant bespeaks also the expansive liveliness of the donor – even, or especially, after his death. Thus, organ removal, the partial alienation-of-self-from-body, turns out to be, in this curious way, a *reaffirmation* of the self's embodiment, thanks to the generous act of donation.

We are now ready to think about buying and selling, and questions regarding the body as property.

II Property

The most common objections to permitting the sale of body parts, especially from live donors, have to do with matters of equity, exploitation of the poor and the unemployed, and the dangers of abuse – not excluding theft and even murder to obtain valuable commodities. People deplore the degrading sale, a sale made in desperation, especially when the seller is selling something *so* precious as a part of his own body. Others deplore the rich man's purchase, and would group life-giving organs with other most basic goods that should not be available to the rich when the poor can't afford them (like allowing people to purchase substitutes for themselves in the military draft). Lloyd Cohen's proposal for a futures market in organs was precisely intended to avoid these evils: through it he addresses only increasing the supply without embracing a market for allocation – thus avoiding special privileges for the rich; and by buying early from the living but harvesting only from the dead he believes – I think mistakenly – that we escape the danger of exploiting the poor. (This and other half-market proposals seeking to protect the poor from exploitation would in fact cheat them out of what their organs would fetch, were the rich compelled to bid and buy in a truly open market.)

I certainly sympathize with these objections and concerns. As I read about the young healthy Indian men and women selling their kidneys to wealthy Saudis and Kuwaitis, I can only deplore the socioeconomic system that reduces people to such a level of desperation. And yet, at the same time, when I read the personal accounts of some who have sold, I am hard-pressed simply to condemn these individuals for electing apparently the only non-criminal way open to them to provide for a decent life for their families. As several commentators have noted, the sale of organs – like prostitution or surrogate motherhood or baby-selling – provides a double-bind for the poor. Proscription keeps them out of the economic mainstream, whereas permission threatens to accentuate their social alienation through the disapproval usually connected with trafficking in these matters.

Torn between sympathy and disgust, some observers would have it both ways: they would permit sale, but ban advertising and criminalize brokering (i.e., legalize prostitutes, prosecute pimps), presumably to eliminate coercive pressure from unscrupulous middlemen. But none of these analysts, it seems to me, has faced the question squarely. For if there were nothing fundamentally wrong with trading organs in the first place, why should it bother us that some people will make their living at it? The objection in the name of exploitation and inequity – however important for determining policy – seems to betray deeper objections, unacknowledged, to the thing itself – objections of the sort I dealt with in the discussion of propriety. For it is difficult to understand why someone who sees absolutely no difficulty at all with transplantation and donation should have such trouble sanctioning sale.

True, some things freely giveable ought not to be marketed because they cannot be sold: love and friendship are prime examples. So, too, are acts of generosity: it is one thing for me to offer in kindness to take the ugly duckling to the dance, it is quite another for her father to pay me to do so. But part of the reason love and generous deeds cannot be sold is that, strictly speaking, they cannot even be *given* – or, rather, they cannot be given *away*. One "gives" one's love to another or even one's body to one's beloved, one does not donate it; and when friendship is "given" it is still retained by its "owner." But the case with organs seems to be different: obviously material, they are freely alienable, they can be given and given away, and, therefore, they can be sold, and without diminishing the unquestioned good their transfer does for the recipient – why, then, should they not be for sale, of course, only by their proper "owner"? Why should not the owner-donor get something for his organs? We come at last to the question of the body as property.

Whose body?

Even outside of law and economics, there are perhaps some common-sense reasons for regarding the body as property. For one thing, there is the curious usage of the possessive pronoun to identify my body. Often I do indeed regard my body as a tool (literally, an organ or instrument) of my soul or will. My organism is organized: for whose use? – why, for my own. My rake is mine, so is the arm with which I rake. The "my-ness" of my body also acknowledges the privacy and unsharability of my body. More importantly, it means also to assert possession against threats of unwelcome invasion, as in the song "My Body's Nobody's Body But Mine," which reaches for metaphysics in order to teach children to resist potential molesters. My body may or may not be mine or God's, but as between you and me, it is clearly mine.

And yet, I wonder. What kind of *property* is my body? Is it mine or is it me? Can it – or much of it – be alienated, like my other property, like my car or even my dog? And on what basis do I claim property *rights* in my body? Is it really "my own"? Have I labored to produce it? Less than did my mother, and yet it is not hers. Do I claim it on merit? Doubtful: I had it even before I could be said to be deserving. Do I hold it as a gift – whether or not there be a giver? How does one possess and use a gift? Are there limits on my right to dispose of it as I wish – especially if I do not know the answer to these questions? Can one sell – or even give away – that which is not clearly one's own?

The word property comes originally from the Latin adjective *proprius* (the root also of "proper" – fit or apt or suitable – and, thus, also of "propriety"), *proprius* meaning "one's own, special, particular, peculiar." Property is both that which is one's own, and also the right – indeed, the exclusive right – to its possession, use, or disposal. And while there might seem to be nothing that is more "my own" than my own body, common sense finally rejects the view that my body is, strictly speaking, my property. For we do and should distinguish among that which is *me*, that which is *mine*, and that which is mine as *my property*. My body is me; my daughters are mine (and so are my opinions, deeds, and speeches); my car is my property. Only the last can clearly be alienated and sold at will.

Philosophical reflection, deepening common sense, would seem to support this view, yet not without introducing new perplexities. If we turn to John Locke, the great teacher on property, the right of property traces home in fact to the body:

> Though the earth and all creatures be common to all men, yet every man has a property in his own person; this nobody has a right to but himself. The labour of his body and the work of his hands we may say are properly his.

The right to the fruits of one's labor seems, for Locke, to follow from the property each man has in his own person. But unlike the rights in the fruits of his labor, the rights in one's person are for Locke surely inalienable (like one's inalienable right to liberty, which also cannot be transferred to another, say, by selling oneself into slavery). The property in my own person seems to function rather to limit intrusions and claims possibly made upon me by others; it functions to exclude me – and every other human being – from the commons available to all men for appropriation and use. Thus, though the right to property stems from the my-own-ness (rather than the in-commons-ness) of my body and its labor, the body itself cannot be, for Locke, property like any other. It is, like property, exclusively mine to use; but it is, unlike property, not mine to dispose of. (The philosophical and moral weakness in the very idea of property is now exposed to view: Property rights stem from the my-own-ness of my labor, which in turn is rooted in the my-own-ness of my body; but this turns out to be only relatively and politically my own.)

Yet here we are in trouble. The living body as a whole is surely not alienable, but parts of it definitely are. I may give blood, bone marrow, skin, a kidney, parts of my liver, and other organs without ceasing to be me, as the by-and-large self-same embodied being I am. It matters not to my totality or identity if the kidney I surrendered was taken because it was diseased or because I gave it for donation. And, coming forward to my cadaver, however much it may be me rather than you, however much it will be my mortal remains, it will not be me; my corpse and I will have gotten divorced, and, for that reason, I can contemplate donating from it without any personal diminution. How much and what parts of the bodily me are, finally, not indispensably me but merely mine? Do they thus become mine as my property? Why or why not?

The analysis of the notion of the body as property produces only confusion – one suspects because there is confusion in the heart of the idea of property itself, as well as deep mystery in the nature of personal identity. Most of the discussion would seem to support the common-sense and common-law teaching that *there is no property in a body* – not in my own body, not in my own corpse, and surely not in the corpse of my deceased ancestor. (Regarding the latter, the common-law courts had granted to next of kin a quasi-property right in the dead body, purely a custodial right for the limited purpose of burial, a right which also obliged the family to protect the person's right to a decent burial against creditors and other claimants. It was this wise teaching that was set aside by the Uniform Anatomical Gift Act.) Yet if my body is not my property, if I have no property right in my body – and here, philosophically and morally, the matter is surely dubious at best – by what *right* do I give parts of it away? And, if it be by right of property, how can one then object – in principle – to sale?

Liberty and its limits

Let us try a related but somewhat different angle. Connected to the notion of private property is the notion of free contract, the permission to transfer our entitlements at will to other private owners. Let us shift our attention from the vexed question of ownership to the principle of freedom. It was, you will recall, something like the principle of freedom – voluntary and freely given donation – that was used to justify the gift of organs, overcoming the presumption against mutilation. In contrast to certain European countries, where the dead body now becomes the property of the state, under principles of escheatage or condemnation, we have chosen to stay with individual rights. But why have we done so? Is it because we want to have the social benefits of organ transplantation without compromising respectful burial, and believe that leaving matters to individual choice is the best way to obtain these benefits? Or is the crucial fact our liberal (or even libertarian) belief in the goodness of autonomy and individual choice *per se*? Put another way, is it the dire need for organs that justifies opening a freedom of contract to dispose of organs, as the best – or least bad – instrument for doing so? Or is the freedom of contract paramount, and we see here a way to take social advantage of the right people have to use their bodies however they wish? The difference seems to me crucial. For the principle of autonomy, separated from specific need, would liberate us for all sorts of subsequent uses of the human body, especially should they become profitable.

Our society has perceived a social need for organs. We have chosen to meet that need not by direct social decision and appropriation, but, indirectly, through permitting and encouraging voluntary giving. It is, as I have argued, generosity – that is, more the "giving" than the "voluntariness" – that provides the moral ground; yet being liberals and not totalitarians, we put the legal weight on freedom – and hope people will use it generously. As a result, it looks as if, to facilitate and to justify the practice of organ donation, we have enshrined something like the notions of property rights and free contract in the body, notions that usually include the possibility of buying and selling. This is slippery business. Once the principle of private right and autonomy is taken as the standard, it will prove difficult – if not impossible – to hold the line between donation and sale. (It will even prove impossible, philosophically, to argue against voluntary servitude, bestiality, and other abominations.) Moreover, the burden of proof will fall squarely on those who want to set limits on what people may freely do with their bodies or for what purposes they may buy and sell body parts. It will, in short, be hard to prevent buying and selling human flesh not only for transplantation, but for, say, use in luxury nouvelle cuisine, once we allow markets for transplantation on libertarian grounds. We see here, in the prism of this case, the limits and, hence, the ultimate insufficiency of rights and the liberal principle.

Astute students of liberalism have long observed that our system of ordered liberties presupposes a certain kind of society – of at least minimal decency, and with strong enough familial and religious institutions to cultivate the sorts of men and women who can live civilly and responsibly with one another, while enjoying their private rights. We wonder whether freedom of contract regarding the body, leading to its being bought and sold, will continue to make corrosive inroads upon the kind of people we want to be and need to be if the uses of our freedom are not to lead to our willing dehumanization. We have, over the years, moved the care for life and death from the churches to the hospitals, and the disposition of mortal remains from the clergy to the family and now to the individual himself – and perhaps, in the markets of the future, to the insurance companies or the state or to enterprising brokers who will give new meaning to insider trading. No matter how many lives are saved, is this good for how we are to live?

Let us put aside questions about property and free contract, and think only about buying and selling. Never mind our rights, what would it mean to fully commercialize the human body even, say, under state monopoly? What, regardless of political system, is the moral and philosophical difference between giving an organ and selling it, or between receiving it as a gift and buying it?

Commodification

The idea of commodification of human flesh repels us, quite properly I would say, because we sense that the human body especially belongs in that category of things that defy or resist commensuration – like love or friendship or life itself. To claim that these things are "priceless" is not to insist that they are of infinite worth or that one cannot calculate (albeit very roughly, and then only with aid of very crude simplifying assumptions) how much it costs to sustain or support them. Rather it is to claim that the bulk of their meaning and their human worth do not lend themselves to quantitative measures; for this reason, we hold them to be incommensurable, not only morally but factually.

Against this view, it can surely be argued that the entire system of market exchange rests on our arbitrary but successful attempts to commensurate the (factually) incommensurable. The genius of money is precisely that it solves by convention the problem of natural incommensurability, say between oranges and widgets, or between manual labor and the thinking time of economists. The possibility of civilization altogether rests on this conventional means of exchange, as the ancient Greeks noted by deriving the name for money, *nomisma*, from the root *nomos*, meaning "convention" – that which has been settled by human agreement – and showing how this fundamental convention made possible commerce, leisure, and the establishment of gentler views of justice.

Yet the purpose of instituting such a conventional measure was to facilitate the satisfaction of *natural* human needs and the desires for well-being and, eventually, to encourage the full flowering of human possibility. Some notion of need or perceived human good provided always the latent non-conventional standard behind the nomismatic convention – tacitly, to be sure. And there's the rub: In due course, the standard behind money, being hidden, eventually becomes forgotten, and the counters of worth become taken for worth itself.

Truth to tell, commodification by conventional commensuration always risks the homogenization of worth, and even the homogenization of things, all under the aspect of quantity. In many transactions, we do not mind or suffer or even notice. Yet the human soul finally rebels against the principle, whenever it strikes closest to home. Consider, for example, why there is such widespread dislike of the pawnbroker. It is not only that he profits from our misfortunes and sees the shame of our having to part with heirlooms and other items said (inadequately) to have "sentimental value." It is especially because he will not and cannot appreciate their human and personal worth and pays us only their market price. How much more will we object to those who would commodify our very being?

We surpass all defensible limits of such conventional commodification when we contemplate making the convention-maker – the human being – just another one of the commensurables. The end comes to be treated as mere means. Selling our bodies, we come perilously close to selling out our souls. There is even a danger in contemplating such a prospect – for if we come to think about ourselves like pork bellies, pork bellies we will become.

We have, with some reluctance, overcome our repugnance at the exploitative manipulation of one human body to serve the life and health of another. We have managed to justify our present arrangements not only on grounds of utility or freedom but also and especially on the basis of generosity, in which the generous deed of the giver is inseparable from the organ given. To allow the commodification of these exchanges is to forget altogether the impropriety overcome in allowing donation and transplantation in the first place. And it is to turn generosity into trade, gratitude into compensation. It is to treat the most delicate of human affairs as if everything is reducible to its price.

There is a euphemism making the rounds in these discussions that makes my point. Eager to encourage more donation, but loath to condone or to speak about buying and selling organs, some have called for the practice of "rewarded gifting" – in which the donor is rewarded for his generosity, not paid for his organ. Some will smile at what looks like double-talk or hypocrisy, but even if it is hypocrisy, it is thereby a tribute paid to virtue. Rewards are given for good deeds, whereas fees are charged for services, and prices are paid merely for goods. If we must continue to practice organ transplantation, let us do so on good behavior.

Anticipating the problem we now face, Paul Ramsey twenty years ago proposed that we copy for organ donation a practice sometimes used in obtaining blood: those who freely give can, when in need, freely receive. "Families that shared in premortem giving of organs could share in freely receiving if one of them needs transplant therapy. This would be – if workable – a civilizing exchange of benefit that is not the same as commerce in organs." Ramsey saw in this possibility of organized generosity a way to promote civilized community and to make virtue grow out of dire necessity. These, too, are precious "commodities," and provide an additional reason for believing that the human body and the extraordinary generosity in the gift of its parts are altogether too precious to be commodified.

III The price of progress

The arguments I have offered are not easy to make. I am all too well aware that they can be countered, that their appeal is largely to certain hard-to-articulate intuitions and sensibilities that I at least believe belong intimately to the human experience of our own humanity. Precious though they might be, they do not exhaust the human picture, far from it. And perhaps, in the present case, they should give way to rational calculation, market mechanisms, and even naked commodification of human flesh – all in the service of saving life at lowest cost (though, parenthetically, it would be worth a whole separate discussion to consider whether, in the longer view, there are not cheaper, more effective, and less indecent means to save lives, say, through preventive measures that forestall end-stage renal disease now requiring transplantation: the definitions of both need and efficiency are highly contingent, and we should beware of allowing them to be defined for us by those technologists – like transplant surgeons – wedded to present practice). Perhaps this is not the right place to draw a line or to make a stand.

Consider, then, a slightly more progressive and enterprising proposal, one anticipated by my colleague, Willard Gaylin, in an essay, "Harvesting the Dead," written in 1974. Mindful of all the possible uses of newly dead – or perhaps not-quite-dead – bodies, kept in their borderline condition by continuous artificial respiration and assisted circulation, intact, warm, pink, recognizably you or me, but brain dead, Gaylin imagines the multiple medically beneficial uses to which the bioemporium of such "neomorts" could be put: the neomorts could, for example, allow physicians-in-training to practice pelvic examinations and tracheal intubations without shame or fear of doing damage; they could serve as unharmable subjects for medical experimentation and drug testing, provide indefinite supplies of blood, marrow, and skin, serve as factories to manufacture hormones and antibodies, or, eventually, be dismembered for transplantable spare parts. Since the newly dead body really is such a precious resource, why not really put it to full and limitless use?

Gaylin's scenario is not so far-fetched. Proposals to undertake precisely such body-farming have been seriously discussed among medical scientists in private. The technology for maintaining neomorts is already available. Indeed, in the past few years, a publicly traded corporation has opened a national chain of large, specialized nursing homes – or should we rather call them nurseries? – for the care and feeding solely of persons in persistent vegetative state or ventilator-dependent irreversible coma. Roughly ten establishments, each housing several hundred of such beings, already exist. All that would be required to turn them into Gaylin's bioemporia would be a slight revision in the definition of death (already proposed for other reasons) – to shift from death of the whole brain to death of the cortex and the higher centers – plus the will not to let these valuable resources go to waste. (The company's stock, by the way, has more than quadrupled in the last year alone; perhaps someone is already preparing plans for mergers and manufacture.) Repulsive? You bet.

Useful? Without doubt. Shall we go forward into this brave new world?

Forward we are going, without anyone even asking the question. In the twenty-five years since I began thinking about these matters, our society has overcome longstanding taboos and repugnances to accept test-tube fertilization, commercial sperm-banking, surrogate motherhood, abortion on demand, exploitation of fetal tissue, patenting of living human tissue, gender-change surgery, liposuction and body shops, the widespread shuttling of human parts, assisted-suicide practiced by doctors, and the deliberate generation of human beings to serve as transplant donors – not to speak about massive changes in the culture regarding shame, privacy, and exposure. Perhaps more worrisome than the changes themselves is the coarsening of sensibilities and attitudes, and the irreversible effects on our imaginations and the way we come to conceive of ourselves. For there is a sad irony in our biomedical project, accurately anticipated in Aldous Huxley's *Brave New World*: We expend enormous energy and vast sums of money to preserve and prolong bodily life, but in the process our embodied life is stripped of its gravity and much of its dignity. This is, in a word, progress as tragedy.

In the transplanting of human organs, we have made a start on a road that leads imperceptibly but surely toward a destination that none of us wants to reach. A divination of this fact produced reluctance at the start. Yet the first step, overcoming reluctance, was defensible on benevolent and rational grounds: save life using organs no longer useful to their owners and otherwise lost to worms.

Now, embarked on the journey, we cannot go back. Yet we are increasingly troubled by the growing awareness that there is neither a natural nor a rational place to stop. Precedent justifies extension, so does rational calculation: We are in a warm bath that warms up so imperceptibly that we don't know when to scream.

And this is perhaps the most interesting and the most tragic element of my dilemma – and it is not my dilemma alone. I don't want to encourage; yet I cannot simply condemn. I refuse to approve; yet I cannot moralize. How, in this matter of organs for sale, as in so much of modern life, is one to conduct one's thoughts if one wishes neither to be a crank nor to yield what is best in human life to rational analysis and the triumph of technique? Is poor reason impotent to do anything more than to recognize and state this tragic dilemma?

Note

† Leon R. Kass, M.D., *Toward a More Natural Science: Biology and Human Affairs* (New York: The Free Press, 1985).

Chapter 21

J. Radcliffe-Richards, A. S. Daar, R. D. Guttmann, R. Hoffenberg, I. Kennedy, M. Lock, R. A. Sells, N. Tilney, for the International Forum for Transplant Ethics

THE CASE FOR ALLOWING KIDNEY SALES

When the practice of buying kidneys from live vendors first came to light some years ago, it aroused such horror that all professional associations denounced it[1,2] and nearly all countries have now made it illegal.[3] Such political and professional unanimity may seem to leave no room for further debate, but we nevertheless think it important to reopen the discussion.

The well-known shortage of kidneys for transplantation causes much suffering and death.[4] Dialysis is a wretched experience for most patients, and is anyway rationed in most places and simply unavailable to the majority of patients in most developing countries.[5] Since most potential kidney vendors will never become unpaid donors, either during life or posthumously, the prohibition of sales must be presumed to exclude kidneys that would otherwise be available. It is therefore essential to make sure that there is adequate justification for the resulting harm.

Most people will recognise in themselves the feelings of outrage and disgust that led to an outright ban on kidney sales, and such feelings typically have a force that seems to their possessors to need no further justification. Nevertheless, if we are to deny treatment to the suffering and dying we need better reasons than our own feelings of disgust.

In this paper we outline our reasons for thinking that the arguments commonly offered for prohibiting organ sales do not work, and therefore that the debate should be reopened.[6,7] Here we consider only the selling of kidneys by living vendors, but our arguments have wider implications.

The commonest objection to kidney selling is expressed on behalf of the vendors: the exploited poor, who need to be protected against the greedy rich. However, the vendors are themselves anxious to sell,[8] and see this practice as the best option open to them. The worse we think the selling of a kidney, therefore, the worse should seem the position of the vendors when that option is removed. Unless this appearance is illusory, the prohibition of sales does even more harm than first seemed, in harming vendors as well as recipients. To this argument it is replied that the vendors' apparent choice is not genuine. It is said that they are likely to be too uneducated to understand the risks, and that this precludes informed consent. It is also claimed that, since they are coerced by their economic circumstances, their consent cannot count as genuine.[9]

Although both these arguments appeal to the importance of autonomous choice, they are quite different. The first claim is that the vendors are not competent to make a genuine choice within a given range of options. The second, by contrast, is that poverty has so restricted the range of options that organ selling has become the best, and therefore, in effect, that the range is too small. Once this distinction is drawn, it can be seen that neither argument works as a justification of prohibition.[7]

If our ground for concern is that the range of choices is too small, we cannot improve matters by removing the best option that poverty has left, and making the range smaller still. To do so is to make subsequent choices, by this criterion, even less autonomous. The only way to improve matters is to lessen the poverty until organ selling no longer seems the best option; and if that could be achieved, prohibition would be irrelevant because nobody would want to sell.

The other line of argument may seem more promising, since ignorance does preclude informed consent. However, the likely ignorance of the subjects is not a reason for banning altogether a procedure for which consent is required. In other contexts, the value we place on autonomy leads us to insist on information and counselling, and that is what it should suggest in the case of organ selling as well. It may be said that this approach is impracticable, because the educational level of potential vendors is too limited to make explanation feasible, or because no system could reliably counteract the misinformation of nefarious middlemen and profiteering clinics. But even if we accepted that no possible vendor could be competent to consent, that would justify only putting the decision in the hands of competent guardians. To justify total prohibition it would also be necessary to show that organ selling must always be against the interests of potential vendors, and it is most unlikely that this would be done.

The risk involved in nephrectomy is not in itself high, and most people regard it as acceptable for living related donors.[10] Since the procedure is, in principle, the same for vendors as for unpaid donors, any systematic difference

between the worthwhileness of the risk for vendors and donors presumably lies on the other side of the calculation, in the expected benefit. Nevertheless the exchange of money cannot in itself turn an acceptable risk into an unacceptable one from the vendor's point of view. It depends entirely on what the money is wanted for.

In general, furthermore, the poorer a potential vendor, the more likely it is that the sale of a kidney will be worth whatever risk there is. If the rich are free to engage in dangerous sports for pleasure, or dangerous jobs for high pay, it is difficult to see why the poor who take the lesser risk of kidney selling for greater rewards – perhaps saving relatives' lives,[11] or extricating themselves from poverty and debt – should be thought so misguided as to need saving from themselves.

It will be said that this does not take account of the reality of the vendors' circumstances: that risks are likely to be greater than for unpaid donors because poverty is detrimental to health, and vendors are often not given proper care. They may also be underpaid or cheated, or may waste their money through inexperience. However, once again, these arguments apply far more strongly to many other activities by which the poor try to earn money, and which we do not forbid. The best way to address such problems would be by regulation and perhaps a central purchasing system, to provide screening, counselling, reliable payment, insurance, and financial advice.[12]

To this it will be replied that no system of screening and control could be complete, and that both vendors and recipients would always be at risk of exploitation and poor treatment. But all the evidence we have shows that there is much more scope for exploitation and abuse when a supply of desperately wanted goods is made illegal. It is, furthermore, not clear why it should be thought harder to police a legal trade than the present complete ban.

Furthermore, even if vendors and recipients would always be at risk of exploitation, that does not alter the fact that if they choose this option, all alternatives must seem worse to them. Trying to end exploitation by prohibition is rather like ending slum dwelling by bulldozing slums: it ends the evil in that form, but only by making things worse for the victims. If we want to protect the exploited, we can do it only by removing the poverty that makes them vulnerable, or, failing that, by controlling the trade.

Another familiar objection is that it is unfair for the rich to have privileges not available to the poor. This argument, however, is irrelevant to the issue of organ selling as such. If organ selling is wrong for this reason, so are all benefits available to the rich, including all private medicine, and, for that matter, all public provision of medicine in rich countries (including transplantation of donated organs) that is unavailable in poor ones. Furthermore, all purchasing could be done by a central organisation responsible for fair distribution.[12]

It is frequently asserted that organ donation must be altruistic to be acceptable,[13] and that this rules out payment. However, there are two problems with this claim. First, altruism does not distinguish donors from vendors. If a father who saves his daughter's life by giving her a kidney is altruistic, it is difficult to see why his selling a kidney to pay for some other operation to save her life should be thought less so. Second, nobody believes in general that unless some useful action is altruistic it is better to forbid it altogether.

It is said that the practice would undermine confidence in the medical profession, because of the association of doctors with money-making practices. That, however, would be a reason for objecting to all private practice; and in this case the objection could easily be met by the separation of purchasing and treatment. There could, for instance, be independent trusts[12] to fix

charges and handle accounts, as well as to ensure fair play and high standards. It is alleged that allowing the trade would lessen the supply of donated cadaveric kidneys.[14] But although some possible donors might decide to sell instead, their organs would be available, so there would be no loss in the total. And in the meantime, many people will agree to sell who would not otherwise donate.

It is said that in parts of the world where women and children are essentially chattels there would be a danger of their being coerced into becoming vendors. This argument, however, would work as strongly against unpaid living kidney donation, and even more strongly against many far more harmful practices which do not attract calls for their prohibition. Again, regulation would provide the most reliable means of protection.

It is said that selling kidneys would set us on a slippery slope to selling vital organs such as hearts. But that argument would apply equally to the case of the unpaid kidney donation, and nobody is afraid that that will result in the donation of hearts. It is entirely feasible to have laws and professional practices that allow the giving or selling only of non-vital organs. Another objection is that allowing organ sales is impossible because it would outrage public opinion. But this claim is about western public opinion: in many potential vendor communities, organ selling is more acceptable than cadaveric donation, and this argument amounts to a claim that other people should follow western cultural preferences rather than their own. There is, anyway, evidence that the western public is far less opposed to the idea, than are medical and political professionals.[15]

It must be stressed that we are not arguing for the positive conclusion that organ sales must always be acceptable, let alone that there should be an unfettered market. Our claim is only that none of the familiar arguments against organ selling works, and this allows for the possibility that better arguments may yet be found.

Nevertheless, we claim that the burden of proof remains against the defenders of prohibition, and that until good arguments appear, the presumption must be that the trade should be regulated rather than banned altogether. Furthermore, even when there are good objections at particular times or in particular places, that should be regarded as a reason for trying to remove the objections, rather than as an excuse for permanent prohibition.

The weakness of the familiar arguments suggests that they are attempts to justify the deep feelings of repugnance which are the real driving force of prohibition, and feelings of repugnance among the rich and healthy, no matter how strongly felt, cannot justify removing the only hope of the destitute and dying. This is why we conclude that the issue should be considered again, and with scrupulous impartiality.

Notes

1 British Transplantation Society Working Party. "Guidelines on living organ donation," *BMJ* 1986; 293: 257–58.

2 The Council of the Transplantation Society. "Organ Sales," *Lancet* 1985; 2: 715–16.

3 World Health Organization. "A Report on Developments Under the Auspices of WHO (1987–1991)," *WHO* 1992 Geneva. 12–28.

4 Hauptman PJ, O'Connor KJ. "Procurement and Allocation of Solid Organs for Transplantation," *N Engl J Med* 1997; 336: 422–31.

5 Barsoum RS. "Ethical Problems in Dialysis and Transplantation: Africa," in Kjellstrand CM, Dossetor JB, (eds.) *Ethical Problems in Dialysis and Transplantation.* Kluwer Academic Publishers, Netherlands. 1992: 169–82.

6 Radcliffe-Richards J. "Nephrarious Goings on: Kidney Sales and Moral Arguments," *J Med Philosph.* Netherlands: Kluwer Academic Publishers, 1996; 21: 375–416.

7 Radcliffe-Richards J. "From Him that Hath Not," in Kjellstrand CM, Dossetor JB, eds. *Ethical Problems in Dialysis and Transplantation.* Netherlands: Kluwer Academic Publishers, 1992: 53–60.

8 Mani MK. "The Argument Against the Unrelated Live Donor," ibid. 164.

9 Sells RA. "The Case Against Buying Organs and a Futures Market in Transplants," *Trans Proc* 1992; 24: 2198–202.

10 Daar AD, Land W, Yahya TM, Schneewind K, Gutmann T, Jakobsen A. "Living-donor Renal Transplantation: Evidence-based Justification for an Ethical Option," *Trans Reviews* 1997.

11 Dossetor JB, Manickavel V. "Commercialisation: the Buying and Selling of Kidneys," in Kjellstrand CM, Dossetor JB, eds. *Ethical Problems in Dialysis and Transplantation.* Netherlands: Kluwer Academic Publishers, 1992: 61–71.

12 Sells RA. "Some Ethical Issues in Organ Retrieval 1982–1992," *Trans Proc* 1992; 24: 2401–03.

13 Sheil R. "Policy Statement from the Ethics Committee of the Transplantation Society," *Trans Soc Bull* 1995; 3: 3.

14 Altshuler JS, Evanisko MJ. *JAMA* 1992; 267: 2037.

15 Guttmann RD, Guttmann A. "Organ Transplantation: Duty Reconsidered," *Trans Proc* 1992; 24: 2179–80.

Chapter 22

C. Cohen

THE CASE FOR PRESUMED CONSENT TO TRANSPLANT HUMAN ORGANS AFTER DEATH

A necessary condition for the lifesaving uses of the organs of a human body after death is the permission of those who have rightful authority over the body in question. Each of us is the possessor, master, of our own body, and therefore (it is almost universally agreed) we have wide moral authority to give or withhold permission for the use of our body after death.

Just rules for the disposition of human bodily remains therefore must respect individual autonomy in that disposition. Autonomous expressions of will regarding the posthumous disposition of one's organs are most often not made while alive, and therefore a decedent's autonomous judgments are rarely known with certainty. After death, authority over the body is commonly thought to rest with the decedent's family, who are likely to represent best the true will of the decedent. But even they cannot exercise the decedent's autonomy, since no one can do that.

Presumed consent

A great change in our national system of organ procurement, a change grounded in the moral foundation of human autonomy, is now called for. Current American practice tacitly assumes that, absent specific notification to the contrary, decedents are best protected if we act as though they had autonomously willed that their organs not be donated for transplantation. Organ procurement therefore now relies utterly upon consent expressly given (by the decedent before death, or by his family after death) that rebuts this presumption. I argue that this system of rules, formal and informal, ought to be wholly reformed; the underlying assumption that ought to be made is the very opposite of the one now made. To protect the autonomous wishes of decedents we ought to assume that they did will or would have willed their organs for beneficent medical uses. Expectations would be reversed under the reformed system; the normal pattern would be one in which lifesaving transplants of cadaveric organs proceed as a matter of course unless consent for such uses had been expressly refused. Absent express refusal, no permission for the donation of organs need be sought from any party, no discussion of any kind being required save that called for by purely medical considerations. Presumed consent is the name commonly given to the system here proposed.[1]

This is certainly not a new idea; it was first brought to wide attention by Dukenminier and Sanders in 1968,[2] and has since been entertained by a number of others, sometimes only halfheartedly or as a possible recourse to which our need for cadaver organs unhappily drives us.[1,3-6] I submit, however, that a system in which consent is presumed is not merely expedient or advantageous; it is also just. Such a system is good because it maximizes benefits for all concerned. Because it best protects the autonomy of decedents it is also right, more right than the system now employed.

Needs and goods

What is good for society is largely a function of what its members need. We need a great many more organs for transplant than are presently procured. The gravity of that need, and the likelihood that it will increase as the years go on, I take to have been established by others, at this conference and elsewhere. Therefore, the existing system of organ procurement depending utterly upon express consent, almost always sought from the decedent's family at the time of the loved one's death, is not very good and certainly is not good enough.

The enlargement of organ supply required to meet these compelling needs does not appear feasible within the current framework of express consent, whether that consent be sought voluntarily or the request for consent be required by law. Whatever the reasons for this – the failure of young persons to express their judgment while healthy, the psychological stress upon families at the time of the dying of a loved one, the reluctance or awkwardness of physicians and administrators in making donation requests as patients are dying, or others – long experience teaches that circumstances commonly conspire to block the needed express consent for the donation of the organs of a decedent. In sum, the need for organs is great and will increase, while the present system of procurement through express consent fails and almost certainly will continue to fail to meet that need.

We may conclude without serious doubt that overall human well-being will be substantially improved if some way were found, within the boundaries of morally right conduct, to increase greatly the supply of human cadaver organs for transplantation. A system of presumed consent would very probably increase the supply of needed organs vastly. This is the outcome one would expect, and it is confirmed by experience with presumed consent in other countries,[4] although even where presumed consent is operative the needs for cadaver organs are not fully met.[7] The disadvantages of presuming consent are minimal, nearly nil. On balance, therefore, an organ procurement system founded upon presumed consent is almost certainly good.

What is right to presume?

For utilitarian moralists, therefore, the justifiability of the proposed reform is plain. But for those of us whose deepest moral principles are not grounded in utility, serious moral questions remain. Is the presumption of consent right? Principles of right conduct ought not be sacrificed to improve the balance of results. Can a system of presumed consent be fully reconciled with the autonomous disposition of human organs by those having rightful authority over them? The answer is yes.

Putting the question in that way, however, suggests that there is some difficulty in the reconciliation. Suppose the critical question were posed in this way: How can we most fully realize most persons' autonomous wishes concerning the disposition of the organs of their own bodies after their death? While healthy, most persons do not confront the matter; while dying most cannot decently be asked their preferences; when dead all are silent. In the great run of cases, therefore, we do not learn what we would most like to know at the time we need to know it: the actual desires of persons whose body organs are at issue. Therefore, under any system whatever, we are forced to make some very general presumptions concerning the wishes of persons about the uses of their organs after their deaths. And under any system we must receive and implement autonomous expressions of preference that do not conform to the general presumption.

That is the form of what we do now, but the present substance is problematic. We presume now for all persons that there was a will not to donate; then under some circumstances we seek express consent to do what we had presumed would not have been wanted. That express consent we most often seek from the family of the decedent, there being no other resource.

The consequences of presuming in this pattern are often gravely unfortunate; familial distress is commonly caused by the mere request for consent, and the refusal of it often results in the loss of human lives that might otherwise have been saved. But in addition to its unfortunate results, the present system is wrong, wrong because it undermines in practice the very principle upon which it was supposed to have been built. As it must, it incorporates into law and practice a presumption about what people generally want to happen to the organs of their bodies after death, but the presumption thus incorporated is one now known to be inconsistent with the actual wishes of most persons, at least, of most persons in these times in the United States. We honestly aim to protect autonomy, but by our current practice we commonly vitiate it.

The majority of people now heartily support the concept of organ donation. Most persons, when asked, express without qualification their willingness to donate their own organs after death. As early as 1968, support in the United States for organ transplantation from cadavers was shown to be strong and widespread.[8] By 1975, still an early date from the perspective of organ transplantation, a majority of those even in rural and relatively unsophisticated areas expressed positive support for organ transplantation.[9] In

regions more sophisticated, in Los Angeles County, for example,[10] and in Houston,[11] those supporting organ donation (even back in 1975) were more than three fourths of the whole. In Liverpool in 1979 that figure was 93 percent.[12]

Some who support transplantation are nevertheless uneasy with the vision of the donation of their own organs. Every person conscious of his own will has some difficulty in picturing his own death. But what most Americans would say in their most rational moments, and (if they were in a position to be asked) would most likely say about the lifesaving uses of their own bodily remains, is clear: Yes, our organs may be used by others, if this will save lives. The present system, depending entirely upon the express consent of the decedent's family after death, thus errs in its empirical underpinning, and by that error promotes a great moral mistake.

All too frequently the spouses and children of dying patients respond negatively to the request for donation. But families are thus being obliged to answer a question terrible, at the moment it is being asked, for the very reason it must then be asked. And they are being asked at that moment to override what is widely presumed. Under such circumstances the responses of families often do not reliably reflect the autonomous wishes of decedents. We ask the wrong persons, at the worst possible times, what they should never have been asked at all.

As a matter of morality, our weightiest obligation here is to decedents, about whose wishes we (usually) must make some presumption. To best realize their autonomy we should presume what we have strong empirical reason to believe was in fact their wish, that if their organs might be used to save another life, they ought to be so used.

Presuming general consent to organ donation is therefore the right thing to do; that is the chief reason, and a very powerful reason, to turn the present system right side up. At the moral core of this matter lies autonomy, of which consent in the disposition of one's body is one manifestation; presuming consent for beneficent transplantation of organs is the best, imperfect but still the best, realization of autonomy in any population like ours that strongly favors the donation of cadaveric organs.

Other good consequences of presumed consent

By this morally right presumption we may improve and prolong the lives of persons who would otherwise have died. But other great goods also ensue. Grieving families are given enormous relief. Many people who, when rational and calm, would donate their own organs without qualm, want not to think about the matter when not obliged to do so, and when forced to make that decision for others at moments of despair and stress, are agonized. At the very moment when the removal of a loved one's vital organs is most dreadful to contemplate, when feelings of guilt or helplessness are most likely to distort calm judgment, grieving families need not confront the matter. Moreover, this resolution of the matter by an earlier, and universally understood presumption, would be a service not only to the bereaved, but also to physicians and nurses who are relieved of the need to ask questions laden with pain and doom, as grief-ridden families are relieved of the need to answer them.

Protecting those who object

The operational details of a system of presumed consent would be important, of course, but they present no insuperable difficulties. Some persons do not wish to have their organs removed for any purpose, even to save lives after their deaths. To them it must be said, without hesitation or rancor, "as you wish." Giving to every person the opportunity, while alive, freely to opt out of the system of general donation is a social obligation entailed by respect for individual autonomy, and an obligation readily fulfilled. Any individuals, *for any reason or without reason*, must have and will have the right and the fullest opportunity to cancel the presumption in their own case, and that simply by registering this wish, without need for justification or argument or delay, or for any other judgment by any other person.

Once the system of presumed consent has become widely understood (and probably even when first introduced) those who opt out will be a minority, and the interests of that minority must be carefully respected. A national computerized registry will be required, to which there will be appropriate access by authorized individuals and by hospitals, so that those who do object may record their wishes in ways that will protect the autonomy of their judgments. In short: any person's exercise of the option to demur, for any reason whatever, must settle the matter for that person's remains. No one else need ever be asked anything.

Transitional matters

The transformation of the present system of organ procurement into one of presumed consent (with ready option out) must take place in ways that will not result in surprise or disadvantage. Of no one may it later be said that objections would have been registered if only the rules had been known. Wide public education must therefore precede the reversing reform, and the revised presumption must be openly and clearly expressed in ways that all may fully grasp. That is surely within our power. The twice-annual shift from standard to daylight saving time and back is infuriating to many, and more coercive than the shift proposed here, but (I observe) very few are they who, the very next day, do not know the correct time. When there is a widespread understanding that human organs are an absolutely priceless resource, wasted only at the cost of life, most persons will be proud of this shift. But for those who remain troubled by it the fullest opportunity to opt out must be given, their recourse made simple, convenient, and always revocable. In less time than we now may think, I hazard, most will wonder how ever it could have been any other way. Did once people need to give express permission in order that another's life be saved by what would otherwise soon rot? Did once people commonly presume that some must die because others were in psychological distress? It was that way once, but that was back in unenlightened times.

Autonomy and the family

Until the presumption of consent has become widely understood and assimilated, objections to the donation of organs by the decedent's next-of-kin will have to be respected. This family veto is likely to reduce the supply of needed organs at first, but authorizing it will smooth passage into the new system. Eventually, I believe, we will come to regard the authority of the family to block donation as more a burden to the decedent's autonomous will than a safeguard of it.

On the other hand, if any individuals while competent register an affirmative expression of their will that their organs be donated if they prove usable after death, by this act converting a presumed consent into consent expressly given, the will so expressed ought not be subject to contravention by the will of any others, even that of family members.

Presumed consent has been criticized by some as insensitive to the ethical demands of bereaved families,[13] but this complaint misconceives the ethical issue. The psychological well-being of bereaved families must be safeguarded, of course, and will be. But at bottom the moral authority for consent to donate organs lies only with the person whose organs they are, or were. Families properly enter, in this as in other proxy contexts, to represent the will of those who cannot speak for themselves; however good their motivation, family members ought not to be permitted to contravene the wills of those with genuine moral authority in the matter.

In spirit and in detail an organ procurement system should aim to realize the will of the donor while alive, and to preclude the frustration of that will. With very few exceptions, imposed in special circumstances, we are properly sovereign over our own bodies while alive, and our wishes concerning the uses of our organs after death deserve continuing respect. Autonomy ought to be the ruling principle in this sphere; under a system of presumed consent it will be, and if not perfectly realized at least autonomy will be more fully realized then than it is now.

Objections and replies

Objections of three kinds have been registered to systems of presumed consent. Some are mechanical, some sociologic, some moral; all fail.

1. Mechanical or technical objections are raised to the workings of the system. The registry of those opting out would become too complicated, it is said, or some might fail to get ready access to it, and so on.[14] Others are troubled by the fact that maintaining uniformity across state boundaries will present a problem.[15] Such objections are essentially insignificant. If a change to presumed consent is fundamentally wise we can surely make it work, and we can devise the needed machinery to smooth its operation in practice. Occasional breakdowns there will be under any system, of course; in a few cases, especially at the outset, the presumption of consent may result in some organs being

used that should not have been. But when the common presumption made is consonant with the common will on the matter, misfire is much less likely than when (as under the present system) it is dissonant. Moreover, while the life-saving use of organs that ought not to have been used cannot be right, it is at least a wrong more tolerable than the wrong of not using organs that should have been used. In choosing between the two approaches, each subject to some operational failure, it is surely wise to implement the one whose failures are likely to be fewer and whose results are certain to be better.

2. What I call "sociologic" objections are rooted in the fear of attitudes or practices that the new system will (allegedly) promote. The fears are various:

(a) That organ farms or other macabre fantasies will be encouraged by a system of presumed consent.
(b) That the spirit of voluntarism will be undermined.[14]
(c) That the distrust of physicians will be increased because, as the routine harvesters of organs, they will come to be viewed as persons in whose hands a very sick person cannot be safe. Or that hospitals will by this change be transformed in the popular imagination into places of bodily mutilation and brutality. May[16] writes:

While the procedure of routine salvaging may, in the short run, furnish more organs for transplants, in the long run, its systemic effect on the institutions of medical care would seem to be depressing and corrosive of that trust upon which acts of healing depend.

Such dismal speculations, and others like them, and the objections to which they give rise, have no good empirical foundation. Anecdotal horror stories, in a context of antipathy toward medical researchers and hospitals, are magnified by imagination into what may be called anticipatory speculative condemnation. Anxiety and mistrust of the same sort long accompanied the development of recombinant DNA technology, we will recall, and commonly arise when changes in old ways are proposed. The real empirical consequences of presuming consent, of which we cannot now be certain of course, are in fact as likely to support humane care as to undermine it. The life-enhancing possibilities of organ transplantation, the change of focus from those dying to those who may yet live, may bring a fuller appreciation of the gratitude of organ recipients and their families, and may prove a great boon for humanity, celebrating as it does the most wholesome and productive of human values. Widespread recognition of the goods achieved by organ transplantation may do more to enhance the spirit of voluntarism than to erode it. All such claims are speculative; the fears have as little foundation as the hopes. We ought not to treat ungrounded speculations as rational objections to a presumption about the common will that we know to be correctly applicable to most people.

Those who find the removal of organs from a cadaver to be frightening, or self-seeking, or otherwise unscrupulous, will of course attack any system that promotes organ transplantation as brutal. Those who make of physicians common objects of abuse will feel threatened by any proposal appearing to enlarge their authority. Ghastly misbehaviors in the handling of dead bodies are of course possible under any system whatever. Avoiding the ugly, the abusive, and the insensitive is a matter of wise and intelligent administration. Wisdom and intelligence may on occasion prove wanting, but their want is in no case likely to be the consequence of the presumption made about the will of decedents. And what is most insensitive after all, even macabre, is watching persons die whose lives could and should have been saved, but were not.

3. Finally, account must be taken of two genuinely moral objections:

(a) The first is based on the moral conviction that the process of harvesting human organs is intrinsically wrong. Some (but not all) Orthodox Jews, seeking to respect Divine command, and some who hope for a resurrection of their bodies in the afterlife have this conviction. The religiosity of such objectors we must respect, but their convictions cannot be allowed to block an otherwise justifiable change in the presumption of consent, so long as these persons and all others are clearly free to reject the presumption of consent effectively in their own case, and are unhindered in doing so. With those who hold such views no contest is in order, nor any effort to persuade or bring pressure. Convictions about the history of the body after death, or other supernatural beliefs causing organ removal to be thought unacceptable, are not the business of the community at large. We may all believe what we please, and all must be free to work our own will regarding our own organs, without objection or obstruction.

(b) A second moral objection rests upon what is claimed to be a critical difference between consenting and not objecting. If we rely upon express consent we realize autonomy (this critic holds), but if we rely only upon the absence of objection we may fail to do so. Hence a procurement system built upon express consent is always morally sound even if clumsy, while one built upon the presumption of consent could not be reliably sound even if it were expeditious.

Wrong. There is a difference between express consent and not objecting, but that difference cannot guide us in a moral choice between the one presumption and the other. If persons do in reality object to the use of their organs (but never register that objection) a system that requires express consent will protect his autonomy more surely than the revised system here defended. But with the presumption reversed, the very same point can be made in reverse: if one does in reality consent to the use of one's organs (as most of us do, although never registering that consent) a system that presumes *consent* will protect his autonomy more surely than the present system can.

The difference in focus here is that between positive acts (expressly consenting or expressly refusing consent) and negative acts (refraining from refusing or refraining from consenting). This is an operational difference, not a moral one. It can have moral consequences, but the merit of proceeding in the one way or the other depends largely upon what we believe to be the general inclination of those about whom one of those presumptions must be made. If we knew that only one or two persons in 10 would autonomously donate their organs, a system that presumed consent, protecting 10 percent automatically but obliging the other 90 percent to register their objections to make their will effective, would be unfair. But if we have good reasons to believe that 7 or 8 of 10, or even 6 of 10, would in fact choose to donate their own organs for lifesaving uses after death, a system that presumes the absence of consent (what we have now) similarly protects a minority and obliges the majority to register their views expressly, and it is then unfair.[6]

Whether we require consent to be expressed, or require refusal to be expressed, should depend upon what we believe the majority would have done in fact, if all had registered their views. We may presume one way, or presume the other, but presume we must. Either way we place a heavier load upon those for whom the presumption made is incorrect. Moral principles by themselves give no indication which presumption is the fairer. Acting justly requires respect for the autonomous judgments people would actually have made, or (if we are unsure of that) what we may most reasonably suppose they would have made. That reasonable supposition we can reliably make. Therefore, presuming that consent would have been given is in fact fairer, more protective, and more likely to realize autonomy than presuming (as we do now) that it would not have been given.

Notes

1 Matas AJ, Veith FJ: *Theor Med* 5:155, 1984

2 Dukenminier J, Sanders D: *N Engl J Med* 179:413, 1968

3 Caplan AL: *Hastings Cent Rep* 13:23, 1983

4 Hull AR: *Nephrol News Issues* October:28, 1990

5 Matas A, Arras J, Muyskens J: *J Health Polit Policy Law* 10:231, 1985

6 Menzel PT: *Strong Medicine.* Oxford University Press, 1990, p 171

7 Childress J: *Hearings before the Subcommittee on Investigations and Oversight of the Committee on Science and Technology,* US House of Representatives, 1983, 281

8 Gallup G: *The Gallup Report.* Princeton, NJ: Gallup, January 17, 1968

9 Kidney Foundation of Eastern Missouri, *Assessment of Public and Professional Attitudes Regarding Organ Transplantation,* St Louis, Mo, 1975

10 Transplantation Council of Southern California, *Public Opinion and Attitudes about Medical Transplantation Among Los Angeles County Residents,* Los Angeles, 1975

11 Cleveland SE: *Psychosom Med* 37:306, 1975

12 Sells RA: *J Royal Soc Med* 72:109, 1979

13 Mahoney J: *J Med Ethics* 1:67, 1975

14 Sadler AM, Sadler BL: *Hastings Cent Rep,* October: 6, 1984

15 Prottas J: *Hearings Before the Subcommittee on Investigations and Oversight of the Committee on Science and Technology,* US House of Representatives, 1983, 745

16 May W: *Hastings Cent Rep* 1:6, 1973

Chapter 23

R.M. Veatch and J.B. Pitt

THE MYTH OF PRESUMED CONSENT: ETHICAL PROBLEMS IN NEW ORGAN PROCUREMENT STRATEGIES

The acute shortage of organs for transplantation has led to considerable interest in laws that are designed to increase the number of organs procured. These laws are often referred to as "presumed consent" laws. Such laws are alleged in many popular and scholarly articles to exist in several European countries and Singapore, among other places. The reasoning behind recent arguments in favor of adopting a so-called "presumed consent" law in the United States is that if we can presume the consent of the deceased to organ procurement there will be a substantial increase in the yield of organs.

The difference between consent and salvaging

The problem with this approach, however, is that, with a few exceptions, the existing laws never actually claim to presume consent, nor can they rightly be said to do so. They simply authorize the state's taking of the organs without explicit permission. It therefore seems wrong to call them presumed consent laws. They are, in effect, what used to be called *routine salvaging* laws.[1] We believe the time has come to be more careful in distinguishing between policies of presumed consent and those of routine salvaging. While the net outcome may be the same under either kind of policy, the underlying assumptions about the relation of the individual to society are radically different.

It is our hypothesis that those who support a societal right to procure organs without consent find it embarrassing to speak bluntly about taking organs without consent, hence they adopt the *language* of presuming consent even when there is no *basis* for such a presumption. In doing so, they preserve the appearance of the preferred gift-mode and the guise of respect for individual choice. (This desire for euphemistic language is also seen in the persistent practice of referring to persons from whom organs are taken as *donors* even in cases, such as small children, in which these people could never have actually made a gift or donation.) We shall suggest that important matters of societal relations are at stake in distinguishing between policies that allow the procurement of organs on the presumption that people would consent and those that simply take organs without consent. One form of society gives central place to the individual, holding that his or her person can be used by the state only with some form of consent. That has been the society of liberal Western culture, particularly the United States. It underlies the gift-mode and the doctrine of consent that has been central not only to organ procurement, but to the practice of medicine in general, for decades.

Another form of society gives more central authority to the state, authorizing it to use the individual for important societal purposes even without individual consent. It underlies routine salvaging, or the taking of organs without consent. For the purposes of this article, we are not pressing for one form of policy or the other. It is possible that the time has come for elevation of the state by adopting a routine salvaging law. That seems to be the rationale behind new movements toward enhancing communitarianism and stressing the common good in social policy.[2] It is also possible that the importance of the individual continues to require procuring organs in the gift or donation mode in which organs may be taken only with proper permission. The conflation of these two is, we shall suggest, a dangerous prospect indeed.

The state of the law

Countries with routine salvaging laws (with no claim of presuming consent)

It is striking that it is so common for commentators to refer to these laws as "presumed consent" laws. For example, according to Gerson, the French law on organ procurement adopted in 1976 is one which presumes the consent of persons who do not, during their lifetime, expressly refuse to have their organs taken upon their death.[3] However, on examination of the law itself, one is hard pressed to find any mention of presuming consent, overt or implicit. The law states that "An organ to be used for therapeutic or scientific purposes may be removed from the cadaver of a person who has not during his lifetime made known his refusal of such a procedure."[4] Although the law offers a provision for those willing and able to record their dissent, it is not clear why we should conclude that the rationale behind the opting out system it establishes is based upon the presumed consent of

the decedent rather than the primacy of the state. Gerson, citing an article by Cantaluppi, also attributes presumed consent laws to Austria, Belgium, (the former) Czechoslovakia, Finland, Italy, Norway, Spain, and Switzerland, among other countries[4] (p 1019, note 35). Not one of these laws mentions anything about presuming consent, directly or indirectly.[5–13] Among the other countries that have laws authorizing organ procurement without claiming to presume consent are Cyprus,[14] Hungary,[15] Singapore,[16] Syria,[17] and the former Yugoslavia.[18] Some of these have been referred to in the literature as countries with presumed consent laws, yet none of them actually claims to presume consent in its legislation.

Laws with an explicit presumption of consent

By contrast we have located a few laws and proposed laws that do actually state a presumption of consent or its equivalent. For example, the Colombian law on organ procurement states that, "... there shall be a legal presumption of donation if a person during his life time [sic] has refrained from exercising his right to object to the removal from his body of anatomical organs or parts during his death ..."[19]

Within the United States at least two states have recently considered laws that would properly be called presumed consent laws. In Maryland, a bill proposed on March 10, 1993, had it not been defeated, would have allowed for the presumption of consent of those who did not opt out. It read, "In the absence of specific objection by an individual expressed during that individual's lifetime, or by any of the individual's next of kin immediately following the individual's death, the individual is deemed to have consented to the donation of the individual's body or any part of the individual's body for and of the purposes specified ..."[20] In Pennsylvania, a subchapter entitled, "Presumed Anatomical

Gifts," of a proposed amendment to an act, reads, "Organs and tissues may be removed, upon death, from the body of any Commonwealth resident by a physician or surgeon for transplantation or for the preparation of therapeutic substances, unless it is established that a refusal was expressed ..."[21] Both of these proposed law changes adopt the language of presuming consent. This, of course, begs the question of whether consent can actually be presumed in these jurisdictions at all.

Why consent cannot validly be presumed

Although the difference between the European laws that do not presume consent and the New World laws and proposals that do presume it may seem small, matters of fundamental importance are at stake. It is important to see why consent cannot validly be presumed in the present cultural environment.

To presume consent is to make an empirical claim. It is to claim that people *would consent* if asked, or, perhaps more precisely, that they would consent to a policy of taking organs without explicit permission. The reasoning behind true presumed consent laws is that it is legitimate to take organs without explicit consent because those from whom the organs are taken would have agreed had they been asked when they were competent to respond.

That, however, is a claim which, if it is to be made with authority, must be corroborated with empirical evidence. Social survey evidence makes clear that if we assume people would agree to having their organs procured if they were asked, we would be wrong at least 30 percent of the time. A recent 1993 Gallup poll shows that only 37 percent of Americans are "very likely" to want their organs transplanted after their death, and only 32 percent are "somewhat likely." Furthermore, only 55 percent are willing to grant formal permission for organ removal. It should

also be noted that although 55 percent are willing to grant permission, only 28 percent have *actually* done so. (The Gallup Organization, Inc, conducted for The Partnership for Organ Donation, Boston, MA, February 1993, pp 4, 15.) In other words, only about half of the Americans who are willing to grant permission have taken the proactive steps necessary to do so, creating a large number of *false negatives*. We might expect that if ours were an opting out system, we might also see a large number of *false positives*. Based even on the larger figure of 69 percent who would be either "very likely" or "somewhat likely" to want their organs to be transplanted, it is clear that there can be no basis for presuming consent. Claiming such a presumption is an ill-informed notion at best; it is an outright deception at worst.

Perhaps even more pertinent to this discussion are the relative proportions of Americans who would agree to the system of presumed consent itself, as the ethos of the presumed consent mode would seem to demand. One recent survey shows that only 38 percent of Americans agree with presumed consent, defined as a system in which doctors routinely remove organs from deceased persons unless the person indicated a wish to the contrary while alive.[22] Another survey shows that number to be only seven percent.[23]

To gain a better understanding of the issues involved, a comparison with the presumption of consent to treatment in an emergency room is helpful. When people suffer accidents or heart attacks that render them incapable of consenting to medical treatment, they are rushed to an emergency room where they are treated by the hospital team. They are treated without explicit consent. This policy is defended on the grounds that consent is presumed.[24] Under these circumstances such a presumption allows us to preserve the notion that people can receive medical treatment only with their consent.

In the case of the emergency room treatment of the patient incapable of giving explicit consent, the presumption of consent is surely valid. Were we to conduct a survey of the population asking its members whether they would want such a presumption made, agreement would be close to unanimous. To be sure, some small group would object. A patient who is a Jehovah's Witness may refuse blood products; a Christian Scientist may refuse treatment altogether. This reveals that on occasions the presumption of consent in the emergency room may be an erroneous presumption (it will, on occasion, yield *false positives*). But it will be accurate an overwhelming percentage of the time, and the presumption is therefore justified.

By contrast if we presume consent in the case of organ procurement, we will be wrong at least 30 percent of the time. (It is interesting to ask exactly what percentage of people would have to agree to a policy when surveyed before we can presume that individuals being treated by that policy would have consented. One's first instinct might be to assume that a majority must indicate endorsement of the policy, but that surely is wrong. It would lead to erroneous presumption of consent as much as half the time. One possibility is to take a figure of 95 percent approval in a survey as sufficient to presume that any one individual would have consented, if asked. That would mean five percent of the time we would have erred in presuming the individual would have consented. Even then the rights of individuals would be violated five percent of the time.) In a society that affirms the right of the individual not to have his or her body invaded without appropriate consent, procuring organs on the basis of a presumption of consent will violate that right at least 30 percent of the time.

What is at stake

What is at stake is something very fundamental: the ethics of the relation of the individual to the society. A pioneer in the study of contemporary medical ethics, Paul Ramsey, introduced the issue in distinguishing between organ procurement in the modes of "giving" and "taking."[25] In liberal Western society certain rights are attributed to the individual. Among these is the right to control what is done with one's body. Hence, in Western culture medical treatment is acceptable only with the consent of the individual or the individual's appropriate surrogate. Research on a human subject is ethically acceptable only when consent is obtained. According to the Nuremberg Code, such voluntary consent is absolutely essential. An individual is in a position whereby he or she has the authority to give to society by authorizing medical research and now by authorizing procurement of organs for transplant, research, therapy, and other purposes.

The alternative is the mode of "taking" or what Dukeminier and Sanders called "routine salvaging." In this model the central authority has claims over the individual without relying on the individual's consent or approval. In the model of presumed consent, the individual is before the state; in the alternative the individual is subordinate. This underscores the problems associated with casual misuse of the term, "presumed consent." Many authors merely confuse routine salvaging for presumed consent, claiming that they are the same thing.[26,27] Others have implied that their versions of so-called "presumed consent" can be justified by the concept of eminent domain.[28,29] However, eminent domain involves the taking of private property for public use, and has no bearing on questions of consent. Clearly a system which validly presumes the consent of persons does not – cannot – rely upon notions of eminent domain.

Choosing the language of legitimating organ procurement is, in effect, choosing how we want to see the individual in relation to the state. Those who use the language of presumed consent are trying to hold on to the liberal model in which gift-giving is the foundation of organ procurement. In cases in which consent can validly be presumed, presumed consent seems consistent with such an orientation. However, in cases in which the evidence makes clear that consent cannot be presumed, this language is simply a disguise for the less acceptable reality of state authority over the individual.

This in itself, of course, does not make routine salvaging wrong; it is, however, deceptive if one advocates such a relationship in the name of the more liberal mode of gift-giving and consenting. Such deception is a moral affront to members of a society built upon respect for the rights of the individual.

It is worth speculating why there is this strong propensity to use the language of presuming consent when the apparent intention is to take organs without consent. One possibility is that, at least in countries reflecting liberal political philosophy's affirmation of the rights of the individual, it is more comforting to use the language of gift-giving and consent. It leaves the impression of the priority of the individual. Thus there is a strong tendency to use the language of the gift mode (words such as "donor"), even in cases in which the source of the organs may be a small child who never could have made an actual donation and in cases in which a medical examiner rather than the individual whose organs are being taken is the one approving the procurement. The language of consent is a more comfortable language, one that may be necessary to win approval of policies that de facto authorize procurement without donation.

The policy implications; alternatives to presuming consent

Routine salvaging

There are a number of alternatives to persisting in calling organ procurement "presumed consent" in cases where there is no explicit individual permission. One would be to follow Dukeminier

and Sanders and bluntly call it "routine salvaging." That at least is an honest policy. That would be appropriate if one wanted to refer to a policy that is grounded not in a presumption of individual consent, but in a belief that the society had a right to procure organs without individual consent.

Routine salvaging with opting out

Most versions of routine salvaging policies include a provision that permits individuals to opt out by executing a document explicitly asking not to have organs procured.[30–33] These policies place the burden on the individual actively to signal a desire not to have organs procured. Such policies, however, are still not presuming the consent of those who fail to opt out (or at least there is no valid basis for presuming that those who fail to opt out would consent if asked). Some people might not even know about the organ procurement policy or its opting out provisions. Surveys show that persons of lower education are far less likely to consent to organ donation, yet it is precisely this group of people who would run the highest risk of not knowing the proper procedures for opting out of the system.[23] Others might be overwhelmed by other concerns so that they simply never execute the necessary document. Many people do not execute economic wills during their lifetimes. If they do not, their assets will be dispersed at the time of their deaths according to default state policy. It would be a mistake, however, to presume that all who die intestate have consented to or favor the state's default provisions. It would be more appropriate to say that because of their inability or lack of willingness to complete a will, such persons have unfortunately not been able to secure their own preferences. A routine salvaging policy with opting out shows more respect for the wishes of the individual than does a straight salvaging law without such a provision, but it still involves the taking of organs. It replaces the gift-mode with one of taking without asking for permission.

Required request of next of kin

Another option is the type of law that requires hospitals to request permission of the next of kin in cases in which the deceased has not explicitly expressed his or her preferences for or against organ procurement.[34,35] Such policies are an uneasy intermediary between the gift and salvaging modes. They imply that the state needs permission, but not the permission of the deceased.

There are both practical and theoretical problems with required request of the next of kin. At the practical level, there are doubts that such required requests significantly increase the yield of organs.[36,37] There may even be backlash refusals of relatives to cooperate if they feel they have been treated crassly in a moment of family crisis. At the theoretical level, if the moral foundation of procuring organs is the consent of the individual whose organs are procured, then next-of-kin approval is, at best, a poor substitute. It may be the best available alternative in cases in which there is no knowledge of the deceased's wishes, but the better course, ethically and practically, for those who rest organ procurement in the approval of the individual whose organs are going to be procured, would be a policy that maximizes the opportunity for the individual to record his or her wishes, not one that rests on the wishes of a family member.

Required response

An option that preserves the gift mode (and the correlative affirmation of respect for the integrity of the individual) is a policy we have called *required request*[38] and routine inquiry,[39] but might better be referred to as *required response*. The United Network for Organ Sharing Ethics Committee's Presumed Consent Subcommittee has recently endorsed "required response" as an alternative to presuming consent.[40] It would provide for organized requests for individuals to consent or refuse to consent to organ procurement and would require some response from the individual. A number of mechanisms have been suggested for obtaining such responses. This is not the place to develop such proposals in detail. This article will simply list some mechanisms that have been proposed. Asking for a decision to donate and requiring a response might be done as in the past during the drivers' license renewal process or could be included on the federal income tax return. The latter policy has the advantages of reaching almost all adults (the only people who can exercise a gift of organs), providing for a single federal list of those willing to donate, and providing for yearly updates. The new feature would be that people would be required to answer the question. For any required response mechanism, it would probably be wise to permit an "I don't know" response. Such responses would trigger requests from next of kin, just as would be necessary for minors and others who are not mentally competent to make a gift of an organ.

If this approach to organizing and encouraging the giving of organs is unacceptable, then the alternative would appear to be some version of routine salvaging, presumably with opting out. However, until there is empirical evidence that the presumption of consent is warranted, those who favor the state's procurement without some explicit permission should affirm their communitarianism and acknowledge that their policy commits them to the view that the state has the right to take organs without permission. They should refrain from the misleading language of "presumed consent," which implies the gift mode, but cannot be justified unless there is evidence that essentially all people (or all people whose explicit opting out cannot be located at the moment of crisis) would consent to such an arrangement if they were asked. That presumption is warranted for life-and-death emergency room treatment, but not presently for procuring organs.

Acknowledgment

The authors gratefully acknowledge the help of the following persons, consulted regarding foreign legislation: Dr Francese Abel S.J., Dr Edwin Bernat, Fr Antonio Puca, Dr Knut W. Ruyter, and Dr Paul Schotsmans.

Notes

1 Dukeminier J, Sanders D: N Engl J Med 279:413, 1968

2 Callahan D: *What Kind of Life:The Limits of Medical Progress*. New York: Simon and Schuster, 1990

3 Cantaluppi. Cited by Gerson W: *NYU J Int Law Politics* 19:1013, 1987

4 Farfor J: Br Med J 1:497, 1977

5 Int Dig Health Legislat 37(1):332, 1986 (Austrian Law of June 1, 1982)

6 Int Dig Health Legislat 38(3):523, 1987 (Belgian Law of June 13, 1986)

7 Int Dig Health Legislat 3(3):477, 1982 (Czechoslovakian Mandatory directives of February 27, 1978)

8 Int Dig Health Legislat 36(4):971, 1985 (Finnish Law of April 26, 1985)

9 Int Dig Health Legislat 28(3):621, 1977 (Italian Law of December 2, 1975)

10 Puca A: *Trapianto di Cuore E Morte Cerebrale del Donatore*. Torino, Italy: Edizione Camilliane, 1993, pp 128–133; 201–224

11 Lov om transplantasjon og avgivelse av Iik m.m. February 9, 1973

12 Ley de Octubre 1979, no. 30/79, art. 5.3

13 Int Dig Health Legislat 36(1):50, 1985 (Swiss Regulation of September 17, 1984)

14 Int Dig Health Legislat 40(4):836, 1989 (Cyprus Law of May 22, 1987)

15 Int Dig Health Legislat 40(3):588, 1989 (Hungarian Law of February 17, 1988)

16 Republic of Singapore, "Human Organ Transplant Bill," Government Gazette Bills Supplement October 31, 1986, pp 1–10

17 Int Dig Health Legislat 38(3):530, 1987 (Syrian Law of December 20,1986)

18 Int Dig Health Legislat 42(1):46, 1992 (Yugoslavian Decree of October 18, 1990)

19 Int Dig Health Legislat 41(3):436, 1990 (Colombian Law of December 20, 1988)

20 Md State Senate Bill 428, § 4–509.2

21 General Assembly of the Commonwealth of Pa, proposed amendment to Title 20, Chapter 86, Subchapter C

22 Organ Donation Study. United Network for Organ Sharing Executive Summary. February 15, 1992

23 Manninen DL, Evans R: *JAMA* 253:3111, 1985

24 Applebaum PS, Lidz CW, Meisel A: *Informed Consent: Legal Theory and Clinical Practice*. New York: Oxford, 1987, pp 66–69

25 Ramsey P: *The Patient as Person*. New Haven: Yale, 1970

26 Silver T: *Boston University Law Review* 68:681, 1988

27 Spital A: N Engl J Med 325:1243, 1991

28 McNeil DR: *Hamline J Public Law Policy* 9:343, 1989

29 Stuart FP, Veith FJ, Cranford RE: *Transplantation* 31:238, 1981

30 Kennedy I: J Med Ethics 5:133, 1979

31 Steinbrook R: J Health Polit Policy Law 16:504, 1981

32 Steuer JD, Bell SK: Crit Care Nurse 9:466, 1989

33 Schotsmans P: In Land W Dossetor JB (eds): *Organ Replacement Therapy: Ethics, Justice and Commerce*. Heidelberg: Springer-Verlag Berlin, 1991

34 NY State Task Force on Life and the Law: The Required Request Law March, 1986

35 Caplan A: N Engl J Med 311:981, 1984

36 Caplan AL, Welvang P: Clin Transplant 3:170, 1989

37 Ross SE, Nathan H, O'Malley KF: J Trauma 30:820, 1990

38 Veatch RM: *Death, Dying and the Biological Revolution* (revised ed). New Haven: Yale, 1989 p 216

39 Veatch R: N Engl J Med 25:1246, 1991

40 Dennis JM, Hanson P, Hodge E, et al: UNOS Update 10(2):16, 1994

Chapter 24

Aaron Spital

MANDATED CHOICE FOR ORGAN DONATION: TIME TO GIVE IT A TRY

The successful development of transplantation is one of the most miraculous accomplishments of modern medicine. Unfortunately, the ability to deliver this medical miracle is limited by a severe and steadily worsening shortage of organs.[1] According to the United Network for Organ Sharing, as of 31 March 1996, more than 45 000 persons in the United States were on the national waiting list for transplantation;[2] this list grows by several hundred each month. It is estimated that eight of these people will die each day while waiting for transplantation.[3] Even more tragic is the realization that many of these deaths are preventable. Because only about 40 percent of potential cadaveric organ donors become actual donors, large numbers of life-saving organs are continuously being lost.[4] Clearly, something is wrong with our current organ procurement system.

Our current organ procurement system

In the United States, explicit consent is required before organs may be removed and used for transplantation. The Uniform Anatomical Gift Act, which has been passed in some form in all states and the District of Columbia, provides the legal framework for this process.[5–7] This statute gives all competent adults legal authority to decide for themselves whether or not they wish to become organ donors after their deaths. Unfortunately, relatively few people take advantage of this law and record their wishes about posthumous organ donation.[8,9] Furthermore, even when such directives are available, organ procurement organizations still ask the family of the deceased for consent, despite the clear stipulation of the Uniform Anatomical Gift Act that the decedent's wishes must be honored.[5–7,10] In effect, the law is simply ignored, and the question of organ donation after death is almost always left for the family to decide.

The need to obtain family consent is a major barrier to organ procurement.[1,8,11–14] Because most organ donors are young people who die unexpectedly, the family is often devastated and in shock. Under these circumstances, clear thinking may be impossible. The need to consider organ donation at such a terrible time places additional stress on the family. Furthermore, family members are often unaware of their loved one's wishes, which makes the question of donation even more difficult for them to answer.[9,13,15,16] The need to ask for permission is also stressful for hospital personnel who fear aggravating the family's pain. Considering this mixture of grief, confusion, and anxiety, it is not surprising that more than 50 percent of families say no.[4] Indeed, despite suggested techniques designed to increase the number of families who say yes, such as delaying the request for organ donation until after the notification of death,[17,18] a recent study concluded that "the major impediment to procurement was the low rate of family consent".[4]

Toward greater individual control through mandated choice

The high rate of family refusal contrasts sharply with public opinion polls that show widespread support for organ donation.[9,15] This suggests that if the stress accompanying the decision-making process could be avoided, the rate of consent would increase. This goal could be accomplished by eliminating the need for families to consider donation at the emotionally charged time of a relative's death. Instead, adults could decide for themselves at their leisure whether or not they wish to become organ donors upon their deaths. By further ensuring that each person's wishes would be known and honored, favorable public sentiment toward organ donation should translate into increased rates of organ procurement.

This proposal to transfer control away from the family and back to the individual is consistent with the intent of the Uniform Anatomical Gift Act, which states that the wishes of the individual are paramount.[5-7] Furthermore, the Council on Ethical and Judicial Affairs of the American Medical Association recently concluded that "the individual's interest in controlling the disposition of his or her own body and property after death suggests that it is ethically preferable for the individual, rather than the family, to decide to donate organs".[14] Similar views have been expressed by many philosophers and ethicists,[19-23] and several surveys suggest that most of the public agrees.[13,24-26]

How can such an individualistic approach be achieved? Mandated choice has been proposed as an alternative method for obtaining consent, which is designed to accomplish precisely this.[1,5,8,10,13,14,22,27-30] Under mandated choice, all competent adults would be required to decide and record whether or not they wish to become organ donors upon their deaths. This could be accomplished by asking about organ donation on driver's license applications, tax returns, or official state identification cards. The application or tax return would not be accepted until the question of donation was answered. A change of mind could easily be communicated with a written directive at any time. However, a person's decision would be binding and could not be overridden by the family unless that person had made a provision granting his or her family veto power.

There are several advantages to this approach.[1,8,13,14,22,27-30] First, by eliminating the need to obtain family approval, the added stress now experienced by many families and health care workers when confronting the question of organ donation would be removed, and the family consent barrier would fall; in fact, many families might be comforted to know that their relatives wishes were clear and would be honoured. Second, mandated choice would take advantage of favorable public attitudes because all competent adults would decide about organ donation for themselves in a relaxed setting, where thinking is likely to be clear. Third, because all adults would be forced to consider this issue, mandated choice might be the most effective method for increasing public awareness of the great value of organ donation, and this might further stimulate participation. Fourth, mandated choice would eliminate occasional delays resulting from the need to obtain family consent that can jeopardize the quality of organs. Finally, mandated choice would preserve altruism and voluntarism, which are the philosophical foundations of our current system for obtaining consent. Indeed, mandated choice would promote autonomy because, more than any other system, it would ensure that a person's wishes would be honored, whatever they may be.

Making good decisions about complicated issues requires careful consideration; therefore, the question of organ donation should not be sprung upon people at motor vehicle bureaus. This issue should be considered in a setting that provides ample opportunity for reflection and discussion with family and friends, and these deliberations should take place before a decision is made. Therefore, regardless of how preferences are recorded under mandated choice, all adults should be informed long before their decisions are recorded that they will have to decide about organ donation for themselves and why. This information should be coupled with ongoing educational programs that outline the great value of organ donation and dispel fears that inhibit participation.[15,18,27] Mandated choice is designed to complement these vital programs, not to replace them.

Several authors have expressed concern that requiring people to decide about organ donation is coercive.[4,31] However, as Katz[28] has pointed out, "since the gain to the public . . . is likely to be substantial . . . we as a society can legitimately decide to tolerate the negligible intrusion on an individual's privacy presented. . . ." In addition, mandated choice is not coercive with regard to the choices a person makes, and it ensures that those choices will be honored[13,14]. This assurance also suggests that the fear that mandated choice would generate public resentment and an actual decline in the rate of consent is unlikely to occur. Two recent studies support this conclusion. Among a random sample of 1000 adults in the United States, 65 percent said they would support mandated choice[8]. In a subsequent national survey, a nearly identical number (63 percent) said they would sign up to donate their organs

under this plan; furthermore, of the 30 percent who had previously decided to donate, 95 percent said they would still do so under mandated choice[13]. These and other studies[9] also refute the claim that a widespread fear of being declared dead too soon would greatly inhibit participation[4].

Role of the family

Mandated choice has been criticized as being insensitive to families.[31,32] In fact, it is kinder to families than is our present system because it eliminates the need for devastated families to confront the emotionally wrenching issue of organ donation during their most trying moments.[1,8,13,14,29,33] The concern that most families would not tolerate being excluded from the final consent process is not well substantiated. As previously noted, surveys have shown that most of the public believes that the individual, rather than the family, is best suited to decide about organ donation and that when advance directives exist, families should not be able to override the wishes of their loved ones[8,13,24-26,29]. These data indicate that most people in the United States believe that with regard to posthumous organ donation, individual autonomy should be respected. Therefore, once family members and medical personnel realize that the best way to protect autonomy, including their own, is to know and honor a person's wishes, they would probably be willing to accept advance directives as binding. It is hoped that self-determination about organ donation would eventually become accepted as routine, just as mandated autopsy is in deaths in which foul play is suspected.

Although mandated choice would give ultimate control regarding organ donation to the individual, this does not mean that the family is unimportant. Family discussions about this sensitive issue have always been of great value, and they always will be. Such discussions may provide useful insights that can help people explore their own feelings as they try to decide whether or not to donate. These exchanges also serve to inform family members of each other's wishes. This knowledge may avoid the distress that might otherwise occur if organs were taken from recently deceased persons who had previously agreed to donate but had not notified their families of their wishes. Furthermore, although most people in the United States seem to believe that adults should decide about organ donation for themselves, a significant minority believe that their families are better suited for this task. Under mandated choice, these people could include a provision granting their families veto power[33]. However, even in these cases, a personal decision should still be made and recorded because this information would be very helpful to families trying to decide. Finally, all families should be informed of any plans for organ retrieval, treated with the utmost respect and sensitivity, and offered as much support as they need to help them deal with the enormous trauma caused by the unexpected loss of a loved one.

Potential effect on public commitment to organ donation

Because the purpose of mandated choice is not only to protect individual autonomy but also to increase public commitment to organ donation, it is important to estimate the potential impact that implementing this system would have on the rate of consent before an actual trial is undertaken. As noted above, in a recent Gallup poll of 1002 randomly selected adults in the United States[13], 63 percent said they would sign up to donate their organs if mandated choice became law. Furthermore, because devoting thought to organ donation

correlates with a positive response[13], and because mandated choice forces everyone to consider this issue, this system may actually encourage the 13 percent who were undecided to say yes. If so, under mandated choice, as much as 75 percent of the U.S. adult population would become committed potential organ donors.

Conclusions

Our current organ procurement system is inadequate. The need to obtain family consent is at least in part to blame and results in a daily loss of potentially life-saving organs. To rectify this tragic situation, we need to redirect our focus away from the family and back to the individual, the one who is usually best suited to decide the disposition of his or her own body after death. Mandated choice appears to be an acceptable method for achieving this goal and was recently endorsed by the Presumed Consent Subcommittee of the United Network for Organ Sharing[27] and the Council on Ethical and Judicial Affairs of the American Medical Association[14]. Of course, whether or not mandated choice would actually increase public commitment to organ donation remains to be seen. There have been no actual trials of this proposal (Cate F. Personal communication), and the problems with public opinion polls are well known. However, the results of these polls are encouraging enough to recommend that a pilot study of mandated choice be undertaken as soon as possible. With so many lives at stake, we cannot afford to simply continue our current insufficient approach to organ procurement. It is time to try something new.

Acknowledgments

The author thanks Sam Spital for his very helpful suggestions.

Notes

1 Spital A. "The shortage of organs for transplantation. Where do we go from here?" N Engl J Med. 1991; 325: 1243–6.

2 "Patients waiting for transplants." United Network for Organ Sharing Update. 1996; 12: 26–7.

3 Fox MD. "The transplantation success story" [Editorial]. JAMA. 1994; 272: 1704.

4 Siminoff LA, Arnold RM, Caplain AL, Viming BA, Seltzer DL. "Public policy governing organ and tissue procurement in the United States. Results from the National Organ and Tissue Procurement Study." Ann Intern Med. 1995; 123: 10–7.

5 "Medical technology and that law organ transplantation." Harvard Law Review. 1990; 103: 1614–43.

6 Lee PP, Kissner P. "Organ donation and the Uniform Anatomical Gift Act." Surgery, 1986; 100: 867–75.

7 Overcast TD, Evans RW, Bowen LE, Hoe MM, Livak CL. "Problems in the identification of potential organ donors. Misconceptions and fallacies associated with donor cards." JAMA. 1984; 751: 1559–67.

8 Spital A. "Consent for organ donation: time for a change." Clin Transplant, 1993; 7: 525–8.

9 The Partnership for Organ Donation. "The American public's attitudes; toward organ donation and transplantation," Boston; 1993.

10 Iserson KV. "Voluntary organ donation: autonomy ... tragedy." JAMA. 1993; 270: 1930.

11 Caplan AL. "Organ transplants: the costs of success." Hastings Cent Rep. 1983; 13: 23–32.

12 Kokkedee W. "Kidney procurement policies in the Eurotransplant region." Soc Sci Med. 1992; 35: 177–82.

13 Spital A. "Mandated choice. A plan to increase public commitment to organ donation." JAMA. 1995; 273: 504–6.

14 "Strategies for cadaveric organ procurement. Mandated choice and presumed consent." Council on Ethical and Judicial Affairs, American Medical Association. JAMA. 1994; 272: 809–12.

15 Prottas JM, Batten HL. "The willingness to give: the public and the supply of transplantable organs." J. Health Polit Policy Law. 1991; 16: 121–34.

16 Tymstra T, Heyink JW, Pruim J. Slooff MJ. "Experience of bereaved relatives who granted or refused permission for organ donation." Fam Pract. 1992; 9: 141–4.

17 Garrison RN, Bently FR, Raque GH, Polk HC Jr, Sladek LC, Evanisko MJ. et al. "There is an answer to the shortage of organ donors." Surg Gynecol Obstet. 1991; 173: 391–6.

18 Perkins KA. "The shortage of cadaver donor organs for transplantation. Can psychology help?" Am Psychol. 1987; 42: 921–30.

19 Childress JF. "Ethical criteria for procuring and distributing organs for transplantation." J Health Polit Policy Law. 1989; 14: 87–113.

20 Cohen C. "The case for presumed consent to transplant human organs after death." Transplant Proc. 1992; 24: 2168–72.

21 Peters DA. "A unified approach to organ donor recruitment, organ procurement, and distribution." Journal of Law and Health 1989–90; 3: 157–87.

22 Veatch RM, Pitt JB. "The myth of presumed consent: ethical problems in new organ procurement strategies." Transplant Proc. 1995; 27: 1888–92.

23 Waithe ME. "Mandated choice for organ donation" [Letter], JAMA, 1995; 273: 1177.

24 Corlett S. "Public attitudes toward human organ donation." Transplant Proc. 1985; 17(6 Suppl 3): 103–10.

25 Marris RJ, Jasper JD, Lee BC, Miller KE. "Conserving to donate organs whose wishes carry the most weight?" J Appl Soc Psychol. 1991; 21: 3–14.

26 Manninen DL, Evans RW. "Public attitudes and behavior regarding organ donation," JAMA. 1985; 253: 3111–5.

27 Dennis JM, Hanson P. Hodge EE, Krom RAF, Veatch RM. "An evaluation of the ethics of presumed consent and a proposal based on required response." United Network for Organ Sharing Update, 1994; 10: 16–21.

28 Katz BJ. "Increasing the supply of human organs for transplantation: a proposal for a system of mandated choice." Beverly Hills Bar Journal. Summer 1984; 18: 152–67.

29 Spital A. "Mandated choice. The preferred solution to the organ shortage?" Arch Intern Med. 1992; 152: 2421–4.

30 Veatch RM. "Routine inquiry about organ donation – an alternative to presumed consent." N Engl J Med. 1991; 325: 1246–9.

31 Prottas J. "Mandated choice for organ donation" [Letter]. JAMA. 1995; 274: 942–3.

32 Murray TH, Youngner SJ. "Organ salvage policies; a need for better data and more insightful ethics." JAMA. 1994; 272: 814–15.

33 Glasson J. Orentlicher D. "Mandated choice for organ donation" [Letter]. JAMA. 1995; 273: 1176–7.

Chapter 25

Linda Beecham

DONORS AND RELATIVES MUST PLACE NO CONDITIONS ON ORGAN USE

Patients and relatives will be unable to impose any conditions on the use of donated organs under new rules due to be introduced by the UK government. The previous health secretary, Frank Dobson, ordered an inquiry in July 1999 after a transplant coordinator agreed to a family's request that the organs of a dead relative be given to a white person.

The inquiry's report says that the dead man's kidneys and liver were "wrongly accepted and wrongly passed through the system." It criticises senior staff in the UK transplant service and the Department of Health for failing to act to stop the practice when details emerged. The report concluded: "To attach any condition to a donation is unacceptable because it offends against the fundamental principle that organs are donated altruistically, and should go to patients in the greatest need."

All NHS staff will shortly receive guidance reminding them that organs offered under racist conditions must be refused. Under the new guidelines, relatives will be unable to stipulate that the organs should go, for example, to a child or a non-smoker. The Department of Health is also to review how transplant services can best be modernised. The UK Transplant Support Services Authority, which coordinates transplants, will be renamed UK Transplantation and asked to procure more organs. More than 6000 people were waiting for a transplant in 1999, and only 212 operations were carried out.

Chapter 26

T. M. Wilkinson

WHAT'S NOT WRONG WITH CONDITIONAL ORGAN DONATION?

In a well known British case, the relatives of a dead man consented to the use of his organs for transplant on the condition that they were transplanted only into white people. The British government condemned the acceptance of racist offers and the panel they set up to report on the case condemned all conditional offers of donation. The panel appealed to a principle of altruism and meeting the greatest need. This paper criticises their reasoning. The panel's argument does not show that conditional donation is always wrong and anyway overlooks a crucial distinction between making an offer and accepting it. But even the most charitable reinterpretation of the panel's argument does not reject selective acceptance of conditional offers. The panel's reasoning has no merit.

There was no clear policy for dealing with this offer, and officials and doctors in effect accepted it. In the resulting outcry, the British government and the panel it set up to report on the case condemned the acceptance of not only racist conditions but any conditional offers of donation. Their recommendations have become National Health Service (NHS) policy. The purpose of this paper is to criticise the panel's reasons.[1]

In the British case, the organs (two kidneys and a liver) (i) saved at least one life (the recipient of the liver) that would otherwise not have been saved and (ii) as it happened went to those who would have received them had the offer been unconditional. Nonetheless, the panel held that acceptance of the offer was wrong *even given those facts*. They said: "to attach *any* condition to a donation is unacceptable, because it offends against the fundamental principle that organs are donated altruistically and should go to patients in the greatest need".[2]

Is it really so bad to attach a condition to an organ donation? Of course it was bad in the case of the racist. The motive there was some mix of hatred and contempt and there is nothing to be said for it. But what about the condition that an organ go to a relative? There seems nothing morally wrong about agreeing to donate a kidney, say, on condition that it go to a sibling, whether the donation is to be from a living person or a dead one. A special concern for one's nearest and dearest is, on all but the most extreme views, at least morally permissible and often obligatory. Note here that we are considering the moral assessment of the donor's behaviour and not the decision by the transplant services to accept or reject the offer. This vital distinction – which is overlooked by the panel – is something I shall come to later.

Setting aside a special concern for one's nearest and dearest, let us consider the panel's explanation of what is wrong with attaching a condition to a donation. The panel claims that conditional donation "offends against the fundamental principle that organs are donated altruistically and should go to patients in the greatest need". Altruism in its normal sense refers roughly to a non-self interested concern for the interests of others. Importantly, a wide variety of other regarding motives can be described as altruistic, such as a special concern for children, or the deaf, or the poor. "Altruism" does not have a specific application. It does not require, for example, that actions be motivated out of adherence to a greatest happiness principle or, saliently here, a greatest needs principle. Consequently, there need be nothing non-altruistic about conditional donation. Wanting organs to go to a child – although also apparently opposed by the panel – is not a violation of altruism any more than donating to a children's charity is.[3] Racial conditions can be altruistic if, for instance, the donor wants the organs to go to groups (say blacks in the US, or Maori and Pacific Islanders in New Zealand) that typically do relatively badly in receiving organs. It is not obvious that even the racist condition in the present case violates the rule that organs should be given altruistically; rather it looks like a mix of altruism towards whites and nastiness to non-whites.

If conditional donation cannot be condemned as invariably a failure of altruism, can it be condemned as a failure to donate according to the greatest need? It is unclear why people would be obliged, if they choose to donate, to donate according to the greatest need. To say that they are is both to deny donors any moral discretion in their donation and to hold that it is the greatest need, and not some other worthy goal, such as helping underrepresented groups, that is the only permissible aim. Perhaps this view about discretion and need is correct, although few apply it elsewhere, for instance to the conditions attached by those who make wills. But it is not defended in any way by the panel.

Bear in mind again that at this stage we have only been talking about the moral assessment of the donor. It is an entirely separate question whether a conditional donation should be accepted, even if the donor does no wrong in attaching the condition *or even if he or she does*. There is a crucial distinction between *attaching* a condition to an offer and *accepting* it. Because the panel's argument deals only with attaching conditions, this distinction makes the panel's reasoning, on the face of it, irrelevant to the question it considers, which is whether the offers should be accepted.

Attaching a condition might be "unacceptable", but that does not yet give us a reason against *accepting* the offer. Perhaps the panel believes an offer that is itself wrong should not be accepted. As a general principle, that cannot possibly be correct. There must be some offers that should be accepted if they do a great deal of good even if the motives are slightly discreditable. If I offer money to Oxfam in order to spite my wife, it would be preposterous for Oxfam to be morally required to turn down money that would save lives on the grounds that I should have given the money unconditionally and my offer was a morally bad one. But although the general principle is unsound, perhaps we can draw something relevant from the panel's claim that would morally forbid accepting conditional offers in the specific context of transplantation.

What is surely doing the work in the panel's argument is the appeal to greatest needs. Accepting a conditional donation might (but might not) prevent the system allocating according to the greatest need. The counterargument here is that when we think about the principle of allocating according to greatest need, it cannot be understood as a fundamental principle and when its foundations are explored, there can be no unequivocal condemnation of accepting conditional donations.

The principle of allocating according to needs cannot be fundamental if it is taken to be a prescription for action. It has instead to be inferred from a principle such as: "it is important that needs are met". Suppose that trying to meet needs by a direct method caused needs to be less well met than by some indirect method. Then the indirect method should be chosen. This is not only not a hairsplitting point, it is of great significance in other fields, such as economics. The most famous defence of a market allocation of goods appeals to its superiority in providing them to any direct method, such as state allocation. That is why, for instance, the vital need for food is largely met through the market. Even those who think the market does not do as well as some alternative do not dispute the idea here which is that, if it did, it should be preferred by a needs principle to trying to meet needs directly.

The application to conditional donation is straightforward. Since the panel's argument is supposed to rule out conditional donation whatever its effects, it begs no question to consider a case where acceptance of the condition causes some organs to become available that otherwise would not, and reduces access to organs for no one. Suppose that if the condition is accepted, the organs would not go to the person in greatest need but to someone who needs them less (but still needs them badly). The principle of allocating according to greatest need would condemn this. But the idea that needs should be met that underlies the principle would require accepting it. This is because whether or not the offer is accepted, the person with the greatest need would not receive those organs, but accepting the offer allows someone's needs to be met while refusing it meets no one's need for organs. A principle of meeting needs should then say that the offer should be accepted, no matter how important it is to meet the greatest need. The conclusion that the offer

should be accepted goes through even more strongly if it is the person in greatest need who gets the organs, as the panel says happened in the case of the British racist.

Note that my arguments here have not been that a needs principle is outweighed by others. It is that whatever underlies a concern for needs merely contingently supports the principle that one should act so as to meet the greatest need, and that conditional donation cannot be categorically rejected out of a concern that needs be met.

What remains of the panel's argument? Here is one way to take their argument: it is wrong to accept organs if one would then be in the position of not being free to allocate them to the person in greatest need. That principle needs much more elaboration and defence than can be wrung out of the panel's report, especially since its effect could be to cause some to die with no gain to anyone else. Even if the principle is sound, it does not support the unconditional rejection of conditional donation, because it does not justify rejecting conditional offers that would lead to the organs going to those in greatest need anyway. Recall that in the British case the liver did go to someone in greatest need; acceptance of the offer thus did leave those involved free to allocate to the person in greatest need. (They must have been free or else they could not have done it.[4]) Moreover, what was an accident here could become policy. Conditional offers could be accepted on the understanding that the organs would only be used when satisfying the condition coincided with allocating to the person in greatest need.

This paper has tried to show that the panel's arguments against conditional donation, or acceptance, have no merit. There are other arguments. A proper discussion of these would take much longer, but let us here close with one often mentioned and which may have been in the back of the mind of the panel (although they did not mention it). This is the objection that a policy of accepting conditional donation could have net bad effects on the supply of organs, perhaps because some refuse to donate in a system they see as compromised, perhaps because some attach conditions they would not otherwise have done. This objection, and particularly the criterion of "net bad effects" needs spelling out, but here are two preliminary replies. First, the objection speculates about the effects, and we might counterspeculate that people would not respond to the policy as predicted, especially if they do not find out about it. The second reply is about the role of speculation. In the present case, at least one life was saved for certain. When it comes to certainties of saving lives, there needs to be more on the other side than mere speculation. Why not try the policy of accepting conditional donations, get the benefits, and then think about discontinuing if the harms materialise?

Notes

1 Department of Health. *An Investigation into Conditional Organ Donation* [report of the panel]. London; Department of Health, 2000.
2 See reference 1: 25 (emphasis added).
3 See reference 1:10.
4 Cohen GA. "Are disadvantaged workers who take hazardous jobs forced to take hazardous jobs?" *History, Labour, and Freedom*. Oxford; Clarendon Press, 1988.

Chapter 27

J.L. Bernat

ARE DONORS AFTER CIRCULATORY DEATH REALLY DEAD, AND DOES IT MATTER? YES AND YES

Case Report

A 45-year-old man had a sudden headache, followed rapidly by unconsciousness. He was taken to a community hospital ED, where he required intubation and ventilation. A brain CT scan showed a large, right-hemispheric, intracerebral, intraventricular, and subdural hemorrhage with early brain herniation. On urgent transfer to the medical center, he was deeply comatose, with a dilated, unreactive right pupil. Following ED insertion of an external ventricular drain, he underwent immediate surgical removal of the intraparenchymal and subdural hemorrhage, removal of a ruptured arteriovenous malformation, and hemicraniectomy. Subsequent CT scans showed large infarctions of both temporal lobes and the right occipital lobe.

On the fifth postoperative day, he remained deeply comatose, with elevated intracranial pressures in the 50s mm Hg. Both pupils were unreactive, and all brainstem reflexes were absent. He was overbreathing the ventilator. A brain CT scan showed further brain herniation. Given the dismal neurologic prognosis, his family requested withdrawal of life-sustaining therapy and consented to organ donation. With his family at the bedside, he was extubated and apnea was observed. Within minutes, he developed pulseless cardiac electrical activity,

followed within a minute by asystole. After 5 min of asystole and apnea, he was declared dead and rushed to the operating room for organ donation. His liver and kidneys were recovered.

Organ donation after the circulatory determination of death (DCDD) has become a common practice in the United States, Canada, and many European countries during the past two decades.[1] DCDD programs are desirable for society because they increase the organ pool and for families because they offer the opportunity for organ donation from patients who are not brain dead. As in this case, families may consent to organ donation after the patient has been declared dead after cessation of circulation and breathing following an earlier decision to discontinue life-sustaining therapy. Programs in the United States and Canada permit such controlled DCDD, but most exclude uncontrolled DCDD on patients declared dead after failed CPR. The generally accepted principles of DCDD[2] are listed in Table 27.1.

In the United States, the proliferation of DCDD programs has been stimulated by endorsements from the Institute of Medicine[3–5] and the Department of Health and Human Services,[6] and by the establishment of DCDD quality criteria for hospitals by the Joint Commission. The Division of Transplantation in

Table 27.1 Principles of DCDD

Respect the Dead Donor Rule.
Determine death using accepted tests and procedures.
Separate the death determination team from the organ procurement team.
Separate the decision to refuse life-sustaining therapy from the decision to donate.
Obtain surrogate consent for withdrawal of life-sustaining therapy.
Obtain surrogate consent for organ donation.
Provide palliative care during dying.
Provide end-of-life family support.
Properly design and scrupulously follow the DCDD protocol.
Provide adequate documentation.

Source: Adapted with permission from the Massachusetts Medical Society.[2] DCDD = donation after the circulatory determination of death.

the Department of Health and Human Services, Health Resources and Services Administration (HRSA) recently convened a multidisciplinary panel to clarify the standards for circulatory death determination in DCDD, whose report was published earlier this year.[7] I summarize several of the panel's arguments here.

Several critics of DCDD claimed that the donor patient is not truly dead at the moment death is declared because if CPR were performed and could restore circulation, its loss would not be irreversible, as required by the statute of death.[8] If this claim were true, DCDD would violate the Dead Donor Rule (DDR) that requires that the vital multiorgan donor first be dead so the donation does not kill the donor.[9] Other DCDD critics argued that this violation is irrelevant because the DDR should be abandoned as unnecessary.[10] In this dialogue, I argue that: (1) DCDD donors are dead once their circulation and respiration have ceased permanently; (2) permanent cessation of circulation and respiration is the medical practice standard for death determination; (3) determining death in DCDD complies with US death statutes as "accepted medical practice" and, therefore, does not violate the DDR; and (4) as a public policy for organ donation, it is unnecessary and unwise to abandon the DDR.

DCDD donors are dead at death declaration

The controversy over circulatory death determination turns on whether the cessation of circulation is irreversible at the moment death is declared. Resolving this issue requires clarifying the distinction between an irreversible and a permanent loss of a function. An irreversible cessation of a function means that the function *cannot* be restored by any known technology. A permanent cessation of a function means that the function *will not* be restored, because it will neither return spontaneously nor return as a result of medical intervention because CPR efforts will not be attempted. Being irreversible is an absolute condition that implies impossibility and does not rely on action, whereas being permanent is a contingent condition that admits possibility and relies on action.[7]

In our patient's case, once breathing, heartbeat, and circulation stop beyond the point when spontaneous resumption of heartbeat and circulation (autoresuscitation) can occur, and given that CPR will not be attempted because the patient has a do-not-resuscitate order, the patient's circulation has ceased permanently. Permanent cessation of circulation constitutes a valid proxy for its irreversible cessation because it quickly and inevitably becomes irreversible and because there is no difference in outcome between using a permanent or irreversible standard.

Michael DeVita[11] called the mandated observation period of mechanical asystole the "death watch." The duration necessary to exclude autoresuscitation is an empirical question. Studies on a few hundred patients have documented no return of cardiac rhythm after 65 s of asystole.[12] A current systematic review of autoresuscitation found no instances of resumption of circulation in extubated patients who did not receive CPR, even among those patients whose cardiac rhythm autoresuscitated to electrical activity.[13] The duration of required mechanical asystole stipulated by most DCDD protocols has been prudently extended to 5 min in light of the reduced CIs resulting from the small patient database of autoresuscitation. This interval will decrease in the future as more cases are recorded. There is a consensus that mechanical asystole is sufficient for cessation of circulation and that electrical asystole is unnecessary because the essential issue in death determination is circulation, not cardiac electrical activity.[14]

Permanent cessation is the medical practice standard for death determination

Physicians have always used the permanent cessation standard when determining circulatory death. When called to the bedside of a patient admitted for terminal palliative care who had a do-not-resuscitate order and who was found pulseless and apneic, physicians declare death immediately without proving that the cessation of breathing and circulation is irreversible. They know that autoresuscitation in this setting does not occur and that CPR will not be administered. It is inconsequential and unnecessary to wait until the cessation of vital signs persists long enough to unequivocally establish irreversibility

or to prove irreversibility by attempting CPR and showing it is impossible to restore circulation.

The medical standard for circulatory death determination among hospitalized patients should be the same (permanent cessation) whether the patient is a DCDD donor. But because of the consequentiality of the determination in the DCDD circumstance, it is prudent to require a higher level of proof of complete circulatory absence in the DCDD donor. The HRSA-recruited panel recommended performing a valid, objective test proving absent circulation in circulatory death determination, such as percutaneous Doppler ultrasound, intraarterial pulse monitoring, or echocardiography in cases in which there is not electrical asystole.[7]

Death determination in DCDD complies with US death statutes

Most death statutes in the United States are based on the Uniform Determination of Death Act (UDDA), the model death statute proposed by the President's Commission for the Study of Ethical Problems in Medicine and Biomedical and Behavioral Research in 1981.[15] Similar death statutes are in effect in Canada and elsewhere. The UDDA (Table 27.2) stipulates "irreversible cessation of circulatory and respiratory functions" and

Table 27.2 The Uniform Determination of Death Act

An individual who has sustained either:

(1) irreversible cessation of circulatory and respiratory functions, or

(2) irreversible cessation of all functions of the entire brain, including the brainstem, is dead.

A determination of death must be made in accordance with accepted medical standards.

Source: Reprinted with permission from the President's Commission for the Study of Ethical Problems in Medicine and Biomedical and Behavioral Research.[15]

that physicians conduct the determination in accordance with "accepted medical standards."

Although the UDDA did not define "irreversible," many commentators assumed that the President's Commission intended the strict definition of "irreversible" as noted above. The discussion in their guiding document, *Defining Death*, however, made no such stipulation. The medical consultants to the President's Commission noted that the prevailing medical practice standard is the permanent cessation of circulation and respiration.[16] Therefore, "permanent" was the intended construal of "irreversible" in the UDDA, as clarified by the term "accepted medical standards" (Alexander M. Capron, executive director of the President's Commission, December 15, 2008, personal communication). Physicians declaring death by permanent cessation of circulation and respiration, therefore, are in compliance with the statutory requirement for death and do not violate the DDR.

It is unwise and unnecessary to abandon the DDR

The DDR, required by the Uniform Anatomic Gifts Act, is often regarded as an ethical axiom of vital multiorgan donation.[2] Robert Truog and colleagues[10,17] have argued that the DDR should be abandoned and replaced by consent for donation when the donor is incipiently dying and beyond harm. Although some educated and motivated donors would continue to donate without a DDR, I believe that maintaining it is a desirable component of sound national policy for programs of organ donation. On at least the societal level, it matters that the organ donor is dead.

Abandoning the DDR, although possibly resolving one problem, could spawn new problems by inadvertently jeopardizing public confidence in organ donation.[18] I believe that the acceptance of organ transplantation as a social practice depends on public confidence that physicians can accurately and reliably determine death and on public certainty that surgeons will not remove their vital organs until after they die. Eliminating the DDR would create a justifiable fear of abuses by physicians or family members and a resultant loss of confidence in the organ donation enterprise. Additionally, doing so is an unnecessary step because death determination in DCDD does not violate the DDR. Similar conclusions were reached by the Institute of Medicine,[3] the HRSA-recruited panel,[7] a Canadian DCDD consensus panel,[19] and the US President's Council on Bioethics.[20]

Notes

1 Steinbrook R. "Organ donation after cardiac death."*N Engl J Med.* 2007; 357(3): 209–13.
2 Bernat JL. "The boundaries of organ donation after circulatory death." *N Engl J Med.* 2008; 359 (7): 669–71.
3 Institute of Medicine. *Non-Heart-Beating Organ Transplantation: Medical and Ethical Issues in Procurement.* Washington, DC: National Academy Press; 1997.
4 Institute of Medicine. *Non-Heart-Beating Organ Transplantation: Practice and Protocols.* Washington, DC: National Academy Press; 2000.
5 Institute of Medicine. *Organ Donation: Opportunities for Action.* Washington, DC: National Academy Press; 2006.
6 Shafer TJ, Wagner D, Chessare J, et al. "US organ donation breakthrough collaborative increases organ donation." *Crit Care Nurs Q.* 2008; 31(3): 190–210.
7 Bernat JL, Capron AM, Bleck TP, et al. "The circulatory-respiratory determination of death in organ donation." *Crit Care Med.* 2010; 38(3): 963–70.
8 Menikoff J. "Doubts about death: the silence of the Institute of Medicine." *J Law Med Ethics.* 1998; 26(2): 157–65.
9 Robertson JA. "The dead donor rule."*Hastings Cent Rep.* 1999; 29(6): 6–14.
10 Miller FG, Truog RD. "Rethinking the ethics of vital organ donations." *Hastings Cent Rep.* 2008; 38(6): 38–46.

11 DeVita MA. "The death watch: certifying death using cardiac criteria."*Prog Transplant.* 2001; 11(1): 58–66.

12 DeVita MA, Snyder JV, Arnold RM, Siminoff LA. "Observations of withdrawal of life-sustaining treatment from patients who became non-heart-beating organ donors."*Crit Care Med.* 2000; 28(6): 1709–12.

13 Hornby K, Hornby L, Shemie SD. "A systematic review of autoresuscitation after cardiac arrest."*Crit Care Med.* 2010; 38(5): 1246–53.

14 Bernat JL, D'Alessandro AM, Port FK, et al. "Report of a national conference on donation after cardiac death."*Am J Transplant.* 2006; 6(2): 281–91.

15 President's Commission for the Study of Ethical Problems in Medicine and Biomedical and Behavioral Research. *Defining Death. Medical, Legal and Ethical Issues in the Determination of Death.* Washington, DC: US Government Printing Office; 1981.

16 "Guidelines for the determination of death. Report of the medical consultants on the diagnosis of death to the President's Commission for the Study of Ethical Problems in Medicine and Biomedical and Behavioral Research." *JAMA.* 1981; 246(19): 2184–86.

17 Truog RD, Miller FG. "The dead donor rule and organ transplantation."*N Engl J Med.* 2008; 359(7): 674–75.

18 D'Alessandro AM, Peltier JW, Phelps JE. "Understanding the antecedents of the acceptance of donation after cardiac death by healthcare professionals." *Crit Care Med.* 2008; 36(4): 1075–81.

19 Shemie SD, Baker AJ, Knoll G, et al. "National recommendations for donation after cardiocirculatory death in Canada: Donation after cardiocirculatory death in Canada." *CMAJ.* 2006; 175 (8 Suppl): S1–S24.

20 President's Council on Bioethics. *Controversies in the Determination of Death: A White Paper by the President's Council on Bioethics.* Washington, DC, 2009. www. bioethics.gov/reports/death/index.html. Accessed January 16, 2010.

Chapter 28

Robert D. Truog and Franklin G. Miller

COUNTERPOINT: ARE DONORS AFTER CIRCULATORY DEATH REALLY DEAD, AND DOES IT MATTER? NO AND NOT REALLY

From at least 1968 until the present, prevailing ethical thinking about organ donation has been shaped by the perceived need to conform to the so-called Dead Donor Rule (DDR), which states that vital organs may be removed from patients only after they have been declared dead. Elsewhere, we have argued against the fiction that individuals diagnosed as brain dead are dead because they maintain an extensive range of biologic functioning of the organism as a whole.[1-3] The determination of death in the case of donation after the circulatory determination of death (DCDD) relies on a more subtle fudging of the truth: Although imminently dying, donors under DCDD protocols are not known to be dead at the time that organs are procured.

We agree with Bernat[4] that organ donation as described in the presented case is ethically acceptable. We disagree sharply, however, over *why* is it acceptable. Whereas Bernat believes it is acceptable because it respects the DDR, we believe it is ethical (despite contravening the DDR) because it respects the rights of patients to make rational choices over how they die and their opportunities to donate vital organs. These differences are profound and have tremendous importance for the future of end-of-life care and organ transplantation.

The practice of DCDD is at an ethical crossroads. Those who believe that it must conform to the DDR must offer convincing evidence that the donors are dead. We believe the evidence shows that DCDD donors are dying but are not yet known to be dead. In other words, we believe Bernat has made a serious logical mistake by conflating a prognosis of imminent death with a diagnosis of death. This mistake is compounded by his view that patients who meet these prognostic criteria can be treated as if they meet the diagnostic criteria.

Is permanence a valid proxy for irreversibility?

The Uniform Determination of Death Act stipulates two criteria for determining death, including the traditional criterion of the irreversible loss of circulatory and respiratory function.[5] Bernat[4] acknowledges that under DCDD protocols we do not know whether patients have suffered the irreversible loss of circulatory function, but he argues that they have permanently lost this function and that permanence is a valid proxy for irreversibility. Bernat and colleagues[6] state, "An 'irreversible' cessation of function means that the function *cannot* be restored by any known technology ... and does not rely on intent or action. In contrast, a 'permanent' cessation of function means that the function will not be restored because it will neither return spontaneously, nor will return as a result of medical intervention because resuscitation efforts will not be attempted."

Bernat offers no explanation for why permanence is a "valid stand-in" for irreversibility other than an appeal to "accepted medical practice." He provides three examples, all based on the observation that in a variety of end-of-life scenarios, physicians commonly declare patients to be dead at the time that their heart stops beating, prior to the time when the physicians could know that the loss of circulatory function has become irreversible.

Although Bernat[4] has accurately described "accepted medical practice" in the context of end-of-life care that does not involve organ donation, we contend that this observation is irrelevant in situations where following the DDR requires knowing whether the patient is merely dying or already dead. In this context, as in many areas of medicine, the accuracy that is required of our assessments depends on the consequences of our assessments being wrong.

For example, if a physician must be absolutely certain that a patient is well hydrated before giving a nephrotoxic drug, he or she will use different criteria of assessment than if he or she is merely interested in knowing if the patient is in good fluid balance outside of a potentially toxic insult. If a physician is about to amputate a limb for what is believed to be a local malignancy, he or she will use more precise criteria to rule out metastatic disease than if amputation is not a feasible therapeutic option. In other words, the criteria physicians use to establish the presence or absence of a clinical condition are context dependent. Because the use of certain criteria is "accepted medical practice" in many common end-of-life situations, we cannot infer that they are adequate in other end-of-life contexts, such as DCDD, in which (assuming the necessity of the DDR) it is crucial to be certain whether the patient is already dead.

In Napoleon's empire, physicians were prohibited from declaring death until they had performed a "death watch" for 24 h.[7] In today's hospitals, we commonly declare death at the moment of asystole, confident that nothing consequential will happen to the patient over the next several minutes and that the routine flow of hospital activity will provide an adequate death watch to ensure that the criteria for irreversibility are satisfied. But when the few minutes following the onset of asystole may involve lethal actions such as surgical incision and the removal of vital organs, the methods of assessment must be correspondingly more precise. In particular, if one holds that the DDR is an inviolate principle of organ donation, then the difference between "dying" and "dead" becomes crucial. Even a very slim chance of a false positive makes a difference. If the patient is, in fact, not dead or not known to be dead at the time of the declaration of death, then the DDR has been violated.

Introducing intentions into the definition of death creates problems

Although Bernat offers no reasons for accepting permanence as a valid stand-in for irreversibility other than a misplaced reference to accepted medical practice, we believe there are sound reasons for rejecting this equivalence. The difference between life and death is an ontologic distinction; it is about two different "states of the world." But under Bernat's formulation, this fundamental distinction becomes dependent on the intentions and actions of those surrounding the patient at the time of death. Consider, for example, the following thought experiment:

Case 1

An otherwise healthy young athlete is involved in a severe automobile accident, leaving him with devastating brain injury but not brain death. He is taken to the operating room for DCDD organ donation, becomes pulseless shortly after the withdrawal of life support, and remains pulseless for 2 min. Is he dead?

Case 2

An otherwise healthy young athlete collapses in pulseless ventricular fibrillation while playing basketball. No one initiates CPR. After 2 min, paramedics arrive. Is he already dead? (Used with permission from Doug B. White's lecture, "The Ethics of Donation After Circulatory Determination of Death," San Diego, 2009.)

Using Bernat's reasoning, and assuming that it would be appropriate for the paramedics to initiate CPR in case 2, the young man is dead in case 1 but alive (or at least not known to be dead) in case 2. The only difference between these cases is whether those who are present intend to initiate CPR. If death is an ontologic state that exists independent of our thoughts, opinions, and intentions, then the status of these two individuals as dead or alive should be treated the same.

Case 3

Consider a variation of case 1 where the young man's parents are present during the withdrawal of life support and when death is declared after 2 min of pulselessness. But imagine that in their overwhelming grief they suddenly change their minds and demand that he be resuscitated. Imagine that the clinicians comply and are successful (a reasonable possibility). Has the patient been raised from the dead?

As Bernat has correctly observed, "Because no mortal can return from being dead, any resuscitation or recovery must have been from a state of dying but not from death."[8] Following this reasoning, this patient was not in fact dead,

but only dying, at the time that death was declared. In sum, current practices of procuring organs through DCDD protocols do not conform to the DDR.

When is it legitimate to procure organs?

If DCDD is incompatible with the DDR, why do Bernat and others continue to insist on the necessity of maintaining the rule to justify the status quo? The reasons seem to hinge on the perceived need to uphold the traditional norm prohibiting doctors from killing patients and the prospect of jeopardizing the trust of the public.

In our view, the pivotal ethical questions in these cases all center on the decision to withdraw life support, which is prior to any decisions about organ donation. As we have argued in detail elsewhere, withdrawing life support causes the death of patients.[9] Given that it is ethically justifiable for doctors to cause their patients' death by stopping a ventilator, with valid consent, we see no good reasons for precluding procuring organs for transplantation prior to withdrawing life support.

The *de facto* violation of the DDR in current practices of organ donation suggests that we should shift the focus of ethical inquiry from when is it legitimate to determine that an organ donor is dead to when is it legitimate to procure vital organs. When a valid decision has been made to stop life-sustaining treatments, no harm or wrong is done by procuring vital organs prior to death because the patient will be dead within a short interval of time as a result of stopping life support, regardless of whether organs are procured. The absence of harm plus appropriate consent legitimates vital organ donation. The DDR does no genuine moral work in current practices of vital organ donation because neither donors who are brain dead nor

290 Robert D. Truog and Franklin G. Miller

donors under DCDD protocols are known to be dead at the time organs are procured. We should be working toward honestly facing the fact that currently we are procuring vital organs from patients who are not known to be dead and that it is ethically legitimate and desirable to do so.

With regard to maintaining the trust of the public, to date the assumption has been that the public is not capable of engaging in a discussion about the ethical complexities of organ transplantation and needs to be reassured that current practices accord with traditional ethical principles. Yet the limited research that has been conducted does not support the notion that the public has settled views on this issue,[10] and multiple anecdotal reports suggest the existence of a diversity of opinion. In the case of a child who donated organs through a DCDD protocol, for example, the parents stated that if they had found out that another child had died because they were not able to donate their daughter's heart, it would have been "like another slap in our faces." They would have permitted simply taking out their daughter's heart under general anesthesia, without the choreographed death. When pressed about the fact that this would have violated the DDR, the father replied, "There was no chance at all that our daughter was going to survive ... I can follow the ethicist's argument, but it seems totally ludicrous."[11] We submit that he gets it exactly right.

1 Truog RD. "Is it time to abandon brain death?" *Hastings Cent Rep.* 1997; 27(1): 29–37.

2 Truog RD, Miller FG. "The dead donor rule and organ transplantation." *N Engl J Med.* 2008; 359(7): 674–75.

3 Miller FG, Truog RD. "Rethinking the ethics of vital organ donations." *Hastings Cent Rep.* 2008; 38(6): 38–46.

4 Bernat JL. "Point: are donors after circulatory death really dead, and does it matter? Yes and yes." *Chest.* 2010; 138(1): 13–16.

5 "Guidelines for the determination of death. Report of the medical consultants on the diagnosis of death to the President's Commission for the Study of Ethical Problems in Medicine and Biomedical and Behavioral Research." *JAMA.* 1981; 246(19): 2184–6.

6 Bernat J, Capron A, Bleck T, et al. "The circulatory-respiratory determination of death in organ donation." *Crit Care Med.* 2010; 38(3): 963–70.

7 Pernick MS. "Back from the grave: recurring controversies over defining and diagnosing death in history." In: Zaner RM, ed. *Death: Beyond Whole-Brain Criteria.* Boston, MA: Kluwer Academic Publishers; 1988: 17–74.

8 Bernat JL. "Are organ donors after cardiac death really dead?" *Clin Ethics.* 2006; 17(2): 122–32.

9 Miller FG, Truog RD, Brock DW. "Moral fictions and medical ethics." *Bioethics.* In press.

10 Siminoff LA, Burant C, Youngner SJ. "Death and organ procurement: public beliefs and attitudes." *Kennedy Inst Ethics J.* 2004; 14(3): 217–34.

11 Sanghavi D. "When does death start?" *The New York Times Magazine.* December 20, 2009: MM38.

Chapter 29

J.L. Bernat

POINT: ARE DONORS AFTER CIRCULATORY DEATH REALLY DEAD, AND DOES IT MATTER?

Rebuttal

Truog and Miller[1] criticize elements of the recent analysis of the circulatory determination of death that my colleagues and I conducted.[2] Their most penetrating criticism is that in DCDD (donation after the determination of cardiac death) donors we blurred the ontologic distinction between death and dying. They correctly note that I have previously analyzed this distinction in depth in the DCDD context[3] and then pose an illustrative case comparison. My rebuttal addresses this distinction and shows that the essential issue in DCDD is not one of ontology but of medical practice.

Truog and Miller[1] ask how we could consider the patient in case 1 to be dead if the analogous patient in case 2, who represents the typical patient we all hope will undergo a successful resuscitation, is not dead. They acknowledge that most hospital death determinations are made at the moment of asystole, which, from a purely ontologic perspective, is before the patient is dead. They observe that this practice is acceptable because "nothing consequential will happen to the patient over the next several minutes" until the patient is truly dead. Our society permits physicians to declare death earlier for social benefits, rather than awaiting signs of rigor mortis or other unequivocal signs of circulatory irreversibility.

We argue that cessation of circulation and respiration is permanent at that point in time after asystole when autoresuscitation cannot occur and if CPR will not be performed. Because the transition to irreversible cessation of circulation and respiration is rapid and inevitable, permanence serves as a valid surrogate marker for irreversibility. We argue that this same situation holds for both organ donation and nondonation circulatory death determinations because identical conditions apply.

However, if an ICU physician declared the patient in case 3 dead after two min of asystole based on the plan not to perform CPR, but then performed successful CPR, it would show that the prevailing practice of early death determination could create errors if the conditions under which it is valid have been violated. Similarly, if an ICU physician were to perform CPR on a DCDD donor (analogous to case 3), the same error would occur. From a purely ontologic perspective, neither patient is dead until irreversibility can be proved or is obvious. But we allow physicians to declare death at the point of permanent cessation without awaiting or proving irreversibility because this is what physicians and society have determined that we mean by death. Death statutes, such as the Uniform Determination of Death Act, accommodate this practice by their language, stating,

"A determination of death must be made in accordance with accepted medical standards." Thus, medical practice issues, not ontologic ones, are paramount in the DCDD argument.

Truog and Miller's claim that the DCDD case is more consequential because of the lethality of removing organs is simply wrong. In fact, organ donation has no impact whatsoever on the inevitable process during which permanent cessation of circulation becomes irreversible.[4] This process parallels the gradual destruction of the brain by circulatory arrest and proceeds completely unaffected by organ removal, including that of the asystolic heart.

Notes

1 Truog RD, Miller FG. "Counterpoint: are donors after circulatory death really dead, and does it matter? No and not really." *Chest*. 2010; 138(1): 16–18.

2 Bernat JL, Capron AM, Bleck TP, et al. "The circulatory-respiratory determination of death in organ donation." *Crit Care Med*. 2010; 38(3): 972–79.

3 Bernat JL. "Are organ donors after cardiac death really dead?" *J Clin Ethics*. 2006; 17(2): 122–32.

4 Menikoff J. "Doubts about death: the silence of the Institute of Medicine." *J Law Mccl Ethics*. 199S; 26(2): 157–65.

What sort of consent does respect for autonomy imply?

INTRODUCTION TO PART SIX

APART FROM ETHICAL challenges created by biomedical science, bioethics addresses more traditional medical ethics issues. A case in point is the so-called 'triumph of autonomy'; i.e., in the post-war period, medical authoritarianism and paternalism – which was based on the assumption that 'doctor knows best' – gave way to respect for the autonomy of patients and research participants. Although autonomy is a contested concept, it is rooted in the idea that the respect persons deserve entails that competent agents should be allowed to decide for themselves. Since this is concordant with the liberal values dominant in Western society – especially the emphasis on individual freedoms – autonomy's 'triumph' is unsurprising. The upshot is that certain rights are now established in medicine, health care and research, including the rights to consent, confidentiality and privacy.

The selections focus on the first of these: competent persons have the right to (voluntary, informed) consent to participate in biomedical research and to undergo medical treatment. The first selection presents the problematic: on the one hand, respect for autonomy entails a right to consent but, on the other, there are numerous 'barriers to informed consent'. In the next piece, Childress suggests a way of negotiating this dilemma. Childress is co-founder (along with Tom Beauchamp) of principlism, which advocates seeing medical ethics problems in the light of certain 'principles of biomedical ethics'. One such is the principle of respect for autonomy. Carefully distinguishing this principle from similar sounding ones, such as 'the principle of autonomy', allows Childress to defend autonomy from misdirected criticisms. It also enables him to meet substantive worries, such as that emphasising autonomy undermines our sense of obligations to others: Childress points out that the principle of respect for autonomy 'involves correlative rights and obligations [and] is thus a principle of obligation, rather than liberation from obligation'. He concludes that autonomy provides 'one among several important moral principles in biomedical ethics'; nonetheless, its role and importance are limited, a point which some critics of autonomy fail to appreciate.

Veatch's more radical response to the problematic Lidz *et al.* report is that consent is a 'transitional concept' that should be abandoned. His depiction of current consenting practices comprises a physician-determined treatment plan which is presented to a patient for approval; the plan is supposed to be in the best interests of the patient. Ironically, this means that 'the consent model buys into more of the traditional, authoritarian understanding of clinical decision making than many people realize'. More importantly, Veatch argues that pluralism and subjectivism about values – i.e., that there can be reasonable disagreement about what is good, and in what well being consists – entail that 'there is no reason to assume that the health professional can be expected to know what will promote the best interest of the patient'. Given this, the current model of consent fails to achieve its purpose – namely, to ensure that a patient's beliefs and values inform their decisions about their treatment – so should be replaced. He proposes 'an institutional framework for pairing [providers and patients] based on deep value convergence'.

Veatch's proposal obviously raises practical problems, but a deeper worry is the presumption that all value systems are equally respectable, the consent related task being to match them up. But at the heart of the debate about autonomy and consent is the patient who decides on their treatment according to dubious values and a questionable worldview. Savulescu and Momeyer address this head-on, arguing that, often, 'what purports to be medical deference to a patient's values is not this at all: rather, it is acquiescence to irrationality'. Specifically, since decisions based on irrational beliefs do not deserve respect, doctors 'should care more about the rationality of patients' beliefs'. This does not justify overriding all ignorant or irrational choices, but it does require doctors to help patients to deliberate more effectively in order to reach decisions based on rational beliefs.

Onora O'Neill has become a leading critic of informed consent due largely to her influential Gifford and Reith lecture series. Her critique starts by describing the 'exceptional numbers of people' to whom the current model of informed consent is simply irrelevant, including the very young and the very ill. Focusing on the relationship between consent and autonomy, O'Neill queries the easy assumption that consent is important because it protects autonomy; after all, a competent patient can consent to options in a way that lacks autonomy. The real value of consenting, she suggests, is in ensuring that one is neither deceived nor coerced. This, in turn, has practical implications: deception is avoided by giving patients and participants control of the amount of information they receive; coercion is avoided by allowing patients and participants to rescind their decisions. Hence, many of our current practices, such as overloading patients with information and ramifying requests for specific consent, are inappropriate.

It is unclear whether expandable information and rescindable consent, as proposed by O'Neill, will provide research subjects and patients with adequate protection. More details about alternative models of consent are required. Whitney points out that there are numerous sorts of treatment decisions: for example, some are more important than others; some involve choosing between competing treatments, whereas in other cases there is only one viable treatment option. Likewise, there are various decision making processes, depending on whether the provider or the patient is preeminent, or whether a process of 'shared decision making' is apposite. Since no one decision making process will be universally appropriate, the task is to develop a typology to match a decision making process to a treatment decision. Whitney's expressly nascent suggestion is that importance and certainty are two pertinent characteristics of a medical decision; for example, 'decisional priority for choices of high importance and low certainty should rest unequivocally with patients'.

Selections

Lidz, C.W., Meisel, A., Osterweis, M., Holden, J.L., Marx, J.H. and Munetz, M.R. 1983. 'Barriers to Informed Consent', *Annals of Internal Medicine*, 99 (4), 539–43.

Childress, J.F. 1990. 'The Place of Autonomy in Bioethics', *The Hastings Center Report*, 20 (1), 12–17.

Veatch, R.M. 1995. 'Abandoning Informed Consent', *Hastings Center Report*, 25 (2), 5–12.

Savulescu, J. and Momeyer, R.W. 1997. 'Should Informed Consent be Based on Rational Beliefs?' *Journal of Medical Ethics*, 23 (5), 282–8.

O'Neill, O. 2003. 'Some Limits of Informed Consent', *Journal of Medical Ethics*, 29 (1), 4–7.

Whitney, S.N. 2003. 'A New Model of Medical Decisions: Exploring the Limits of Shared Decision Making', *Medical Decision Making*, 23 (4), 275–80.

Chapter 30

Charles W. Lidz, Ph.D., Alan Meisel, J.D., Marian Osterweis, Ph.D., Janice L. Holden, R.N., John H. Marx, Ph.D. and Mark R. Munetz, M.D.

BARRIERS TO INFORMED CONSENT

Over the last 30 years, the legal doctrine of informed consent has emerged as the focus of increased regulation of the decision-making process for medical treatment.[1,2] The common law always has required the patient's consent to medical treatment in all but exceptional circumstances.[3-5] Recently this requirement has metamorphosed into a more elaborate requirement of "informed consent" under which a physician is obligated not only to obtain the patient's consent but to [disclose relevant information about the treatment before obtaining consent and beginning treatment.[3] In the absence of informed consent or a recognized exception to the requirement,[4] a physician who provides treatment to a patient may incur civil liability.[6]

The legal model of informed consent requires the following elements: Information is "disclosed" to patients by their physician or other health care personnel designated to do so. Physicians make reasonable efforts to ascertain whether patients understand the treatment decision at hand and provide the information necessary for them to do so. Patients make a decision either to undergo or forego the procedure in question. Patients make their decisions voluntarily. No coercion, duress, or undue influence is imposed by health care personnel.

When patients are incapable of understanding the issues in the treatment decision – that is, when they are "incompetent" – health care personnel should invoke some sort of proxy decision-making process.[6] All of these factors are supposed to lead to patients actively and autonomously participating in a decision-making process with the physician. The final decision is supposed to be made by the patient. Most published research on informed consent has raised serious questions about the extent to which these elements are found in real medical decision making. Physicians often do not fully inform[1,7] (Benson P. Unpublished observations), patients do not fully understand[8,9] (Benson P. Unpublished observations), and physicians actually make decisions[1,7]. The reasons the ideal decision-making process envisioned by law is not realized in practice have not been fully explored. This article reports some of the barriers to the implementation of the doctrine that emerged from extensive social scientific observations of medical and surgical decision making. The study was done for the President's Commission for the Study of Ethical Problems in Medicine and Biomedical and Behavioral Research.

Methods

The study was primarily based on "participant observation," a general research strategy that uses a series of different methods when the research goal is to gain new knowledge about the everyday world.[10-13] Because we were trying to learn about a routine part of hospital life, we followed the basic outline of participant observation and chose the least obtrusive method that would still gain the necessary data.

The two observers studied two inpatient wards (a medical cardiology service and surgical service) and a surgical outpatient clinic in a university teaching hospital. After obtaining the cooperation of the staff, the observers spent a few days watching the routine procedures, getting to know the staff, and becoming a part of each of the settings. After this period the observers systematically began to record staff and patient behavior related to the treatment decision-making process, and to interview patients after each decision, using a semi-structured interview schedule.

The observers used speed writing to take verbatim notes of all observed patient-staff interactions; interviews with patients were recorded. The observers conducted in-depth recorded interviews with 101 patients (43 in cardiology and 58 in surgery). More details of the research methods can be obtained in the appendix to the President's Commission Report.[14]

General barriers to informed consent

Although there were differences in the degree to which decision making approximated the informed consent model in different circumstances, some barriers to informed consent existed in almost all situations. Although the barriers were of many different types, they all interfered with the patient's playing an active role in decision making.

Treatment decision making as a process

The first general barrier is a fundamental failure of the legal doctrine of informed consent to deal with the realities of medical decision making. Medical decisions are processes that emerge and evolve over a period of time, not discrete events that occur only once. A typical surgical decision, for example, starts with the patient being referred to the surgeon by a general practitioner, who may have told the patient about the problem and its treatment. In the clinic that we observed, the patient was usually seen by a nurse who recorded the patient's basic symptoms, followed by an intern or junior resident who obtained the patient's history and did a physical examination. The intern or resident then made a general determination of what should be done, and often expressed an opinion about this to the patient. The next step was for either the senior resident or chief resident to repeat much of the work the intern had done, and make a recommendation.

If surgery was required, the chief resident would often tell the patient a little about the procedure, and the patient would be given a chance to ask questions. Some risks of the procedure were often described in a general way. After the patient was hospitalized, further discussion with the patient would occur during both morning and evening rounds, and the nurses often spoke informally with the patient about aspects of the treatment. However, the physician's major effort to formally comply with the informed consent doctrine typically occurred the night before surgery, before the nurse brought in the consent form for the patient to sign. Just before the consent form was signed, the nurses often described to the patient in some detail the preparation for, and immediate results of, the operation itself (preoperative teaching).

During this time, the patient was forming impressions of the risks, benefits, alternatives, and nature of the surgery. However, the formal effort to describe any risks or disadvantages of the surgery and get a signed consent form only occurred the night before surgery. By this time the patient was both socially and emotionally committed to the surgery, and such disclosure never, in our observations, caused a patient to refuse the procedures. In both surgical and nonsurgical cases, the formal disclosure and consent were done after the decision had been made to have the treatment.

Patient attitudes

Another major barrier to patient participation in decision making was the attitudes of the patients themselves. First, with a few exceptions that we will discuss below, the patients believed that decisions about their treatment should be primarily or completely up to their physicians because of their technical expertise and commitment to the best interests of the patients. However, this does not mean that patients did not want information about their treatment. Indeed, many patients raised this issue with us, and some complained bitterly that they had not been told enough. But a close examination of their comments shows that patients rarely wanted information in order to direct their treatment. There were four reasons that patients wanted information.

Information for Compliance: Most patients wanted information to facilitate compliance with treatment decisions. They typically transferred responsibility for decision making to their physician and wanted information in order to get the most from the physician's decisions.

A 70-year-old diabetic woman was taken off oral medication and given insulin for her diabetes without any consultation, and was given no information relevant to that decision. The patient was very interested in what she was told about the new nutritional requirements that were necessary because of the switch to insulin treatment. She was satisfied with the amount and kind of information she had received about her treatment — despite the fact that she was informed about the treatment decisions only after they were

implemented — but said she would have liked more information about the new dietary requirements so she could comply with them exactly. About the change in her treatment, she had little to say except "Whatever reasons they had ... I'm sure that they're doing what's best for me."

Information as a Courtesy: Patients wanted information about their treatment as a sign of respect for them as persons, so they could be prepared for what was to happen, and not feel that they were being moved around like inanimate objects. A patient was brought into the hospital with atrial flutter and a chronic lung disease. He had little curiosity about what was being done to him.

Interviewer: Can you tell me how you came to (get the electro-cardio conversion)?

Patient: The doctors decided it. They had me on medication from Monday noon when I was admitted until Tuesday morning. Tuesday afternoon they decided to take me down there.

Interviewer: What kind of medication?

Patient: I don't know. Whatever they shoot into you to try to get the heart back in rhythm. They said something like the medication didn't work. ...

Interviewer: Who do you think should be making the decisions about (your treatment)?

Patient: Well, I'm a firm believer that there should always be two opinions with anything, and that's why I like what they do here where they got three or four doctors that make the decision.

Interviewer: And what about yourself? Do you have any involvement in the decision?

Patient: No, not particularly.

Although this patient was uninterested in information necessary for deciding about his treatment, he was indignant about the information he was not getting.

Interviewer: Well, do you feel, then, that you're getting enough information about . . .?

Patient: Well, I'm getting information about what's wrong and everything. The only thing I don't like is that you never know from one minute to the next what's coming next. Just like this morning in here, they were talking about this test and that, but they don't tell you they're going to do it. And I didn't even know it. Then the nurse came up and she said, "Well, you're going to go down there."

The information that was of interest to him and of which he felt deprived concerned not the rationale for or risks associated with various tests and procedures but rather their scheduling. The information that he saw as significant was when and what was going to be done. These patients were comfortable relinquishing all decision-making responsibility. What they were not pleased with was a lack of consideration and courtesy manifested by physicians' failure to give them information as to when "you're going to go down there," which kept the patient from preparing for the test.

Information for Veto: A third and less frequent conception of necessary information involved exercising a veto over a decision the physician had already made. These patients expected their physician to tell them briefly what procedure would be done. They thought it appropriate that their physician would decide what would best facilitate recovery and wanted to be informed in case this decision involved something they profoundly wanted to avoid. As one patient said, "Just tell me what you want to do so I know something about it, and I'll almost always go along with it because I know that you know best for me. But I guess if it were something I really, for some reason, didn't want to do, I guess then I'd just tell them not to go ahead." These patients were not altogether clear about what were legitimate grounds for rejecting a proposed course of treatment. However, they clearly did not expect to be given a recitation of all the information necessary for a rational choice.

Information for Decision Making: Only about 10 percent of the patients we interviewed saw themselves as having an active role in decision making. They wanted information so they could keep up with what was going on in their treatment, and often participated in the decisions. As a group these patients were distinguished by the fact that most of them had chronic diseases whose treatment usually required their active participation. Their discussions with their physicians involved a give and take that contrasted sharply with the passivity of the other patients. A patient who had had a life-threatening disease of uncertain origin for two years was told by his surgeon that the surgeon wanted to remove a large mass in his intestine, despite negative biopsy findings. The patient responded, "Well, I want to talk to my regular doctors about this. I also think I've got to talk to my family and make a decision with them. . . . And, quite frankly, unless it is very important, I have to get home for awhile. I have a lot of personal and business things that I've got to arrange." This patient clearly did not feel the decision was simply up to his surgeon.

A 45-year-old high school graduate receiving welfare had had bowel bypass surgery for weight loss that caused periodic bowel obstructions. This patient's description of her most recent trip to the emergency room is indicative of her role in decision making.

Interviewer: What did the doctors in the emergency room tell you so far is going on in there?

Patient: They didn't tell me. They ... asked me what the problem was and I told them. And like they said to me, how did I know it was a bowel obstruction and, like I told you, if you throw up stuff that is awful smelly green and smells like shit, it probably is ... The (doctor) I had when I first went in ... told me what we'll do. He's gonna send me over for x-rays and whatever, and I said to him, put the stomach tube in first. Because that is my biggest problem. I can go over and get my x-rays done, I can stand on my head, but get the swelling out, because the pain was unbearable.

Interviewer: Did he do that?

Patient: Yeah.

Except when dealing with physicians who were expert on this type of condition, this patient directed much of her treatment herself.

Physician attitudes

Most of the physicians we observed did not view informed consent as an integral part of good patient care. Although they generally supported the idea of giving patients more information, few physicians gave it much attention. Most physicians who were interested in informed consent seemed primarily concerned with it as a legal problem. Indicative of this attitude was a senior member of the house staff who, when asked how the presence of observers had changed the service, said that he was more inclined to discuss the patients' treatment with them on rounds when he knew the observers would be

there. Ordinarily, he said, he would often discuss patients' treatment with them privately because "The intern(s) think it's real boring to stand there and listen to me tell somebody about what their treatment is going to be." Even at this early stage of their careers, the interns found the dialogue with the patient uninteresting.

The disinclination of many physicians to take informed consent seriously reflects the basic logic of medical decision making. Except for the goal of relieving acutely problematic symptoms, medical treatment is usually done only when the physician feels that a clear diagnosis has been arrived at. In almost every case, the physician believes that there is a preferred treatment for the condition. Because the diagnosis is based on the physician's knowledge and experience, and the best treatment has been determined by a combination of medical research and clinical experience, most physicians find it hard to see how the patient can choose differently except by sacrificing his or her health. Whereas the doctrine of informed consent implies that there are a series of alternative treatments from which the patient may choose, physicians usually see only superior and inferior treatments. Physicians see little to be gained by allowing a patient to choose the treatment. Thus discussions with patients that we observed very rarely involved a presentation of alternatives; instead, the physician told the patient what was going to be done.

Specific barriers to informed consent

Although there are various general barriers to informed consent, decision making processes differ depending on the patient's problem and the structure of the medical care setting.

Inpatients versus outpatients

In general, inpatients were much less likely to question their physicians or to participate in decision making. Outpatients were more likely to assert a need for a specific treatment or to have a specific problem dealt with and asked more questions.

This difference may reflect a feeling on the part of inpatients that the major decision had been made before they entered the hospital. However, even when substantial changes were made in the treatment after the patient was admitted, inpatients were generally passive. Perhaps this problem shows the regression and emotional dependency that develops in patients when they enter the hospital.[15]

Acute versus chronic illness

The most striking differences in decision-making patterns we saw were between patients with acute illness and those with chronic, long-standing disorders. This difference can be seen by considering two different cases. The first patient had acute myocardial infarction. Although his intern was probably the most inclined to talk with patients of all interns we saw, and although the patient was a talkative and lively person, the patient was not told much about his treatment. He was told that he had a severe heart attack and would have to stay in the hospital. Aside from that, his conversations with his doctors focused on symptom reports and discussions of when he could leave the coronary care unit and the hospital. We interviewed him when he had been in the hospital for 10 days.

Interviewer: What sort of treatment have they given you? What have they done for you?

Patient: What have they done for me? I don't know all of the medications I'm taking. I would say I am on at least six or seven different kinds of medications for the heart. Probably some blood thinners and probably also the, I just don't have any idea. I haven't even bothered. This is their business. It's none of mine. It is not my field of work. And however . . . I put my faith and trust in them that they know what they are giving me.

This sort of passivity and distance from the treatment decisions was typical of acute patients. This patient, like many patients with serious but acute problems, was denying the seriousness of his illness and trying to avoid any responsibility for its treatment. Patients with chronic diseases were usually more active.

A 28-year-old divorced woman with chronic renal failure had been on hemodialysis for six years. She was admitted to University Hospital with a clotted fistula. The surgeon knew her well.

Physician: Well, what happened?

Patient: My blood pressure dropped.

Physician: When did it happen?

Patient: Well, after dialysis and after I gave myself some lactulose.

Physician: (Examining the fistula) Remind me, is this the first graft on this site?

Patient: Yes.

Physician: The question is if we should even bother to try to open this up.

Patient: It was working just fine until my blood pressure dropped.

Physician: Well, let's try to open it up again. We'll make a small incision under a local and see if we can fix it.

The doctor seems to have decided to repair the graft, at least in part, on the basis of the patient's observation.

The differences between patients' chronic and acute disorders seem to reflect basic differences in orientation to their situation in life. Patients with acute disorders usually take on a set of behaviors and attitudes that has been called "the sick role," including temporarily giving up control of the care of their bodies to the physician.[15] In return they expect to be made well and to return to their normal lives. However, patients with serious chronic diseases cannot so easily give up responsibility for their treatment, deny the reality or seriousness of their illness, and wait to be cured. To do so means giving up responsibility for larger parts of their life while waiting for something that will probably never come.

The complexity of the medical system

Although many factors were involved in the differences in the informed consent process on the surgical and medical cardiology wards, one of the main ones was the organization of the medical staff.

On the surgical ward, it was clear that the chief resident had the most authority in an unambiguous chain of command. This authority was communicated to the patients by his interns in the morning when they told patients that the chief resident would talk to them later during afternoon rounds. Patients quickly learned that the chief resident would directly or indirectly determine their fate.

The chain of command on the cardiology unit was more complicated, partly because of the large size of the service. The house staff was divided into two teams, each consisting of one resident and two interns. Patients were assigned to an intern who was, according to hospital policy, the patient's "primary physician." The intern was directly responsible for maintaining the patient on a day-to-day basis, making decisions about laboratory work and routine diagnostic tests, and watching clinical symptoms for changes in status.

The resident's task was to supervise the interns and to guide their decision making. Residents were available formally in rounds and informally throughout the day as needed. The degree of resident participation in decision making differed in practice from taking primary charge of the patient's care to serving as a consultant to the intern.

Other members of house staff also had substantial input into decisions. These included various faculty and visiting cardiologists who conducted rounds, the faculty in their roles as directors of various testing laboratories, the cardiology fellows who were receiving postresidency training, and the entire nursing staff. Furthermore, every patient had an attending physician. Although some of the attending physicians had been assigned to patients upon admission to the hospital and had only a formal relationship with the patient, in other cases the attending physician was also involved in the decision-making process.

The differences in authority structure had two effects on the informed consent process, first in terms of who was responsible for telling the patient what was going on, and second in terms of patients knowing who was making decisions. On the surgical ward it was clear that the chief resident was in charge. It was his responsibility to inform patients, and in turn the patients went to him with their questions. On the cardiology ward the responsibility for informing patients was so diffuse that it was sometimes completely overlooked. Furthermore, the responsibility for decision making was ambiguous and many patients did not know where to turn.

The number of decisions

Most discussions of informed consent have implicitly assumed two things about the decision-making process that have led to the widespread use of consent forms. The first assumption is that the decision about treatment is made at a specific point in time. We noted above that decision making is usually a complex process that takes place over a long period. A second assumption underlying the use of consent forms is that there is typically only one or a small number of important decisions to be made about any patient's treatment.

On the surgical service, the second assumption was typically true. Decisions focused on whether to operate. However, medical treatments were often more complex, involving many smaller decisions with major consequences to the patient's well being. Patients with cardiomyopathy, for example, often had their medications changed or adjusted several times a day. Each of these adjustments not only had potential for improving the patient's condition but often risked permanent injury or death. Each of these decisions was based on a complex interplay between the physician's diagnosis of the overall problem, the acute need to keep various blood chemistry levels in the proper range, the apparent physical condition of the patient, and the patient's symptoms. Although the ward physicians made little effort to tell the patient anything, it is hard to see how all of these considerations could have been discussed with the patient each time the medications were modified.

Conclusion

We have sought to describe some of the obstacles to the legal doctrine of informed consent that envisions patients as active participants in the decision-making process for their treatment. The barriers to this type of role for the patient are more extensive than noncooperation on the part of the physicians or patient ignorance. The barriers include organizational factors, the complexity of the decisions, patient inexperience, and faith in their physicians.

Having reviewed both the general difficulties of implementing the doctrine and the specific situations in which informed consent is difficult, one may think that informed consent is unrealistic in terms of treatment and hospital organization. However, just as specific situations undercut the possibility of informed consent as a meaningful reality, so other specific situations seem to make an active dialogue between patient and physician more likely. In essence, these features are the reverse of the ones that inhibit this dialogue. In situations where treatment involves chronically ill patients who have to make a small number of decisions about treatments that they have experienced previously, and in which the center of medical responsibility is clear, there is a substantial chance that a real dialogue will take place. It was in these situations that we found most of the patients who actively sought to play a role in their treatment and understood the issues involved to help make decisions.

If there is any hope of encouraging patient participation through the doctrine of informed consent, the implementation of the doctrine must become more flexible and responsive to the complexities of medical care. Our findings suggest that with certain types of chronic patients and in certain types of organizational structures, an active patient role is feasible. In other types of settings, less can be expected. In either case, developing procedures to encourage true informed consent cannot consist solely of designing more carefully worded consent forms.

Notes

1 President's Commission for the study of Ethical Problems in Medicine and Biomedical and Behavioural Research: Making Health Care Decisions. *The Ethical and Legal Implications of Informed Consent in the Patient-Practitioner Relationship.* Washington, D.C.: U.S. Government Printing Office, 1982.

2 Restructuring informed consent. *Yale Law* 1970; 79: 1533–76.

3 Meisel A. "The expansion of liability for medical accidents." *Nebraska Law Rev.* 1977; 56: 51–152.

4 Meisel A. "The 'exceptions' to the informed consent doctrine: striking a balance between competing values in medical decisionmaking." *Wisconsin Law Rev.* 1979; 2: 413–88.

5 Katz J. "Informed consent – a fairy tale?: law's vision." *Univ Pittsburgh Law Rev.* 1977; 39: 137–74.

6 Meisel A., Roth L., Lidz C. "Toward a model of the legal doctrine of informed consent." *Am J Psychiatry.* 1977; 134: 285–9.

7 Lidz C.W., Meisel A., Zerubavel E., Ashley M., Sestak R.S., Roth L.H. *Informed Consent: A Study of Psychiatric Decisionmaking.* New York: Guilford Press; 1983.

8 Roth L.H., Lidz C.W., Meisel A. et al. "Competency to decide about treatment or research: an overview of some empirical data." *Int. J Law Psychiatry.* 1982; 5: 29–50.

9 Leonard C.O., Chase G.A., Childs B. "Genetic counseling: a consumer's view." *N Engl J Med.* 1972; 287: 433–9.

10 Denzin N. *The Research Act: A Theoretical Introduction to Sociological Methods.* Chicago: Aldine Publishing Co.; 1970.

11 Schwartz H., Jacobs J. *Qualitative Sociology.* New York: Free Press; 1979.

12 Bogdan R., Taylor S. *Introduction to Qualitative Research Methods.* New York: Wiley-Interscience; 1975.

13 Whye W.F. *Street Corner Society.* Chicago: University of Chicago Press; 1955.

14 Lidz C.W., Meisel A. "Informed consent and the structure of medical care." Appendix to PRESIDENT'S COMMISSION FOR THE STUDY OF ETHICAL PROBLEMS IN MEDICINE AND BIOMEDICAL AND BEHAVIORAL RESEARCH. *The Ethical and Legal Implications of Informed Consent in the Patient-Practitioner Relationship.* Washington, D.C., U.S. Government Printing Office, 1983.

15 Parsons T. *The Social System.* Glencoe. Illinois: Free Press; 1951.

Chapter 31

James F. Childress

THE PLACE OF AUTONOMY IN BIOETHICS

I come not to bury autonomy, but to praise it. Yet my praise is somewhat muted; for autonomy merits only two cheers, not three. Five years ago at the fifteenth anniversary of the founding of The Hastings Center the general theme was "Autonomy–Paternalism–Community." Hearing several sharp criticisms – indeed, virtual rejections of autonomy – I stressed in my oral remarks and later in my published paper that we "need several independent moral principles, such as individual and communal beneficence and respect for personal autonomy." It is "unfortunate and even pernicious," I continued, to suggest that "biomedical ethics is allegedly moving beyond autonomy to community and paternalism," for such an approach would reduce "ethical reflection to a mere mirror of societal concerns at a particular time, when in fact the task for serious ethical reflection is to indicate the importance and relative weight of several moral considerations that should be maintained in some tension or balance."[1]

Reaffirming that statement five years later, I want to defend the principle of respect for personal autonomy as one among several important moral principles in biomedical ethics. My defense will proceed by sketching and clarifying some presuppositions and implications of this principle in light of several major criticisms.

Many of those criticisms are misplaced, because they are (perhaps deliberately) not directed at the most defensible conceptions of the principle of respect for autonomy. I will contend that an adequate conception of the principle of respect for autonomy can meet the main criticisms levelled by various critics, whether communitarians, narrativists, virtue theorists, traditionalists, or religionists. My main argument focuses on the *principle of respect for autonomy as an important moral limit and as limited*. As a moral limit, it constrains actions; but it is also limited in scope and in weight, in addition to being complex in its application. Both critics and defenders tend to neglect these senses of limit in their focus on an oversimplified, overextended, overweighted principle of respect for autonomy.

Misdirected criticisms

In several ways, the principle of respect for autonomy has been misunderstood and misinterpreted, in part as a result of flawed formulations and defenses by its supporters. Critics have often supposed that they were attacking the concept of autonomy when in fact they were aiming their fire at particular conceptions of autonomy, often the least defensible ones.

It has been a mistake to use the term "autonomy" or even the phrase "principle of autonomy" as a shorthand expression for "the principle of respect for autonomy."[2] It is important to correct this mistake because many critics seem to suppose that proponents of this principle have an ideal of personal autonomy and believe that we ought to be autonomous persons and make autonomous choices. However, the ideal of personal autonomy is neither a presupposition nor an implication of the principle of respect for personal autonomy, which obligates us to respect the autonomous choices and actions of others.

The ideal of autonomy must be distinguished from the *conditions* for autonomous choice. It is important for the moral life that people be competent, be informed, and act voluntarily. But they may choose, for example, to yield their first-order decisions (that is, their decisions about the rightness and wrongness of particular modes of conduct). For example, they may yield to their physicians when medical treatment is proposed or to their religious institution in matters of sexual ethics. Abdication of first-order autonomy appears to involve heteronomy, that is, rule by others. However, if a person autonomously chooses to yield first-order decision-making to a professional or to a religious institution, that person has exercised what may be called second-order autonomy.[3] People who are subservient to a professional or to a religious institution may lack first-order autonomy – self-determination regarding the content of their first-order decisions and choices – because they have exercised and continue to exercise second-order autonomy in selecting the professional or institution to which they choose to be subordinate. Hence, in those cases, respect for their second-order autonomy is central, even though their first-order choices are heteronomous. This point is important because of the common supposition that the principle of (respect for) autonomy is at odds with all forms of heteronomy, authority, tradition, etc.

The term "respect" also requires amplification. One meaning of respect is to refer to or have regard for or to consider. For example, a boxer may respect his opponent's right hook. A second meaning is more relevant – to consider worthy of high regard, to esteem, or to value. This meaning reflects the attitude that is proper in relation to autonomous choices. Although this attitude does not depend on the content of those choices, it is not inconsistent with criticism of them. In a third sense, respect is more than an attitude, it is an act of refraining from interfering with, or attempting to interfere with the autonomous choices and actions of others, through subjecting them to controlling influence, usually coercion or manipulation of information.[4]

The principle of respect for autonomy can be stated negatively as "it is [prima facie] wrong to subject the actions (including choices) of others to controlling influence." This principle provides the justificatory basis for the right to make autonomous decisions. This right in turn takes the form of specific autonomy-related (if not autonomy-based) rights, such as liberty and privacy. This negative formulation focuses on avoidance of controlling influences, including coercion and lying. However, the principle of respect for autonomy also has clear positive implications in the context of certain relationships, including health care relationships. For example, in research, medicine, and health care, it engenders a positive or affirmative obligation to disclose information and foster autonomous decisionmaking. Nevertheless, it is important to distinguish negative and positive rights based on or related to the principle of respect for autonomy, and the limits on positive rights may be greater than the limits on negative rights. For example, the positive right to request a particular treatment may be severely limited by research protocols and by just allocation schemes.

Finally, the principle of respect for autonomy is ambiguous because it focuses on only one aspect of personhood, namely self-determination, and defenders often neglect several other aspects, including our embodiment. A strong case can be made for recognizing a principle of "respect for persons," with respect for their autonomous choices being simply one of its aspects – though perhaps its main aspect. But even then we would have to stress that persons are embodied, social, historical, etc. Some of these issues emerge when we try to explicate the principle of respect for autonomy by noting its complexity.

Complexity of respect for personal autonomy

In determining what the principle of respect for autonomy requires, it is important to recognize its complexity, which is widely neglected by both defenders and critics. Some of my earlier remarks highlighted aspects of this complexity – for example, the distinction between first-order and second-order choices. Because of the complexity of persons, judgment is required, rather than the mechanical application of a clear-cut moral principle.

One difficulty in respecting people's choices is determining what they are choosing, what preferences they are expressing, etc. This complexity is magnified because people communicate not only through written statements (such as signed consent forms) or through words, but through nonverbal signs as well.[5]

Furthermore, patients may be ambivalent or even express contradictory preferences. In the maze of signals, the professional may have to make a judgment about whether a patient really wants full or only partial disclosure, or really wants to undergo a test to determine whether he could donate a kidney to a sibling, etc.

Another major difficulty in respecting personal autonomy stems from the fact that

people exist in and through time and their choices and actions occur over time. Consent itself is given and withdrawn over time and a patient's present statements should not always be taken at face value. Hence in discharging our obligations under the principle of respect for autonomy, we not only have to determine whether a patient is autonomous and just what he or she is choosing, we also have to put that patient's present consents and dissents in a broad temporal context encompassing both the past and the future. As temporal beings through and through, people may express different preferences at different times. Often discussion of the principle of respect for autonomy focuses on the present moment – for example, is there an informed consent or refusal at this time? Respecting persons becomes very complex when their temporality is properly included. Which choices and actions should we respect? In particular, is it justifiable to override a patient's present autonomous choices and actions in the light of his/her past or (anticipated) future choices and actions? And is a decision to do so respect for personal autonomy or a paternalistic breach of the principle of respect for autonomy?[6]

Past or prior consent/refusal poses no problem if the patient cannot currently autonomously express his or her wishes. As in the case of advance directives, we respect personal autonomy by acting on that past or prior statement. Matters become more problematic, however, when a person's present choices appear to contradict those previous choices, which may have even been made with a view to preventing future change. For example, in one case a twenty-eight-year-old man decided to terminate chronic renal dialysis because of his restricted lifestyle and the burdens on his family – he had diabetes, was legally blind, and could not walk because of progressive neuropathy. His wife and physician agreed to provide him medication to relieve his pain while he died and agreed not to put him back on dialysis even if he requested it under the influence of uremia, morphine sulfate, and ketoacidosis (the last resulting from the cessation of insulin). While dying in the hospital, the patient awoke complaining of pain and asked to be put back on dialysis. The patient's wife and physician decided to act on the patient's earlier request that he be allowed to die, and he died four hours later.[7] In my judgment, the spouse and physician should have put the patient back on dialysis in view of his current request and the irreversibility of the decision to let him die in accord with his earlier statements. After putting him back on dialysis, they could have determined if he had autonomously revoked his prior choice; if he then persisted in his prior decision, they could have proceeded again with more confidence.

A critical question in this case and others is whether people have autonomously revoked their previous consents/refusals. Thus, it is necessary to continue to assess a person's degree of autonomy over time to determine whether he or she is autonomously revoking previous consents or dissents. The principle of respect for autonomy requires that we attend to both a person's prior consent/refusal and present revocation, but the present revocation takes priority if it is autonomous.

What is the role of authenticity in judgments about which actions respect personal autonomy? The consistency or inconsistency of a present choice or action with a person's life plan and risk budget over time may help us determine whether the revocation is genuine. For Bruce Miller, authenticity means that "an action is consistent with the attitudes, values, dispositions and life plans of the person."[8] Its intuitive idea is "acting in character." We wonder whether actions are autonomous if they are out of character (for example, a sudden and unexpected decision to discontinue dialysis by a woman who has displayed considerable courage and zest for life despite years of disability). Similarly, we are less likely to challenge actions as nonautonomous if they are in character (a Jehovah Witness's refusal of a blood transfusion, for example). Nevertheless, as important as the idea of character is, it would be a mistake to make authenticity a criterion of autonomy. At most, actions apparently out of character and inauthentic can be caution flags that warn others to request explanations and justifications to determine whether the actions are autonomous. It is important, however, not to rule out in advance the possibility of a change or even a conversion in basic values.

In some situations the health care professional may have good reasons to believe that if a patient is kept alive, for example, by a particular treatment that she is now refusing, she will eventually ratify the coercive or deceptive treatment on her behalf, perhaps even thanking the professional. Such a ratification does occur in some cases. Can anticipation of future consent justify present actions against a patient's express choices, in part on the grounds that the present actions respect what the person will be rather than what she now is? My response is that actual or predicted future consent is neither necessary nor sufficient to justify interventions against current choices. At most, a patient's probable future consent may provide evidence that the criteria for justified paternalistic interventions have been met.[9]

Finally, respecting personal autonomy is complex because there are several varieties of consent and refusal. Although express consent (or refusal) is the primary model, consent (or refusal) may also be implicit, tacit, or presumed. To take one example, solid organ procurement in the U.S. is structured around express consent or donation, whether by the individual while alive or by the family after the individual's death. But there is also presumed consent in the donation of corneas in a dozen states. Presumed donation is not necessarily a breach of the principle of respect for autonomy. In some circumstances, silence or a failure to refuse donation could appropriately be construed as donation. For presumed donation – perhaps better viewed as tacit donation – to be autonomous and valid, society needs to make sure that the conditions of understanding and voluntariness have been met. Otherwise, the appeal to presumed donation may only be expropriation.

Scope or range of respect for autonomy

In explicating the principle of respect for autonomy as limited, I want to focus on its limited scope or range, and on its limited weight or strength. If these limits are not recognized, it is too easy to dismiss the principle as extending too far or as outweighing or overriding too much. Deflation of claims for and about the principle of respect for autonomy is essential to its preservation.

Respect for persons who are autonomous may legitimately differ from respect for persons who are not autonomous. The presence, absence, or degree of autonomy is a morally relevant characteristic (though hardly the only morally relevant characteristic) in shaping our actions and attitudes toward others. When people are autonomous, respect for them requires (or prohibits) certain actions that may not be required (or prohibited) in relation to non-autonomous persons. Several principles may establish minimum standards of conduct, such as noninfliction of harm in relation to all persons whatever their degree of autonomy. But what the principle of respect for autonomy requires (and prohibits) in relation to autonomous persons and in relation to nonautonomous persons will differ. Thus, Kant excluded children and the insane from his discussion of the principle of respect for persons and Mill applied his

discussion of liberty only to those in the "maturity of their faculties."

Nevertheless, it is appropriate to operate with a presumption in favor of adults' autonomy, unless and until they are determined to be substantially nonautonomous. Several factors of autonomy are relevant; these include incompetence, i.e., an inability to perform certain tasks, lack of understanding, and lack of voluntariness (both internal and external). When these signs of nonautonomy occur, and people are at risk of harm or loss of benefits to themselves, interventions based on beneficence can be justified, and they do not violate the principle of respect for autonomy even if the person refuses. This is limited beneficence or limited paternalism.[10]

However, the principle of respect for autonomy can be overextended in ways that are misleading and even dangerous. One simple but risky overextension is to refer to the cadaveric *source* of organs for transplantation as a *donor* even if he or she never "donated," perhaps because the individual never had autonomy or never chose to donate. The donor is one who autonomously decides to donate, whether an individual while alive or a family member after the individual's death. If the decedent never made a decision to donate while alive, the family is the donor. A more troubling example can be found in presumed (consent) donation for corneas; as noted above, it often appears to be a fiction for expropriation.

Another troubling example is the appeal to substituted judgment in circumstances where it does not plausibly apply. If a person has previously (and competently) expressed preferences with sufficient clarity, that person's autonomous preferences can and should be extended to periods of lack of autonomy. However, for patients who have never been autonomous or for previously autonomous patients whose prior preferences and values cannot be reliably traced, it is more defensible to rely on a best-interests standard, based on nonmaleficence and beneficence, rather than on a substituted judgment standard, based on autonomy. The standard of substituted judgment should be rejected in such situations as an illegitimate fiction.

A final point needs to be made about scope or range. The "principle of autonomy" has been criticized as minimalist and perhaps even egoistic in nature or at least in application in our sociocultural context.[11] This criticism focuses on a person's claim to have his or her autonomy respected rather than on a person's obligation to respect the autonomy of others. The principle of respect for autonomy, however, involves correlative rights and obligations. And it is thus a principle of obligation, rather than liberation from obligation. Here again the confusion may stem in part from the misleading language of "principle of autonomy," which should be replaced by the "principle of respect for autonomy."

Even as a principle of obligation, respect for autonomy does not exhaust the moral life. Other principles are important, not only where autonomy reaches its limits. For example, focusing narrowly on the principle of respect for autonomy can foster indifference; thus principles of care and beneficence are necessary. But without the limits set by the principle of respect for autonomy, these other principles may support arrogant enforcement of "the good" for others. Nevertheless, these and other principles sometimes outweigh or override the principle of respect for personal autonomy.

Limits of weight or strength

The principle of respect for autonomy is more than a maxim. Yet it is not absolutely binding and does not outweigh all other principles at all times. Two major alternatives remain. It could be viewed as serially ordered, taking absolute priority over some other principles; or it could be viewed as prima facie binding, competing equally with other prima facie principles in particular circumstances. I take the latter approach. Even though this avoids a priori rankings and is thus case-oriented or situational, it is different from some perspectives on casuistry, because the logic of prima facie principles dictates a procedure of reasoning or justification for infringements of principles in particular circumstances. For example, the prima facie principle of respect for autonomy can be overridden or justifiably infringed when the following conditions are satisfied: Proportionality – when in the circumstances there are *stronger* competing principle(s); Effectiveness – when infringing the principle of respect for autonomy would *probably protect* the competing principle(s); Last Resort – when infringing the principle of respect for autonomy is *necessary* to protect the competing principle(s); Least Infringement – when the infringement of the principle of respect for autonomy is the *least intrusive or restrictive* in the circumstances, consistent with protecting the competing principle(s).[12]

In addition, wherever possible and appropriate, we should explain and justify the infringement of the principle of respect for autonomy to those agents whose autonomy has been infringed.

The question of mandatory screening or testing for HIV infection instructively illustrates the reasoning required when moral values conflict. As the first public health crisis in an era of firmly established civil rights and liberties, AIDS poses important questions about the place and significance of the principle of respect for autonomy, especially in relation to the community as well as to other individuals. The needs of the community in public health may well override the rights related to the principle of respect for autonomy of some individuals under some circumstances to reduce the spread of HIV infection. Consider, for example, the principles or rules of liberty, privacy, and confidentiality.

These may be derived from the principle of respect for autonomy, but even if they have independent standing, they are nevertheless closely related to the principle of respect for autonomy, for individuals may exercise or waive their rights to liberty, privacy, and confidentiality and thereby remove the constraints on actions by others in particular cases. But even when individuals do not waive their rights, their rights and their autonomous choices regarding those rights may sometimes be overridden.

Even in actions to protect the community, it is important to start with a presumption in favor of the principle of respect for autonomy, as expressed in liberty and privacy, and then to determine whether that presumption can be rebutted by arguments for mandatory screening or testing. Critics sometimes doubt whether it is appropriate for the community to have to bear the burden of proof for overriding respect for autonomy, but in view of the community's power and tendency to abridge autonomy, along with the importance of the principle of respect for autonomy, this is not an inappropriate burden and it can sometimes be met. For example, if we apply the conditions identified above for overriding prima facie obligations, it would be necessary to consider the proportionality and effectiveness of any proposed mandatory screening or testing; the absence of an alternative; the least infringement of autonomy and privacy (the least restrictive and intrusive options) consistent with achieving the end; and finally, an explanation and justification to those whose autonomy and liberty are infringed on behalf of a communal good.[13] In view of what we now know about HIV and its transmission, very few types of mandatory screening and testing would meet these conditions – donations of blood, semen, and organs and perhaps a few others.

This pattern of justification holds in efforts to protect the community or other individuals, including health care professionals, within the community. Whatever the target, it is important to recognize when the principle of respect for autonomy – and associated principles – are being overridden, rather than camouflaging the justification as one of respect for autonomy. The wrong approach appears in recent Virginia legislation that appeals to "deemed consent" to justify HIV testing and release of test results in certain situations. The legislation provides that

> whenever any health care provider, or any person employed by or under the direction and control of a health care provider, is directly exposed to body fluids of a patient in a manner which may, according to then current guidelines of the Centers for Disease Control, transmit human immunodeficiency virus, the patient whose body fluids were involved in the exposure *shall be deemed to have consented to testing* for infection with human immunodeficiency virus. Such patient shall also be *deemed to have consented to the release* of such test results to the person who was exposed. In other than emergency situations, it shall be the responsibility of the health care provider to inform patients of this provision prior to providing them with health care services which create a risk of such exposure.
>
> Virginia Code § 32.1–45.1;
> emphasis added

The danger of both overextending and overweighting the principle of respect for autonomy is evident in this move to "deemed consent." It is an inappropriate fiction to construe testing and release of information as based on the principle of respect for autonomy in situations where individuals did not consent and perhaps even explicitly refused to consent. Whatever the rationale for the Virginia legislation, it is better to face directly the conflict between the principle of respect for autonomy and other principles rather than to reinterpret the principle of respect for autonomy by extending it to circumstances where it does not apply. Then we can address whether the principle of respect for autonomy can be outweighed by competing principles in the circumstances.

The principle of respect for autonomy is very important in the firmament of moral principles guiding science, medicine, and health care. However, it is not the only principle, and it cannot be assigned unqualified preeminence. A clear example of overconcentration on the principle of respect for autonomy and its implications can be seen in research involving human subjects, where for years the subject's voluntary, informed consent tended to overshadow all other ethical issues. As a consequence, there was neglect of other important moral considerations that must be met prior to soliciting the potential subject's consent to participate – e.g., research design, probability of success, risk-benefit ratio, and selection of subjects.[14] To be sure, if researchers do not receive the potential subject's voluntary, informed consent, they may not enlist that subject. However, the right of the potential subject to refuse to participate in research became for many the only moral constraint worthy of attention, even though this issue should not be addressed until other prior important ethical issues have been resolved.

In addition, concentration on the principle of respect for autonomy invited inadequate reasons for rejecting or redirecting some research on some populations. For example, critics of research involving prisoners tended to argue that the principle of respect for autonomy cannot be met in an inherently coercive environment. However, a more defensible ethical criticism emerges from the principle of justice – the unfair imposition of the burdens of research on a captive and vulnerable population many of whom have already suffered serious deprivations in the society.

Yes, we should go beyond the principle of respect for autonomy – in the sense of going beyond its misconceptions and distortions and in the sense of incorporating other relevant moral principles. But going beyond should not mean abandoning. Despite its complexity in application, despite its limits in scope or range and in weight or strength, and despite social changes, the principle of respect for personal autonomy has a critical role to play in biomedical ethics in the 1990s. But that role requires a sense of limits; we must not overextend or overweight respect for autonomy.

Notes

1 James F. Childress, "Ensuring Care, Respect, and Fairness for the Elderly," *Hastings Center Report* 14:5 (1984), 27–31. For criticisms of autonomy, see essays by Daniel Callahan, Eric Cassell, and Robert Morison in the same issue.

2 In the third edition of *Principles of Biomedical Ethics* (New York: Oxford University Press, 1989), Tom L. Beauchamp and I reformulate what we had earlier called "the principle of autonomy" as "the principle of respect for autonomy."

3 See, for example, Gerald Dworkin, "Autonomy and Behavior Control," *Hastings Center Report* 6:1 (1976), 23–8.

4 This formulation is influenced by Ruth R. Faden and Tom L. Beauchamp, *A History and Theory of Informed Consent* (New York: Oxford University Press, 1986).

5 See, for example, Eric J. Cassell, *Talking with Patients*, 2 vols. (Boston: MIT Press, 1985).

6 See James F. Childress, *Who Should Decide?: Paternalism in Health Care* (New York: Oxford University Press, 1982).

7 Childress, *Who Should Decide?*, 224–25. This case was prepared by Gail Povar, MD.

8 Bruce Miller, "Autonomy and the Refusal of Life-Saving Treatment," *Hastings Center Report* 11:4 (1981), 22–28.

9 Childress, *Who Should Decide?*.

10 Childress, *Who Should Decide?*.

11 See, for example, Daniel Callahan, "Minimalist Ethics," *Hastings Center Report* 11:5 (1981), 19–25.

12 For a somewhat different formulation, see Beauchamp and Childress, *Principles of Biomedical Ethics*, 3rd ed., 53.

13 See James F. Childress, "An Ethical Framework for Assessing Policies to Screen for Antibodies to HIV," *AIDS and Public Policy Journal* 2 (Winter 1987), 28–31.

14 See, for example, James F. Childress, *Priorities in Biomedical Ethics* (Philadelphia: The Westminster Press, 1981), 51–73.

Chapter 32

Robert M. Veatch

ABANDONING INFORMED CONSENT

Consent has emerged as a concept central to modern medical ethics. Often the term is used with a modifier, such as *informed* or *voluntary* or *full*, as in loosely used phrases like "fully informed and voluntary consent." In some form or another, modern ethics in health care could hardly function without the notion of consent.

While we might occasionally encounter an old-guard retrograde longing for the day when physicians did not have to go through the process of getting consent, by and large consent is now taken as a given, at least at the level of theory. To be sure, we know that actual consent is not obtained in all cases and even when consent is obtained, it may not be adequately informed or autonomous. For purposes of this discussion, we shall not worry about the deviations from the ideal; rather the focus will be on whether consent ought to be the goal.

This consensus in favor of consent may turn out to be all too facile. *Consent* may be what can be called a transition concept, one that appears on the scene as an apparently progressive innovation, but after a period of experience turns out to be only useful as a transition to a more thoroughly revisionary conceptual framework.

This paper will defend the thesis that consent is merely a transitional concept. While it emerged in the field as a liberal, innovative idea, its time may have passed and newer, more enlightened formulations may be needed. Consent means approval or agreement with the actions or opinions of another; terms such as *acquiescence* and *condoning* appear in the dictionary definitions. In medicine, the physician or other health care provider will, after reviewing the facts of the case and attempting to determine what is in the best interest of the patient, propose a course of action for the patient's concurrence. While a few decades ago it might have been considered both radical and innovative to seek the patient's acquiescence in the professional's clinical judgment, by now that may not be nearly enough. It is increasingly clear if one studies the theory of clinical decisionmaking that there is no longer any basis for presuming that the clinician can even guess at what is in the overall best interest of the patient. If that is true, then a model in which the clinician guesses at what he or she believes is best for the patient, pausing only to elicit the patient's concurrence, will no longer be sufficient. Increasingly we will have to go beyond patient consent to a model in which plausible options are presented (perhaps with the professional's recommendation regarding a personal preference among them, based on the professional's personally held beliefs and values), but with no rational or "professional" basis for even guessing at which one might truly be in the patient's best interest.

To demonstrate that the concept of consent will no longer be adequate for the era of contemporary medicine, some work will be in order. After briefly summarizing the emergence of the consent doctrine, we will look at what we learn from axiology – the philosophical study of the theory of the good – that calls into question the adequacy of *consent* as a way of legitimating clinical decisions. This, I suggest, will provide a basis for demonstrating why experts in an area such as medicine ought not to be expected to be able to guess correctly what course is in the patient's interest, and therefore should not be able to propose a course to which the patient's response is mere consent or refusal.

The history of consent

The history of consent reveals that it is a relatively recent phenomenon.[1] None of the classical documents in the ethics of medicine had anything resembling a notion of consent, informed or no. Autonomy of decision making, especially lay decisionmaking, was not in the operating framework. For example, neither the Hippocratic oath nor any of the other Hippocratic writings says anything about consent or any other form of patient participation in decisionmaking.[2] The oath explicitly prohibits even disclosure of information to patients. Up until the revision of 1980, the American Medical Association's published *Current Opinions* did not include any notion of consent either. To this day the AMA *Opinions* permit physicians to treat without consent when the physician believes that consent would be "medically contraindicated."[3] Consent is essentially a twentieth-century phenomenon, but one that has its roots in post-Reformation affirmation of the individual and the liberal political philosophy and related judicial system derived from it rather than in professional physician ethics.[4]

While classical medical ethics had no doctrine of informed consent, what can be called modern medicine did begin making limited room for the notion. Wide recognition of the importance of patient concurrence in a medical intervention first arose in research involving human subjects. The Nuremberg Code gives pride of place to the consent requirement.[5] In clinical medicine explicit consent has, at least until very recently, been reserved for more controversial treatments and for choices that are perceived as "ethically exotic." Consent is frequently invoked for treatment decisions in which the patient is seen as drawing on ethical and other values coming from outside medicine. Often this arises more in the refusal of consent than approvals of treatment. For instance, patients are now sometimes given the opportunity to refuse consent for certain death-prolonging interventions during a terminal illness.

Sometimes the consent notion gets a bit muddled – as when patients are asked to "consent" to DNR (do-not-resuscitate) orders. The idea of consenting to an "order" is strange, but the idea of consenting to nontreatment is even more so. A more appropriate language would refer to refusal of consent to resuscitation rather than consenting to nonresuscitation.

Modern medicine has reluctantly made room for the consent doctrine and has recognized, at least in theory, the right of patients to consent and refuse consent to certain kinds of treatment. Usually explicit consent is reserved for these more complex and exotic decisions. It is still common to hear people distinguish between treatments for which consent is required and those for which it is not. Surely it would be better to speak of those for which consent must be explicit and others that still require consent even though the consent can be implied or presumed. For example, many would probably say that routine blood drawings of modest amounts of blood can be done without consent. This would more appropriately be described as being done without explicit consent and with no specific information needing to be transmitted. The mere extending of the arm should count as an adequate consent.

Likewise, when a physician writes a prescription, he or she is supposed to review the alternatives and choose the best medication, select a brand name or generic equivalent, choose a route of administration, a dosage level, and length of use of the medication. The patient may signal "consent" simply by accepting the prescription and getting it filled at the local pharmacy.

Up until now no one has seriously questioned the adequacy, from the left, of an approach that permits explicit consent for special and complex treatment, including research and surgery, and implicit or presumed consent for more routine procedures. More careful analysis reveals that, in fact, the consent model buys into more of the traditional, authoritarian understanding of clinical decisionmaking than many people realize. As in the days prior to the development of the consent doctrine, the clinician is still supposed to draw on his or her medical knowledge to determine what he or she believes is in the best interest of the patient and propose that course of treatment. Terms such as "doctor's orders" may have been replaced by more appropriate images, but the physician is still expected to determine what is "medically indicated," the "treatment of choice," or what in the "clinical judgment" of the practitioner is best for the patient. The clinician then proposes that course, subject only to the qualification that through either word or action, the patient signals approval of the physician-determined plan.

Consent and the theory of the good

Current work on the theory of medical decision-making and in axiology makes increasingly clear that this pattern no longer makes sense. It still rests on the outdated presumption that the clinician's moral responsibility is to do what is best

for the patient, according to his or her ability and judgment, and that there is some reason to hope that the clinician can determine what is in the patient's best interest. The idea in medical ethics of doing what is best for the patient has achieved the status of an unquestioned platitude, but like many platitudes, it may not stand the test of more careful examination. On several levels the problems are beginning to show.

The best interest standard in surrogate decisions

The "best interest standard" has become the standard for surrogate decisionmaking in cases in which the wishes of the patient are not known and substituted judgment based on the patient's beliefs and values is not possible. But the best interest standard, if taken literally, is terribly implausible. In fact, no decision-maker is held to it in practice.

Two problems arise. First, since such judgments are increasingly recognized to be terribly complex and subjective, it is now widely accepted that the surrogate need not choose literally what is best. It would be extremely difficult to determine whether the absolute best choice has been made. Surely, the opinion of the attending physician cannot serve as a definitive standard. A privately appointed, parochial ethics committee might be better, but still surely is not definitive. If every surrogate decision were taken to court, we still would not have an absolute assurance that the best choice had been made.

Fortunately, we generally do not hold parents and other surrogates to a literal best interest standard when they make decisions for their wards. We expect, tolerate, even encourage a reasonable range of discretion. That is why it makes sense to replace the best interest standard with a "standard of reasonableness" or what could be called a "reasonable interest standard."[6]

There is a second reason why the best interest standard is inappropriate for surrogate decisions. Often surrogates have legitimate moral obligations to people other than the patient. Parents, for example, are pledged to serve the welfare of their other children. When best interests conflict, it is logically impossible to fulfill simultaneously the best interest standard for more than one child at the same time. Surely, all that is expected is that a reasonable balance of the conflicting interests be pursued.

Problems with best interest in clinician judgments

Although the problems with the best interest standard in surrogate decisions are more immediately apparent, a more fundamental and important problem with best interest arises when clinicians are held to the best interest standard in an ethic of patient care. For a clinician to guess at what is the best course for the patient, three assumptions must be true regarding a theory of the good. First, the clinician must be expected to determine what will best serve the patient's medical or health interest; second, the clinician must be expected to determine how to trade off health interests with other interests; and third, the clinician must be expected to determine how the patient should relate the pursuit of her best interest to other moral goals and responsibilities, including serving the interests of others and fulfilling any moral duties she may have that happen to conflict with her interest. An examination of the theories of the good and the morally right will reveal that it is terribly implausible to expect a typical clinician to be able to perform any one of these tasks completely correctly, let alone all three of them. If the clinician cannot be expected to guess at what serves the well-being of the patient and determine when patient well-being should be subordinated to other moral requirements, then there is no way that he or she can be expected to propose a course of treatment to which the patient would offer their consent.

Alternative theories of the good

To understand the limits on the ability of the clinician to propose a course that will maximize patient well-being, we need to examine briefly current theories of the good. Axiology – the study of theories of the good or valuable – is a field of normative philosophical ethics that is in considerable turmoil. Fortunately, for our purposes, any plausible contemporary theory leads to the same radical conclusion: there is no reason to believe that a physician or any other expert in only one component of well-being should be able to determine what constitutes the good for another being.

Determining what it means to say something is in someone's best interest turns out to be a very difficult task. Establishing the proper criteria for determining that one course or another maximizes the good for an individual is even harder. One major philosophical contributor to this debate suggests that there are at least three major groups of answers to the question of what is in someone's best interest.[7] Hedonistic theories hold that what best serves someone's interest is that which makes the person's life happiest. Desire-fulfillment theories hold that what is best for someone is what would in one way or another fulfill his or her desires (recognizing that one's desires may include many ends other than happiness). There is no reason to assume that a clinician – a specialist in one relatively narrow aspect of well-being and a relative stranger to the patient – should be able to guess either at what would make the patient most happy or what would fulfill a patient's desires.

The third group of theories of the good is collectively referred to as objective list theories. As summarized by Parfit, "According to this theory, certain things are good or bad for people, whether or not these people would want to have the good things, or to avoid the bad things. The good things might include moral goodness, rational activity, the development of one's abilities, having children and being a good parent, knowledge, and the awareness of true beauty."[8] Other people's lists of objective goods may differ, but, as Bernard Gert has argued, there is a remarkable convergence on the items for the list, as long as the items are kept quite general.

Most uses of the concept of best interests in health care ethics appear to rely on some theory of objective goods. If a physician is expected to determine what is in the patient's best interest and then present it to the patient for consent, there must be a presumption of a good that is, in some sense, objective, knowable by one who is committed to pursuing the patient's welfare. When the standard of best interest is used by a court or a theorist dealing with surrogacy decisions for patients whose personal wants, desires, preferences, and beliefs are not known, surely there is operating some notion that the good of the patient is objective and external to the patient.

What is striking here is that even with objective list theories, there is an enormous gap between what it would take to know what is "objectively in a patient's interest" and what the usual clinician can be expected to know about the patient. For example, many objective list theories include among the things that are good for people such items as spiritual well-being, freedom, sense of accomplishment, and "deep personal relationships."[9] Holders of such objective list theories claim that these are good for people regardless of whether they make people happy and, in contradistinction to desire-fulfillment theories, whether the individual desires these or even knows they are possible.

Certainly a physician is normally not in a good position to determine whether a medical intervention will contribute to the patient's sense of accomplishment or to that individual's "deep personal relationships." If this is true even for an objective theory of the good, there will be no way for the health professional to know whether the patient's good is served with a medical intervention without asking the patient. To put it bluntly, the only way to know whether an intervention is good medicine is to ask the patient.

Our conclusion seems clear: regardless of the theory of the good chosen, there is no reason to assume that the health professional can be expected to know what will promote the best interest of the patient. This conclusion seems clear on its face for hedonistic and desire-fulfillment theories of the good. It is less obvious but equally true even for objective theories of the good.

Elements of well-being

The problem can be made more clear by formulating an account of what could be called the spheres or elements of well-being. Regardless of the theory of the good, we can understand what promotes the good for persons better by asking what the elements are that contribute to one's well-being. Another way of putting the question would be to ask in what areas one's limited amount of personal resources – time, money, energy, and material – ought to be invested in order to maximize well-being.

The main elements of well-being

Several elements can be identified. These would surely include some concern with medicine or what could be called one's organic well-being. Closely related, but distinct, would be psychological well-being. It would be a terrible distortion to assume that well-being involved only the organic and psychological, however. Reasonable persons would devote considerable attention and resources to other elements, including the social, legal, occupational, religious, aesthetic, and other components that together make up one's total well-being. There is no reason to assume that each of these components is the same size. By trading off emphasis on different components one should be able to increase or decrease the size of the whole. Well-being is not a zero-sum game.

The problem is central to the concern about the concept of consent. It is unrealistic to expect experts in any one component to be able to speak knowledgeably about well-being in its other components. If this is true, then it makes no sense to expect them to come up with a proposed intervention that will promote the total well-being of the individual.

The subcomponents of organic or medical well-being

One obvious response is to back off from the claim that the goal of medicine is to promote total well-being. A more modest formulation of the end of medicine would be that it should promote not the total well-being of the patient, but the health, the medical good, of the patient.

While this is surely more realistic, it still raises two serious problems. First, even if the health professional were to limit concern to the component of health narrowly construed to mean organic well-being, this component of well-being cannot be thought of as a single, univocal good. There are several, often competing goods in the medical realm. Indeed, organic well-being has subcomponents, including the preservation of life, cure of disease, relief of suffering, and promotion of health.

If a physician is to attempt to use "clinical judgment" to choose what will serve the medical good of the patient, he or she will have to have a definitively correct account of what the relationship is among these medical goods and how they should be traded off against one another. Unless someone is prepared to argue not only that there is a definitively correct relationship

among these various medical goods, but also that the physician ought to be able to identify that definitively correct relationship, even the more modest goal of recommending what is *medically* best for the patient in order for the patient to consent to it will be illusory.

There is a second, even more serious problem with the strategy of expecting clinicians to recommend the medically best course and having the patient consent or refuse consent for that recommendation. At best such a strategy will produce what is *medically* best for the patient. However, a realistic goal for a person is to maximize total well-being (subject to possible moral constraints requiring one to take into account the well-being of others and to act in ways that are morally required). Assuming that resources are not infinite, no rational person wants to maximize his or her health. Assuming decreasing marginal utility, it would be irrational to expect that one could maximize the amount of one's total well-being by maximizing the amount of one of its components. Health professionals not only have to figure out how to balance the various medical goods against one another; they also have to realize that rational patients do not want that goal pursued, at least if it comes at the expense of goods in other spheres and lowers the patient's total well-being.

The example of V-tach

An example may help illustrate the difficulty. A forty-year-old man is diagnosed as having moderately severe polyfocal ventricular arrhythmia consisting of PVCs at a rate of over 600 per hour with occasional runs of v-tach. Other than being able to perceive the arrhythmias, he is asymptomatic. There is some reason to believe that these unstable heart rhythms could lead to a serious episode of ventricular fibrillation and even death, but many people are known to suffer such arrhythmias indefinitely without any adverse effect. There are four classes of medications on the market that can be used to suppress the arrhythmias: membrane-stabilizing agents, beta-blockers, repolarization inhibitors, and calcium channel blockers. Some of the classes have subclasses. Although they all suppress PVCs, there is no definitive evidence that suppressing them lowers the mortality risk. In each class of agents there are many different drugs, each with slightly different desired effects and side effects. There are many different dose levels and routes of administration. There are different manufacturers, different costs, and different risk profiles involved. Some clinicians are sufficiently concerned about the side effects that they recommend to their patients that no medication be used. Other clinicians use modest pharmacological intervention with low risk and proportionally low suppression of the arrhythmias. Others more aggressively suppress the PVCs, believing that on balance more good is done. The choices involve matters that are not trivial. The side effects range from nausea, vomiting, diarrhea, headache, and ringing in the ears, to unwanted changes in heart rhythm, a lupus erythematosus-like syndrome, coma, and even death. The dose forms range from a simple one-a-day capsule to more complex combinations taking different medications at several different times a day. The costs range from nothing to several hundred dollars a year for the rest of the patient's life. There are easily over a hundred possible courses of action, each of which has its own unique combination of advantages and disadvantages and might be favored depending on the patient's idiosyncratic circumstances and values and what one considers to be the correct theory of the good.

A patient will have his arrhythmia discovered by a physician, who will assess the options and recommend either no medication, or some variant of conservative or more aggressive medication. That physician will opt for one (or more)

of the classes of drugs, some route of administration, dose range, cost, and profile of side effects and will write the prescription, subject, whether the patient realizes it or not, to the consent of the patient. Perhaps some of the variables in the decision will be discussed with the patient and the patient will be given a chance to object.

What the patient may not know is that another equally competent clinician with different instincts about how aggressive to be, which risks are worth taking, and what tradeoffs with nonmedical well-being should be made, will choose some other course and propose it for patient consent. Many of the hundred therapeutic regimens could be perceived as plausible by responsible clinicians. The correct choice will depend on the tradeoffs among the various medical goods and between the medical and nonmedical goods, including a judgment about how the patient should spend marginal dollars. In such a situation it is simply hubristic for clinicians to believe that, out of the hundreds of subtle value tradeoffs to be made, they can come up with just the course that will maximize the patient's well-being.

Why experts should not propose a course for patient consent

It should now be clear why it makes no sense to continue to rely on consent as the mode of transaction between professionals and their clients. In order for a physician to make an initial estimate of which treatment best served the patient's interest, he or she would first have to develop a definitive theory of the relationship among various medical goods and pick the course that best served the patient's medical good. Then the clinician would have to estimate correctly the proper relationship between the patient's medical good and all other components of the good so that the patient's overall well-being was served.

Even if this could be done, there is a final problem. In virtually any moral theory the well-being of the individual is only one element. Plausible consequentialist theories (such as utilitarianism) also insist that the good of other parties be taken into account. Plausible nonconsequentialist theories, including Kantian theories, natural law theories, much of biblical ethics, and all other deontological theories, hold that knowing what will be in the best interests of persons does not necessarily settle the question of the right thing to do. Many patients may purposely want to consider options that do not maximize their well-being. A patient may acknowledge, for example, that his well-being would be served if he lived longer, but choose to sacrifice his interests to conserve resources for his offspring. A pregnant woman might conclude, for another example, that her interests would be served if she had an abortion, but that such a course would still be morally wrong. Both of these people would rationally not choose the course of action that admittedly maximized their personal well-being. Even if physicians can figure out what maximizes medical well-being and how medical well-being should be related to other elements of well-being, that still does not necessarily lead to the course that is right, all things considered. To know what is "good medicine" and what should be recommended for the patient's assessment and consent, one needs to know how to answer all three of these questions. There is no basis for assuming that physicians have any special expertise in answering any of them.

Why physicians should guess wrong

This may simply establish that physicians are no better than the rest of us at guessing what counts as the medical good, how the medical good relates to total good, and whether the patient's total good should be promoted. A defender of the emerging process of informed consent might respond by pointing out that physicians should at least be considered to be as good as any other lay person in answering these value theory questions. Since physicians clearly have technical expertise, if they are as good as anyone else at answering the value theory questions, would not efficiency suggest that the physician could combine her technical knowledge of the patient's case with the physician's ordinary level of wisdom when it comes to making value judgments? This could lead to a recommended course of action for the patient. As long as the patient can decline, would that not be the most efficient way to maximize the patient's welfare?

There are two problems here. First, this democratic theory of value which holds that professional experts are as good as others at guessing what maximizes the good can, at best, lead to a general estimate of what the typical person would consider to be good. If, however, any plausible value theory includes elements that are unique to the individual, then merely figuring out what would be good for the general person will not be good enough. We need to know what is good for this particular person. We need not hold to the liberal position that the individual is the best predictor of what serves his or her interest. All we need to acknowledge is that experts in a particular area are not the best at this task.

This leads to a second problem with relying on the claim that experts in one component of well-being should be as good as anyone else at determining the total good for the patient. Experts in any one component of well-being predictably will answer the questions about what maximizes interests in an atypical fashion. This is not because they are self-serving, trying to increase utilization of their services. I assume, perhaps naively, that most health professionals are altruistic and do a good job most of the time of holding self-interest in check.

The problem is more theoretical and more fundamental. Anyone who has given his or her life to an area of professional specialization ought to be expected to value the contributions of that area in an atypical way. Cardiologists ought to believe atypically that cardiology does good for people. Presumably that is one reason why people choose to go into a professional field. But this means if they ask themselves what medical intervention will best promote the overall well-being of the patient, they predictably will give an answer different from the one that would be given by lay people, including the patient himself. They often, but not always, will overvalue the benefits of their field. If one were seriatim to ask experts in each of the components of well-being what portion of an individual's total resources should be invested in their component and then sum the collective recommendations, it is likely that the total would exceed 100 percent.

Specialists not only make the value tradeoffs atypically, they also make the moral tradeoffs atypically. In short, if clinicians are asked to guess at what will serve the best interest of the patient, they ought to come up with the wrong answer. Insofar as the consent process involves a clinician using personal judgment to guess at what would be best for the patient, it still rests on the false assumption that clinicians can, at least tentatively, estimate what specific treatment interventions are in the individual patient's interests.

This should make clear why relying on the consensus of medical experts does not provide an adequate way to figure out what the "treatment of choice" should be for the patient with the cardiac arrhythmia. One might make the move of holding the clinician to the "professional standard" of choosing to recommend the drug that the majority of her colleagues similarly situated would have chosen. The most this move could do is tell us what people with the dominant value profile of the medical profession would choose under similar circumstances. It does not tell us what persons with the typical value profile of the average lay person would choose, let alone what one with the value profile of the individual patient would choose. The consensus of the profession does not provide an adequate basis for deciding what the "treatment of choice" is and what should be recommended to patients.[10] The idea that a clinician can determine what is a "medically indicated" treatment for a patient rests on a confused understanding of what the skills of a physician can be expected to provide.[11] If the clinician cannot be expected to figure out what is best for the patient, then he or she cannot obtain a valid consent to a recommended treatment of choice.

Choice: the liberal alternative

If consent is no longer adequate as a mechanism for assuring that the patient's beliefs and values will help shape decisions about what a patient ought to do, what are the alternatives? Adherents to medical ethical systems that emphasize autonomy may prefer the concept of choice to that of consent. In this alternative the patient would be presented with a list of plausible treatment options, together with a summary of the potential benefits and risks of each. It is important to emphasize that choice is conceptually different from consent and potentially could replace consent as the basis for patient involvement in health care decisions.

This "liberal" solution, however, faces serious, probably insurmountable problems. First, if the choices that are plausible for the patient are contingent on the beliefs and values of the patient, then the professional cannot be sure that all plausible options are being presented unless he or she has knowledge of the patient's beliefs and values − knowledge that we have argued is normally unavailable. Second, some options (for example, suicide) may be so offensive to some practitioners that they ought not to present them. Third, it is increasingly recognized that even the description of the "facts" necessarily must incorporate certain value judgments, such that even the clinician of good will cannot give a value-free account of the likely outcomes of the alternatives. In short, while the choice alternative may go part of the way toward giving the patient more active control, it is naive to believe it will be able to solve the problems with the consent model.

Pairing based on "deep values"

There is another alternative worth considering. If a clinician is skilled and passionately committed to maximizing the patient's welfare, and knows the belief and value structure and socioeconomic and cultural position of the patient quite well, there would be some more reason to hope for a good guess. Unfortunately, not only is that an ever-vanishing possibility, even knowing the value system of the patient well probably would not be sufficient. The value choices that go into a judgment about what is best for another are so complex and subtle that merely knowing the other's values and trying to empathize will probably not be enough. There is ample evidence that unconscious value distortions will not only influence the clinician's judgment about what is best, but even influence the very interpretation of the scientific data.

There might be more hope if the patient were to choose her cadre of well-being experts (lawyers, accountants, physicians) on the basis of their "deep" value systems. That way when unconscious bias and distortion occur, as inevitably they must, they will tip the decision in the direction of the patient's own system.

I say "deep" value system because I want to make clear that I am not referring to the cursory assessment of the professional's personality, demeanor, and short-term tastes. That would hardly suffice. If, however, there were alignments, "value pairings," based on the most fundamental worldviews of the lay person and professional, then there would be some hope. This probably would mean picking providers on the basis of their religious and political affiliations, philosophical and social inclinations, and other deeply penetrating worldviews. To the extent that the provider and patient were of the same mind set, then there is some reason that the technically competent clinician could guess fairly well what would serve the patient's interest.

The difficulty in establishing a convergence of deep values cannot be underestimated. Surely it would not be sufficient, for instance, to pair providers and patients on the basis of their institutional religious affiliations. Not all members of a religious denomination think alike. But there is reason to hope that people can establish an affinity of deep value orientations, at least for certain types of medical services. For example, certain institutionalized health care delivery systems are now organizing around identifiable value frameworks, recruiting professional and administrative staff on the basis of commitment to that value framework, and then announcing that framework to the public so as to attract only those patients who share the basic value commitment of the institution. A hospice is organized around such a constellation of values. It recruits staff committed to those values and attracts patients who share that commitment. When hospice-based health care providers present options to patients they should admit that they do not present all possible options. (They do not propose an aggressive oncology protocol, for instance; most would not present physician-assisted suicide or active mercy killing.) They should also admit that when they explain options and their potential benefits and harms they do so in ways that incorporate a tone of voice or body language that reflects their value judgments. Patients, however, need be less concerned about this value encroachment than if they were discussing options with a provider who was deeply, instinctively committed to maximally aggressive life preservation. There will be biases, but they will be less corrupting of the patient's own perspective.

Other delivery systems are beginning to organize around deep value orientations: feminist health centers, holistic health clinics, and the National Institutes of Health Clinical Center all announce at least their general value orientations to potential patients.

Providing an institutional framework for pairing based on deep value convergence in more routine health care may be more difficult, but not impossible. HMOs could be organized by social and religious groups that could formally articulate certain value commitments. A Catholic HMO, like a Catholic hospital, could articulate to potential members not only a set of values pertaining to obstetrical and gynecological issues, but also a framework for deciding which treatments are morally expendable as disproportionally burdensome. A liberal Protestant health care system would announce a different framework; a libertarian secular system still another. A truly Protestant health care system, for example, would probably reflect the belief that the lay person is capable of having control over the "text." The medical record, accordingly, would plausibly be placed in the patient's hands just as the Bible is.

Such value pairings will obviously not be a total matching, but they should at least place provider and patient in the same general camp. Moreover, organizing health care delivery on the basis of explicit value pairings would put both provider and patient on notice that values are a necessary and essential part of health care decisionmaking, a part that cannot be avoided and cannot be handled adequately by merely obtaining the consent of the patient to a randomly assigned provider's guess about what would be best.

With such an arrangement the problems that arise with use of consent for the normal random pairing of lay people and professionals are mitigated. The clinician has a more plausible basis for guessing what would serve the interests of the patient and, more importantly, will let a system of beliefs and values influence the presentation of medical information in a way that is more defensible. To be sure, such deep value pairing will not eliminate the problem of the necessary influence of beliefs and values on communication of medical facts, but it will structure the communication so that the inevitable influence will resemble the influence that the patient would have brought to the data were he or she to become an authority in medical science.

Barring such radical adjustment in the basis for lay-professional pairings, there is no reason to believe that the process of consent will significantly advance the lay person's role in the medical decisionmaking process. The concept of consent will have to be replaced with a more radical, robust notion of active patient participation in the choice among plausible alternatives – either by getting much greater information to the patient or by actively selecting the professional on the basis of convergence of "deep" value systems.

Notes

1 Robert M. Veatch, "Three Theories of Informed Consent: Philosophical Foundations and Policy Implications," *The Belmont Report: Ethical Principles and Guidelines for the Protection of Human Subjects of Research* (Washington, D.C.: National Commission for the Protection of Human Subjects of Biomedical and Behavioral Research, DHEW Publication No. (05)78–0014), pp. 26–1 through 26–66;

President's Commission for the Study of Ethical Problems in Medicine and Biomedical and Behavioral Research, *Making Health Care Decisions: A Report on the Ethical and Legal Implications of Informed Consent in the Patient-Practitioner Relationship*, vol. 1 (Washington, D.C.: U.S. Government Printing Office, 1982); Ruth Faden and Tom L. Beauchamp in collaboration with Nancy P. King, *A History and Theory of Informed Consent* (New York: Oxford University Press, 1986); Jay Katz, *The Silent World of Doctor and Patient* (New York: The Free Press, 1984).

2   Ludwig Edelstein, "The Hippocratic Oath: Text, Translation and Interpretation," *Ancient Medicine: Selected Papers of Ludwig Edelstein*, ed. Owsei Temkin and C. Lilian Temkin (Baltimore: Johns Hopkins University Press, 1967), pp. 3–64.

3   American Medical Association, *Current Opinions of the Council on Ethical and Judicial Affairs of the American Medical Association: Including the Principles of Medical Ethics and Rules of the Council on Ethical and Judicial Affairs* (Chicago: American Medical Association, 1989), p. 32.

4   Schloendorff v. New York Hospital (1914), discussed in Jay Katz, *Experimentation with Human Beings: The Authority of the Investigator, Subject, Professions,* and State in the Human Experimentation Process (New York: Russell Sage Foundation, 1972), p. 526.

5   *Encyclopedia of Bioethics*, s.v. "Nuremberg Code, 1946."

6   Robert M. Veatch, "Limits of Guardian Treatment Refusal: A Reasonableness Standard," *American Journal of Law & Medicine* 9, no. 4 (Winter 1984): 427–68; Robert M. Veatch, *Death, Dying, and the Biological Revolution*, rev. ed. (New Haven: Yale University Press, 1989).

7   Derek Parfit, *Reasons and Persons* (Oxford: Clarendon Press, 1984), pp. 493–503.

8   Parfit, *Reasons and Persons*, p. 499; see also Bernard Gert, "Rationality, Human Nature, and Lists," *Ethics* 100 (1990): 279–300.

9   See David DeGrazia, "Sketch of a Tentative Value Theory for Human Persons," photocopy, 11 November 1991, for an objective list that includes deep personal relationships.

10   Robert M. Veatch, "Consensus of Expertise: The Role of Consensus of Experts in Formulating Public Policy and Estimating Facts," *Journal of Medicine and Philosophy* 16 (1991): 427.

11   Robert M. Veatch, "The Concept of 'Medical Indications,'" in *The Patient-Physician Relation: The Patient as Partner* (Bloomington: Indiana University Press, 1991), pp. 54–62.

Chapter 33

Julian Savulescu and Richard W. Momeyer

SHOULD INFORMED CONSENT BE BASED ON RATIONAL BELIEFS?

I Introduction

Medical ethics places great emphasis on physicians respecting patient autonomy. It encourages tolerance even towards harmful choices patients make on the basis of their own values. This ethic has been defended by consequentialists and deontologists.

Respect for autonomy finds expression in the doctrine of informed consent. According to that doctrine, no medical procedure may be performed upon a competent patient unless that patient has consented to have that procedure, after having been provided with the relevant facts.

We have no quarrel with these principles. We do, however, question their interpretation and application. Our contention is that being autonomous requires that a person hold rational beliefs. We distinguish between rational choice and rational belief. Being autonomous may not require that one's choices and actions are rational. But it does require that one's beliefs which ground those choices are rational. If this is right, what passes for respecting autonomy sometimes consists of little more than providing information, and stops short of assessing whether this information is rationally processed. Some of what purports to be medical deference to a patient's values is not this at all: rather, it is acquiescence to irrationality. Some of what passes for respecting

patient autonomy may turn out to be less respect than abandonment. Abandonment of patients has never been regarded as a morally admirable practice.

We will outline three ways in which patients hold irrational beliefs: (1) ignorance, (2) not caring enough about rational deliberation, and (3) making mistakes in deliberation. We argue that it is the responsibility of physicians not only to provide relevant information (which addresses 1), but to improve the rationality of belief that grounds consent (2 and 3).

II Rationality and autonomy

II.I True belief and autonomy

The word, "autonomy", comes from the Greek: autos (self) and nomos (rule or law).[1] Autonomy is self-government or self-determination. Being autonomous involves freely and actively making one's own evaluative choices about how one's life should go.

It is a familiar idea that it is necessary to hold true beliefs if we are to get what we want. For example, John loves Northern Indian dishes and loathes Southern Indian dishes. Yet he is very confused about which dishes belong to which area. He consistently orders Southern Indian dishes thinking he is ordering Northern Indian dishes. His false beliefs cause him to fail to get what he wants.

However, true beliefs are important for evaluative choice in a more fundamental way: we cannot form an idea of *what* we want without knowing what the options on offer are *like*. Consider a person with gangrene of the foot. She is offered an amputation. In evaluating "having an amputation" she is attempting to evaluate a complete state of affairs: how much pain she will experience, whether she will be able to live by herself, visit her grandchildren, and so on. (Importantly, knowing the name of one's disease and the nature of the operation are less important facts.)

II.II True belief and practical rationality

Practical rationality is concerned with what we have reason to care about and do. Let's distinguish between what there is good reason to do and what it is rational to do. Paul sits down after work to have a relaxing evening with his wife. She gives him a glass of what he believes is wine, but is in fact poison. There is a good reason for Paul not to drink it, even if this is not known to Paul. However, if he believes that it is wine, it is rational for him to drink it.[2] Thus:

> It is rational for a person to perform some act if there would be a good reason to perform that act if the facts were as he/she believes them to be.

Thus holding true beliefs is important in two ways: (1) it promotes our autonomy and (2) allows us to see what there is good reason to do. This does not collapse autonomous choice with rational choice. Even holding all the relevant true beliefs, a person may autonomously choose some course which he or she has no good reason to choose. For example, assume that the harms of smoking outweigh the benefits. Jim

has good reason to give up smoking. However, he may choose to smoke knowing all the good and bad effects of smoking. His choice is then irrational but his beliefs may be rational and he may be autonomous. His choice is not an expression of his autonomy if he believes that smoking is not only pleasurable, but good for your health.

II.III Coming to hold true beliefs

One important way to hold true beliefs is via access to relevant information. For example, one way to get Paul to believe that the wine is poisoned is to provide him with evidence that it is poisoned.

We can never know for certain that our beliefs are true. We can only be confident of their truth. Confidence is the likelihood that a belief is true. Beliefs which are based on evidence (rational beliefs) are more likely to be true than unfounded (irrational) beliefs. The likelihood that our beliefs are true is a function both of how informed they are and of how we think about that information.

Theoretical rationality is concerned with what it is rational to believe.

> It is rational for a person to believe some proposition if he/she ought to believe that proposition if he/she were deliberating rationally about the evidence available and his/her present beliefs, and those beliefs are not themselves irrational.

Let's say that a person is "deliberating rationally" if[3]:

1   She holds a degree of belief in a proposition which is responsive to the evidence supporting that proposition. For example, the firmer the evidence, the greater the degree of belief ought to be.

2   She examines her beliefs for consistency. If she detects inconsistency, she ought appropriately to contract her set of beliefs or adjust her degree of belief in the relevantly inconsistent beliefs.

3   She exposes her reasoning to the norms of inductive and deductive logic. Valid logic is important because it helps us to have the broadest range of true beliefs.

Consider the following example. Peter is trying to decide whether to have an operation. Suppose that he is provided with certain information and reasons in the following way.

(1)   There is a risk of dying from anaesthesia, (true)

(2)   I will require an anaesthetic if I am to have this operation, (true)

Therefore, if I have this operation, I will probably die.

The conclusion does not follow from the premises. Peter comes to hold an irrational belief because he commits a logical error. Irrational beliefs are less likely to be true than rational beliefs. Since knowledge of truth is elusive for subjective beings like us, the best we can hope for is informed, rational belief.

If we are right that information is important to evaluative choice because of its contribution to a person holding the relevant true beliefs necessary for evaluation, then deliberating rationally is as important as being informed, since this also affects the likelihood that one's beliefs are true. Being fully autonomous requires not only that we are informed, but that we exercise our theoretical rationality.

III An example of irrational belief: Jehovah's Witnesses and blood

Jehovah's Witnesses (JWs) who refuse life-saving blood transfusions for themselves are often taken to be paradigm cases of autonomous, informed choice based on different (non-medically shared) values that require respect and deference.

Jehovah's Witnesses refuse life-saving blood transfusions because they believe that if they die and have received blood, they will turn to dust. But if they refuse blood (and keep Jehovah's other laws) and die, they will enjoy eternal life in Paradise.[4]

Jehovah's Witnesses interpret the Bible as forbidding the sustaining of life with blood in any manner. They base this belief on passages such as:

> Every creature that lives and moves shall be food for you . . . But you must not eat the flesh with the life, which is the blood, still in it.[5]

Anyone eating the blood of an animal would be "cut off" or executed.[6] The only legitimate use of animal blood was as a sacrifice to God. Leviticus 17: 11 states:

> . . . the life of the creature is the blood, and I appoint it to make expiation on the altar for yourselves: it is the blood, that is the life, that makes expiation.[7]

Jehovah's Witnesses believe these views concerning blood were important to the early Christian Church. At a meeting of the apostles and older men of Jerusalem to determine which laws would continue to be upheld in the new Church, blood was again proscribed:

> . . . you are to abstain from meat that has been offered to idols, from blood, from anything that has been strangled, and from fornication.[8]

Jehovah's Witnesses believe that these passages imply more than a dietary proscription. They attach great symbolic significance to blood: it represents the life or soul. Thus they claim that the exhortation "abstain from blood" applies to all forms of blood, at all times. They argue that there is no moral difference between sustaining life by taking blood by mouth ("eating blood") and taking blood directly into the veins.

Relative to their beliefs, JWs are practically rational. Any (practically) rational person would choose to forgo earthly life if this ensured that one would enjoy a blissful eternal existence in the presence of God. If JWs are irrational, it is because their beliefs are irrational. A failure of theoretical rationality causes them to do what there is good reason not to do and frustrates their autonomy.

We believe that the beliefs of JWs are irrational. One way to show this is to question the rationality of belief in the existence of God or in the truth of some religious version of morality. For argument's sake, we will accept theism. However, the vast majority of those in the Judaeo-Christian tradition have not interpreted these passages from the Bible as proscribing blood transfusion. The beliefs of JWs are irrational on at least two counts: their particular beliefs are not responsive to evidence nor are their interpretations of Biblical text consistent. These failures of rationality are shared with other forms of religious "fundamentalism" and so-called "literal" interpretations of religious texts. It is worth noting that many JWs are also Creationists, believing all of Genesis to be literally true. Ignorance of historical context, the diverse intentions and circumstances of Biblical peoples and authors, oral and written traditions in the Middle East, other religious traditions and interpretations of Biblical texts, inconsistencies between different canonised works and the like all help ground an unduly simplistic interpretation of the Bible.

Mere ignorance, however, is not to be equated with irrationality. Wilful ignorance is. And wilful ignorance is what lies behind grounding understanding of the Bible on faith rather than the kinds of knowledge suggested above. This sort of wilful ignorance cuts across educational levels as it is rooted in dogmatism and closed-mindedness rather than degrees of education.

However, we believe that JWs' beliefs are irrational even in terms that should be acceptable to JWs.

Firstly, their interpretation is inconsistent with other passages of the Bible and Christian practices. It is inconsistent with the Christian practice of communion. Communion is the holy ceremony of the Last Supper. At the Last Supper

> Jesus took bread, and having said the blessing he broke it . . . with these words: 'Take this and eat; this is my body.' Then he took a cup [of wine], and having offered thanks to God . . . [said] . . . 'Drink from it . . . For this is my blood, the blood of the covenant, shed for many, for the forgiveness of sins'.[9]

Secondly, Paul warns against slavish obedience to law:

> . . . those who rely on obedience to the law are under a curse . . .[10]
>
> Christ bought us freedom from the curse of the law by becoming . . . an accursed thing.[11]

The answer is not obedience to law but faith.

> . . . the law was a kind of tutor in charge of us until Christ should come, when we should be justified through faith; and now that faith has come, the tutor's charge is at an end.[12]

Paul himself does not understand the Bible to be literally true, as evidenced when he speaks of the story of the origin of Abraham's sons being "an allegory".[13] He goes on to say:

Mark my words: I, Paul, say to you that if you receive circumcision Christ will do you no good at all ... [E]very man who receives circumcision is under obligation to keep the entire law. When you seek to be justified by way of law, your relation with Christ is completely severed ... [O]ur hope of attaining that righteousness ... is the work of the Spirit through faith ... the only thing that counts is faith active in love.[14]

If the beliefs of JWs are irrational, why are they are irrational?

IV Three examples of holding a false belief

In all three of the following cases, the person lacks a true belief which is relevant to choice. We describe how to help a person come to hold true beliefs, drawing out the parallels with patients and JWs.

Case 1. Lack of information

Arthur 1 is burning rubbish in the garden. The fire grows rapidly. It begins to threaten surrounding buildings. They are not in imminent danger but Arthur wants to douse the fire with water before it gets out of hand. He goes to the shed where he keeps a jerry can of water for just such a situation. He has a high degree of belief that this can contains water. Unbeknownst to him, someone has substituted water for petrol in the can. He throws the liquid on the fire and the petrol ignites, causing an explosion. He is badly burnt. Was Arthur irrational?

We need a more complete description of the state of affairs.

Arthur always locks the shed. There had been no signs of forced entry. There was only one jerry can in the shed. It was in the position where Arthur always kept it, next to the shovel.

He had only the previous weekend refilled it with water after using it to put out another garden fire. If Arthur simply had no reason to suspect that the can contained anything but water, it was rational to believe that it contained water. A person who unavoidably lacks relevant information is neither theoretically nor practically irrational.

What should we do if we see Arthur about to throw the liquid onto the fire?

Arthur is rational, but he lacks a relevant true belief that he could have. In this case, the solution is simple. Provide information. Tell him, "Stop. The can contains petrol." If there were no time to provide this information, we ought to grab the can from his hands.

Many patients who hold false beliefs are like Arthur 1: uninformed. What we ought to do is provide them with information. If this is not possible, we should do what is best for them.

Are there any JWs like Arthur 1? Jehovah's Witnesses are remarkably well informed about blood transfusion, the effects of refusing it, and the Biblical context of their belief. But some may be unaware of the conflicting Biblical passages. These ought to be treated like Arthur 1. However, many are not like him. The provision of information is not alone an adequate response. What is required is rational argument.

Case 2. Not engaging in rational deliberation

Arthur 2 is the same as Arthur 1, but in this case Arthur goes to the shed and finds it unlocked. He is not sure whether he left it locked last weekend. He thinks he probably did. The jerry can is next to the lawn mower. Arthur thinks that he normally keeps it next to the shovel. But, again, he is not sure. Is he irrational if he believes the can contains water?

Arthur clearly ought to believe that the door is open and the can is next to the lawnmower. But for these propositions to constitute evidence for the conclusion that the contents of the can are not water, Arthur must believe that the position of the can and door have changed. Should Arthur believe that he left the jerry can next to the shovel? This depends on the degree of belief Arthur has in his recollection of how things were. If he is vague, then there is no evidence.

Arthur may not lack information as much as a context for that information because he fails to remember relevant facts. This may be beyond control. In this case, Arthur 2 is like Arthur 1. But in some cases, a person fails to remember because he fails to think about the issue. And he may fail to think about the issue because he fails to care enough about the truth of his beliefs or the consequences of his actions.

Arthur could be directed to think more carefully about what he sees and of the possible implications of his actions. There may be other evidence he would find, if he looked, for believing the propositions that the door was locked and the can was next to the shovel. He may notice other items in the shed have been moved.

It is often thought that consultation in medicine involves presenting information so that it is understood. But even understanding is not enough. Facts must be assembled to tell a story or to construct an argument which stands in the foreground of deliberation. The arrangement and form of the facts is as important as their content.

Are there any JWs who are like Arthur 2? There is, we are assuming, evidence that their beliefs are false. However, being informed of these facts is not sufficient to cause them to hold the relevant true beliefs. They also need to care about thinking about that information in a rational way. The hallmark of faith is a stubbornness to respond to the evidence for a proposition. While this may be necessary for belief in God, it cannot be the appropriate paradigm for interpretation of God's word. the Bible, as a guide to how to live, aims to sanction some ways of living and proscribe others. Faith in any *interpretation* of God's word cannot be acceptable.[15] When interpreting Biblical text, the appropriate paradigm for theists is rationality and not faith. Indeed, the efforts of JWs to argue for their interpretation of the Bible indicates that they subscribe to this paradigm. What they are required to do by that paradigm is to care more about the proper exercise of rationality.

Intervention in this case would include trying to persuade JWs to care more about rationality by showing how they themselves appeal to rational argument and why the Bible must be interpreted rationally.

We are often like Arthur 2 and some JWs: we fail to care enough about what we believe and what we commit to memory. This failing is at the interface of practical and theoretical rationality: we fail to care enough (a practical failing) about the rationality of our beliefs (a theoretical failing).

Case 3. Theoretical irrationality

Arthur 3 is the same as Arthur 2 but in this case, Arthur is sure he left the door to the shed locked and sure that the jerry can is in a different position from where he left it. On entering the shed, he smells petrol. He doesn't normally keep petrol in the shed. None the less, he throws the fluid on the fire.

As the evidence mounts up, Arthur becomes more theoretically irrational if he fails to consider the possibility that the can contains petrol. At the limit, if the evidence is overwhelming, he is like a person who believes that p, and that if p then q, but fails to believe that q.

Why might Arthur be theoretically irrational?

He may simply fail to believe that what he smells is petrol. This would be an error of perception.

He may fail to examine his beliefs for inconsistency. He may fail to compare what he believes to what the evidence suggests is the case.

Most importantly of all, Arthur may not be very talented at theoretical reasoning. He may not be good at assembling the evidence and drawing conclusions from it. It is not enough for a person to throw up any explanation for evidence presented to him. To move from "I saw a light on the water" to "I saw a ghost at Dead Man's Bluff" is to make an unjustified and irrational leap. Ideally, we should infer to the best explanation.[16]

Physicians, concerned to promote theoretical rationality, may assemble facts in a way which together suggest a conclusion. But patients may still fail to draw the right conclusion. Telling a patient that he has "advanced cancer" may imply that he will die. But the patient may not conclude this. Indeed, even telling a patient that he will die may not convey "the message" that the physician intends to give: perhaps that the patient ought to sort out his affairs, that he will not offer any more curative treatments, and so on.

How would a person who is in a similar epistemic position to Arthur, but who is more theoretically rational than Arthur 3, convince him that the can contains petrol? He would engage Arthur in argument. He would provide reasons. He might say something like, "The can seems to be in a different position from how you left it. That might suggest that someone has used it. I can smell petrol. Perhaps someone has used the can to carry petrol."

The reason why most JWs hold an irrational belief is because they make mistakes in their theoretical reasoning. What is the best way to correct these mistakes? For many, it is a matter of someone versed in the relevant texts taking them through the argument.

In other cases, the route may be more indirect. A JW may be presented with some information, call it, I, which should or would cause

him to conclude that C if he also held other beliefs, B. However, he may fail to believe B or utilise B. He may have forgotten B in the urgent search for salvation, or had it drummed out of his head, or failed to see any longer its relevance. Intervention requires that we tap into these other beliefs. For an argument to be convincing for him may require the construction of the appropriate context: to show him that his belief should be rejected in his own terms.

Our object is the beliefs of JWs, not necessarily their choices. In some circumstances, JWs might autonomously choose to reject blood. We can autonomously adopt a course of action with a low probability of success, provided that we hold the relevant rational beliefs. Neither risk-takers nor the exceedingly cautious are necessarily non-autonomous, nor are they necessarily doing what there is good reason not to do.[17]

If JWs were to hold the relevant informed, rational beliefs, they might then autonomously choose to reject blood. But from their revised epistemic position, many would no doubt accept blood.

V Summary and implications

Where most rational agents differ from JWs is that they do not hold all of the following beliefs:

1   There is a God.
2   Divinely conferred immortality is possible for human beings after death.
3   God forbids eating blood.
4   Accepting a blood transfusion is no different from eating blood.
5   If one eats blood when alive, one turns to dust upon death.
6   We know 3–5 to be true based on faith that a (selectively) literal interpretation of the Bible reveals God's will.
7   If one lives a faithful life in accord with Jehovah's laws, eternal life is assured.

336   Julian Savulescu and Richard W. Momeyer

Many health care workers no doubt believe 1 and 2 and some variation on 7; it is 3, 4, 5, or especially 6 that is rejected. But this is a difference not in moral beliefs or values but about the structure of reality. This is a difference of opinion about metaphysics.

Hence if we are to respect JWs' refusals of life-prolonging blood transfusions, it is not on the grounds that we are obliged to respect decision-making that is based on a different value system from ours. Their values are the same as many other theists and atheists. They value earthly life and immortality as much as others do.

We often hear that we should allow people to do what there is no good reason to do out of respect for their nature as autonomous beings, as ends in themselves. But many such instances are something else entirely. They are cases in which people hold irrational beliefs. They are cases of theoretical irrationality. We do not respect autonomy when we encourage people to act on irrational beliefs. Rather, such beliefs limit a person's autonomy.

Rational deliberation

Our aim has been to expand the regulative ideal governing consent. We have argued that true beliefs are necessary for evaluation. Information is important to choice insofar as it helps a person to hold the relevant true beliefs. But in order to hold the relevant true beliefs, competent people must also think rationally. Insofar as information is important, rational deliberation is important. Just as physicians should aim to provide relevant information regarding the medical procedures prior to patients consenting to have those procedures, they should also assist patients to think more clearly and rationally. They should care more about the rationality of patients' beliefs.

Since holding true beliefs is necessary to be autonomous, we do not respect autonomy when

we allow patients to act on irrational beliefs. Should physicians override choices based on irrational beliefs? Should life-saving blood transfusions be given to JWs against their wishes?

When we look at how informed medical decisions must be, we see that a requirement of informedness functions as an ideal to be striven for, and not as a requirement to be enforced. Society generally accepts that patients should be informed of all the relevant facts, but not that they must be compelled to accept information which they do not want. To force information on a person would be coercive.

The requirements of theoretical rationality should be on a par with requirements of informedness. This raises the question whether we should override both choices made in ignorance of relevant information or on the basis of irrational beliefs. We believe that there are reasons against taking this radical departure from the notion of informed consent as a regulative ideal.

The first reason is consequentialist: if we allow doctors to override choices based on some species of irrationality, then other JWs will be distressed at the thought their decisions will be overridden. The general misery and distrust of medicine that would result would reduce the value of such a policy. In the vast majority of cases, JWs' refusal of blood does not compromise their care. In fact, many may receive better care. Given the small numbers of people who would be saved by such a policy, it is not clear that it would be for the best. As is usually the case, education is better than compulsion.

Secondly, though a practice of allowing people to act out of wilful ignorance or irrationality may not promote their autonomy in the short term, respect for autonomy is not the only ground for non-interference in another person's life. It is surely enough that it is his life, and that he ought to be allowed to do what there is no good reason to do, if he chooses. Respect for persons is not restricted to respect for wholly rational persons.

In some cases, irrationality is so gross that it calls into question a person's competence. In these cases, intervention may be justified. But at lesser degrees of irrationality, we encourage the development of autonomy for all people in the long term by adopting a policy of empowering people to make their own choices.

Thirdly, *requiring* that choice be grounded on rational beliefs before it is respected is fraught with dangers. Those who claim to know Truth with certainty are at least as dangerous as those who claim to know Right and Good with certainty. Dogmatic ideologues of either sort show a lamentable propensity to use their "knowledge" to oppress others, sometimes "benignly" as paternalists, more often tyrannically as authoritarians. Hence a measure of epistemic scepticism about our own rationality or the lack of rationality in others is highly desirable.

In the end, deferral to irrationality, to partial autonomy, to imperfect consent and to unexplored values and metaphysical beliefs in patients may be necessary, even morally required. But before reaching this point, a physician committed to the highest standard of care will exercise her talents as an educator to promote greater rationality in patients. Not to make the effort to promote rational, critical deliberation is to risk a very contemporary form of patient abandonment: abandonment to human irrationality.

Duties as educators

In important ways, physicians have always been expected to be educators: about how bodies work, do not work, and go awry; about how to care for our bodies in sickness and health; about, in the end, how to live a mortal embodied existence. Our discussion suggests, however, that physician duties as educators are more

extensive. For in order genuinely to respect autonomy and patients' values, physicians must be prepared to do more than provide patients with information relevant to making evaluative choices. They must attend to how that information is received, understood and used. Good education is not restricted to providing information. It requires encouraging in others the requisite skills for dealing with information rationally.

If an ethic of respect for persons in contemporary medicine rules out – except in the most extreme cases – coercion as a response to patient irrationality, it also makes more imperative a "critical educator" response to patient irrationality. One caveat, however: effective educators know when to promote critical enquiry. Physicians, whose primary obligations are to the medical wellbeing of patients, will do well to resist the secondary obligation to promote rational criticism of deeply held beliefs at a time when their patients are impaired and suffering greatly. Thus the time to engage a hypothetically irrational JW in a critical enquiry about her convictions on "eating blood" is not the time at which she might benefit from an immediate blood transfusion because her life is in jeopardy.

It may be a very contemporary form of physician abandonment of patients in need to accept wilfulness as autonomy, the mere provision of information as adequate for informed consent, and acceptance of any morally or metaphysically bizarre view held by patients as grounds for not pursuing a medically beneficial course of treatment. But if physicians are to promote autonomy, if they are to respect patients as persons, if they are to help patients to choose and do what there is good reason to do, they should care more about the rationality of their patients' beliefs. Physicians must concern themselves with helping patients to deliberate more effectively and, ultimately, must themselves learn to care more about theoretical rationality. To do any less is to abandon patients to autonomy-destroying theoretical irrationality.

Acknowledgement

Thanks to Derek Parfit, Michael Lockwood, and David Malyon for many helpful comments.

Notes

1   Dworkin G. *The Theory and Practice of Autonomy.* Cambridge: Cambridge University Press, 1988: 12.
2   Adapted from Parfit D. *Reasons and Persons.* Oxford: Clarendon Press, 1984: 153.
3   Forrest P. *The Dynamics of Belief: A Normative Logic.* Oxford: Basil Blackwell, 1986.
4   Watch Tower Bible and Tract Society of Pennsylvania. *Family Care and Medical Management for Jehovah's Witnesses.* New York: Watch Tower Bible and Tract Society of New York, 1995.
5   *The New English Bible:* Genesis 9: 3–4.
6   *The New English Bible:* Leviticus 17: 10, 13, 14; 7: 26, 7; Numbers 15: 30, 31; Deuteronomy 12: 23–5.
7   *The New English Bible:* Leviticus 17: 11.
8   *The New English Bible:* Acts 15: 29.
9   *The New English Bible:* Matthew 26: 26–9.
10   *The New English Bible:* Galatians 3: 10.
11   *The New English Bible:* Galatians 3: 13.
12   *The New English Bible:* Galatians 3: 24–5.
13   *The New English Bible:* Galatians 4: 24.
14   *The New English Bible:* Galatians 5: 2–6.
15   Belief in God may be a basic belief: a belief which does not rest on other beliefs. (Swinburne R. *Faith and Reason.* Oxford: Clarendon Press, 1981: 33.) A basic rational belief is a belief in a proposition that is (1) self-evident or fundamental, (2) evident to the senses or memory, or (3) defensible by argument, inquiry or performance (Kenny A. *Faith and Reason.* New York: Columbia University Press, 1983: 27). Beliefs about eating blood are secondary beliefs. They must be justified in terms of other beliefs. It is precisely this that cannot be rationally done.
16   Armstrong DM. *What is a Law of Nature?* Cambridge: Cambridge University Press, 1983: 59.
17   Pascal gave a rationalist argument for belief in God: we have more to lose if we do not believe in God, and we are wrong (eternal torment), than we have to lose if we do believe in God, and we are wrong (living under an illusion). So we ought to believe that God exists (Pascal B. *Pensées.* Geneve: Pierre Cailler, 1947: fragment 223). Theoretically rational JWs could give a similar justification for refusing blood.

Chapter 34

O. O'Neill

SOME LIMITS OF INFORMED CONSENT

Abstract

Many accounts of informed consent in medical ethics claim that it is valuable because it supports individual autonomy. Unfortunately there are many distinct conceptions of individual autonomy, and their ethical importance varies. A better reason for taking informed consent seriously is that it provides assurance that patients and others are neither deceived nor coerced. Present debates about the relative importance of generic and specific consent (particularly in the use of human tissues for research and in secondary studies) do not address this issue squarely. Consent is a propositional attitude, so intransitive: complete, wholly specific consent is an illusion. Since the point of consent procedures is to limit deception and coercion, they should be designed to give patients and others control over the amount of information they receive and opportunity to rescind consent already given.

Across the last 25 years informed consent has been central to discussions of ethically acceptable medical practice. It is seen as necessary (and by some as sufficient) ethical justification for action that affects others, including medical treatment, research on human subjects, and uses of human tissues. Some commonly cited reasons for thinking that informed consent is of great importance are quite unconvincing: informed consent has been supported by poor arguments and lumbered with exaggerated claims. My intention is not to deny its importance, or to argue for any return to medical paternalism, but to take it sufficiently seriously to identify some of its limitations as well as its strengths.

Informed consent is nothing strange. It is a familiar and ethically important aspect of everyday transactions. Shopping and borrowing a book from the library, taking one's clothes to the cleaners and buying a train ticket are ethically acceptable if, but only if, all parties to the transaction take part willingly in awareness of ways in which others' proposed action will bear on them. It may seem pompous to speak of giving informed consent to these everyday transactions. Traditionally we emphasise informed consent only in more formal contexts, typically involving documents, signatures, and legal requirements and other rituals of consent, such as those used in signing a contract or getting married. But in everyday as in more formal contexts we accept that transactions are ethically and legally questionable, or even void, unless all parties are aware of the essential features of the transaction and take part willingly.

There is broad agreement that informed consent has become more important in medicine in the last 25 years because medical practice too has become more formalised.[1] The largely tacit understandings and trust which (we at least imagine) used to be found in everyday, one to one, face to face relations between doctors and patients have given way (as the title of one book rather ominously puts it) to relations between patients and *Strangers at the Bedside*.[2] Of course, medicine is not the only part of life in which formality, bureaucracy, and explicit ways of seeking, giving, recording, and respecting informed consent have multiplied. They are also more prominent in education, financial services, consumer protection, and other fields in which social relations have become less personal, more bureaucratic, and more complex, displacing traditional relations of trust. The change has been accelerated because institutions and professionals increasingly see obtaining informed consent as protection against accusation, litigation, and compensation claims. As one sociologist of medicine aptly writes, informed consent has become "the modern clinical ritual of trust".[3] As often with rituals, there is disagreement both about its real meaning and about its proper performance.

Before turning to these disagreements I note several reasons why rituals of informed consent cause more difficulty in medicine than in almost any other area of life. The first reason is very familiar: we can give informed consent only if we are competent to do so. Informed consent has its place in relationships "between consenting adults"; it is possible only when we are, as John Stuart Mill puts it, "in the maturity of our faculties".[4] But medical practice constantly has to deal with exceptional numbers of people who are (temporarily or permanently) not in the maturity of their faculties. Innumerable discussions of informed consent in medicine and medical ethics focus on these hard cases; there are lots of them.

We cannot give informed consent when we are very young or very ill, mentally impaired, demented or unconscious, or merely frail or confused. Often people cannot give informed consent to emergency treatment. Even in the maturity of our faculties we may find it quite taxing to give informed consent to complex medical treatment when feeling lousy.

These hard cases provide a staple diet for medical ethics. Some writers look for ways to make consent easier for those who find it hard. Others seek alternative criteria for permissible treatment of patients who cannot consent, and concede that many patients have to be treated with a degree of paternalism.

A second limitation of informed consent procedures in medicine is that they are useless for selecting public health policies. Public policies, including public health policies, have to be uniform for populations. We cannot adjust water purity levels or food safety requirements to individual choice, or seek informed consent for health and safely legislation or quarantine restrictions.

Vaccination policies are an interesting and possibly hybrid case: in so far as we think of them as a matter of public health policy they cannot be based on individual choice, or on informed consent. In the United Kingdom, however, we have treated vaccination only partly as a public health matter. We allow parents to refuse to have their children vaccinated without medical reason. Some have done so at little or no cost or risk to their children by sheltering behind protection provided by others' vaccinated children. The proportion of children vaccinated with measles, mumps, and rubella (MMR) has fallen, and free riders now face a problem. They still do not want to expose their children to the risk of measles, but can no longer do so by refusing vaccination. Their current ambition – well stoked by parts of the media[5] – is to use an alternative vaccine which they claim (evidence is not provided) would be safer for their children, but which will not provide the same level of protection for the population – including for infants below the age of vaccination. Public health policies can be undermined if their implementation depends on individual informed consent.

A third limitation of informed consent is that medical treatment of individuals uses personal information about third parties that is disclosed without their consent. For example, family history information, genetic information and information about exposure to infections are often disclosed to medical practitioners without the consent of all to whom the information pertains. We do not expect patients to obtain prior consent to disclosure of such information from their relatives and contacts, and this would often be impractical or impossible. This humble but pervasive fact about the way medical information is sought and used cannot be reconciled with the claim that informed consent is necessary for all ethically acceptable medical practice.

A fourth limitation of informed consent emerges when people with adequate competence to consent are under duress or constraint, so less able to refuse others' demands. Prisoners and soldiers, the vulnerable, and dependent often have ordinary capacities to consent but cannot refuse, so undermining any "consent" they offer. These cases have traditionally been seen as problematic in recruiting subjects for experiments; they are no less problematic in obtaining informed consent to medical treatment.

Behind the ritual of informed consent

Evidently informed consent cannot be relevant to all medical decisions, because it cannot be provided by patients who are incompetent to consent, cannot be used in choosing public health policies, cannot be secured for all disclosure of third party information, and cannot be obtained from those who are vulnerable or dependent. Informed consent might nevertheless

be important for the ethically acceptable treatment of individual patients who are competent and free to consent in cases where no information about third parties is needed. Indeed, it is a commonplace of medical ethics that informed consent is indispensable in these cases. The reasons offered for this view are varied and perplexing.

Informed consent in medical ethics is commonly viewed as the key to respecting patient autonomy. This claim is endlessly repeated but deeply obscure. There are many distinct conceptions of *individual autonomy* in circulation, and even more views of the value and importance of these various conceptions. In a survey of views of autonomy, Gerald Dworkin noted that it has been equated with:

> Liberty (positive or negative) ... dignity, integrity, individuality, independence, responsibility and self knowledge ... self assertion ... critical reflection ... freedom from obligation ... absence of external causation ... and knowledge of one's own interests.[6]

Other writers have equated autonomy with "privacy, voluntariness, self mastery, choosing freely, choosing one's own moral position and accepting responsibility for one's choices".[7] The list could be extended in many ways, and the feasibility and the value of all conceptions of individual autonomy are hotly contested. It seems to me, however, that if informed consent is ethically important, this *cannot* be because it secures some form of individual autonomy, however conceived. Informed consent procedures protect choices that are timid, conventional, and lacking in individual autonomy (variously conceived) just as much as they protect choices that are self assertive, self knowing, critically reflective, and bursting with individual autonomy (variously conceived).

Contemporary accounts of autonomy have lost touch with their Kantian origins, in which the links between autonomy and respect for persons are well argued; most reduce autonomy to some form of individual independence, and show little about its ethical importance.[8]

The ethical importance of informed consent in and beyond medical practice is, I think, more elementary. It provides reasonable assurance that a patient (research subject, tissue donor) has not been deceived or coerced. I shall not rehearse the deeper theoretical reasons for thinking that we have obligations not to deceive or to coerce. I believe there are convincing reasons for thinking that we have such obligations, which provide good reasons not to impose treatment or action on patients – or on others – without their informed consent.

In saying this I do not mean to suggest that informed consent is the *only* ethically important consideration, in medicine or elsewhere. The libertarian tendency in medical ethics sees informed consent as *necessary* and *sufficient* justification for action. For libertarians everything is morally permissible "between consenting adults". Most other ethical positions do not view consent as *sufficient* justification. Even if there is informed consent, we may judge surgery without medical purpose, medical practice by the unqualified, or unnecessarily risky treatment unacceptable and may think it wrong to use human tissues as commodities, as inputs to industrial processes, or as items for display.[9] Informed consent is one tip of the ethical iceberg: those who think otherwise overlook the rest of the iceberg.

Performing the ritual: consent procedures

How can consent show that there is neither deception nor coercion? What makes a ritual of informed consent effective? Events at Alder Hey

Hospital and the Bristol Royal Infirmary have made these questions urgent and controversial in the UK. Is the task to ensure that patients, research subjects, and tissue donors sign up to specific propositions set out in explicit consent forms? Or can a single signature – or a gesture of assent – imply consent to a range of distinct propositions? Proponents of *specific* and *generic* consent are at work up and down the land drafting regulations, codes of practice, and guidelines, consent forms, and information leaflets. How are these disputes to be settled? Should they be settled in the same way for treating patients, for recruiting research subjects and for removing tissues (including postmortem removal)? What should be done given that it is seldom feasible to get specific consent to future uses of donated tissue? Is it necessary to seek further consent whenever new research purposes are envisaged? If so, what is to be done if donors cannot be found or are dead?[10] Can agreement on these issues be achieved in time to shape reform of the Human Tissues Act 1961, which the government promised in their response (or reaction?) to the Redfern Report on events at Alder Hey?[11]

A reasonable starting point is to note that consent is a *propositional attitude*, given in the first instance not to another's action, but to a proposition describing the action to be performed (other propositional attitudes include *knowing, desiring, hoping, expecting, believing*). Propositions may be more or less specific, and some limit has to be drawn to the amount of detail included. The inclusion of excessive or technical detail, for example, will eventually overtax even the most energetic, and undermine the possibility of informed consent. On the other hand, consent that is too vague and general may also fail to legitimate action.

It is commonly assumed that in consenting to a description of what is to be done patients also consent to other descriptions of the treatment or procedure that are, for example, *entailed by* or *logically equivalent to* the description to which consent is given. It is also commonly assumed that in consenting to a description of what is to be done the patient consents to the *likely consequences* of its being done. Both assumptions are evident in the thinking that assumes that implied consent will reach the parts that generic consent does not reach; but proponents of specific consent procedures also assume that consent travels beyond the propositions to which it is explicitly and literally given in signing a consent form.

Yet strictly speaking, consent (like other propositional attitudes) is not transitive. I may consent to A, and A may entail B, but if I am blind to the entailment I need not consent to B. Consent is said to be *opaque* because it does not shadow logical equivalence or other logical implications: when I consent to a proposition its logical implications need not be transparent to me. Transitivity fails for propositional attitudes. Consent and other propositional attitudes also do not shadow most causal connections. I may consent to C, and it may be well known that C causes D, but if I am ignorant of the causal link I need not consent to D. Again, transitivity fails for propositional attitudes. When I consent to a proposition describing an intended transaction, neither its logical implications nor the causal links between transactions falling under it and subsequent events need be transparent to me: *a fortiori* I may not consent to them.

Events at Alder Hey illustrate the *opacity of consent*. Some parents consented to removal of tissue, but objected that they had not consented to the removal of organs – although, of course, organs are composed of tissues. They did not agree that their consent to removal of tissue implied their consent to the removal of organs. As a point of logic the parents were right.

These simple facts create a dilemma. The real limits of patient and donor comprehension

suggest that it is unreasonable to seek consent for every detail of a proposed treatment, or of a proposed research protocol, or of a proposed use of tissues. Yet the logic of propositional attitudes suggests that we cannot simply assume that implied consent will spread from one proposition to another, or from one proposition to the expected consequences of that which it covers, making any further consent unnecessary. There are many ways of skinning this cat. I conclude by sketching one approach that I think plausible.

Our aim in seeking others' consent should be not to deceive or coerce those on the other end of a transaction or relationship: these are underlying reasons for taking informed consent seriously. It follows that consent is not always improved by trying to ensure that it is given to more, or more specific, propositions: more specific consent is not invariably better consent. Complex forms that request consent to numerous, highly specific propositions may be reassuring for administrators (they protect against litigation), and may have their place in recruiting research subjects: yet they will backfire if patients or practitioners come to see requesting and giving consent as a matter of ticking boxes. Our aim should, I suggest, be to achieve *genuine* consent, and this may not always be best done by seeking specific consent to a great many propositions.

Patients, research subjects, and tissue donors give genuine consent only if they are neither coerced nor deceived, and can judge that they are not coerced or deceived; yet they must not be overwhelmed with information. This balance can perhaps be achieved by giving them a limited amount of accurate and relevant information and providing user friendly ways for them to extend this amount (thereby checking that they are not deceived) as well as easy ways of rescinding consent once given (thereby checking that they are not coerced). Genuine consent is apparent where patients can control the amount of information they receive, and what they allow to be done.

Genuine consent is not a matter of overwhelming patients with information, arrays of boxes to tick or propositions for signature. The quest for perfect specificity is doomed to fail since descriptions can be expanded endlessly, and there is no limit to a process of seeking more specific consent. It is not, however, difficult to give patients control over the amount of information they choose to receive, by offering easy access to more specific information that lies behind an initial, or second, or third layer of information provided. Accurate information of varying degrees of specificity can be provided by offering fact sheets, explanatory leaflets, discussion, and (with care) by counselling – and time to absorb further information. If additional accurate information is reliably available as demanded, patients will not be deceived: even a patient who decides on the basis of limited information has judged that the information was enough to reach a decision, and is not deceived.

Nor is it difficult to give patients greater control over what happens by making sure that their consent is rescindable, and that they know it is rescindable. Of course, consent to treatment is not always rescindable: I cannot have my appendix put back in once removed. But I can decide that I want no further chemotherapy, or refuse recommended medication. And consent to participate in clinical trials or in research, or to give tissues (for purposes other than transplantation) can be rescinded. Patients and others who know they can at any time change their mind about continuing a treatment, about participating in research, or about use of tissues they have given are not coerced and know that they are not coerced.

Patients who know they have access to extendable information and that they have given rescindable

consent have in effect a veto over what is done. It is true that exercising the veto may come at a price for patients: if I do not consent to surgery I do not get it. But for research subjects the cost of refusal is only exclusion from a study, and for tissue donors only the loss of an opportunity to be generous. This way of looking at informed consent seems to me not only to reduce possibilities of deception and coercion, but to make it plain to patients, research subjects, and tissue donors that they may determine how far they will be informed, and that (when it is technically possible) they remain free to rescind their initial choice. Where these standards are met, there are reasonable assurances that nobody is coerced or deceived.

[...]

Notes

1 Faden R, Beauchamp T. *A History and Theory of Informed Consent.* New York: Oxford University Press, 1986.

2 Rothman D. *Strangers at the Bedside: A History of How Law and Ethics Transformed Medical Decision Making.* New York: Basic Books, 1991.

3 Wolpe P. "The Triumph of Autonomy in American Bioethics: A Sociological View." In: Devries R, Subedi J, eds. *Bioethics and Society: Sociological Investigations of the Enterprise of Bioethics.* Englewood Cliff, NJ: Prentice Hall, 1998: 38–59.

4 Mill JS. "On Liberty," in Warnock M, ed. *Utilitarianism, On Liberty and Other Essays.* London: Fontana, 1962: 135.

5 Jenkins S. "A Sad Case of Media Meddling not Reason." *The Times* 2000 Feb 8.

6 Dworkin G. *The Theory and Practice of Autonomy.* Cambridge: Cambridge University Press, 1988: 6.

7 See reference 1: 8.

8 O'Neill O. *Autonomy and Trust in Bioethics.* Cambridge: Cambridge University Press, 2002.

9 "Nuffield Council on Bioethics." *Human Tissue: Ethical and Legal Issues.* London: Nuffield Council on Bioethics, 1995. www.nuffield.org.uk/bioethics/index.html. O'Neill O. "Medical and Scientific

Uses of Human Tissues." *Journal of Medical Ethics* 1996; 22: 1–3.

10 House of Lords' Select Committee on Science and Technology. *Report on Human Genetic Databases: Challenges and Opportunities.* HL57; written evidence, 2000, HL

115. www.parliament.the-stationery-office.co.uk/ pa/ld/ldsctech.htm

11 *The Report of The Royal Liverpool Children's Inquiry.* London: The Stationery Office, 2001: www.rlcin-quiry.org.uk/

Chapter 35

Simon N. Whitney

A NEW MODEL OF MEDICAL DECISIONS: EXPLORING THE LIMITS OF SHARED DECISION MAKING

Shared decision making is widely accepted as an ethical imperative[1-5] and as an important part of reasoned clinical practice.[6] Major texts in decision analysis,[7] medical ethics,[8] and evidence-based medicine[9] all encourage physicians to include patients in the decision-making process.

One reason for this emphasis on collaboration is the inequality between patients and physicians. In theory, a patient has near-absolute control over his or her own body and treatment; in practice, the physician is more powerful in many ways.[2] Shared decision making is, among other things, a way to ensure that the patient's voice is heard as choices are made. Because patients have differing preferences for both the processes and outcomes of care, their participation is vital for decisions such as lumpectomy with radiation therapy compared to mastectomy for early breast cancer[2] or systemic methotrexate compared to laparoscopic salpingostomy for tubal pregnancy.[10,11]

These two examples have an important commonality: both are close cases. Yet there is a second class of decisions in which there is only one realistic option. If a woman has inflammatory breast cancer rather than infiltrating ductal carcinoma, her values are less relevant, because lumpectomy is much less likely to cure her than chemotherapy followed by mastectomy.

Similarly, if a woman with a tubal pregnancy experiences sudden abdominal pain and develops unstable vital signs, her consent to surgery represents a kind of collaboration; however, she has no real choice, so it is not the kind of collaboration usually envisioned in the literature on shared decision making. Discussions of shared decision making pay scant heed to these obvious decisions, precisely because they are obvious. Nevertheless, they are common in clinical practice, and the appropriate allocation of decisional priority within them should not be ignored. It is also noteworthy that these are major decisions that will have a substantial impact on a patient's life. There is a second group of more modest importance that may invite a different approach to decision making. This article proposes a typology or model to simplify the complex universe of different decision types.

Although the model presented here focuses on the central decision makers, the patient and the physician, they do not deliberate alone. Patients have friends, families, and coworkers to consider, and physicians have colleagues and consultants. Their decisions may also be influenced by a wide range of outside factors, such as an insurance company's policies or a coach's wish to have an injured player returned to the field quickly. These influences do not lessen the validity of the preferences of the patient and the physician.

This model uses two key characteristics – importance and certainty – to array medical choices on a decision plane, so that specific types of decisions populate identifiable zones of the plane. These zones have distinct features that may be used to predict how fully decisions will be shared and the type of conflict that will arise if a patient and a physician disagree. This model owes a debt to both Deber and coworkers,[12,13] who have contended that most patients wish to defer to physicians when a problem has a single correct solution, but many wish to participate in decisions when one of several alternatives must be chosen, and to Braddock and his colleagues,[14] who proposed that two dominant characteristics of a medical decision are its effect on the patient (herein called "importance") and its degree of medical consensus (herein called "certainty"). As developed here, this model is a theoretical construct informed only by the ordinary observations of everyday practice, not structured research. Its empirical validity remains to be demonstrated.

Decisional authority and priority

Throughout a clinical encounter, a patient maintains decisional authority – the right to accept or reject any reasonable intervention (e.g., a diagnostic test, psychotherapy, a procedure, or a medication). However, the physician is the logical decision maker in some types of decisions. Consequently, decisional priority may lie with the patient, with the physician, or with both. For instance, a patient with severe hypothyroidism would ordinarily accept a physician's judgment that exogenous thyroid hormone is advisable. Here, the physician has decisional priority, meaning that he or she is better situated than the patient to make this decision. Therefore, the physician will normally take the lead, even though the final decision is still the patient's. In contrast, the patient should have decisional priority for other types of decisions, including,

for instance, the choice of whether or not to undergo a coronary artery bypass graft procedure for angina.

This model depicts a monotonic shift in who should have decisional priority from one participant to the other as importance and certainty change. It is possible, however, that patients' and physicians' preferences follow a U-shaped curve that does not reflect this linear transition. As an example, most patients might believe that physicians should make minor decisions but would want greater participation in the decision-making process when the decisions have moderate importance. But as the stakes continue to rise, some patients may wish their physicians to reassume the role of decision maker, so that if the chosen therapeutic course goes badly, the patient need not suffer blame – from self or family – for a poor outcome.[13]

Key characteristics of decisions

Importance

Some decisions are major, others are minor. The importance or seriousness of a medical decision begins with its probable effect on a patient's health and well-being, but it also reflects the moral, financial, social, legal, and esthetic repercussions of the decision for the patient and others. The importance of a decision reflects both medical facts and personal values and may be viewed differently by different parties. As an example, a patient and a physician may disagree about whether it is more important to treat hypertension or maintain normal sexual function. Because the patient's perspective is always at the core of good decision making, physicians are well advised to understand patients' priorities. Importance is a continuous characteristic, but for the purposes of discussion, it may be mapped onto a 4-point scale (Table 35.1). Of course, any choice may have profound

Table 35.1 Levels of Importance of Medical Decisions

Importance	Example	Rationale for Choice of Level
Major	Delivery of an infant via postmortem cesarean section performed on a woman who has died of a gunshot wound	Performing the surgery may save the infant's life, but at the cost of severe hypoxic damage; a tough choice with major consequences
Important	Initiating evaluation for abuse in a child with suspicious injuries	Significant legal and clinical consequences if abuse is present but unreported
Routine	Choice of anesthetic for a patient with impaired liver function	An error would be harmful but is not difficult to avoid
Minor	Timing of blood work to confirm response of anemia to iron	Whether this is checked in 1 month or 2 is usually unimportant

consequences, but this happens less often for minor than for major decisions.

Major decisions grip our attention (and attract ethicists), but minor decisions fill physicians' workdays. Primary-care physicians probably confront major decisions rarely, important decisions occasionally, and routine and minor decisions frequently.

Certainty

For a physician, certainty reflects the degree to which a decision-analytic approach using good-quality data would demonstrate that there is a single preferred intervention. In the absence of suitable data, certainty is present in practical terms if expert opinion holds one intervention to be superior. Clinicians do not always agree with one another, of course, and in fact, two clinicians may each be quite confident that his or her own approach is superior.[15-17] When a physician feels that one choice is better but knows that other clinicians disagree, the patient should be informed of the controversy and offered a second opinion from someone holding the other view.

A decision that is high in certainty has a widely accepted clinical response (e.g., treating neonatal sepsis with parenteral antibiotics). In contrast, a decision is low in certainty if the available interventions are very similar, if there are scant relevant data, if there is controversy over the optimal treatment, or if there is good evidence that suggests little difference in outcomes between treatments. Patients should have maximal decisional priority in situations in which medical certainty is low. When the decision is also of major importance, physicians should educate their patients to help them synthesize the available information and decide on the best course of action. Kassirer and Pauker[18] call uncertain decisions "toss-ups" and comment,

> Even when the [decision] analysis shows a slight benefit of one option [such as surgery], factors such as a preference for long-term medical therapy or an unwillingness to be hospitalized and away from one's family for several weeks may well sway the decision toward a competing choice (such as medical therapy).

An example of how levels of certainty might be arrayed is provided in Table 35.2.

One example of a low-certainty decision is the choice of a generic compared to a brand name drug. In the absence of a therapeutic difference, a patient may reasonably request one form or the other on the basis of cost, size, or prior reaction to a dye or preservative. As another example, two low-certainty choices may use different

Table 35.2 Levels of Certainty of Medical Decisions

Certainty	Physician's Preference	Example	Rationale for Choice of Level
High	Strong	Exploratory thoracotomy for a patient with penetrating trauma to the chest and shock	Experts agree that the patient will probably die without surgery
Intermediate	Some	Limiting the use of premarin/progesterone therapy after menopause	There is good evidence showing some overall health risk, yet symptomatic benefits may be significant
Low	None	Trimethoprim-sulfamethoxazole compared to nitrofurantoin for an uncomplicated urinary tract infection	Both medications are effective, so prior experience, cost, the size of the pill, or similar factors may tip the scales

interventions to reach the same outcome (e.g., the use of an arm sling compared to a figure-of-eight brace for a midshaft fracture of the clavicle) or different interventions to reach different outcomes (e.g., mastectomy compared to lumpectomy and radiation for a small breast cancer). The patient should have decisional priority for these choices, although the patient's preferences may have to compete with cost, the physician's convenience, and resource availability.

Patients' preferences become less determinative, however, when the medical preference for one choice is compelling. So, for instance, a woman in labor might reasonably reject operative delivery even if she has a marginal placenta previa; if she has a complete placenta previa, however, her physician is likely to vigorously recommend a cesarean section. Here, as elsewhere, even when decisional priority shifts to the physician, decisional authority remains with the patient, and she may accept or reject her physician's recommendation.

The decision plane

All medical decisions may be mapped onto a plane on the basis of their importance and certainty (Figure 35.1).

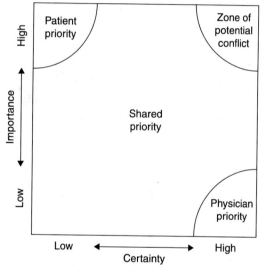

Figure 35.1 Decision plane for medical decisions.

The location of a particular decision will vary with the circumstances of an individual patient, and if those circumstances change, its location will also change. Consider, for example, the treatment of streptococcal pneumonia with antibiotics. All physicians administer antibiotics to previously healthy patients with pneumonia, but many physicians would not recommend an antibiotic for a patient with pneumonia who already has severe dementia and advanced cancer with intractable suffering. In each case, there is a consensus (high certainty) about the best management of the pneumonia: in the former case, antibiotics are recommended, and in the latter, they are not. As a patient's underlying condition progresses by stages from good health to the last hours of life, the characteristics of the decision will move from the certainty of the first case (provide antibiotics), through an intermediate zone of low certainty, and on to the certainty of the last case (withhold antibiotics). A patient in the intermediate zone might, for example, have advanced cancer with dyspnea and pain that is only partly relieved by medication. In view of this patient's suffering, if he or she develops pneumonia, the decision of whether or not to administer antibiotics would be characterized by substantial medical uncertainty. The choice of whether to treat the pneumonia or to allow it to progress unhindered should be made by the patient if he or she is capable of making the decision.

The decision plane may be divided into specific zones, as discussed below.

Zone of patient priority

Decisions that have high importance and low certainty are, or should be, the patient's (or the surrogate's) to make (Figure 35.1). One classic preference-sensitive decision is the choice between mastectomy and lumpectomy with radiation for localized breast cancer; decisions like these have been of particular value in honing our respect for patients' autonomy.[2]

Zone of physician priority

Patients retain authority over all decisions, but few patients are likely to want to make decisions that are clearly minor and for which there is one best choice. Consequently, these decisions are customarily made by physicians. For example,

adding oblique views to an X-ray of the finger will sometimes reveal a fracture that was not visible on a two-view study, and the only way to prevent Rh sensitization after an Rh-negative woman miscarries is to administer Rho immune globulin. Sometimes these choices are discussed with patients and sometimes they are not, but patients are not ordinarily asked to provide their views or explore their values with regard to these choices.

Zone of shared priority

The zone of shared priority constitutes the remainder of the plane. Decisions in this zone should be negotiated, although the way in which decision making should be shared will depend on the specific choice at hand. For some choices, a very detailed informed consent process that pays careful attention to a patient's preferences is appropriate, as, for instance, when a patient with an ischemic leg may be better served by amputation than by a third attempt at revascularization. For other choices, a concise description of the proposed plan is adequate, as when a physician tells the mother of a child who is fatigued, "I'd like to check a blood count" and obtains her approval with a nod. A physician who provides only a cursory explanation of the pros and cons of amputation flouts fundamental principles of consent and patient autonomy; one who spontaneously provides a careful description of the merits of checking a blood count violates common norms of appropriateness and economy of time.

Zone of potential conflict

Decisional priority reposes with patients for decisions that have major importance and with physicians for decisions that have high certainty. Some decisions have both features; this is the zone of potential conflict. Because these decisions are important, patients have every right to make them against their physicians' advice; because they are not only important, but have only one medically acceptable choice, physicians may feel strongly that such patients are making a critical mistake.

Fortunately, the usual dynamic in this situation is for physicians to make strong recommendations and for patients to accept them. However, serious conflicts may arise when patients reject physicians' recommendations, no matter what the reason for patients' preferences. Consider, for example, a pregnant woman who has three young children and has aggressive cervical cancer. Her physician might believe strongly that she should undergo a hysterectomy, which would maximize her chance of cure but at the cost of the fetus's life; the woman might be unwilling to terminate her pregnancy, preferring to endanger her own life rather than sacrifice that of her fetus. Although each case like this must be judged on its own merits, the physician generally can rely only on persuasion to prevail, because decisional authority remains with the patient. This zone is the wellspring of classic bioethics cases like those of Dax Cowart[19] and Baby K,[20] as well as entire genres of ethical dilemmas, like the rejection of lifesaving treatment by patients on religious grounds.

Implications

Improving Theory

Shared decision making is much advocated but inconsistently achieved; studies have shown patient involvement in decisions in both inpatient and outpatient settings to be less than ideal.[14,21,22] Domineering physicians, meek patients, a lack of time, and limited patient education and understanding certainly form part of the explanation. This model suggests an

additional explanation: some types of choices are inherently ill suited to shared decision making. Minor decisions that have one medically preferred choice are probably quite common, and it is entirely possible that even intelligent, self-aware patients would yield decisional priority for these decisions to their physicians. This hypothesis may be tested by asking patients their preferences with regard to this type of decision; if the hypothesis is confirmed, decisions of this type would be an exception to the general recommendation that patients and physicians share decision making.

Improving clinical practice

Although this model suggests that physicians should make many medical decisions, decisional priority for choices of high importance and low certainty should rest unequivocally with patients. These decisions may also be made by physicians if that is their patients' unambiguous and adequately informed preference. Clues that indicate a decision to be of this type include a clinician's feeling that a decision is difficult or that different physicians might make different recommendations.[18]

Conclusion

This model provides a new perspective on medical decision making, one in which the characteristics of an individual decision suggest whether decisional priority should remain with a patient, be assumed by his or her physician, or be shared. It reserves an important place for shared decision making and suggests an explanation for why many decisions are shared minimally or not at all. Structurally, this model is a form of typology – a two-dimensional, conceptual classification.[23] It was developed from an ideal of what ought to be, not from empirical demonstrations of what is.

This model is exploratory, not definitive. Although it describes some characteristics of medical decisions, it does not do full justice to the complex interactions that exist between patients, physicians, and the environment, nor does it integrate the impact of culture and the structure of care, including such factors as resources available and insurance. This is therefore but a first step.

Notes

1 President's Commission for the Study of Ethical Problems in Medicine and Biomedical and Behavioral Research. Making Health Care Decisions: The Ethical and Legal Implications of Informed Consent in the Patient-Practitioner Relationship. Washington (DC): President's Commission; 1982.

2 Katz J. The Silent World of Doctor and Patient. New York: Free Press; 1984.

3 Brody H. The Healer's Power. New Haven (CT): Yale University Press; 1992.

4 Emanuel EJ, Emanuel LL. "Four models of the physician-patient relationship." JAMA. 1992; 267(16): 2221–6.

5 Quill TE, Brody H. "Physician recommendations and patient autonomy: finding a balance between physician power and patient choice." Ann Int Med. 1996; 125(9): 763–9.

6 Sox HC Jr, Blatt MA, Higgins MC, Marton KI. Medical Decision Making. Boston: Butterworth-Heinemann; 1988.

7 Weinstein MC, Fineberg HV. Clinical Decision Analysis. Philadelphia: Saunders; 1980.

8 Beauchamp TL, Childress JF, Principles of Biomedical Ethics. 4th ed. New York: Oxford University Press; 1994.

9 Sackett DL, Straus SE, Richardson WS, Rosenberg W, Haynes RB. Evidence-Based Medicine: How to Practice and Teach EBM. 2nd ed. New York: Churchill Livingstone; 2000.

10 Nieuwkerk PT, Hajenius PJ, van der Veen F, Ankum WM, Wijker W, Bossuyt PM. "Systemic methotrexate therapy versus laparoscopic salpingostomy in tubal pregnancy. Part II. Patient preferences for systemic methotrexate." Fertil Steril. 1998; 70(3): 518–22.

11 Nieuwkerk PT, Hajenius PJ, Ankum WM, van der Veen F, Wijker W, Bossuyt PM. "Systemic methotrexate therapy versus laparoscopic salpingostomy in patients with tubal pregnancy. Part I. Impact on patients' health-related quality of life." Fertil Steril. 1998; 70(3): 511–17.

12 Deber RB, Baumann AO. "Clinical reasoning in medicine and nursing: decision making versus problem solving." Teach Learn Med. 1992; 4(3): 140–6.

13 Deber RB, Kraetschmer N, Irvine J. "What role do patients wish to play in treatment decision making?" Arch Intern Med. 1996: 156(13): 1414–20.

14 Braddock CH III, Edwards KA, Hasenberg NM, Laidley TL, Levinson W. "Informed decision making in outpatient practice: time to get back to basics." JAMA. 1999; 282(24): 2313–20.

15 Taylor KM, Margolese RG, Soskolne CL. "Physicians' reasons for not entering eligible patients in a randomized clinical trial of surgery for breast cancer." N Engl J Med. 1984; 310: 1363–7.

16 Freedman B. "Equipoise and the ethics of clinical research." N Engl J Med. 1987; 317(3): 141–5.

17 Baumann AO, Deber RB, Thompson GG. "Overconfidence among physicians and nurses: the 'micro-certainty, macro-uncertainty' phenomenon." Soc Sci Med. 1991; 32(2): 167–74.

18 Kassirer JP, Pauker SG. "The toss-up" [editorial]. N Engl J Med. 1981; 305(24): 1467–9.

19 Kliever LD, editor. Dax's Case: Essays in Medical Ethics and Human Meaning. Dallas (TX): Southern Methodist University Press; 1989.

20 Annas GJ. "Asking the courts to set the standard of emergency care: the case of Baby K." N Engl J Med. 1994; 330: 1542–5.

21 Lidz CW, Meisel A, Zerubavel E, Carter M, Sestak RM, Roth LH. Informed Consent: A Study of Decision making in Psychiatry. New York: Guilford; 1984.

22 Braddock CH III, Fihn SD, Levinson W, Jonsen AR, Pearlman RA. "How doctors and patients discuss routine clinical decisions. Informed decision making in the outpatient setting." J Gen Int Med. 1997; 12(6): 339–45.

23 Bailey KD. Typologies and Taxonomies: An Introduction to Classification Techniques. Thousand Oaks (CA): Sage; 1994.

Received 7 March 2002 from the Department of Family and Community Medicine, Baylor College of Medicine, Houston, TX. This project was supported by grant K08 HS11289 from the US Agency for Healthcare Research and Quality. I would like to express my gratitude to Anna Kieken, PhD, and Pamela Paradis Tice, BA, ELS(D), for their skillful editing and to Robert D. Canning, PhD, Irene Easling, DrPH, Judy Levison, MD, Robert Volk, PhD, and the anonymous reviewers for their thoughtful comments on earlier drafts of the manuscript. Revision accepted for publication 26 January 2003.

Is it permissible to impose on individuals for the sake of the public's health?

INTRODUCTION TO PART SEVEN

WHEREAS CLINICAL MEDICINE centres on encounters between doctors and individual patients, public health comprises activities such as immunisation, screening, and health promotion, which are intended to protect and promote the health of populations of people. This 'population perspective' creates distinctive ethical challenges. One such arises because public health brings into view inequalities in health status, and uneven access to good quality health care services, within and between national populations: what causes such injustice and how can it be ameliorated? The selections comprising this chapter focus on another public health ethics problematic, one most pertinent to developed health care systems. The public's health can be protected and promoted by programmes that impose on individuals, such as mandatory testing and screening, quarantining during outbreaks, and compelling people to adopt healthier lifestyles. This generates dilemmas between individual rights and communal benefits and – since public health typically enjoys the backing of the state – introduces themes from political philosophy, such as the grounds and limits of state interference in the lives of individuals.

One response is to construct a framework within which to consider public health ethics dilemmas. Childress et al. suggest that the relevant terrain 'includes a loose set of general moral considerations ... arguably relevant to public health. Public health ethics, in part, involves ongoing efforts to specify and to assign weights to these'. This sounds formulaic, but Childress et al. suggest that these considerations inform *prima facie* cases for and against public health interventions, the real work being to adjudicate disputes over their salience and weight in particular circumstances. They focus on conflicts with a certain structure, i.e. between 'general moral considerations that are generally taken to instantiate the goal of public health – producing benefits, preventing harms, and maximizing utility – and those that express other moral commitments'. Since 'other moral commitments' include respecting autonomy and preserving justice, such conflicts are serious. Childress et al. suggest 'five "justificatory conditions" ... to help determine whether promoting public health warrants overriding such values as individual liberty or justice in particular cases'.

The next two selections represent two perspectives on this conflict between individual liberty and public health goals. The first perspective is societal: infringing on individual liberty is justified in order to protect society as a whole. Gostin discusses the extent to which the protective legal powers of the state should be extended in the post 9/11 context, arguing that public health infrastructure and legislation need updating: 'The balance between individual interests and common goods needs to be recalibrated in an age of terrorism'. His suggestion that 'resource allocations, ethical values, and law should transform to reflect the critical importance of the health, security, and well-being of the populace' resonates beyond his controversial defence of the Model State Emergency Health Powers Act. Nonetheless, like Childress *et al.*, Gostin is clear that public health ethics requires sensitive balancing of competing, weighty considerations, concluding that neither personal liberty nor health and security are fundamental because 'both … are important to human flourishing'.

The second perspective is that of the individual: limiting an individual's liberty is justified for their own sake. Important here is Mill's 'harm principle' that the State is justified in coercing the competent against their will only to avoid harms to third parties. Also important are libertarianism, the political philosophy that aims to maximise individual freedom of thought and action, and paternalism, the view that individuals can be coerced for their own good (for example, to make them healthier). Given these definitions, 'libertarian paternalism' sounds oxymoronic. But Thaler and Sunstein suggest not. Drawing on empirical work in behavioural economics, they argue from three main premises: agents do not always choose in their own best interests; there is no viable alternative to acting paternalistically towards others; and paternalism does not always involve coercion. Given these, libertarian paternalism is perfectly possible; it involves establishing 'choice architecture' – such as the design of a company cafeteria – so as to 'nudge' people towards the better options. Although paternalistic in as much as they are intended to make agents benefit themselves, such policies do not limit freedom of choice.

Societal and individual perspectives on the conflict between individual liberties and public health goals recur in the case study discussions in the remaining four selections. Holm relates Thaler and Sunstein's notion of libertarian paternalism to soft paternalism, the shared theme being that public health goals are achieved by making it easier for people to choose freely the healthy option. The paper demonstrates how much more there is to discuss, even in the limited case of obesity interventions. For example, the ethics of 'early-life intervention' centres on balancing 'the best interests of the child' and parents' rights to bring up their child however they think fit. This theme re-emerges in Isaac *et al.*'s discussion. Whilst mass immunisation neatly encapsulates the trade-off between individual rights and public health goals, childhood immunisation is further complicated by societal assumptions about parental rights: 'how far should we interfere with parental choices about child rearing?'

The final two pieces centre on the trade-off between individual liberties and the public health goal of reducing smoking related disease. In his discussion of extending legislation to ban smoking outdoors, Chapman recalls Mill's harm principle and the limits of paternalism by pointing out that it is one thing to restrict the freedom of individuals to stop them harming others, but quite another to impose on them for their own sake. For example, the official justification for criminalising smoking in many public places in the UK is to protect third parties, such as bar staff, from the effects of secondary smoking; but a ban on smoking outdoors would be overly paternalistic in being aimed at benefiting the smoker. Bayer and Fairchild advocate the development of an ethics of public health surveillance. They query the World Bank's insistence that public health surveillance does not comprise research, suggesting that 'the guidelines governing research and clinical practice, so focused on protecting individuals, cannot be imported to the public health setting'.

Selections

Childress, J.F., Faden, R.R., Gaare, R.D., Gostin, L.O., Kahn, J., Bonnie, R.J., Kass, N.E., Mastroianni, A.C., Moreno, J.D. and Nieburg, P. 2002. 'Public Health Ethics: Mapping the Terrain', *Journal of Law, Medicine and Ethics*, 30 (2), 170–8.

Gostin, L.O. 2002. 'Public Health Law in an Age of Terrorism: Rethinking Individual Rights and Common Goods', *Health Affairs*, 21 (66), 79–93.

Thaler, R.H. and Sunstein, C.R. 2003. 'Libertarian Paternalism', *The American Economic Review*, 93 (2), 175–9.

Holm, S. 2007. 'Obesity Interventions and Ethics', *Obesity Reviews*, 8 (Supplement 1), 207–10.

Isaacs, D., Kilham, H.A. and Marshall, H. 2004. 'Should Routine Childhood Immunizations be Compulsory?' *Journal of Paediatrics and Child Health*, 40 (7), 392–6.

Chapman, S. 2000. 'Banning Smoking Outdoors is Seldom 'Ethically Justifiable', *Tobacco Control*, 9 (1), 95–7.

Fairchild, A.L. and Bayer, R. 2004. "Ethics and the Conduct of Public Health Surveillance', *Science*, 303 (5658), 631–2.

Chapter 36

James F. Childress, Ruth R. Faden, Ruth D. Gaare, Lawrence O. Gostin, Jeffrey Kahn, Richard J. Bonnie, Nancy E. Kass, Anna C. Mastroianni, Jonathan D. Moreno, and Phillip Nieburg

PUBLIC HEALTH ETHICS: MAPPING THE TERRAIN

Public health ethics, like the field of public health it addresses, traditionally has focused more on practice and particular cases than on theory, with the result that some concepts, methods, and boundaries remain largely undefined. This paper attempts to provide a rough conceptual map of the terrain of public health ethics. We begin by briefly defining public health and identifying general features of the field that are particularly relevant for a discussion of public health ethics.

Public health is primarily concerned with the health of the entire population, rather than the health of individuals. Its features include an emphasis on the promotion of health and the prevention of disease and disability; the collection and use of epidemiological data, population surveillance, and other forms of empirical quantitative assessment; a recognition of the multidimensional nature of the determinants of health; and a focus on the complex interactions of many factors – biological, behavioral, social, and environmental – in developing effective interventions.

How can we distinguish public health from medicine? While medicine focuses on the treatment and cure of individual patients, public health aims to understand and ameliorate the causes of disease and disability in a population. In addition, whereas the physician-patient relationship is at the center of medicine, public health involves interactions and relationships among many professionals and members of the community as well as agencies of government in the development, implementation, and assessment of interventions. From this starting point, we can suggest that public health systems consist of all the people and actions, including laws, policies, practices, and activities, that have the primary purpose of protecting and improving the health of the public.[1] While we need not assume that public health systems are tightly structured or centrally directed, we recognize that they include a wide range of governmental, private and non-profit organizations, as well as professionals from many disciplines, all of which (alone and together) have a stake in and an effect on a community's health. Government has a unique role in public health because of its responsibility, grounded in its police powers, to protect the public's health and welfare, because it alone can undertake certain interventions, such as regulation, taxation, and the expenditure of public funds, and because many, perhaps most, public health programs are public goods that cannot be optimally provided if left to individuals or small groups.

The Institute of Medicine's landmark 1988 definition of public health provides additional insight: "Public health is what we, as a society, do collectively to assure the conditions in which people can be healthy."[2] The words "what we, as a society, do collectively" suggest the need for cooperative behavior and relationships built on overlapping values and trust. The words "to assure the conditions in which people can be healthy" suggest a far-reaching agenda for public health that focuses attention not only on the medical needs of individuals, but on fundamental social conditions that affect population levels of morbidity and mortality. From an ethical standpoint, public health activities are generally understood to be teleological (end-oriented) and consequentialist – the health of the public is the primary end that is sought and the primary outcome for measuring success.[3] Defining and measuring "health" is not easy, as we will emphasize below, but, in addition, "public" is a complex concept with at least three dimensions that are important for our discussion of ethics.

First, public can be used to mean the "numerical public," i.e., the target population. In view of public health's goal of producing net health benefits for the population, this meaning of public is very important. In measurement and analysis, the "numerical public" reflects the utilitarian view that each individual counts as one and only one. In this context, ethical analysis focuses on issues in measurement, many of which raise considerations of justice. For example, how should we define a population, how should we compare gains in life expectancy with gains in health-related quality of life, and whose values should be used in making those judgments?

Second, public is what we collectively do through government and public agency – we can call this "political public." Government provides much of the funding for a vast array of public health functions, and public health professionals in governmental roles are the focal point of much collective activity. In the United States, as Lawrence Gostin notes, government "is compelled by its role as the elected representative of the community to act affirmatively to promote the health of the people," even though it "cannot unduly invade individuals' rights in the name of the communal good."[4] The government is a central player in public health because of the collective responsibility it must assume and implement. The state's use of its police powers for public health raises important ethical questions, particularly about the justification and limits of governmental coercion and about its duty to treat all citizens equally in exercising these powers. In a liberal, pluralistic democracy, the justification of coercive policies, as well as other policies, must rest on moral reasons that the public in whose name the policies are carried out could reasonably be expected to accept.[5]

Third, public, defined as what we do collectively in a broad sense, includes all forms of social and community action affecting public health – we can call this "communal public." Ethical analysis on this level extends beyond the political public. People collectively, outside of government and with private funds, often have greater freedom to undertake public health interventions since they do not have to justify their actions to the political public. However, their actions are still subject to various moral requirements, including, for instance, respect for individual autonomy, liberty, privacy and confidentiality, and transparency in disclosure of conflicts of interest.

General moral considerations

In providing a map of the terrain of public health ethics, we do not suggest that there is a consensus about the methods and content of public health ethics.[6] Controversies persist about theory and method in other areas of applied or practical ethics, and it should not be surprising that variety also prevails in public health ethics.[7]

The terrain of public health ethics includes a loose set of general moral considerations – clusters of moral concepts and norms that are variously called values, principles, or rules – that are arguably relevant to public health. Public health ethics, in part, involves ongoing efforts to specify and to assign weights to these general moral considerations in the context of particular policies, practices, and actions, in order to provide concrete moral guidance.

Recognizing general moral considerations in public health ethics does not entail a commitment to any particular theory or method. What we describe and propose is compatible with several approaches. To take one major example, casuistical reasoning (examining the relevant similarities and differences between cases) is not only compatible with, but indispensable to our conception of public health ethics. Not only do – or should – public health agents examine new situations they confront in light of general moral considerations, but they should also focus on a new situation's relevant similarities to and differences from paradigm or precedent cases – cases that have gained a relatively settled moral consensus. Whether a relatively settled moral consensus is articulated first in a general moral consideration or in precedent cases does not constitute a fundamental issue – both are relevant. Furthermore, some of the precedents may concern how general moral considerations are interpreted, specified, and balanced in some public health activity, especially where conflicts emerge.

Conceptions of morality usually recognize a formal requirement of universalizability in addition to a substantive requirement of attention to human welfare. Whatever language is used, this formal feature requires that we treat similar cases in a similar way. This requirement undergirds casuistical reasoning in morality as well as in law. In public health ethics, for example, any recommendations for an HIV screening policy must take into account both past precedents in screening for other infectious diseases and the precedents the new policy will create for, say, screening for genetic conditions. Much of the moral argument will hinge on which similarities and differences between cases are morally relevant, and that argument will often, though not always, appeal to general moral considerations.[8] We can establish the relevance of a set of these considerations in part by looking at the kinds of moral appeals that public health agents make in deliberating about and justifying their actions as well as at debates about moral issues in public health. The relevant general moral considerations include:

- producing benefits;
- avoiding, preventing, and removing harms;
- producing the maximal balance of benefits over harms and other costs (often called utility);
- distributing benefits and burdens fairly (distributive justice) and ensuring public participation, including the participation of affected parties (procedural justice);
- respecting autonomous choices and actions, including liberty of action;
- protecting privacy and confidentiality;
- keeping promises and commitments;
- disclosing information as well as speaking honestly and truthfully (often grouped under transparency); and
- building and maintaining trust.

Several of these general moral considerations – especially benefiting others, preventing and removing harms, and utility – provide a *prima facie* warrant for many activities in pursuit of the goal of public health. It is sufficient for our purposes to note that public health activities have their grounding in general moral considerations, and that public health identifies one major broad benefit that societies and governments ought to pursue. The relation of public health to the whole set of general moral considerations is complex. Some general moral considerations support this pursuit; institutionalizing several others may be a condition for or means to public health (we address this point later when we discuss human rights and public health); and yet, in particular cases, some of the same general moral considerations may limit or constrain what may be done in pursuit of public health. Hence, conflicts may occur among these general moral considerations.

The content of these various general moral considerations can be divided and arranged in several ways – for instance, some theories may locate one or more of these concepts under others. But, whatever theory one embraces, the whole set of general moral considerations roughly captures the moral content of public health ethics. It then becomes necessary to address several practical questions. First, how can we make these general moral considerations more specific and concrete in order to guide action? Second, how can we resolve conflicts among them? Some of the conflicts will concern how much weight and significance to assign to the ends and effects of protecting and promoting public health relative to the other considerations that limit and constrain ways to pursue such outcomes. While each general moral consideration may limit and constrain public health activities in some circumstances, for our purposes, justice or fairness, respect for autonomy and liberty, and privacy and confidentiality are particularly noteworthy in this regard.

Specifying and weighting general moral considerations

We do not present a universal public health ethic. Although arguably these general moral considerations find support in various societies

and cultures, an analysis of the role of cultural context in public health ethics is beyond the scope of this paper. Instead, we focus here on public health ethics in the particular setting of the United States, with its traditions, practices, and legal and constitutional requirements, all of which set directions for and circumscribe public health ethics. (Below we will indicate how this conception of public health ethics relates to human rights.)

General moral considerations have two major dimensions. One is their meaning and range or scope; the other is their weight or strength. The first determines the extent of conflict among them – if their range or scope is interpreted in certain ways, conflicts may be increased or reduced. The second dimension determines when different considerations yield to others in cases of conflict.

Specifying the meaning and range or scope of general moral considerations – the first dimension – provides increasingly concrete guidance in public health ethics. A common example is specifying respect for autonomy by rules of voluntary, informed consent. However, it would be a mistake to suppose that respect for autonomy requires consent in all contexts of public health or to assume that consent alone sufficiently specifies the duty to respect autonomy in public health settings. Indeed, specifying the meaning and scope of general moral considerations entails difficult moral work. Nowhere is this more evident in public health ethics than with regard to considerations of justice. Explicating the demands of justice in allocating public health resources and in setting priorities for public health policies, or in determining whom they should target, remains among the most daunting challenges in public health ethics.

The various general moral considerations are not absolute. Each may conflict with another and each may have to yield in some circumstances. At most, then, these general moral considerations identify features of actions, practices, and policies that make them *prima facie* or presumptively right or wrong, i.e., right or wrong, all other things being equal. But since any particular action, practice, or policy for the public's health may also have features that infringe one or more of these general moral considerations, it will be necessary to determine which of them has priority. Some argue for a lexical or serial ordering, in which one general moral consideration, while not generally absolute, has priority over another. For instance, one theory might hold that protecting or promoting public health always has priority over privacy, while another might hold that individual liberty always has priority over protecting or promoting public health. Neither of these priority rules is plausible, and any priority rule that is plausible will probably involve tight or narrow specifications of the relevant general moral considerations to reduce conflicts. From our standpoint, it is better to recognize the need to balance general moral considerations in particular circumstances when conflicts arise. We cannot determine their weights in advance, only in particular contexts that may affect their weights – for instance, promises may not have the same moral weights in different contexts.

Resolving conflicts among general moral considerations

We do not believe it is possible to develop an algorithm to resolve all conflicts among general moral considerations. Such conflicts can arise in multiple ways. For example, it is common in public health practice and policy for conflicts to emerge between privacy and justice (for instance, the state collects and records private information in disease registries about individuals in order to allocate and provide access to resources for appropriate prevention and treatment services), or between different conceptions of justice (for

instance, a government with a finite public health budget must decide whether to dedicate resources to vaccination or to treatment of conditions when they arise). In this paper, however, we focus on one particular permutation of conflicts among general moral considerations that has received the most attention in commentary and in law. This is the conflict between the general moral considerations that are generally taken to instantiate the goal of public health – producing benefits, preventing harms, and maximizing utility – and those that express other moral commitments. For conflicts that assume this structure, we propose five "justificatory conditions": effectiveness, proportionality, necessity, least infringement, and public justification. These conditions are intended to help determine whether promoting public health warrants overriding such values as individual liberty or justice in particular cases.

Effectiveness: It is essential to show that infringing one or more general moral considerations will probably protect public health. For instance, a policy that infringes one or more general moral considerations in the name of public health but has little chance of realizing its goal is ethically unjustified.

Proportionality: It is essential to show that the probable public health benefits outweigh the infringed general moral considerations – this condition is sometimes called proportionality. For instance, the policy may breach autonomy or privacy and have undesirable consequences. All of the positive features and benefits must be balanced against the negative features and effects.

Necessity: Not all effective and proportionate policies are necessary to realize the public health goal that is sought. The fact that a policy will infringe a general moral consideration provides a strong moral reason to seek an alternative strategy that is less morally troubling. This is the logic of a *prima facie* or presumptive general moral consideration. For instance, all other things being equal, a policy that provides incentives for persons with tuberculosis to complete their treatment until cured will have priority over a policy that forcibly detains such persons in order to ensure the completion of treatment. Proponents of the forcible strategy have the burden of moral proof. This means that the proponents must have a good faith belief, for which they can give supportable reasons, that a coercive approach is necessary. In many contexts, this condition does not require that proponents provide empirical evidence by actually trying the alternative measures and demonstrating their failure.[9]

Least infringement: Even when a proposed policy satisfies the first three justificatory conditions – that is, it is effective, proportionate, and essential in realizing the goal of public health – public health agents should seek to minimize the infringement of general moral considerations. For instance, when a policy infringes autonomy, public health agents should seek the least restrictive alternative; when it infringes privacy, they should seek the least intrusive alternative; and when it infringes confidentiality, they should disclose only the amount and kind of information needed, and only to those necessary, to realize the goal.[10] The justificatory condition of least infringement could plausibly be interpreted as a corollary of necessity – for instance, a proposed coercive measure must be necessary in degree as well as in kind.

Public justification: When public health agents believe that one of their actions, practices, or policies infringes one or more general moral considerations, they also have a responsibility, in our judgment, to explain and justify that infringement, whenever possible, to the relevant parties, including those affected by the infringement. In the context of what we called "political public," public health agents should offer public justification for policies in terms that fit the overall social contract in a liberal, pluralistic democracy. This transparency stems in part from the requirement to treat citizens as equals and with respect by offering moral reasons, which in principle they could find acceptable, for policies that infringe general moral considerations. Transparency is also essential to creating and maintaining public trust; and it is crucial to establishing accountability. (Below we elaborate a process-oriented approach to public accountability that goes beyond public justification to include, as an expression of justice and fairness, input from the relevant affected parties in the formulation of policy.)

Screening program example

An extended example may illustrate how these moral justificatory conditions function in public health ethics. Let us suppose that public health agents are considering whether to implement a screening program for HIV infection, tuberculosis, another infectious or contagious disease, or a genetic condition (see Figure 36.1 for some morally relevant features of screening programs).

The relevant justificatory conditions will require public health agents to consider whether any proposed program will be likely to realize the public health goal that is sought (effectiveness), whether its probable benefits will outweigh the infringed general moral considerations (proportionality), whether the policy is essential to realize the end (necessity), whether it involves the least infringement possible consistent with realizing the goal that is sought

		Degree of Voluntariness	
		Voluntary	Mandatory
Extent of Screening	Universal		
	Selective		

Figure 36.1 *Features of public health screening programs.*

(least infringement), and whether it can be publicly justified. These conditions will give priority to selective programs over universal ones if the selective programs will realize the goal (as we note below, questions may arise about universality within selected categories, such as pregnant women), and to voluntary programs over mandatory ones if the voluntary programs will realize the goal.[11]

Different screening programs may fail close scrutiny in light of one or more of these conditions. For instance, neither mandatory nor voluntary universal screening for HIV infection can meet these conditions in the society as a whole. Some voluntary and some mandatory selective screening programs for HIV infection can be justified, while others cannot. Mandatory screening of donated blood, organs, sperm, and ova is easily justified, and screening of individuals may also be justified in some settings where they can expose others to bodily fluids and potential victims cannot protect themselves. The question of whether and under what conditions screening of pregnant women for HIV infection should be instituted has been particularly controversial. Even before the advent of effective treatment for HIV infection and the identification of zidovudine (AZT) as effective in reducing the rate of perinatal transmission, there were calls for mandatory screening of pregnant women, especially in "high risk" communities. These calls were defeated by sound arguments that such policies entailed unjustifiable violations of autonomy, privacy, and justice.[12] In effect, the recommended policies failed to satisfy any of the justificatory conditions we have proposed here.

However, once it was established that zidovudine could interrupt maternal-fetal transmission of HIV, the weight of the argument shifted in the direction of instituting screening programs of some type. The focus of the debate became the tensions between the public health interests in utility and efficiency, which argued for mandatory, selective screening in high-risk communities, and considerations of liberty, privacy, and justice, which argued for voluntary, universal screening.[13]

In many situations, the most defensible public health policy for screening and testing *expresses* community rather than *imposes* it. Imposing community involves mandating or compelling testing through coercive measures. By contrast, expressing community involves taking steps to express solidarity with individuals, to protect their interests, and to gain their trust. Expressing community may include, for example, providing communal support, disclosing adequate information, protecting privacy and confidentiality, and encouraging certain choices. This approach seeks to make testing a reasonable, and perhaps moral, choice for individuals, especially by engendering public trust, rather than making it compulsory. Several diseases that might be subjected to screening for public health reasons involve stigma, and breaches of privacy and confidentiality may put individuals' employment and insurance at risk. Expressing community is often an appropriate strategy for public health, and, *ceteris paribus*, it has priority over imposing community through coercive policies.

Processes of public accountability

Our discussion of the fifth justificatory condition – public justification – focused on providing public reasons for policies that infringe general moral considerations; this condition is particularly applicable in the political context. While public accountability includes public justification, it is broader – it is prospective as well as retrospective. It involves soliciting input from the relevant publics (the numerical, political, and communal publics) in the process of formulating public health policies, practices, and actions, as well as justifying to the relevant publics what is being undertaken. This is especially, but not only, important when one of the other *prima facie* general moral considerations is infringed, as with coercive protective measures to prevent epidemics. At a minimum, public accountability involves transparency in openly seeking information from those affected and in honestly disclosing relevant information to the public; it is indispensable for engendering and sustaining public trust, as well as for expressing justice.[14]

Public accountability regarding health promotion or priority-setting for public health funding additionally might involve a more developed fair process. Noting that in a pluralistic society we are likely to find disagreement about which principles should govern issues such as priority-setting in health care, Norman Daniels calls for a fair process that includes the following elements: transparency and publicity about the reasons for a decision; appeals to rationales and evidence that fair-minded parties would agree are relevant; and procedures for appealing and revising decisions in light of challenges by various stakeholders. He explains why this process can facilitate social learning: "Since we may not be able to construct principles that yield fair decisions ahead of time, we need a process that allows us to develop those reasons over time as we face real cases."[15]

Public accountability also involves acknowledging the more complex relationship between public health and the public, one that addresses fundamental issues such as those involving characterization of risk and scientific uncertainty. Because public health depends for its success on the satisfaction of deeply personal health goals of individuals and groups in the population, concepts such as "health" and "risk" cannot be understood or acted upon on the basis of *a priori*, formal definitions or scientific analysis. Public accountability recognizes that the fundamental conceptualization of these terms is

a critical part of the basic formulation of public health goals and problems to be addressed. This means that the public, along with scientific experts, plays an important role in the *analysis* of public health issues, as well as in the development and assessment of appropriate *strategies* for addressing them.

Risk characterization provides a helpful example. A National Research Council report, *Understanding Risk: Informing Decisions in a Democratic Society*, concluded that risk characterization is not properly understood if defined only as a summary of scientific information; rather, it is the outcome of a complex analytic-deliberative process – "a decision-driven activity, directed toward informing choices and solving problems."[16] The report explains that scientific analysis, which uses rigorous, replicable methods, brings new information into the process, and that deliberation helps to frame analysis by posing new questions and new ways of formulating problems, with the result that risk characterization is the output of a recursive process, not a linear one, and is a decision-driven activity.

Assessment of the health risks of dioxin illustrates this process. While scientific analysis provides information about the dose-response relationship between dioxin exposure and possible human health effects, public health focuses on the placement of waste incinerators and community issues in which dioxin is only one of many hazardous chemicals involved and cancer only one of many outcomes of concern. The critical point is that good risk characterization results from a process that "not only gets the science right," but also "gets the right science."[17]

Public health accountability addresses the responsibility of public health agents to work with the public and scientific experts to identify, define, and understand at a fundamental level the threats to public health, and the risks and benefits of ways to address them. The appropriate level of public involvement in the analytic-deliberative process depends on the particular public health problem.

Public accountability requires an openness to public deliberation and imposes an obligation on decision-makers to provide honest information and justifications for their decisions. No ethical principle can eliminate the fact that individual interests must sometimes yield to collective needs. Public accountability, however, ensures that such trade-offs will be made openly, with an explicit acknowledgment that individuals' fundamental well-being and values are at stake and that reasons, grounded in ethics, will be provided to those affected by the decisions.[18] It provides a basis for public trust, even when policies infringe or appear to infringe some general moral considerations.

Public health interventions vs. paternalistic interventions

An important empirical, conceptual, and normative issue in public health ethics is the relationship between protecting and promoting the health of individuals and protecting and promoting public health. Although public health is directed to the health of populations, the indices of population health, of course, include an aggregation of the health of individuals. But suppose the primary reason for some restrictions on the liberties of individuals is to prevent harm to those whose actions are substantially voluntary and do not affect others adversely. The ethical question then is, when can paternalistic interventions (defined as interventions designed to protect or benefit individuals themselves against their express wishes) be ethically justified if they infringe general moral considerations such as respect for autonomy, including liberty of action?

Consider the chart in Figure 36.2: An individual's actions may be substantially voluntary

		Adverse Effects of Individuals' Actions	
		Self-regarding	Other-regarding
Voluntariness of Individuals' Actions	Voluntary	1	2
	Non-voluntary	3	4

Figure 36.2 *Types of individual action.*

(competent, adequately informed, and free of controlling influences) or non-voluntary (incompetent, inadequately informed, or subject to controlling influences). In addition, those actions may be self-regarding (the adverse effects of the actions fall primarily on the individual himself or herself) or other-regarding (the adverse effects of the actions fall primarily on others).

Paternalism in a morally interesting and problematic sense arises in the first quadrant (marked by the number "1" in Figure 36.2) — where the individual's actions are both voluntary and self-regarding. According to John Stuart Mill, whose *On Liberty* has inspired this chart, other-regarding conduct not only affects others adversely, but also affects them directly and without "their free, voluntary, and undeceived consent and participation."[19] If others, in the maturity of their faculties, consent to an agent's imposition of risk, then the agent's actions are not other-regarding in Mill's sense.

Whether an agent's other-regarding conduct is voluntary or non-voluntary, the society may justifiably intervene in various ways, including the use of coercion, to reduce or prevent the imposition of serious risk on others. Societal intervention in non-voluntary self-regarding conduct is considered weak (or soft) paternalism, if it is paternalistic at all, and it is easily justified. By contrast, societal interference in voluntary self-regarding conduct would be strong (or hard) paternalism. Coercive intervention in the name of strong paternalism would be insulting and disrespectful to individuals because it would override their voluntary actions for their own benefit, even though their actions do not harm others. Such interventions are thus very difficult to justify in a liberal, pluralistic democracy.

Because of this difficulty, proponents of public health sometimes contend that the first quadrant is really a small class of cases because individuals' risky actions are, in most cases, other-regarding or non-voluntary, or both. Thus, they insist, even if we assume that strong or hard paternalism cannot be ethically justified, the real question is whether most public health interventions in personal life plans and risk budgets are paternalistic at all, at least in the morally problematic sense.

To a great extent, the question is where we draw the boundaries of the self and its actions; that is, whether various influences on agents so determine their actions that they are not voluntary, and whether the adverse effects of those actions extend beyond the agents themselves. Such boundary drawing involves empirical, conceptual, and normative questions that demand attention in public health ethics. On the one hand, it is not sufficient to show that social-cultural factors influence an individual's actions; it is necessary to show that those influences render that individual's actions substantially non-voluntary and warrant societal interventions to protect him or her. Controversies about the strong influence of food marketing on diet and weight (and, as a result, on the risk of disease and death) illustrate the debates about this condition.

On the other hand, it is not sufficient to show that an individual's actions have some adverse effects on others; it is necessary to show that those adverse effects on others are significant enough to warrant overriding the individual's liberty. Controversies about whether the state should require motorcyclists to wear helmets illustrate the debates about this condition. These controversies also show how the inclusion of the financial costs to society and the emotional costs to, say, observers and rescue squads can appear to make virtually any intervention non-paternalistic. But even if these adverse financial and emotional effects on others are morally relevant as a matter of social utility, it would still be necessary to show that they are significant enough to justify the intervention.

Either kind of attempt to reduce the sphere of autonomous, self-regarding actions, in order to warrant interventions in the name of public health, or, more broadly, social utility, can sometimes be justified, but either attempt must be subjected to careful scrutiny. Sometimes both may represent rationalization and bad faith as public health agents seek to evade the stringent demands of the general moral consideration of respect for autonomy. Requiring consistency across an array of cases may provide a safeguard against rationalization and bad faith, particularly when motives for intervention may be mixed.

Much of this debate reflects different views about whether and when strong paternalistic interventions can be ethically justified. In view of the justificatory conditions identified earlier, relevant factors will include the nature of the intervention, the degree to which it infringes an individual's fundamental values, the magnitude of the risk to the individual apart from the intervention (either in terms of harm or lost benefit), and so forth. For example, even though the authors of this paper would disagree about some cases, we agree that strong paternalistic interventions that do not threaten individuals' core values and that will probably protect them against serious risks are more easily justifiable than strong paternalistic interventions that threaten individuals' core values and that will reduce only minor risks. Of course, evaluating actual and proposed policies that infringe general moral considerations becomes very complicated when both paternalistic and public health reasons exist for, and are intertwined in, those policies.

Social justice, human rights, and health

We have noted potential and actual conflicts between promoting the good of public health and other general moral considerations. But it is important not to exaggerate these conflicts. Indeed, the societal institutionalization of other general moral considerations in legal rights and social-cultural practices generally contributes to public health. Social injustices expressed in poverty, racism, and sexism have long been implicated in conditions of poor health. In recent years, some evidence suggests that societies that embody more egalitarian conceptions of socioeconomic justice have higher levels of health than ones that do not.[20] Public health activity has traditionally encompassed much more than medicine and health care. Indeed, historically much of the focus of public health has been on the poor and on the impact of squalor and sanitation on health. The focus today on the social determinants of health is in keeping with this tradition. The data about social determinants are impressive even though not wholly uncontroversial. At any rate, they are strong enough to warrant close attention to the ways conditions of social justice contribute to the public's health.

Apart from social justice, some in public health argue that embodying several other general moral considerations, especially as articulated in human rights, is consistent with and may even contribute to public health. For example, Jonathan Mann contended that public health officials now have two fundamental responsibilities – protecting and promoting public health and protecting and promoting human rights. Sometimes public health programs burden human rights, but

human rights violations "have adverse effects on physical, mental, and social well-being" and "promoting and protecting human rights is inextricably linked with promoting and protecting health."[21] Mann noted, and we concur, that, ultimately, "ethics and human rights derive from a set of quite similar, if not identical, core values," several of which we believe are captured in our loose set of general moral considerations.[22] Often, as we have suggested, the most effective ways to protect public health respect general moral considerations rather than violate them, employ voluntary measures rather than coercive ones, protect privacy and confidentiality, and, more generally, express rather than impose community. Recognizing that promoting health and respecting other general moral considerations or human rights may be mutually supportive can enable us to create policies that avoid or at least reduce conflicts.

While more often than not public health and human rights – or general moral considerations not expressed in human rights – do not conflict and may even be synergistic, conflicts do sometimes arise and require resolution.[23] Sometimes, in particular cases, a society cannot simultaneously realize its commitments to public health and to certain other general moral considerations, such as liberty, privacy, and confidentiality. We have tried to provide elements of a framework for thinking through and resolving such conflicts. This process needs to be transparent in order to engender and sustain public trust.

Acknowledgments

This work was supported by a grant from The Greenwall Foundation. Other project participants were John D. Arras and Paul A. Lombardo, both of the University of Virginia, and Donna T. Chen of the National Institute of Mental Health and Department of Bioethics, National Institutes of Health.

372 James F. Childress et al.

Notes

1 Our definition builds on the definition of health systems offered by the World Health Organization: Health systems include "all the activities whose primary purpose is to promote, restore, or maintain health." See *World Health Report 2000 Health Systems: Improving Performance* (Geneva: World Health Organization, 2000): at 5.

2 Committee for the Study of the Future of Public Health, Division of Health Care Services, Institute of Medicine, *The Future of Public Health* (Washington, D.C.: National Academy Press, 1988): at 1.

3 We recognize that there are different views about the ultimate moral justification for the social institution of public health. For example, some communitarians appear to support public health as an instrumental goal to achieve community. Others may take the view that the state has a duty to ensure the public's health as a matter of social justice. Although these different interpretations and others are very important for some purposes, they do not seriously affect the conception of public health ethics that we are developing, as long as public health agents identify and inform others of their various goals.

4 L.O. Gostin, *Public Health Law: Power, Duty, Restraint* (Berkeley: University of California Press; New York: The Milbank Memorial Fund, 2000): at 20.

5 T. Nagel, "Moral Epistemology," in R.E. Bulger, E.M. Bobby, and H.V Fineberg, eds., Committee on the Social and Ethical Impacts of Developments in Biomedicine, Division of Health Sciences Policy, Institute of Medicine, *Society's Choices: Social and Ethical Decision Making in Biomedicine* (Washington, D.C.: National Academy Press, 1995): 201–14.

6 For some other approaches, see P. Nieburg, R. Gaare-Bernheim, and R. Bonnie, "Ethics and the Practice of Public Health," in R.A. Goodman et al., eds., *Law in Public Health Practice* (New York: Oxford University Press, in press), and N.E. Kass, "An Ethics Framework for Public Health," *American Journal of Public Health*, 91 (2001): 1776–82.

7 We do not explore here the overlaps among public health ethics, medical ethics, research ethics, and public policy ethics, although some areas of overlap and difference will be evident throughout the discussion. Further work is needed to address some public health activities that fall within overlapping areas – for instance, surveillance, outbreak investigations, and community-based interventions may sometimes raise issues in the ethics of research involving human subjects.

8 Recognizing universalizability by attending to past precedents and possible future precedents does not preclude a variety of experiments, for instance, to determine the best ways to protect the public's health. Thus, it is not inappropriate for different states, in our federalist system, to try different approaches, as long as each of them is morally acceptable.

9 This justificatory condition is probably the most controversial. Some of the authors of this paper believe that the language of "necessity" is too strong. Whatever language is used, the point is to avoid a purely utilitarian strategy that accepts only the first two conditions of effectiveness and proportionality and to ensure that the non-utilitarian general moral considerations set some *prima facie* limits and constraints and establish moral priorities, *ceteris paribus*.

10 For another version of these justificatory conditions, see T.L. Beauchamp and J.F. Childress, *Principles of Biomedical Ethics*, 5th ed. (New York: Oxford University Press, 2001): at 19–21. We observe that some of these justificatory conditions are quite similar to the justificatory conditions that must be met in U.S. constitutional law when there is strict scrutiny because, for instance, a fundamental liberty is at stake. In such cases, the government must demonstrate that it has a "compelling interest," that its methods are strictly necessary to achieve its objectives, and that it has adopted the "least restrictive alternative." See Gostin, *supra* note 4, at 80–81.

11 Of course, this chart is oversimplified, particularly in identifying only voluntary and mandatory options. For a fuller discussion, see R. Faden, M. Powers, and N. Kass, "Warrants for Screening Programs: Public Health, Legal and Ethical Frameworks," in R. Faden, G. Geller, and M. Powers, eds., *AIDS, Women and the Next Generation* (New York: Oxford University Press, 1991): 3–26.

Public health ethics 373

12 Working Group on HIV Testing of Pregnant Women and Newborns, "HIV Infection, Pregnant Women, and Newborns," *Journal of the American Medical Association*, 264, no. 18 (1990): 2416–20.

13 See Faden, Geller, and Powers, *supra* note 11; Gostin, *supra* note 4, at 199–201.

14 In rare cases, it may be ethically justifiable to limit the disclosure of some information for a period of time (for example, when there are serious concerns about national security; about the interpretation, certainty, or reliability of public health data; or about the potential negative effects of disclosing the information, such as with suicide clusters).

15 N. Daniels, "Accountability for Reasonableness," *British Medical Journal*, 321 (2000): 1300–01, at 1301.

16 P.C. Stern and H.V Fineberg, eds., Committee on Risk Characterization, Commission on Behavioral and Social Sciences and Education, National Research Council, *Understanding Risk: Informing Decisions in a Democratic Society* (Washington, D.C.: National Academy Press, 1996): at 155.

17 *Id.* at 16–17, 156.

18 See, for example, N. Daniels and J. Sabin, "Limits to Health Care: Fair Procedures, Democratic Deliberation, and the Legitimacy Problem for Insurers," *Philosophy and Public Affairs*, 26 (Fall 1997): 303–50, at 350.

19 J.S. Mill, *On Liberty*, ed. G. Himmelfarb (Harmondsworth, England: Penguin Books, 1976): at 71. For this chart, see J.F. Childress, *Who Should Decide? Paternalism in Health Care* (New York: Oxford University Press, 1982): at 193.

20 See, for example, the discussion in I. Kawachi, B.P. Kennedy, and R.G. Wilkinson, eds., *Income Inequality and Health*, vol. 1 of *The Society and Population Health Reader* (New York: The New Press, 2000).

21 J.M. Mann, "Medicine and Public Health, Ethics and Human Rights," *The Hastings Center Report*, 27 (May–June 1997): 6–13, at 11–12. Contrast Gostin, *supra* note 4, at 21. For a fuller analysis and assessment of Mann's work, see L.O. Gostin, "Public Health, Ethics, and Human Rights: A Tribute to the Late Jonathan Mann," S.P Marks, "Jonathan Mann's Legacy to the 21st Century: The Human Rights Imperative for Public Health," and L.O. Gostin, "A Vision of Health and Human Rights for the 21st Century: A Continuing Discussion with Stephen P. Marks," *Journal of Law, Medicine, and Ethics*, 29, no. 2 (2001): 121–40.

22 Mann, *supra* note 21, at 10. Mann thought that the language of ethics could guide individual behavior, while the language of human rights could best guide societal-level analysis and response. See Mann, *supra* note 21, at 8; Marks, *supra* note 21, at 131–38. We disagree with this separation and instead note the overlap of ethics and human rights, but we endorse the essence of Mann's position on human rights.

23 See Gostin, *supra* note 4, at 21.

Chapter 37

L.O. Gostin

PUBLIC HEALTH LAW IN AN AGE OF TERRORISM: RETHINKING INDIVIDUAL RIGHTS AND COMMON GOODS

[. . .]

Public and scholarly discourse in the late twentieth century became highly oriented toward "rights." The political community stressed the importance of individual freedoms rather than the health, security, and well-being of the community. The salience of individualism could be seen on both sides of the political spectrum. The ideological left favored a set of personal interests, principally autonomy, privacy, and liberty. This meant that individuals should be free to make choices, restrict the flow of health information, and have unfettered movement, without regard to the needs and desires of the wider community. The ideological right favored a set of proprietary interests, principally the freedom to contract, conduct business, use and develop property, and pursue a profession. This meant that entrepreneurs should be permitted to engage in free enterprise without the fetters of, for example, occupational health and safety regulations, inspections and products liability, zoning and nuisance abatements, and licenses.

In this civil and property rights society, the tone has been distinctly antigovernment. The State has been perceived as inefficient, bureaucratic, and burdensome. Citizens have opposed taxation and broad health and welfare spending as well as oppressive regulation. From a funding perspective, this has meant that health dollars have been allocated primarily to advanced biotechnology and health care, which serve the needs of individual patients, particularly those who can afford private health insurance. Funding for traditional prevention and population-based services represents only a small fraction of health spending, estimated at around 1 percent at the state level and less than 5 percent at the federal level.[1]

As a result of chronic underspending, the public health infrastructure is badly deteriorated.[2] Public health agencies lack the capacity to conduct essential public health services at a level of performance that matches the constantly evolving threats to the health of the public. Critical components of that infrastructure include a well-trained workforce, electronic information and communications systems, rapid disease surveillance and reporting, laboratory capacity, and emergency response capability.[3]

The public health law infrastructure is equally deficient. The law establishes the mission, functions, and powers of public health agencies. Yet public health laws are highly antiquated, after many decades of neglect. Very little consideration has been given to modernizing these laws to reflect advances in public health practice and constitutional law. Reform of public health law is essential to ensure that public health agencies have clear missions and functions, stable sources of financing, adequate powers to avert or manage health threats, and restraints on powers to maintain respect for personal rights and liberties.

The balance between individual interests and common goods needs to be recalibrated in an age of terrorism. The attacks on the World Trade Center and Pentagon on 11 September 2001 and the subsequent dispersal of anthrax spores through the U.S. postal system reawakened the public to the importance of public health, safety, and security.[4] The president's 2003 budget reflects changing priorities, with an influx of funding to prevent and respond to bioterrorism.[5] However, even in this budget, disproportionate funding is devoted to biotechnology rather than to basic prevention and population-based services.[6]

This paper explores the appropriate balance between individual interests and common goods. The current focus on individualism should be seen not as fixed and authoritative, but rather as transient and culturally derived. There is, of course, an alternative philosophical tradition that sees individuals primarily as members of communities. This communitarian tradition views individuals as part of social and political networks, with each individual reliant on the others for health and security.[7] Individuals, according to this tradition, gain value from being a part of a well-regulated society that seeks to prevent common risks.

In legal terms, this communitarian tradition is expressed in the "police power" to protect the health, safety, and security of the population. In fact, the linguistic and historical origins of the concept of "police" demonstrate a close association between government and civilization: *politia* (the state), *polis* (city), and *politeia* (citizenship).[8] The word had a secondary usage as well: cleansing or keeping clean. This use resonates with early-twentieth-century public health connotations of hygiene and sanitation.

First, this paper explains modern efforts at public health law reform. Even before September 11, the Robert Wood Johnson Foundation's (RWJF's) Turning Point initiative supported comprehensive reform of antiquated public health laws – the Public Health Statute Modernization Collaborative. After the anthrax outbreak, the Centers for Disease Control and Prevention (CDC) asked the Center for Law and the Public's Health (CLPH) at Georgetown and Johns Hopkins Universities to draft the Model State Emergency Health Powers Act (MSEHPA). Thirty-five states and the District of Columbia have introduced legislative bills or resolutions based in whole or part on this model act. Nineteen states and D.C. have enacted MSEHPA or a version of it.[9]

Next, the paper shows why existing public health laws provide a weak foundation for public health practice. They are obsolete, inconsistent, and inadequate from a public health and civil liberties perspective. State legislation does not facilitate, and may even impede, the critical variables for public health preparedness: planning, coordination, surveillance, management of property, and protection of persons. Finally, the paper offers a systematic defense of MSEHPA. The model act has galvanized the public debate around the appropriate balance between public goods and individual rights. Although it has had success in many state legislatures, it has been criticized by both ends of the political spectrum. Many of the critiques offer a rationale for protecting property and personal rights against state incursions. This defense shows how MSEHPA creates strong public health powers while safeguarding individual freedoms – adopting clearer standards and more rigorous procedures than existing statutes do.

Two national projects for public health law reform

The Institute of Medicine (IOM), in its foundational 1988 report, *The Future of Public Health*, acknowledged that law was essential to public health but cast serious doubt on the soundness of public health's legal basis. Concluding that "this nation has lost sight of its public health goals and has allowed the system of public health activities to fall into disarray," the IOM recommended reform of an obsolete and inadequate body of enabling laws and regulations.[10] In its 2002 report, *The Future of the Public's Health in the Twenty-first Century*, the IOM notes that little progress has been made in implementing its 1988 proposal. The committee recommends that "public health law be reformed so that it conforms to modern, scientific and legal standards, is more consistent within and among states, and is more uniform in its approach to different health threats."[11] The U.S. Department of Health and Human Services (HHS) in *Healthy People 2010* similarly argued that strong laws are a vital component of the public health infrastructure and recommended that states reform their outdated statutes.[12]

Public Health Statute Modernization Collaborative

In response to a sustained critique of the crumbling public health infrastructure, the RWJF, in partnership with the W.K. Kellogg Foundation, initiated the Turning Point project in 1996: "Collaborating for a New Century in Public Health." Turning Point launched five National Excellence Collaboratives in 2000, including the Public Health Statute Modernization Collaborative. The collaborative's mission was "to transform and strengthen the legal framework for the public health system through a collaborative process to develop a model public health law."

The Public Health Statute Modernization Collaborative is led by a consortium of states, in partnership with federal agencies and national organizations. The collaborative contracted with the author to draft a model public health act under the guidance of a national expert advisory committee. It has published a comprehensive assessment of state public health laws, demonstrating the inadequacies of existing law to support modern public health functions.[13] The objective is to ensure that state public health law is consistent with modern constitutional principles and reflects current scientific and ethical values underlying public health practice. This model public health act will focus on the organization, delivery, and funding of essential public health services and functions. It is scheduled for completion by October 2003, and current drafts are available on the Internet.[14]

Model State Emergency Health Powers Act

The law-reform process took on new urgency after the terrorist attacks of late 2001. In response, the CLPH drafted MSEHPA at the request of the CDC. The act was written in collaboration with members of national organizations representing governors, legislators, public health commissions, and attorneys general.[15] There was also an extensive consultative process involving the major stakeholders such as businesses, public health and civil liberties organizations, scholars, and practitioners. MSEHPA, therefore, was written following a broad dialogue regarding the purpose of emergency public health law, its proper reach, and the protection of civil liberties and private property.

The act is explained in greater detail elsewhere, but the following brief description provides the background for a more sustained defense of the act's approach.[16] MSEHPA is structured to reflect five basic public health functions to be facilitated by law: preparedness, surveillance, management of property, protection of persons, and public information and communication. The preparedness and surveillance functions take effect immediately upon the act's passage. However, the compulsory powers over property and persons take effect only once a state's governor has declared a "public health emergency." A public health emergency is defined as the occurrence of imminent threat of an illness or health condition caused by bioterrorism or a novel or previously controlled or eradicated infectious agent or biological toxin. The health threat must pose a high probability of a large number of deaths or serious disabilities in the population.

The act facilitates systematic planning for a public health emergency. The state Public Health Emergency Plan must include coordination of services; procurement of vaccines and pharmaceuticals; housing, feeding, and caring for affected populations (with appropriate regard for their physical and cultural/social needs); and the proper vaccination and treatment of individuals.

The act provides authority for surveillance of health threats and continuing power to follow a developing public health emergency. For example, it requires prompt reporting by health care providers, pharmacists, veterinarians, and laboratories. It also provides for the exchange of relevant data among lead agencies such as public health, emergency management, and public safety.

MSEHPA provides comprehensive powers to manage property and protect persons, to safeguard the public's health and security. Public health authorities may close, decontaminate, or procure facilities and materials to respond to a public health emergency; safely dispose of infectious waste; and obtain and deploy health care supplies. Similarly, the model act permits public health authorities to physically examine or test individuals as necessary to diagnose or to treat illness; vaccinate or treat individuals to prevent or ameliorate an infectious disease; and isolate or quarantine individuals to prevent or limit the transmission of a contagious disease. The public health authority also may waive licensing requirements for health care professionals and direct them to assist in vaccination, testing, examination, and treatment of patients. Finally, MSEHPA provides for a set of postdeclaration powers and duties to ensure appropriate public information and communication. The public health authority must provide information to the public regarding the emergency, including protective measures to be taken and information regarding access to mental health support.

In summary, MSEHPA requires the development of a comprehensive plan to provide a coordinated, appropriate response in the event of a public health emergency. It facilitates the early detection of a health emergency by authorizing the reporting as well as collection and exchange of data. During a public health emergency, state and local officials are authorized to use and appropriate property as necessary for the care, treatment, and housing of patients and to destroy contaminated facilities or materials. They are also empowered to provide care, testing, treatment, and vaccination to persons who are ill or who have been exposed to a contagious disease and to separate affected individuals from the population at large to interrupt disease transmission. At the same time, the act recognizes that a state's ability to respond to a public health emergency must respect the dignity and rights of persons. Guided by principles of justice, state and local governments have a duty to act with fairness and tolerance toward individuals and groups.

Inadequacy of existing public health legislation

Before beginning a detailed defense of the model act, it is important to show why current law provides a weak foundation for the effective identification and control of serious health threats. Critics attack MSEHPA as if it were proposed in a regulatory vacuum. Yet public health is practiced under a voluminous set of laws and regulations. The issue is not whether the act provides an ideal solution to perennially complex problems. Rather, the issue is whether the act does a much better job than existing legislation does. As the following discussion demonstrates, existing state law is obsolete, fragmented, and inadequate. Outdated state laws do not support, and even thwart, effective public health surveillance and interventions.[17]

Public health legislation is so old that it tells the story of communicable diseases through time, with new layers of regulation with each page in history – from plague and smallpox to tuberculosis and polio, and now HIV/AIDS and West Nile virus. Many laws have not been systematically updated since the early to mid-twentieth century.[18] State laws predate modern public health science and practice. Research demonstrates that existing public health law does not conform to modern ideas relating to the mission, functions, and services of public health agencies.[19] Existing state laws also predate advances in constitutional law and civil liberties (such as privacy and antidiscrimination). For example, many public health laws do not provide rigorous procedural due process protections. Existing laws are so obtuse that few public health practitioners, or even legal counsel, fully understand them. Discussion of law reform, therefore, must take account of the obsolescence and complexity of current legislation.

Public health laws are inconsistent both within states and among them. Within states, different rules apply depending on the particular disease in question. Public health officers may legally exercise compulsory powers (screening, vaccination, directly observed therapy, or isolation) for one health threat but not another, and they may have a duty to assure privacy, nondiscrimination, and procedural due process in some cases but not in others. Inconsistencies among the states and territories lead to profound variation in the structure, substance, and procedures for detecting, controlling, and preventing disease. A certain level of consistency is important in public health because infectious diseases usually occur regionally or nationally, requiring a coordinated approach to surveillance and control.

Many current laws fail to provide necessary authority for each of the key elements for public health preparedness: planning, coordination, surveillance, management of property, and protection of persons. States have not devised clear methods of planning, communication, and coordination among the various levels of government (federal, tribal, state, and local), responsible agencies (public health, law enforcement, and emergency management), and the private sector (food, transportation, and health care). Indeed, because of privacy concerns, many states actually proscribe the exchange of vital information.[20]

Current statutes also do not facilitate surveillance and may even prevent monitoring. For example, many states do not require timely reporting for Category A agents of bioterrorism.[21] At the same time, states do not require public health agencies to monitor data held by hospitals, managed care organizations, and pharmacies and may even prohibit them from doing so.[22]

Extant laws usually do provide powers over property and persons, but their scope is limited. Some statutes permit the exercise of certain powers (such as quarantine) but not others (such as directly observed therapy). Other statutes permit the exercise of powers in relation to certain diseases (such as smallpox and tuberculosis) but not others (such as hemorrhagic fevers). There are numerous circumstances that might require management of property in a public health emergency: shortages of vaccines, medicines, hospital beds, or facilities for disposal of corpses. It may even be necessary to close facilities or destroy property that is contaminated or dangerous. There similarly may be a need to exercise powers over individuals to avert a serious threat to the public's health. Vaccination, testing, physical examination, treatment, isolation, and quarantine each may help to contain the spread of communicable diseases.

In summary, existing public health laws introduce two kinds of error that require correction. On the one hand, many statutes fail to provide adequate powers to deal with the full range of health threats. On the other hand, when they do authorize coercion, statutes rarely provide clear standards and fair procedures for decision making.

A Defense of the Model Act

There have been several specific objections to MSEHPA: *federalism* – federal, not state, law is implicated in a health crisis; *emergency declarations* – the scope of a public health emergency is overly broad; *abuse of power* – governors and public health officials will act without sufficient justification; *personal libertarianism* – compulsory powers over nonadherent individuals are rarely, or never, necessary; *economic libertarianism* – regulation of businesses is counterproductive; and *safeguards of property and persons* – MSEHPA fails to provide strong protection of individual and economic freedoms.

Federalism

Critics argue that acts of terrorism are inherently federal matters, so there is no need for expansion of state public health powers. It is certainly true that federal authority is extraordinarily important in responding to catastrophic public health events: Bioterrorism may trigger national security concerns, require investigation of federal offenses, and affect geographic regions or even the entire country. Consequently, the federal government often takes the lead, in responding to a public health emergency, as it did in the anthrax outbreaks. Indeed, the federal government, under the national defense or commerce powers of the Constitution, is entitled to act in the context of multistate threats to health and security.

The assertion of federal jurisdiction, of course, does not obviate the need for adequate state and local public health power. States and localities have been the primary bulwark of public health in America. From a historical perspective, local and state public health agencies predated federal agencies. Local boards of health were in operation in the late eighteenth century, and state agencies emerged after the Civil War. Federal health agencies, however, did not develop a major presence until Franklin Roosevelt's New Deal. State and local agencies have played a crucial role in infectious disease control from colonial and revolutionary times, through the industrial revolution, to the modern times.

From a constitutional perspective, states have "plenary" authority to protect the public's health under their reserved powers in the Tenth Amendment. The Supreme Court has made it clear that states have a deep reservoir of public health powers, conceiving of state, police powers as "an immense mass of legislation . . . Inspection laws, quarantine laws, and health laws of every description . . . are components of this mass."[23] The Supreme Court, moreover, regards federal police powers as constitutionally limited and has curtailed the expansion of national public health authority.[24]

From an economic and practical perspective, most public health activities take place at the state and local levels: surveillance, communicable disease control, and food and water safety. States and localities probably would be the first to detect and respond to a health emergency and would have a key role throughout. This requires states to have effective, modern statutory powers that enable them to work alongside federal agencies.

Declaration of a public health emergency

Critics express concern that the model act could be triggered too easily, creating a threat to civil liberties. Community-based organizations objected to the idea that a governor might declare a public health emergency for an endemic disease such as HIV/AIDS or influenza. Although this may have been a problem with the act's initial version, the current version expressly states that a governor may not declare a public health emergency for an endemic disease.

Legal scholars express concerns that a governor could declare an emergency for a theoretical or low-level risk. However, the drafters set demanding conditions for a governor's declaration, clearly specifying the level of risk. A public health emergency may be declared only in the event of bioterrorism or a naturally occurring epidemic that poses a high probability of a large number of deaths or serious disabilities. Indeed, the drafters rejected arguments from high-level federal and state officials to set a lower threshold for triggering a health emergency.

Finally, commentators suggest that governors retain too much discretion to declare a public health emergency. However, the act specifies clear criteria for triggering gubernatorial powers and uses language that fetters the exercise of discretion. The act also allows the legislature and judiciary to intervene if the governor has acted outside the scope of his or her authority. Taken as a whole, the drafters carefully limited the circumstances under which the act's more robust powers can be invoked.

Governmental abuse of power

Critics argue that governors and public health authorities would abuse their authority and exercise powers without justification. This kind of generalized argument could be used to refute the exercise of compulsory power in any realm, because executive branch officials may overreach. However, such general objections have never been a reason to deny government the power to avert threats to health, safety, and security. The answer to such general objections is to introduce into the law careful safeguards to prevent officials from acting outside the scope of their authority. The model act builds in effective protection against governmental abuse. It adopts the doctrine of separation of powers, so that no branch wields unchecked authority. These checks and balances offer a classic means of preventing abuse.

MSEHPA creates several hedges against abuse: (1) The governor may declare an emergency only under strict criteria and with careful consultation with public health experts and the community; (2) the legislature, by majority vote, can override the governor's declaration at any time; and (3) the judiciary can terminate the exercise of power if the governor violates the act's standards or procedures or acts unconstitutionally. No law can guarantee that the powers it confers will not be abused. But MSEHPA counterbalances executive power by providing a strong role for the legislature and judiciary. Therefore, it sets clear criteria for the exercise of power, requires a consultative process, and imposes checks and balances. There is little more that any law could do to prevent abuse of power.

Personal libertarianism

Critics imply that the model act should not confer compulsory power at all. In particular, they object to compulsory powers to vaccinate, test, medically treat, isolate, and quarantine. Commentators reason that services are more important than power; that individuals will comply voluntarily with public health advice; and that trade-offs between civil rights and public health are not required and even are counterproductive. Certainly the HIV/AIDS epidemic has demonstrated that public health and civil liberties can be mutually reinforcing – respect for individual freedoms can promote the public's health. Nevertheless, the arguments that law should not confer compulsory power are misplaced.

First, although the provision of services may be more important than the exercise of power, the state undoubtedly needs a certain amount of authority to protect the public's health. Government must have the power to prevent individuals from endangering others. It is only common sense, for example, that a person who has been exposed to an infectious disease should be required to undergo testing or medical examination and, if infectious, to be vaccinated, treated, or isolated.

Second, although most people can be expected to comply willingly with public health measures because it is in their own interests or desirable for the common welfare, not everyone will comply. Individuals may resist loss of autonomy, privacy, or liberty even if their behavior threatens others. Provided that public health powers are hedged with safeguards, individuals should be required to yield some of their interests to protect the health and security of the community.

Finally, although public health and civil liberties may be mutually enhancing in many instances, they sometimes come into conflict. When government acts to preserve the public's health, it can interfere with property rights (for example, freedom of contract, to pursue a profession, or to conduct a business) or personal rights (for example, autonomy, privacy, and liberty). The history of public health is littered with illustrations of trade-offs between public health and civil liberties.[25] It may be fashionable to argue that there is no tension, but public health officials need to make hard choices, particularly in public health emergencies.

Individuals whose movements pose a serious risk of harm to their communities do not have a "right" to be free of interference necessary to control the threat. There simply is no basis for this argument in constitutional law, and perhaps little more in political philosophy. Even the most liberal scholars accept the harm principle – that government should retain power to prevent individuals from endangering others.[26]

The Supreme Court has been equally clear about the limits of freedom in a constitutional democracy. The rights of liberty and due process are fundamental but not absolute. Justice Harlan in the foundational Supreme Court case of *Jacobson v. Massachusetts* (1905) wrote: "There are manifold restraints to which every person is necessarily subject for the common good. On any other basis organized society could not exist with safety to its members."[27] Critics argue, without support from any judicial authority, that the Supreme Court's landmark decision in *Jacobson*, reiterated by the Court over the past century, is no longer apposite. There is, according to this line of argument, a constitutional right to refuse interventions even if the individual poses a public risk. Yet the courts have consistently upheld compulsory measures to avert a risk, including the power to compulsorily test, report, vaccinate, treat, and isolate, provided there are clear criteria and procedures.[28]

Economic libertarianism

Civil libertarians have not been the only group to criticize MSEHPA. Businesses, as well as law and economic scholars, have complained that it interferes with free enterprise. Most economic stakeholders, including the food, transportation, pharmaceutical, and health care industries, have lobbied CLPH faculty and legislators. These groups argue that they may have to share data with government, abate nuisances, destroy property, and provide goods and services without their express agreement.[29]

Generally speaking, the model act provides several kinds of powers to regulate businesses: destruction of dangerous or contaminated property, nuisance abatements, and confiscation of property for public purposes. All of these powers have been exercised historically and comply with constitutional and ethical norms. If businesses have property that poses a public threat, government has always had the power to destroy that property. For example, if a rug were contaminated with anthrax or smallpox, government should have the power to order its destruction. Similarly, if businesses are engaged in an activity that poses a health threat, government has always had the power to abate the nuisance. Businesses must comply with all manner of health and safety regulations that interfere with economic freedoms. Those who believe in the undeterred entrepreneur may not agree with health regulations, but the regulations are necessary to ensure that business activities do not endanger the public. Finally, government has always had the power to confiscate private property for the public good. In the event of bioterrorism, for example, it may be necessary for government to have adequate supplies of vaccines or pharmaceuticals. Similarly, government may need to use health care facilities for medical treatment or quarantine of persons exposed to infection.

Businesses argue that government should not have broad powers to control enterprise and property. If these powers have to be exercised, businesses want to ensure that they are compensated according to market values. The model act follows a classical approach to the issue of property rights. Compensation is provided if there is a "taking" – that is, if the government confiscates private property for public purposes (such as the use of a private infirmary to treat or isolate patients). No compensation would be provided for a "nuisance abatement" – that is, if the government destroys property or closes an establishment that poses a serious health threat. This comports with the extant constitutional "takings" jurisprudence of the Supreme Court.[30] If the government were forced to compensate for all nuisance abatements, it would greatly chill public health regulation.

In American history and constitutional law, private property has always been held subject to the restriction that it not be used in a way that posed a health hazard. As Lemuel Shaw of the Massachusetts Supreme Judicial Court observed as early as 1851: "We think it settled principle, growing out of the nature of well ordered civil society, that every holder of property . . . holds it under the implied liability that it shall not be injurious to the right of the community."[31]

Safeguards of persons and property

The real basis for debate over public health legislation should not be that powers are given, because it is clear that power is sometimes necessary. The better question is whether the powers are hedged with appropriate safeguards of personal and economic liberty. The core of the debate over MSEHPA ought to be whether it appropriately protects freedoms by providing clear and demanding criteria for the exercise of power and fair procedures for decision making. It is in this context that the attack on MSEHPA is particularly exasperating, because critics rarely suggest that the act fails to provide crisp standards and procedural due process. Nor do they compare the safeguards in the model act to those in existing public health legislation.

It is important to note that compulsory powers over individuals (testing, physical examination, treatment, and isolation) and businesses (nuisance abatements and seizure or destruction of property) already exist in state public health law. These powers have been exercised since the founding of the Republic. MSEHPA, therefore, does not contain new, radical powers. Most tellingly, the model contains much better safeguards of individual and economic liberty than appear in communicable disease statutes enacted in the early to mid-twentieth century.

Unlike older statutes, MSEHPA provides clear and objective criteria for the exercise of powers, rigorous procedural due process, respect for religious and cultural differences, and a new set of entitlements for humane treatment. First, the criteria for the exercise of compulsory powers are based on the modern "significant risk" standard enunciated in constitutional law and disability discrimination law. The act also requires public health officials to adopt the "least restrictive alternative." Second, the procedures for intervention are rigorous, following the most stringent requirements set by the Supreme Court, including the right to counsel, presentation and cross-examination of evidence, and reasons for decisions. Third, the act shows tolerance of groups through its requirements to respect cultural and religious differences whenever consistent with the public's health. Finally, the act provides a whole new set of rights to care and treatment of persons subject to isolation or quarantine. These include the right to treatment, clothing, food, communication, and humane conditions.

In summary, MSEHPA provides a modern framework for effective identification of and response to emerging health threats, while demonstrating respect for individuals and tolerance of groups. Indeed, the CLPH agreed to draft the law only because a much more draconian approach might have been taken by the federal government and the states acting on their own and responding to public fears and misapprehensions.

Rethinking the public good

Values in the United States at the turn of the twenty-first century could be characterized fairly as individualistic. There was a distinct orientation toward personal and proprietary freedoms and against a substantial government presence in social and economic life. The homeland terrorist attacks in 2001 reawakened the political community to the importance of public health. Historians will look back and ask whether 11 September 2001 was a fleeting scare with temporary solutions or whether it was a transforming event.

There are good reasons for believing that resource allocations, ethical values, and law should transform to reflect the critical importance of the health, security, and well-being of the populace. It is not that individual freedoms are unimportant. To the contrary, personal liberty allows people the right of self-determination, to make judgments about how to live their lives and pursue their dreams. Without a certain level of health, safety, and security, however, people cannot have well-being, nor can they meaningfully exercise their autonomy or participate in social and political life.

My purpose is not to assert which are the more fundamental interests: personal liberty or health and security. Rather, my purpose is to illustrate that both sets of interests are important to human flourishing. The Model State Emergency

Health Powers Act was designed to defend personal as well as collective interests. But in a country so tied to rights rhetoric on both sides of the political spectrum, any proposal that has the appearance of strengthening governmental authority was bound to travel in tumultuous political waters.

Notes

1   K.W Eilbert et al., *Measuring Expenditures for Essential Public Health Services* (Washington: Public Health Foundation, 1996). For state-level spending, see Centers for Disease Control and Prevention, "Effectiveness in Disease and Injury Prevention Estimated National Spending on Prevention – United States, 1988," *Morbidity and Mortality Weekly Report* 41, no. 29 (1992): 529–36. For federal-level spending, see J.I. Boufford and P.R. Lee, *Health Policies for the Twenty-first Century: Challenges and Recommendations for the U.S. Department of Health and Human Services* (New York: Milbank Memorial Fund, 2001).

2   Institute of Medicine, *The Future of Public Health* (Washington: National Academy Press, 1988).

3   CDC, *Public Health's Infrastructure: A Status Report* (Atlanta: CDC, 2001).

4   L.M. Bush et al., "Index Case of Fatal Inhalational Anthrax Due to Bioterrorism in the United States," *New England Journal of Medicine* (29 November 2001): 1607–11.

5   Office of Management and Budget, Executive Office of the President of the United States, *Budget of the U.S. Government, Fiscal Year 2003*, 4 February 2002, www.whitehouse.gov/omb/budget/fy2003/pdf/budget.pdf (22 August 2002).

6   M.H. Cooper, "Weapons of Mass Destruction," *Congressional Quarterly* (8 March 2002): 195–215.

7   D.E. Beauchamp and B. Steinbock, eds., *New Ethics for the Public's Health* (New York: Oxford University Press, 1999).

8   *Webster's Third New International Dictionary, Unabridged* (1986).

9   The nineteen states are Arizona, Delaware, Florida, Georgia, Hawaii, Maine, Maryland, Minnesota, Missouri, New Hampshire, New Mexico, Oklahoma, South Carolina, South Dakota, Tennessee, Utah, Vermont, Virginia, and Wisconsin.

10   IOM, *The Future of Public Health.*

11   IOM, *The Future of the Public's Health in the Twenty-first Century* (Washington: National Academy Press, November 2002).

12   Department of Health and Human Services, *Healthy People 2010* (Washington: DHHS, 2000).

13   L.O. Gostin and J.G. Hodge Jr., *State Public Health Law Assessment Report*, April 2002, www.turningpoint-program.org/Pages/phsc%20statute%20assmt.pdf (22 August 2002).

14   See "Model State Public Health Act, Draft Document as of May 31, 2002," www.turningpointprogram.org/Pages/phsc_MSPH%20Act3.pdf (22 August 2002); and Turning Point, "Public Health Statute Modernization National Collaborative," www.hss.state.ak.us/dph/deu/turningpoint/nav.htm (22 August 2002).

15   J. Gillis, "States Weighing Laws to Fight Bioterrorism," *Washington Post*, 19 November 2001.

16   See L.O. Gostin et al., "The Model State Emergency Health Powers Act: Planning for and Response to Bioterrorism and Naturally Occurring Infectious Diseases," *Journal of the American Medical Association* 288, no. 5 (2002): 622–28. Readers should refer to the specific language of the MSEHPA for the most accurate account. Center for Law and the Public's Health, "The Model State Emergency Health Powers Act, as of December 21, 2001," www.publichealthlaw.net/MSEHPA/MSEHPA2.pdf (22 August 2002).

17   L.O. Gostin, "Public Health Law Reform," *American Journal of Public Health* 91, no. 9 (2001): 1365–68.

18   L.O. Gostin, S. Burris, and Z. Lazzarini, "The Law and the Public's Health: A Study of Infectious Disease Law in the United States," *Columbia Law Review* 99, no. 1 (1999): 59–128.

19   K.M. Gebbie, "State Public Health Laws: An Expression of Constituency Expectations," *Journal of Public Health Management Practice* 6, no. 2 (2000): 46–54.

20   L.O. Gostin et al., "The Public Health Information Infrastructure: A National Review of the Law on Health Information Privacy," *Journal of the American Medical Association* 275, no. 24 (1996): 1921–27.

21 H.H. Horton et al., "Critical Biological Agents: Disease Reporting as a Tool for Bioterrorism Preparedness," *Journal of Law, Medicine, and Ethics* 30, no. 2 (2002): 262–66.

22 Gostin et al., "The Public Health Information Infrastructure."

23 *Gibbons v. Ogden*, 22 U.S. (9 Wheat.) 1 (1824).

24 *United States v. Lopez* 514 U.S. 549 (1995).

25 L.O. Gostin, *Public Health Law: Power, Duty, Restraint* (Berkeley and New York: University of California Press and Milbank Memorial Fund, 2000); and L.O. Gostin, ed., *Public Health Law and Ethics: A Reader* (Berkeley and New York: University of California Press and Milbank Memorial Fund, 2002).

26 J. Feinberg, *The Moral Limits of the Criminal Law*, 4 vols. (New York: Oxford University Press, 1987–1990).

27 *Jacobson v. Massachusetts*, 197 U.S. 11, 26 (1905).

28 On averting risk, see *Washington v. Harper*, 494 U.S. 210, 227 (1990) (upholding forced administration of antipsychotic medication if the inmate is dangerous to himself or others and the treatment is in the inmate's medical interest). On compulsory testing, see *Skinner v. Railway Labor Executives' Ass'n*, 489 U.S. 601 (1989). On compulsory reporting, see *Whalen v. Roe*, 429 U.S. 589 (1977). On compulsory vaccination, see *Zucht v. King*, 260 U.S. 174 (1922). On compulsory treatment, see *McCormick v. Stalder*, 105 F.3d 1059, 1061 (5th Cir. 1997) (finding that the state's compelling interest in reducing the spread of tuberculosis justifies involuntary treatment). On isolation, see *Greene v. Edwards*, 263 S.E.2d 661 (1980).

29 S. Lueck, "States Seek to Strengthen Emergency Powers: Movement Is Raising Privacy and Civil-Liberties Concerns," *Wall Street Journal*, 7 January 2002.

30 *Lucas v. South Carolina Coastal Council*, 505 U.S. 1003 (1992).

31 *Commonwealth v. Alger*, 7 Cush. 53, 84–85 (1851).

Chapter 38

Richard H. Thaler and Cass R. Sunstein

BEHAVIORAL ECONOMICS, PUBLIC POLICY, AND PATERNALISM: LIBERTARIAN PATERNALISM

Many economists are libertarians and consider the term "paternalistic" to be derogatory. Most would think that the phrase libertarian paternalism is an oxymoron. The modest goal of this essay is to encourage economists to rethink their views on paternalism. We believe that the anti-paternalistic fervor expressed by many economists is based on a combination of a false assumption and at least two misconceptions. The false assumption is that people always (usually?) make choices that are in their best interest. This claim is either tautological, and therefore uninteresting, or testable. We claim that it is testable and false – indeed, obviously false.

The first misconception is that there are viable alternatives to paternalism. In many situations, some organization or agent must make a choice that will affect the choices of some other people. The point applies to both private and public actors. Consider the problem facing the director of a company cafeteria who discovers that the order in which food is arranged influences the choices people make. To simplify, consider three alternative strategies: (1) she could make choices that she thinks would make the customers best off; (2) she could make choices at random; or (3) she could maliciously choose those items that she thinks would make the customers as obese as possible. Option 1 appears to be paternalistic, which it is, but would anyone advocate options 2 or 3?

The second misconception is that paternalism always involves coercion. As the cafeteria example illustrates, the choice of which order to present food items does not coerce anyone to do anything, yet one might prefer some orders to others on paternalistic grounds. Would many object to putting the fruit before the desserts at an elementary school cafeteria if the outcome were to increase the consumption ratio of apples to Twinkies? Is this question fundamentally different if the customers are adults? If no coercion is involved, we think that some types of paternalism should be acceptable to even the most ardent libertarian. We call such actions *libertarian paternalism*.

In our understanding, a policy counts as "paternalistic" if it is selected with the goal of influencing the choices of affected parties in a way that will make those parties better off. We intend "better off" to be measured as objectively as possible, and we clearly do not always equate revealed preference with welfare. That is, we emphasize the possibility that in some cases individuals make inferior choices, choices that they would change if they had complete information, unlimited cognitive abilities, and no lack of willpower. Once it is understood that some organizational decisions are inevitable, that a form of paternalism cannot be avoided, and that the alternatives to paternalism (such as choosing options to make people sick, obese, or generally worse off) are unattractive, we can abandon the less interesting question of whether to be paternalistic or not and turn to the more constructive question of how to choose among paternalistic options.[1]

I. Are choices rational?

The presumption that individual choices should be free from interference is usually based on the assumption that people do a good job of making choices, or at least that they do a far better job than third parties could do. As far as we can tell, there is little empirical support for this claim. Research by psychologists and economists over the past three decades has raised questions about the rationality of the judgments and decisions that individuals make. People do not exhibit rational expectations, fail to make forecasts that are consistent with Bayes' rule, use heuristics that lead them to make systematic blunders, exhibit preference reversals (that is, they prefer A to B *and* B to A) and make different choices depending on the wording of the problem (for many examples, see the two recent collections of papers by Daniel Kahneman and Amos Tversky [2000] and by Thomas Gilovich et al. [2002]). Furthermore, in the context of intertemporal choice, people exhibit dynamic inconsistency, valuing present consumption much more than future consumption. In other words, people have self-control problems (see the other papers in this session [James Choi et al., 2003b; Ted O'Donoghue and Matthew Rabin, 2003] for details and references).

Many economists are skeptical of some of these findings, thinking that people may do a better job of choosing in the "real world" than they do in the laboratory. However, studies of actual choices for high stakes reveal many of the same problems. For example, the Surgeon General reports that 61 percent of Americans are either overweight or obese. Given the adverse effects obesity has on health, it is hard to claim that Americans are eating optimal diets.

Another illustration comes from the domain of savings behavior. Shlomo Benartzi and Thaler (2002) investigate how much investors like the portfolios they have selected in their defined-contribution savings plans. Employees volunteered to share their portfolio choices with the investigators (by bringing a copy of their most recent statement to the lab). They were then shown the probability distributions of expected retirement income for three investment portfolios just labeled A, B, and C. Unbeknownst to the subjects, the three portfolios were their own and portfolios mimicking the average and median choices of their fellow employees. The distributions of expected returns were computed using the software of Financial Engines, the financial information company founded by William Sharpe. On average, the subjects rated the average portfolio equally with their own portfolio, and they judged the median portfolio to be significantly more attractive than their own. Indeed, only 20 percent of the subjects preferred their own portfolio to the median portfolio. Apparently, people do not gain much by choosing investment portfolios for themselves.

II. Is paternalism inevitable?

As the cafeteria line example discussed above illustrates, planners are forced to make some design choices. A simple and important example is the selection of a "default option" to determine what happens if an agent fails to choose for himself. In a fully rational world such design choices would have little effect (at least in high-stakes situations) because agents would simply choose the best option for them regardless of the default. However, numerous experiments illustrate that there is a very strong "status quo" bias (see William Samuelson and Richard Zeckhauser, 1988; Kahneman et al., 1991). The existing arrangement, whether set out by private institutions or by government, tends to stick.

One illustration of this phenomenon comes from studies of automatic enrollment in 401(k) employee savings plans. Most 401(k) plans use an opt-in design. When employees first become eligible to participate in the 401(k) plan, they receive some plan information and an enrollment form that must be completed in order to join. Under the alternative of automatic enrollment, employees receive the same information but are told that unless they opt out, they will be enrolled in the plan (with some default options for savings rates and asset allocation). In companies that offer a "match" (the employer matches the employee's contributions according to some formula, often a 50 percent match up to some cap), most employees eventually do join the plan, but enrollments occur much sooner under automatic enrollment. For example, Brigitte Madrian and Dennis Shea (2001) found that initial enrollments jumped from 49 percent to 86 percent, and Choi et al. (2002) find similar results for other companies.

Should the adoption of automatic enrollment be considered paternalistic? And, if so, should it therefore be seen as a kind of officious meddling with employee preferences? We answer these questions yes and no respectively. If the employer thinks (correctly, we believe) that most employees would prefer to join the 401 (k) plan if they took the time to think about it and did not lose the enrollment form, then by choosing automatic enrollment they are acting paternalistically. They are attempting to steer employees' choices in directions that will promote employees' welfare. But since no one is forced to do anything, we think this steering should be considered unobjectionable to libertarians. The employer must choose some set of rules, and

either plan affects employees' choices. No law of nature says that, in the absence of an affirmative election by employees, zero percent of earnings will go into a retirement plan. Because both plans alter choices, neither one can be said, more than the other, to count as a form of objectionable meddling.

Quick-minded readers might be tempted to think that there is a way out of this dilemma. Employers could avoid choosing a default if they *required* employees to make a choice, either in or out. But some thought reveals that this is not at all a way out of the dilemma; rather, it is simply another option among many that the employer can elect. In fact, Choi et al. (2003a) find that this rule increases enrollments (relative to the opt-in rule) though by not as much as automatic enrollment. Furthermore, the very requirement that employees make a choice has a paternalistic element. Many employees do not want to have to make a choice (and would choose not to have to do so). Should employers really force them to choose?

Why, exactly, does the setting of defaults have such large effects? With respect to savings, the designated default plan apparently carries a certain legitimacy for many employees, perhaps because it seems to have resulted from some conscious thought about what makes most sense for most people. But there is a separate explanation, involving inertia. For any employee, a change from any status quo entails time and effort, and many people seem to prefer to avoid both of these, especially if they are prone to procrastination. When default rules are "sticky" and affect choices as a result, inertia might be the major reason.

For present purposes, the choice among these various explanations does not much matter. The point is only that paternalism, in the form of effects on individual choices, is often unavoidable. When paternalism seems absent, it is usually because the starting point appears so natural and obvious that its preference-shaping effects are invisible to most observers. But those effects are nonetheless there. Of course it is usually good not to block choices, and we do not mean to defend non-libertarian paternalism here. But in an important respect, the anti-paternalistic position is incoherent.

III. Beyond the inevitable (but still libertarian)

The inevitability of paternalism is most clear when the planner has to choose default rules. It is reasonable to ask whether the planner should go beyond the inevitable. Take the cafeteria example discussed above. Putting the fruit before the desserts is a fairly mild intervention. A more intrusive step would be to place the desserts in another location altogether, so that diners have to get up and get a dessert after they have finished the rest of their meal. This step raises the transactions costs of eating dessert, and according to a standard economic analysis the proposal is unattractive: it seems to make dessert-eaters worse off and no one better off. But once self-control costs are incorporated, we can see that some diners would prefer this arrangement, namely, those who would eat a dessert if it were put in front of them but would resist temptation if given a little help. To the extent that the dessert location is not hard to find, and no choice is forbidden, this approach meets libertarian muster.

In the domain of employee savings, Thaler and Benartzi (2003) have proposed a method of increasing contributions to 401(k) plans that also meets the libertarian test. Under this plan, called Save More Tomorrow, employees are invited to sign up for a program in which their contributions to the savings plan are increased annually whenever they get a raise. Once employees join the plan, they stay in until they opt out or reach the maximum savings rate in the plan. In the first company to use this plan,

the employees who joined increased their savings rates from 3.5 percent to 11.6 percent in a little over two years (three raises). Very few of the employees who join the plan drop out. This is successful libertarian paternalism in action.

IV. How to choose: the toolbox of the libertarian paternalist

How should sensible planners (a category we mean to include anyone who must design plans for others, from human-resource directors to bureaucrats to kings) choose among possible systems, given that some choice is necessary? We suggest two approaches to this problem.

If feasible, a comparison of possible rules should be done using a form of cost-benefit analysis. The goal of a cost-benefit study would be to measure the full ramifications of any design choice. To illustrate, take the example of automatic enrollment. Under automatic enrollment some employees will join the plan who otherwise would not. Presumably, some are made better off (especially if there is an employer match), but some may be made worse off (e.g., those who are highly liquidity-constrained). If the issue were just enrollment, we would guess that the gains would exceed the losses. We base this guess partly on revealed choices. Most employees do join the plan eventually, and very few who are automatically enrolled opt out when they figure out what has happened to them. We also judge that the costs of having too little saved up for retirement are typically greater than the costs of having saved too much.

In many cases, however, the planner will be unable to make a direct inquiry into welfare, either because too little information is available or because the costs of doing the analysis are not warranted. The committed anti-paternalist might say, in such cases, that people should simply be permitted to choose as they see fit. We hope that we have said enough to show why this response is unhelpful. What people choose often depends on the starting point, and hence the starting point cannot be selected by asking what people choose. In these circumstances, the libertarian paternalist would seek indirect proxies for welfare: methods that test whether one or another approach is welfare-promoting without relying on unreliable guesswork about that question. We suggest three possible methods.

First, the libertarian paternalist might select the approach that *the majority would choose if explicit choices were required and revealed*. Useful though it is, this market-mimicking approach raises its own problems. Perhaps the majority's choices would be insufficiently informed. Perhaps those choices, in fact, would not promote the majority's welfare. At least as a presumption, however, it makes sense to follow those choices if the planner knows what they would be. A deeper problem is that the majority's choices might themselves be a function of the starting point or the default rule. If so, the problem of circularity dooms the market-mimicking approach. But in some cases, at least, the majority is likely to go one way or the other regardless of the starting point; and to that extent, the market-mimicking strategy seems quite workable.

Second, the libertarian paternalist might select the approach that *would force people to make their choices explicit*. This approach might be chosen if the market-mimicking strategy fails, either because of the circularity problem or because the planner does not know which approach would in fact be chosen by the majority. We have seen the possibility of forced choices in the context of retirement plans; it would be easy to find other examples. Here too, however, there is a risk that the choices that are actually elicited will be inadequately informed or will not promote welfare. In the case of retirement plans, for example, forced choices have been found to produce higher participation rates than requiring opt-ins, but lower rates than requiring opt-outs. If it is

likely that automatic enrollment is welfare-promoting, perhaps automatic enrollment should be preferred over forced choices. The only suggestion is that, where the social planner is unsure how to handle the welfare question, he might devise a strategy that requires people to choose.

Third, the libertarian paternalist might select the approach that *minimizes the number of opt-outs*. For example, very few employees opt out of the 401(k) plan when they are automatically enrolled, though many opt in under the standard enrollment procedure. This is an *ex post* inquiry into people's preferences, in contrast to the *ex ante* approach favored by the market-mimicking strategy. With those numbers, there is reason to think that automatic enrollment is better, if only because more people are sufficiently satisfied to leave it in place.

V. Conclusion

Our goal here has been to defend libertarian paternalism, an approach that preserves freedom of choice but that authorizes both private and public institutions to steer people in directions that will promote their welfare. Some kind of paternalism is likely whenever such institutions set out arrangements that will prevail unless people affirmatively choose otherwise. In these circumstances, the goal should be to avoid random, arbitrary, or harmful effects and to produce a situation that is likely to promote people's welfare, suitably defined.

Note

1 Readers interested in this topic should also consult Colin Camerer et al. (2001) for an illuminating discussion of related issues. That paper shares with the papers in this session the common goal of devising policies that help some agents who are making some mistake, while minimizing the costs imposed on others.

References

Benartzi, Shlomo and Thaler, Richard H. "How Much Is Investor Autonomy Worth?" *Journal of Finance*, August 2002, 57(4), pp. 1593–1616.

Camerer, Colin; Issacharoff, Samuel; Loewenstein, George; O'Donoghue, Ted and Rabin, Matthew. "Regulation for Conservatives: Behavioral Economics and the Case for Asymmetric Paternalism." Working paper, California Institute of Technology, 2001.

Choi, James; Laibson, David; Madrian, Brigitte C. and Metrick, Andrew. "Defined Contribution Pensions: Plan Rules, Participant Decisions, and the Path of Least Resistance," in James M. Poterba, ed., *Tax Policy and the Economy*, Vol. 16. Cambridge, MA: MIT Press, 2002, pp. 67–113.

——. "Benign Paternalism and Active Decisions: A Natural Experiment." Working paper, University of Chicago, Graduate School of Business, 2003a.

——. "Optimal Defaults." *American Economic Review*, May 2003b (*Papers and Proceedings*), 93(2), pp. 180–85.

Gilovich, Thomas; Griffin, Dale and Kahneman, Daniel, eds. *Heuristics and biases: The psychology of intuitive judgment.* Cambridge, U.K.: Cambridge University Press, 2002.

Kahneman, Daniel; Knetsch, Jack L. and Thaler, Richard H. "The Endowment Effect, Loss Aversion, and Status Quo Bias." *Journal of Economic Perspectives*, Winter, 5(1), pp. 193–206.

Kahneman, Daniel and Tversky, Amos, eds. *Choices, Values, and Frames.* Cambridge, U.K.: Cambridge University Press, 2000.

Madrian, Brigitte and Shea, Dennis. "The Power of Suggestion: Inertia in 401(k) Participation and Savings Behavior." *Quarterly Journal of Economics*, November 2001, 116(4), pp. 1149–87.

O'Donoghue, Ted and Rabin, Matthew. "Studying Optimal Paternalism, Illustrated by a Model of Sin Taxes." *American Economic Review*, May 2003 (*Papers and Proceedings*), 93(2), pp. 186–91.

Samuelson, William and Zeckhauser, Richard J. "Status Quo Bias in Decision Making." *Journal of Risk and Uncertainty*, March 1988, 1(1), pp. 7–59.

Thaler, Richard H. and Benartzi, Shlomo. "Save More Tomorrow: Using Behavioral Economics to Increase Employee Saving." *Journal of Political Economy*, 2003 (forthcoming).

Chapter 39

S. Holm

OBESITY INTERVENTIONS AND ETHICS

Background

The purpose of ethical reflection is to help us decide how to act in the real world, all things considered. This means ethical reflection must take into account all aspects of a proposed course of action, it must rely on good factual evidence and on an understanding of the social and legal contexts in which action has to take place.

In the present public health context, the main actors under consideration are, on the one hand, the state and doctors and others in their roles as agents of the state, and, on the other hand, the individual. The main principles or ethical values in play are respect for self-determination, the pursuit of a good life, the promotion of the common good, the obligations between parents and children and justice.[1,2]

The basic problems

Two basic ethical problems recur in discussions about obesity interventions. The first is whether and when it is justifiable to intervene to promote a person's own health or well-being, if they do not want the intervention. When can paternalism be justified?

The second is whether and when it is justifiable to negatively affect a person or people's well-being in order to benefit others or to promote the common good. In this context, it is important to note that pure economic gain is usually not considered sufficient reason to claim that the common good is being promoted.

These two problems do not occur on a neutral background. The current nutritional and physical activity landscape is already shaped by numerous historical factors, by intentional and unintentional effects of government regulation and by the actions of commercial actors. The way society is organized is already affecting people's well-being in different ways.

Hard and soft paternalism

A useful distinction has been made by the Finnish philosopher Heta Häyry between three different forms of paternalism: hard paternalism, soft paternalism and maternalism. Hard paternalism involves direct coercion, soft paternalism involves giving unwanted information or foreclosing some options for action and maternalism involves control by inducing a guilty conscience.[3]

An example that combines soft paternalism and maternalism is the practice of telling pregnant women that they are potentially harming their babies if they drink alcohol during pregnancy, whereas the more traditional 'if you smoke, you die' message on cigarette packs and posters is an example of a more pure form of soft paternalism.

It is generally accepted that hard paternalism requires a stronger justification than soft paternalism. Whereas hard paternalism by definition involves the denial of self-determination, soft paternalism still allows the expression of choice. For instance, someone can be told that certain foods are unhealthy, yet they are still able to buy and eat them.

With regard to the foreclosing of options, there is a gradation from soft to hard paternalism, depending on how difficult it becomes for a person to still pursue the discouraged choice. A general practitioner who gives unsolicited advice on diet and weight loss to a patient who has only come to his surgery for a travel vaccination is engaging in soft paternalism, but may be engaging in hard paternalism if, for instance, he refuses to give the shot until his patient has begun to listen more attentively.

If smoking bans in public places had been justified in terms of their health benefits to smokers, it would have been an example of soft paternalism which gradually edged into hard paternalism as more and more areas became non-smoking. However, the official justification for smoking bans are their health benefits to non-smokers, which actually makes them examples in which the well-being of smokers has been sacrificed for the sake of the well-being of non-smokers.

Choice, rationality and information

Can we dispel some of the worries about hard paternalism by making a distinction between informed and/or rational choices that ought to be respected, and uninformed and/or irrational choices that we can legitimately interfere with because they are in some sense not the person's 'real' choices?[4] Should all choices be respected?

Making such a distinction is more complicated than it might initially seem, because, although we may be able to define the extreme poles of the decision-making spectrum with some precision, it is difficult to say where on the spectrum a given decision becomes so ill informed or so irrational that we can overrule it. True, most decisions made by most people most of the time are not terribly well informed or well thought through. Nevertheless, we are usually quite happy to accept that the choices made are theirs and that we should not use any stronger means than advice or mild persuasion to alter that choice, even if we think they are making the wrong decision. Think, for example, of the parents of teenage children intent on body piercings. This holds true even if the decisions in question are significant ones like changing employment, getting married or buying a house.

If we are worried about the quality of decision-making in the context of food and exercise, and if we believe that some people do not have the necessary information to make good choices, it is much less problematic to provide them with that information than to take their choices away from them.

On the other hand, the need for consumers to have enough information to base informed choices on is one of the strongest reasons for requiring the producers, manufacturers and retailers of food to provide unbiased and easily understandable information about their products to consumers. Such requirements are not limited to information about nutrition, and include many other items that are important to choice, such as weight, country of origin, dolphin-friendliness and so on. Giving nutritional information does not necessarily produce nutritionally optimal choices. However, this is not a reason for not giving the information, as its purpose is to improve the consumer's choice process.

Targeting high-risk groups

What are the ethical issues raised by interventions that target identified high-risk groups? First, there are potential problems of social stigmatization and changed self-perception if the targeting results in individuals being identified as belonging to a high-risk group. Insofar as obesity is already a stigmatized condition, being identified as being at risk of obesity may also amount to or produce stigma. This problem may be obviated to some degree if the targeting is performed by self-identification, but such a strategy will often lead to very inefficient targeting.

Second, there are potential problems of justice. The reason for targeting high-risk groups is usually cost-effectiveness. A given expenditure of resources will result in more obesity being prevented than if we had chosen to employ a 'broader' strategy. But do individuals at similar high risk, who are outside the identified high risk groups, not have the same claim on our preventive resources? A basic principle of formal justice, originally outlined by Aristotle is that equal cases should be treated equally, and this indicates that it is at least potentially problematic to focus prevention only on high-risk groups, unless such groups can be so precisely defined, identified and targeted that all high-risk individuals are included.

A different problem occurs in health promotions that target the whole population. In order to achieve a reduction in the prevalence of obesity, it may be necessary to achieve behaviour change far beyond simply the obese. It might be necessary to shift the whole weight curve downwards (or shift the whole energy intake or physical activity curve). Thus, all those who are overweight, or simply perceive themselves to be so, will have to change their lifestyles in order to reduce the number of obese people. Many of those who are made to feel concerned and are induced to change their lifestyle might never have become obese or have had any significant health problems related to their weight. This means that we are affecting some individuals negatively in order to benefit others or to create healthcare savings at the societal level. This can be justified if the bad thing that is averted (obesity and its health effects) clearly outweigh the negative effects experienced by the non-obese, but if this is truly the case, it needs to be substantiated.

Choice and responsibility

It is a commonplace that people are responsible for the foreseeable consequences of their choices and actions. This seems to indicate that the obese are responsible for being obese and for the health and other consequences of their obesity. There are, however, a number of complications in making a claim of personal responsibility.

The first is that it is not clear to what degree lifestyle is actually a matter of conscious choice. There are at least large elements of socialization and social construction involved in acquiring and maintaining a given lifestyle, and the possibility to break with or significantly alter one's lifestyle is significantly influenced by a range of socioeconomic factors.

The second is that the lifestyle in question may contribute significantly to the person's sense of well-being. They may simply like to live the way they live and may accept future negative effects as a reasonable trade-off for their current pleasure and well-being. We may well think that they have either miscalculated the trade-off over time, or that they could get the same amount of well-being now by changing their lifestyle. But whether a given individual will have a better quality of life as a burger-eating smoker or a vegetarian fitness enthusiast is probably not a question that can be objectively decided by a third party.

The final problem is that, even if we could demonstrate that the obese had chosen to be obese, and are therefore responsible for their own obesity and its consequences, that still would not settle the question of whether we should hold them responsible and treat them differently. Unless we treat all similar, non-obesity-related instances of negative health effects of personal choice in the same way, the obese would have a strong claim that they are being treated unjustly.

We should also note that there is an inherent tension between a focus on personal choice and responsibility and the justification for large-scale interventions that affect some non-obese individuals negatively (e.g. by inducing unnecessary lifestyle changes). If the obese are personally responsible for their condition, it becomes doubtful whether others should sacrifice their interests to benefit the obese. For these reasons, it is ethically problematic to pursue a policy primarily emphasizing choice and responsibility.

This raises further the question of whether it is a condition, in expecting personal responsibility, that society does its part. Can we legitimately expect members of society to show responsibility if society does not shoulder its responsibilities or is not perceived to be shouldering its responsibilities? Is there a requirement of 'reciprocity of responsibility'? It may, for instance, be argued that, if society could reduce obesity by prohibiting specific marketing strategies, but declines to do so, it is hypocritical to put much emphasis on personal responsibility.

Many arguments in the debate about obesity try to determine exactly who has responsibility, but it is important to remember that it is not a zero-sum game. There is more than enough responsibility to go round for all parties; the state, the food industry and the individual. We have no reason to believe that only one of these parties is responsible, and even less reason to believe that only one party should be held responsible.

Early-life intervention

It is generally accepted that many instances of obesity can be traced to early childhood or possibly even to the intrauterine environment. A potential, and on the face of it, very attractive intervention strategy is therefore to aim for optimal nutrition in this period. This would probably involve both intensive dietary advice to all parents, more intensive monitoring of weight and body shape of all children, and targeted intervention in families where monitoring reveals problems. Such an intervention could be seen as an extension of current antenatal and health visiting initiatives.

The main ethical arguments for the intervention would be that it has good consequences for the child and for society, and that parents already have an ethical and legal obligation to promote the 'best interests of the child' (in law the obligation is for the child's interests to be paramount). A general problem with the 'best interests of the child' argument is that we do and must allow parents to act against the best interests of the child, or to act on their conception of the best interest in many contexts. We do not think it is wrong for young parents to save for pensions, even if spending the money now on their child could benefit the child. Furthermore, if there is more than one child in the family, it will sometimes be impossible to promote the best interests of both children at the same time when these interests are in conflict.[5]

It is therefore unclear whether parents actually have a strong obligation to promote optimal nutrition in their children, or just a strong obligation to promote 'good enough' nutrition. Achieving optimal nutrition for the child would, in the prenatal period, physically require that the mother changes her nutritional pattern, and may later require that the whole family begins to eat differently to the way they did before they had the child. This may well be beneficial for everyone, but it is not obvious that there's a strong obligation to do this.

The scepticism about obligation is reinforced by two further factors. First is the observation that dietary advice is not constant and that it is not all based on good evidence. 'Optimal nutrition' therefore actually means 'nutrition currently believed to be optimal'.

Second is the fact is that not all overweight children go on to become obese adults. Targeting risk groups in childhood therefore also necessarily leads to some families being the target of unnecessary interventions.

Promoting body image

Another set of questions arises in interventions aimed at promoting a specific body image or range of body images (i) over the strength of the evidence base for promoting this specific body image; (ii) over the possible side effects (e.g. an increase in anorexia or other eating problems) and (iii) over the risk of stigmatization of those with body shapes deemed not ideal.

In one way, discussing possible future interventions in this area is strange because it is quite obvious that our culture and many other cultures already promote a specific body image, not through public health campaigns but through ordinary advertising.[6] Based on our experiences, we can justifiably claim that it is difficult to promote one body shape as good without implying that other shapes are bad, and it is unclear whether it is possible to prevent people from linking bad body shape to personal and moral badness. This is particularly the case when obesity is already stigmatized.

Conclusion

It will be evident from the discussion so far, that two interlocking discussions underlie many of the ethical controversies in the obesity intervention area.

The first is a discussion concerning the correct analysis of value. Are all values subjective or personal, or are there objective values? Is health good for everyone whether or not they value it, or is it only good for those who choose to pursue it? The second is a discussion concerning the degree to which society or the state is justified in interfering in personal choices, and the legitimacy of different kinds of interventions.

Although these two discussions are conceptually distinct, they are linked because it may be easier to justify paternalistic intervention if there are truly objective values.

However, even if there are objective values, there are probably more than one (e.g. health is not the only value). This means that there might be a conflict of values and that they will have to be balanced or prioritized. It is unclear that health will always come out on top when conflicting values are weighed up.

This means that policies involving hard paternalism will always be difficult to justify. However, even the most ardent libertarians can support many policies based on soft paternalism, especially policies where choice is not curtailed, but choosing healthier options is made easier, for instance, by making the healthy choice the default. Such policies could be labelled 'libertarian paternalist' (7) and are unobjectionable as long as they are based on solid evidence that the healthy choice is really the healthiest option. Such evidence does not necessarily have to amount to 'scientific proof' − a notion that is controversial in itself − but it has to be more than just deduction from theory. All interventions have costs, both ethical and economic/ opportunity costs, and it is only when we have some evidence for their effectiveness that we get a handle on whether the costs outweigh the benefits.

Notes

1   Downie RS, Tannahill C. *Health Promotion — Models and Values*, 2nd edn. Oxford University Press: Oxford, 1998.

2   Dawson A, Verweij M (eds). *Ethics, Prevention and Public Health*. Oxford University Press: Oxford, 2006.

3   Häyry H. *The Limits of Medical Paternalism*. Routledge: London, 1991.

4   Holm S. "Autonomy." In: Chadwick R (ed.). *Encyclopedia of Applied Ethics*, Vol. 1. Academic Press: San Diego, CA, 1997, pp. 267–74.

5   Harris J, Holm S. "Should we presume moral turpitude in our children? Small children and consent to medical research." *Theor Med* 2003; 24: 121–29.

6   Beaufort I, Hilhorst M, Holm S (eds). In *The Eye of the Beholder: the Ethics of Appearance Changing Medical Interventions*. Scandinavian University Publishers: Oslo, 1996.

7   Thaler RH, Sunstein CR. "Libertarian Paternalism." *Am Econ Rev* 2003; 93: 175–79.

Chapter 40

D. Isaacs, H.A. Kilham and H. Marshall

SHOULD ROUTINE CHILDHOOD IMMUNIZATIONS BE COMPULSORY?

Childhood immunization is one of the most important and cost-effective public health measures in our armoury. There is no doubt that immunization reduces the incidence and severity of infectious diseases and saves lives.[1–3] But vaccines are unique among public health measures, in that their administration occasionally causes injury. Should routine childhood immunizations be compulsory? In this paper we consider primarily those immunizations currently recommended in the Australian routine childhood schedule.[2]

Historical perspectives

Compulsory immunization: past and present

There is nothing new about compulsory immunization, nor about vociferous anti-vaccination movements. The British Vaccination Act of 1853 made smallpox vaccination compulsory for all infants in the first three months of life, and made defaulting parents liable to a fine or imprisonment.[4] This was the first time that public laws potentially infringed civil liberties, and the Act spawned an Anti-Compulsory Vaccination League and an anti-vaccination demonstration in Leicester attended by 100 000 protesters.[4] Vaccination rates fell and, in 1898, 'conscientious objectors' were excused from having their children vaccinated.

The United States of America has an even longer history of compulsory immunization. The State of Massachusetts introduced compulsory smallpox vaccination in 1809, while in 1922 the Supreme Court upheld laws requiring vaccination for school entry.[5]

Routine childhood immunization, against at least some infectious diseases, is compulsory in a number of countries, including Croatia, France, Italy, Poland, Slovakia and Taiwan. The United States of America has had school immunization laws, requiring compulsory immunization at entry to licensed day care and to school, for a number of years, although enforcement has been variable. In the 1970s unimmunized children were excluded from school during measles outbreaks. In 1977, a measles outbreak occurred in Los Angeles County. After immunizing thousands of students, 50 000 of the 1.4 million students remained unimmunized and were excluded from school. Most returned within days with proof of immunity, and the number of measles cases plummeted.[5] Recently there has been greater emphasis on enforcing immunization requirements at school entry. There is no national immunization law, all regulations being State-based. The exact requirements vary but all States require diphtheria, measles, polio and rubella immunization. Sanctions for non-compliance also vary, and some States threaten to take child care proceedings if there is persistent failure to immunize. Forty-eight of the 50 States allow exemptions for those with deeply held religious beliefs opposed to immunization, but only 15 States allow parents to decline immunization for 'philosophic' reasons.[5] It is argued that US school immunization laws mainly act as a safety net to ensure that under-privileged children are immunized, while offending 'very few', although up to 2.5 per cent of students are exempted in States which allow philosophic exemption.[5]

Factual considerations

Vaccines save lives; failure to immunize costs lives

The eradication of smallpox in the 1970s, by targeted use of smallpox vaccine, has not only prevented many thousands of deaths, but is estimated to have saved US$1.2 billion annually in the 25 years since the last case was reported.[3] Poliomyelitis has almost been eradicated from the world.[6]

More recently conjugate vaccines have reduced the annual number of cases of Haemophilus influenzae type b (Hib) infection, previously the commonest cause of meningitis in industrialized countries, by over 95 per cent.[7] Most routine childhood vaccines protect against communicable diseases, which can be transmitted person-to-person, and which we will term **transmissible.** Immunization against transmissible infections, for example, diphtheria, pertussis, polio, Hib, hepatitis B, meningococcus, pneumococcus, varicella, protects the child, but also reduces spread to other children and adults, resulting in **herd immunity** (see below). Universal rubella immunization is a special case, where the direct benefit is primarily to persons other than the recipient, by reducing the incidence of congenital rubella syndrome. In contrast, tetanus immunization protects only the recipient, because tetanus is not transmissible.

It is sometimes argued that vaccine-preventable diseases are no longer as serious, and that modern medical treatment would prevent the high morbidity and mortality once seen. A recent example illustrates the fallacy of this belief:

The break-up of the Soviet Union caused enormous disruption to health services. As a result rates of childhood immunization fell drastically. Between 1991 and 1996 there was an outbreak of diphtheria, with over 140 000 cases notified and over 4000 deaths.[8]

Vaccine-preventable diseases remain life-threatening, and outbreaks will recur if immunization levels fall.

Immunizations can be harmful

Although the commonest adverse events following immunization are relatively minor and self-limiting, such as local reactions, fever and irritability, immunizations can occasionally cause severe irreversible complications and rarely, even death.

Vaccine-associated paralytic poliomyelitis (VAPP) is estimated to occur once in every 2.4 million doses of oral polio-virus vaccine (OPV).[9] Measles vaccine causes an acute encephalitis with an incidence of one in a million doses, although in contrast the incidence of acute encephalitis after wild-type measles infection is about one in a thousand.[10] Yellow fever vaccine has caused yellow fever in a small number of recipients, and six deaths have been reported from fulminant yellow fever acquired from the vaccine.[11]

Ethical considerations

Risks versus benefit

We take calculated risks every day of our lives. Travel is a good example. The speed and convenience of road, rail and air travel mean that most persons accept the slight risk of an accident in favour of the benefits offered by quicker travel.

In general the benefits of immunization far outweigh the risks. The risk of vaccine-induced injury is hundreds to thousands of times lower than the risk of similar complications of the natural, wild-type infection.[1-3]

People who are afraid of harming their children by immunization tend to over-emphasize the risks of vaccine injury and to minimize the risk of wild-type disease.[2] This reflects a general tendency to be more worried about causing damage to one's child by doing something to them than by not doing it. This is referred to as the fear of commission rather than of omission.[2] In his autobiography, Benjamin Franklin wrote with tragic eloquence:

In 1736, I lost one of my sons (Francis Folger) a fine boy of 4 years old by the smallpox. I long regretted bitterly and still regret that I had not given it to him by inoculation. This I mention for the sake of parents, who omit the operation on the supposition that they should never forgive themselves if the child died under it: my example shows the regret may be the same either way, and that therefore, the safer should be chosen.

Public health and paternalism

Some public health interventions that have been shown to prevent injury or death have been made compulsory, because the public cannot be trusted to comply unless there is a degree of coercion. Examples include seat-belt legislation, motor-cycle helmets, bicycle helmets and swimming pool fences. Sanctions for disobeying regulations are usually fines, and possible loss of motor vehicle licence for frequent offenders regarding seat belts and motor-cycle helmets. Those who oppose such legislation do so for different reasons, such as because it is paternalistic, but they also try to argue that the intervention may itself be harmful. Seat belts can occasionally cause crush injuries to the chest or spine, and while being thrown from a car is likely to result in injury or death, being thrown out occasionally avoids injury, for example from fire. Helmets decrease the risk of head injury, but may rarely inflict damage. Those opposed to swimming pool fences even try to argue, contrary to the evidence, that pool fences may give a false sense of security and increase the risk of drownings.

Are the above public health interventions comparable to immunization, because the benefits outweigh the risks, or is immunization different? It seems to us that there is an important difference between immunization, which involves the injection of foreign material, even though with the intention of protecting the recipient and the community, and the compulsory use of seat belts, crash helmets or pool fences. Compelling someone against their will to have an immunization could be seen as constituting a physical **assault**, whereas the other interventions are substantially less invasive.

Herd immunity and the paradox of the 'free riders'

Herd immunity is the phenomenon that, once a critical proportion of a population is immune to a particular transmissible disease, through infection or immunization, the disease can no longer circulate in the community.[12] The concept only applies to diseases such as diphtheria, measles and pertussis, which are confined to humans and transmissible person-to-person. If there is an animal reservoir, and no transmission from person-to-person, such as for tetanus or rabies, then an individual derives no benefit from the immunization of others in the community. An individual is only protected against tetanus if that individual is immunized.

The critical level of population immunization to achieve herd immunity varies from disease to disease. For Hib disease, rates fall rapidly once 85 per cent of infants are immunized.[1,2] Measles requires approximately 95 per cent immunization rates to stop any outbreaks.[2,3] Pertussis continues to circulate, although at much reduced intensity, even when high levels of immunization are retained; the reason is thought to be waning immunity in adults, who then infect babies. One major benefit of high rates of immunization is to protect, from diseases like whooping cough, babies too young to have been immunized (almost all whooping cough deaths are of babies under three months old).[7]

An important implication of herd immunity is that failure to immunize a child against a transmissible infection may not only render that child susceptible to infection, but may imperil other children. Unimmunized school children in Colorado had a greatly increased risk of catching measles (22-fold) and pertussis (6-fold).[13] In addition, pertussis outbreaks were more likely in schools with a higher percentage of unimmunized children.[12] Immunization against pertussis is not 100 per cent protective, so fully immunized children were catching pertussis, and possibly transmitting it to their infant siblings, yet the pertussis was circulating largely because some of their fellow school children were not immunized.

When the population is highly immunized against a disease subject to herd immunity, then a parent may elect not to have their child immunized, and the child is protected by the herd. Such parents are sometimes referred to as 'free riders'.[14] If the number of free riders increases, the population becomes more susceptible, and the disease will start to circulate.

What this description also illustrates is that the risk-benefit equation of immunization against a transmissible infection varies for any single child in the community according to community rates of immunization. If almost all other children are immunized, then a child can be unimmunized and benefit from herd immunity. If vaccine-preventable diseases like measles and pertussis are circulating, because of low levels of immunization, the benefit of immunization for any individual far outweighs the risk. A corollary is that immunization of a child against a transmissible infection protects the community, as well as protecting that individual.

Arguments in favour of compulsory immunization

(i) Communitarian

Communitarianism is a modern term for a philosophical theory that insists that we recognize the value not only of individual freedom but also of the common good.[15] Although communitarianism is a modern term, it is an ancient concept: philosophers such as Aristotle and David Hume espoused the importance of the community.[15]

A communitarian may well argue that immunization benefits the whole community and protects the common good of society, and that since its significance in protecting the common good outweighs its significance in limiting individual freedom, immunization should be compulsory. An extreme communitarian might say that everyone in the community should be immunized (unless there is a medical contraindication) and that anyone who declined immunization was effectively declining to be part of the community and should be forced to leave the community. A moderate communitarian would find a less draconian sanction for non-compliance.

(ii) Consequentialist

Consequentialists or utilitarians argue that actions or policies are good or bad according to the balance of their good and bad consequences. Compulsory immunization would be preferable to voluntary immunization if it produced the best overall result, from a perspective that gives equal weight to the interests of each affected party. Compulsory childhood immunization would almost certainly result in less disease and hence less suffering, which would outweigh vaccine adverse events. A bad consequence to consider, however, is the limitation to personal freedom occasioned by coercing people to immunize their children. If compulsory immunization caused concern about coercive government control, yet voluntary immunization could achieve almost equally high rates, then a consequentialist might prefer voluntary immunization. If compulsory immunization was the best way to protect children and was acceptable to the community, then a consequentialist should favour compulsory immunization.

(iii) Rights-based: rights of the child and the community

An advocate of children's rights may well argue that, because children need to be protected from dangerous infectious diseases, they have a right to the protection afforded by immunization.[16] Such a right, it might be argued, generates a duty on the part of parents (or, if they are negligent in the fulfilment of that duty, on the part of the State) to immunize the child. Since it is well known that some parents will be negligent with respect to this duty, the State must accept that it has the duty to ensure that each child is immunized, and if it is objected that parents have a right to decide how best to look after their offspring's health and well-being, the advocate of the child's right may well claim that the child's right to protection has priority over the parents' right to decide.

A communitarian might say that the community's interests should take preference over individual rights. How do we decide which rights should be paramount: the child's, the parents' or the community's? One answer is the degree of risk.[17] If the risk to the child or the community is high, then it may be necessary to over-ride the parents' right to choose. A child bitten by a rabid dog will almost certainly die unless given rabies vaccination. If a parent refused rabies vaccination in this circumstance, the situation would be a child protection issue, and the child's right to protection would be the paramount consideration. This situation is analogous to a child of Jehovah's Witness parents who is bleeding to death: the child is too young to choose, and the child's safety becomes pre-eminent.

If there was an outbreak of a vaccine-preventable disease, which was devastatingly severe and children could not be protected simply by exclusion from school, it might be argued that compulsory immunization would be justified. An example might be an outbreak of smallpox due to a bioterrorist attack.

Arguments against compulsory immunization

(i) Respect for parental autonomy

Respect for the autonomous choices of other persons is one of the most deep-rooted concepts in moral thinking. It is tempting for proponents of immunization to say that a child cannot make an autonomous decision about immunization and we should over-rule parents who decline to have their children immunized. But how far should we interfere with parental choices about child rearing? In any society, particularly a pluralist or multicultural society, there are many views on what is acceptable in rearing children. In general, parents have to live with their choices for their children and it is usual to respect such parental choices. The only exceptions to this are when the parents' actions or choices result in serious harm or neglect, i.e. child protection issues.

(ii) Rights-based: rights of the parents

A rights-based approach can also be used to argue against compulsory immunization, because the child's parents also have rights. These rights derive from the fact that they conceived, bore and reared the child and have a significant emotional and financial investment in the child's current and future well-being. This creates an obligation on others to respect parents' right to bring up their children as they see fit, unless they cause serious harm to the child. To argue that parents should be compelled to immunize their children in the child's 'best interest' is to ignore the fact that a child is part of a family. The child of parents who are religiously or philosophically opposed to immunization is quite likely to grow up opposed to immunization. To have been forcibly immunized in childhood will then be viewed by the adult as a societal assault.

(iii) Variable risk-benefit of different vaccines

Even if protection of the community is a compelling communitarian argument for compulsory immunization, it only applies to transmissible infections, and not tetanus. Furthermore, the risk-benefit equation varies from disease to disease and varies over time for a single disease, depending on incidence. To make all routine childhood immunizations compulsory risks ignoring these important intrinsic differences.

(iv) Trust versus State coercion

The State already applies coercion to many of our daily activities. Do we want to live in the sort of society that extends coercion to routine immunization? At present, many industrialized countries achieve high levels of immunization without the need for compulsion. If such high levels can be maintained through encouragement and incentives, this effectively achieves the aims of the moderate communitarian, without the need for legislation. Compulsory immunization would be certain to inflame those who already believe that their Government interferes too much with their freedom. What is more, coercion may alter perception of risk. People who are coerced into an action may be more likely to perceive the action as being risky than if they are persuaded into it. Recent examples, albeit adult rather than child, have been the mandatory immunization of military personnel against anthrax and smallpox, which led to many protests and loss of confidence. Most parents trust the assurances of health care professionals that the benefits of immunizing their child outweigh the risks. Making immunizations compulsory renders trust redundant. If State coercion can be avoided in the area of routine childhood immunization, so much the better.

(v) Practical issues

Even if it was decided that routine childhood immunization should be compulsory, there are potential practical difficulties in enforcement. We often physically restrain a young child to immunize them, but with parental consent. To physically restrain a child and immunize them against their parents' wish could constitute an assault, which only seems justifiable in a situation of extreme risk, such as post-rabies exposure. The alternative is to introduce sanctions for non-compliance, such as fines or even draconian measures like child care proceedings or imprisonment.

Alternatives to compulsory immunization

Most countries do not have compulsory routine childhood immunization. Instead they employ one or more of the following strategies:

(i) Education

If education of the community and of health care providers about the benefits of immunization achieves levels of vaccine uptake that prevent circulation of infectious diseases, then it is unnecessary to introduce legislation to compel parents to conform.

(ii) Inducements

Inducements may be offered to parents or to providers, such as general practitioners. Inducements to parents usually take the form of linking child care benefit payments and/or maternity benefits to immunization status. Could this be seen as a form of coercion, particularly to poorer families who are far more dependent on such welfare payments? A communitarian might argue that if society provides child and family payments, it is reasonable for society to expect and even demand that children be immunized to help protect the whole community. A comparable situation might be taxes on cigarettes and alcohol. To ban cigarettes or alcohol infringes autonomy and is too coercive. Taxation is less coercive and is proportional (the more you smoke and drink the more you pay). Both taxation of cigarettes and financial penalties for non-immunization follow principles of distributive justice. Smoke if you must, but your taxes will offset the cost to society of smoking-related illnesses. If you choose not to immunize your child, the benefit payments saved will help pay for the cost of infectious diseases.

(iii) School exclusion during outbreaks

In New Zealand and some states of Australia, evidence of children's immunization status must be presented at school entry.[18] Immunization status rather than immunization is compulsory. Unimmunized children are excluded from school during outbreaks.

(iv) Outbreak legislation

It is possible to enact emergency legislation to compel immunization in the event of an outbreak such as an influenza pandemic or a bioterrorist smallpox attack. On the other hand, compulsion is scarcely likely to be necessary when the threat of death is very high.

(v) No fault vaccine injury compensation schemes

If the State makes immunization compulsory, then it seems mandatory that the State should compensate the few children who are injured by vaccines. Compensation should be for medical costs, pain and suffering, disability benefits and, if necessary, benefits for loss of earning and death.[19]

It could be argued that, because parents have their children immunized in good faith, and because no-one is to blame for the rare, severe, unpredictable vaccine injuries that occur, then Governments should introduce no-fault compensation schemes even when immunization is voluntary. Thus no-fault vaccine injury compensation schemes probably ought to be in place regardless of, rather than as an alternative to, compulsory immunization laws. There are at least 13 vaccine injury compensation programmes in the world, and immunization is compulsory in only four of those countries.[19]

Conclusion

Compulsory immunization will be regarded by many as justifiable in terms of the benefit to the individual child and to the community. But, in order to respect autonomy, State coercion should be kept to a minimum. We believe that, in general, children should not be compulsorily immunized when similar results can be achieved by education and inducements. Australia is in the happy position of having achieved very high rates of routine childhood immunization, over 90 per cent, without the need for compulsion.[20]

The case for compulsion might be stronger if immunization levels fell, but might not be necessary, because in that case epidemics would occur and the public would quickly recognize the value of immunization.

Whether or not childhood immunization is compulsory, a strong ethical case can be made for introducing a no-fault compensation scheme in Australia, and indeed in other countries.

Acknowledgements

We would like to express our profound gratitude to Bernadette Tobin of the Plunkett Centre for Ethics in Health Care, St Vincent's Hospital, Sydney for her insightful comments on the manuscript. We also thank Peter McIntyre and Kathy Currow of CHW, David Neil and Justin Oakley of the Department of Bioethics, Monash University, Miles Little and Chris Jordens of the Centre for Values, Ethics and the Law in Medicine, University of Sydney for helpful discussion.

Notes

1 Plotkin SA, Orenstein WA. *Vaccines*, 3rd edn. Philadelphia: WB Saunders, 1999.

2 NHMRC. *The Australian Immunisation Handbook*, 8th edn. Canberra: Australian Government Publishing, 2003.

3 Ada G, Isaacs D. *Vaccination: the Facts, the Fears, the Future.* Sydney: Allen & Unwin, 2000.

4 Wolfe RM, Sharp LK. 'Anti-vaccinationists past and present.' *BMJ* 2002; 325: 430–2.

5 Orenstein WA, Hinman AR. 'The immunization system in the United States – the role of school immunization laws.' *Vaccine* 1999; 17: S19–S24.

6 Drutz JE, Ligon BL. 'Polio: its history and its eradication.' *Semin. Pediatr. Infect. Dis.* 2000; 11: 280–6.

7 McIntyre P, Gidding H, Gilmour R *et al.* 'Vaccine preventable diseases and coverage in Australia, 1999–2000.' *Communicable Dis. Intelligence* 2002; 26: S1–111. Available from URL: www.healthgov.au/pubhlth/cdi/pubs/pdf/vpd99–00.pdf

8 Vitek CR, Wharton M. 'Diphtheria in the former Soviet Union: re-emergence of a pandemic disease.' *Emerg. Infect. Dis.* 1998; 4: 539–50.

9 Centres for Disease Control and Prevention. Poliomyelitis prevention in the United States: introduction of a sequential vaccination schedule of inactivated poliovirus vaccine followed by oral polio-virus vaccine. Recommendations of the Advisory Committee on Immunization Practices (ACIP). [erratum appears in MMWR Morb Mortal Wkly Rep 1997 February 28; 46 (8): 183], *MMWR – Morbidity Mortality Weekly Report* 1997; 46: 1–25.

10 Weibel RE, Caserta V, Benor DE, Evans G. 'Acute encephalopathy followed by permanent brain injury or death associated with further attenuated measles vaccines: a review of claims, submitted to the National Vaccine Injury Compensation Program.' *Pediatrics* 1998; 101: 383–7.

11 Chan RC, Penney DJ, Little D *et al.* 'Hepatitis and death following vaccination with 17D-204 yellow fever vaccine.' *Lancet* 2001; 358: 121–2.

12 Anderson RM, May RM. 'Vaccination and herd immunity to infectious diseases.' *Nature* 1985; 318: 323–8.

13 Feiken DR, Lezotte DC, Hamman RF, Salmon DA, Chen RT, Hoffman RE. 'Individual and community risks of measles and pertussis associated with personal exemptions to immunization.' *JAMA* 2000; 284: 3145–50.

14 Hershey JC, Asch DA, Thumasathit T, Meszaros J, Waters VV. 'The roles of altruism, free riding, and bandwagoning in vaccination decisions.' *Org. Behav. Hum. Decision Process* 1994; 59: 177–87.

15 Beauchamp TL, Childress JF. *Principles of Biomedical Ethics*, 5th edn. New York: Oxford University Press, 2001.

16 Bradley P. 'Should childhood immunisation be compulsory.' *J. Med. Ethics* 1999; 25: 330–4.

17   Hodges FM, Svoboda JS, Van Howe RS. 'Prophylactic interventions on children: balancing human rights with public health.' J. Med. Ethics 2002; 28: 10–16.

18   Dare T. 'Mass immunisations: some philosophical issues.' Bioethics 1998; 12: 125–49.

19   Evans G. 'Vaccine injury compensation programs worldwide.' Vaccine 1999; 17: S25–S35.

20   Hull B, Lawrence G, MacIntyre CR, McIntyre P. Immunisation Coverage: Australia 2001. Canberra: Commonwealth Department of Health and Ageing, 2002; Available from URL: www. immunise.health.gov.au/report.pdf/

Chapter 41

Simon Chapman

BANNING SMOKING OUTDOORS IS SELDOM ETHICALLY JUSTIFIABLE

Several Australian hospitals are proposing to extend their indoor smoking bans outdoors. In 1996, the mayor of Friendship Heights, Maryland similarly sought to ban smoking in municipal parks and on sidewalks. A leading non-smokers' rights advocate in Sydney has attracted publicity for his proposal to take civil action to have his suburban tennis club ban smoking in outdoor spectator areas, typically occupied by a handful of people waiting for courts to become vacant. In Kerala and Goa in India, smoking is banned in public spaces such as beaches, attracting fines.[1]

Such proposals "push the envelope" of tobacco control into areas where questions need to be asked to ensure tobacco control policies are firmly anchored to scientific evidence and especially concern those who value the freedom of individuals to do what they please to the extent that this does not harm others.[2] They invite consideration of whether zero tolerance of public exposure to toxic agents is a reasonable policy for civil societies and whether the loudly proclaimed exquisite sensitivities of a small minority should drive public policy.

Further, they invite us to reflect on the extent to which these policies risk alienating a large number of people who might otherwise be supportive of efforts to reduce environmental tobacco smoke (ETS) exposure in situations where there is significant risk or reduced amenity. In short, we need to ask whether efforts to prevent people smoking outdoors risk besmirching tobacco control advocates as the embodiment of intolerant, paternalistic busybodies, who, not content at protecting their own health, want to force smokers to not smoke even in circumstances where the effects of their smoking on others are immeasurably small. Such alienation may undermine support for other tobacco policies which, if implemented, may bring profound public health benefits to communities.

Why should smokers not be allowed to smoke in hospital grounds or other designated outdoor locations, well away from any reasonable prospect of harming (as distinct from visually offending) anyone? Advocates for outdoor bans advance a number of arguments, each of which raise significant ethical concerns. In this paper, I will rehearse their arguments, comment on why I believe they are ethically unsustainable, and conclude that there is justification for banning smoking in outdoor settings only in circumstances where exposure is sustained and significant such as in crowded spectator stadia.

"Any exposure to ETS is harmful"

At the heart of this debate is the question of at what level of exposure an agent might be reasonably deemed to be harmful. The 1997 Californian Environmental Protection Agency report on the health effects of exposure to ETS concluded that chronic, cumulative exposure – such as experienced by those living for years with smokers or who work in indoor environments where smoking is permitted – can increase the probability of dying from lung cancer by 20 percent, from heart disease by 30 percent, and can exacerbate asthma in children by 60–100 percent.[3] Acute exposure to ETS in healthy young adults has been shown to be associated with dose related impairment of endothelium dependent dilatation, suggesting early arterial damage.[4] However, the transitory and fleeting exposure to others' smoking in open outdoor settings is not remotely comparable to that experienced in confined indoor settings such as were involved in the study cited above.

Comment: Risk aversion research shows that the risks individuals declare unacceptable are not necessarily those which have a high probability of a serious adverse outcome.[5] Equally, people are often prepared to take risks that have high probabilities of negative outcomes. Of all the factors which have been identified as tending to increase public outrage, risks which are imposed rather than voluntary explain much of the variance in public perceptions.[6] Passive smoking represents a quintessential imposed risk and, together with the possibility of dreaded outcomes (like lung cancer), often incites public demand for zero exposure. This explains why many will get incensed about exposure to a mere whiff of tobacco smoke, but will not hesitate to sit around a romantic smoky campfire where they will, by choice, be exposed to a large range and volume of carcinogenic particulates and gases.

The question thus remains as to whether zero tolerance of ETS throughout communities is reasonable public policy. The answer to this question can be only be guided by science.

Science could attempt to quantify the health consequences of brief outdoor exposures and it would seem certain that for some rare individuals with exquisite sensitivity, an acute exposure at such a level might precipitate an adverse episode. Similar claims are sometimes made about a large range of environmental agents, but in general public policy is not based on cocooning such people from exposures that are inconsequential to nearly everyone.

"Outdoor bans send an 'important message' to the community"

Proponents of outdoor bans sometimes argue that such policies send an important "message" to communities that smoking is inconsistent with the sort of healthy environments that hospital directors might wish to promote. Smoking is not permitted anywhere on school grounds, so hospitals ought to set a similar example.

Comment: Schools are different to hospitals. Children do not have the same legal rights or autonomy as adults because they are not regarded as mature enough to make decisions such as choosing to not go to school, to buy alcohol or tobacco, or drive with safety. School authorities ban smoking by teachers on school grounds because of the important exemplar role teachers play with children. Such rules sometimes apply as conditions of employment in a hospital (such as requiring that health care staff should not smoke within sight of patients or visitors, just as some employers require for image reasons that their staff observe dress standards, do not smoke on duty, and so on). However, outdoor smoking does not harm staff or other patients, and because health care workers do not employ or somehow control patients and their visitors, there can be no justification for requiring them to not smoke outdoors.

If any message is sent to the community by an outdoor smoking ban, it may well be one that says health policy makers do not care about evidence of harm, but are more concerned to impose standards cut loose from any evidence base and indifferent to a vital ethical principle of respect for autonomy.

"Other freedoms are curtailed in hospitals . . . why not smoking too?"

We do not allow patients to go outdoors to drink alcohol or use illicit drugs. All patients already voluntarily forgo many freedoms while in hospitals, such as sexual activity. Why then should we allow them to go out to smoke?

Comment: Alcohol in moderation is not only harmless to health but demonstrably beneficial.[7] When alcohol is restricted in hospitals it is because of the risk of anti-social intoxication. In fact, it is not uncommon to allow patients to drink alcohol while in hospital. Narcotic dependent people may be prescribed legal narcotics such as methadone in an attempt to mollify their cravings and allow treatment for other presenting problems. It is unlikely that there is ever any formal ban on sexual activity in hospitals. That sexual activity does not occur much is due to patients mostly not feeling sexually alert, and for lack of privacy. However, the real point here is why "forgoing freedoms" should be construed as virtuous and an argument for adding further restrictions. I would argue that hospitals should rather strive to create an atmosphere that while respecting the rights of others to a healthy and peaceful environment, allows as many normal freedoms as possible for often distressed patients and their visitors.

"An enforced ban will be good for people's health"

Patients would benefit by taking a break from smoking as a necessary part of their admission.

Even a temporary period of restriction on smoking may improve a patient's prognosis and shorten their length of stay. In a publicly funded health system, there are obligations on patients to not unnecessarily complicate their recovery. They should therefore expect to be able not to smoke at all while in hospital.

Comment: Restrictions on smoking certainly do reduce smoking frequency and may also promote cessation.[8] However, while this is an undoubted positive benefit, it cannot be used as a front end justification to restrict smoking. It is a fortunate byproduct of bans introduced because of Millean based concerns about stopping smokers harming others.[9] The decision to bring benefit to oneself is a decision that should be up to the individual, not for others to impose.

Health care workers have a duty to offer assistance to those patients who want to quit (and I'd add, to try and *persuade* them). But there are many smokers who do not want to stop. When they are in hospital, they should not forfeit their rights to smoke if they want to, as long as they are not inflicting it on others. If they are unfortunate enough to be unable to get themselves outside to smoke, it would be a callous sort of health care that would say "you might want to smoke, but we so disapprove, we will not lift a finger to help you".

"Escorting patients outside can cause staff shortages leading to patient neglect"

In times when there are staff shortages, or a generally low staff:patient ratio, if staff need to escort incapacitated patients outdoors to smoke, this could cause neglect of patients remaining on wards. Also, if staff were to not remain with such patients while they smoked outdoors, liability might arise if, for example, the patient fell or had a coronary event.

Comment: This is an argument that should be irrelevant to the smoking status of the patients escorted outdoors. Most hospitals allow, when time permits, staff to escort patients outdoors to get fresh air, sunshine, and a change of scenery. Such attention to patient wellbeing is regarded as compassionate. While such assistance to patients is not a "right", if we exorcised every aspect of patient care down to the level of their "rights", a stay in hospital would be an even more dehumanising experience than it often is. If such practices endanger patients who remain inside, the issue of absences per se should be addressed and not confounded by arguing that some reasons for going outdoors are somehow more deserving than others. Similarly, if there are risks in leaving patients unattended outdoors, such risks obviously exist for patients regardless of their smoking status.

"Escorting staff will be exposed to patients' ETS"

If staff have to remain near an incapacitated patient who is taken outside to smoke, this may force them to be exposed to ETS, which violates their occupational health and safety rights.

Comment: The level of ETS exposure a nurse would encounter in an outdoor setting while escorting a smoking patient would be minimal. If downwind of the plume, the nurse could move upwind, and so on. Concern about such levels of exposure borders on fumophobia. Nurses would be daily exposed to far greater risks from infectious diseases, despite infection control protocols.

"Any form of air pollution should be stopped"

There have been several attempts to quantify the contribution of cigarette smoke to air pollution. For example, Sanner estimated that a pack a day

smoker pollutes the air with the same amount of mutagenic substances as when driving a car with a catalytic converter 35 km (22 miles).[10]

Comment: This startling conclusion appears counterintuitive, not least because the carcinogen benzene has been used to replace lead additives in fuel so that it can be used in catalytic converters. However, ASH UK's Clive Bates has pointed out that tobacco use is probably carbon neutral on a global level. Whatever carbon is put into the atmosphere during burning, equivalent carbon is sequestrated during tobacco growing as tobacco leaf fixes carbon from the air through photosynthesis. In a stable market, for each cigarette smoked, there is a tobacco leaf growing to produce the next cigarette. Non-carbon greenhouse emissions from tobacco burning such as oxides of nitrogen and methane are likely to be minute in the global scheme.[11]

Conclusion

A minority of people in tobacco control do not like to even *see* people smoking. Australian non-smokers' rights activist Brian McBride wrote recently to some of his colleagues about outdoor smoking: "We must be prepared to fight the aesthetics and personal standards argument as well as the health argument, and that is what I intend to do. We should not underestimate the public awareness value of having smokers found guilty of negligent actions in all situations indoors or outdoors. The more cases we run the better."

I would argue that the two need to be kept thoroughly apart. Mixing "aesthetics" arguments with health arguments risks infecting tobacco control with the accusation that it is fundamentally the province of people with capricious authoritarian proclivities, caring little for the scientific bedrock on which public health ought to stand.

Health promotion campaigns have often sought to portray smoking in ways different to those typically portrayed in tobacco advertising: smoking has been framed as desperate, disgusting, and slovenly. Don't such efforts also appeal to aesthetic rather than health concerns? Yes, but for my own part, I am comfortable with portrayals that seek to counterbalance the distortions of tobacco advertising with alternative definitions of reality. If a tobacco company can describe a carcinogenic product as "fresh", I am comfortable in countering that kissing a smoker is like licking an ashtray.

The world is full of people who do not like the "aesthetics" of others' different religions, race, sexual expression, modes of dress, or music. Too often these doctrines have found expression in paternalistic or downright oppressive regimes. We do not need authoritarian doctrines in tobacco control.

Notes

1 Anon. "Goa to ban smoking in public from Oct 2." *DeccanHerald* (India). 1999; September 12.
2 Mill JS. "On Liberty (1859)." In: Wollheim R, ed. *Three Essays*. Oxford: Oxford University Press, 1975: 1–141.
3 Environmental Protection Agency, California. Office of Environmental Health Hazard Assessment. "Health Effects of Exposure to Environmental Tobacco Smoke." Final Report 1997, www.oehha.org/scientific/ets/finalets.htm
4 Celermajer DS, Adams MR, Clarkson P et al. "Passive smoking and impaired endothelium-dependent arterial dilatation in healthy young adults." *N Engl J Med* 1996; 334: 150–4.
5 Chapman S, Wutzke S. "Not in our backyard: media coverage of community opposition to mobile phone towers: an application of Sandman's outrage model of risk perception." *Aust NZ J Public Health* 1997; 21: 614–20.
6 Covello VT. "Informing people about risks from chemicals, radiation, and other toxic substances: a review of obstacles to public understanding and effective risk communication." In: Leiss W, ed. *Prospects and Problems in Risk Communication*. Waterloo, Ontario: University of Waterloo Press, 1989: 1–49.
7 Rimm EB, Klatsky A, Grobbee D, et al. "Review of moderate alcohol consumption and reduced risk of coronary heart disease: is the effect due to beer, wine, or spirits." *BMJ* 1996; 312: 731–6.
8 Chapman S, Borland R, Brownson R, et al. "The impact of workplace smoking bans on declining cigarette consumption in Australia and the USA." *Am J Public Health* 1999; 89: 1018–23.
9 Goodin RE. "The ethics of smoking." *Ethics* 1989; 99: 574–624.
10 Sanner T. "Air pollution from cigarette smoking and gasoline cars with catalytic converter." Abstract for the 8th World Conference on Tobacco or Health, Buenos Aires, Argentina; 30 March–3 April 1992. Abstract 330.
11 Bates C. "Tobacco and greenhouse." Message posted to Globalink 26 June 1999.

Chapter 42

Amy L. Fairchild and Ronald Bayer

ETHICS AND THE CONDUCT OF PUBLIC HEALTH SURVEILLANCE

"**Surveillance is not research**. Public health surveillance is essentially descriptive in nature. It describes the occurrence of injury or disease and its determinants in the population. It also leads to public health action. . . . If we confuse surveillance with research, we may be motivated to collect large amounts of detailed data on each case. The burden of this approach is too great for the resources available. . . ."

World Bank Group[1]

Does the collection and analysis of data always constitute research and therefore require ethical oversight? This question has been raised in the context of quality assurance, program evaluation, oral history, and public health surveillance. In October 2003, for example, the Office for Human Research Protection (OHRP) in the Department for Health and Human Services (HHS) sought to resolve a longstanding controversy over whether those conducting oral history studies had to subject their work to institutional review board (IRB) review by a definitional sleight of hand: "Oral history interviewing activities, in general, are not designed to contribute to generalizable knowledge and, therefore, do not involve research as defined by HHS regulations and do not need to be reviewed by an institutional review board".[2] The effort to distinguish radically between research and non-research-related data gathering

was also reflected in a 2002 report by the World Bank designed to underscore the central importance of public health surveillance[1] (see quotation, above).

The efforts of the World Bank to draw a sharp distinction between research and surveillance were fundamentally rooted in economics, not ethics. The financial resources required to collect valid research data, it argued, were overwhelming. It was concern for human rights accompanying the AIDS epidemic within the U.S., and internationally, that fostered recent efforts on the part of the Centers for Disease Control and Prevention (CDC) and World Health Organization (WHO) to extend to surveillance ethical considerations heretofore restricted to research.[3,4] WHO endeavors represent a response to cumulative years of experience working in different countries and settings, which made clear the need for a systematic response to the many ethical questions that arose when implementing HIV surveillance.[5] Examination of the history leading to these undertakings will reveal, in our view, that tortured efforts based on definitions of these activities can only lead to inconsistencies.

In response to federal research protections promulgated in the 1970s, epidemiologists and ethicists began to discuss whether the principle of informed consent extended to the use of medical records and whether the insistence on individual consent would render epidemiological research virtually impossible.[6–8] In 1981, HHS regulations for the protection of human subjects explicitly exempted epidemiological research involving already existing data from informed consent requirements, provided the risk to subjects was minimal, the research did not record data in a way that was individually identifiable, and the research could not otherwise be conducted.[9] But the discussion did not extend to public health surveillance, which includes not only name-based reporting for conditions like TB and HIV, but also monitoring of food poisonings and blood lead levels.

The 1991 *International Guidelines for Ethical Review of Epidemiological Studies* issued by the Council for International Organizations of Medical Sciences (CIOMS) stated that although the vast majority of surveillance should be subject to approval by ethical review committees, "[a]n exception is justified when epidemiologists must investigate outbreaks of acute communicable diseases. Then they must proceed without delay to identify and control health risks".[10]

The CIOMS guidelines did indicate there were still areas of uncertainty, such as "when both routine surveillance of cancer and original research on cancer are conducted by professional staff of a population-based cancer registry." To resolve difficult situations, CIOMS called for the guidance of ethical review committees[10].

In the early 1990s, in response to the charge that the CDC's blinded HIV sero-prevalence studies among childbearing women constituted research conducted without informed consent, HHS's Office for Protection from Research Risks (OPRR) began to advance the notion that all surveillance was research and might require particular kinds of review.[11] This stance alarmed the CDC as well as the Council of State and Territorial Epidemiologists (CSTE). According to the CDC, "The implications of calling public health surveillance research are broad and far reaching. . . . If all surveillance activities were research, it might mean each local health department would have to form institutional review boards (IRBs)".[12] To the CDC, this was more than a bureaucratic consideration: if surveillance activities were designated research, the CDC feared that "people with TB could prevent their names from being reported to the health department or refuse to provide information about their 'contacts',"[13] thus inhibiting disease prevention efforts.

A first set of CDC recommendations drafted in 1996 in response to an OPRR mandate made the case that public health surveillance is differentiated from pure research by intent: Whereas "[t]he intent of research is to contribute to or generate generalizable knowledge; the intent of public health practice is to conduct programs to prevent disease and injury and improve the health of communities".[14]

The CDC is rarely involved in the use of surveillance for public health interventions like contact tracing. However practical their research might be, it remained research requiring IRB review in the minds of CDC officials. As the CDC began to work with state health departments to refine their guidelines, the profoundly different ways in which the states and the federal government drew the boundary between research and public health surveillance became clear.

States, typically operating under statutes mandating departments of health to collect and act on individual-level morbidity data for the explicit purpose of controlling disease, tended to view most public health surveillance activities as practice.[15] The New York City Department of Health, for example, maintained that "we derive knowledge that may protect the particular 'victims' before us. However, it may also be that it is too late to help the particular victims, but that the activity, or the information derived from it, becomes generalizable so as to protect the general population." CSTE concurred, arguing that "we are rarely able to conduct an investigation that provides any medical benefit to those already infected" (15). For example, in the case of food-borne outbreak investigations, the "major benefit has been to others than those we identified and obtained data and specimens from".[16]

However, it was also the provenance of surveillance undertakings that determined whether an activity was research or practice for the states. CSTE explained that its members consistently collected data with an eye not only to the present but also to the future. By review of previous investigations of trichinosis outbreaks, for example, the Alaska state health department not only developed an early diagnostic test for the disease but also identified an animal species not previously known to harbor Trichinella, as well as a new subspecies of the disease-causing organism: "We do not view these activities as research" when conducted by the state; "if conducted by an entity other than the state public health agency," however, "we would define it as research and require an outside researcher to obtain IRB approval"[16]. By definition, then, what public health departments did was not research.

In the end, the CDC maintained that "activities can be viewed differently at federal and state levels".[17] That is, the same initiative might be designated research at the federal level and require IRB review and practice at the state level, requiring no ethical review. But even at the federal level, some health officials have found that they do not distinguish between research and practice consistently and that they sometimes face political pressure to define an activity as practice rather than research.[18]

The recent efforts to provide definitional solutions to the question of research and public health practice involve twists and turns that inevitably produce results that are riddled with inconsistencies and that are conceptually unsatisfying. It is time to resolve the matter by acknowledging the necessity of ethical review of public health surveillance activities at both state and federal levels, whether such activities fall neatly under the classification of research or practice or exist in a gray borderland.

Those involved in public health efforts appear increasingly ready to embrace ethical principles to govern the practice of public health surveillance.[19] Not yet have they greeted with enthusiasm proposals to establish mechanisms to assure ethical review. In the history of research ethics, it was only when forums were created to assure consideration of the rights of subjects that guiding principles were given any meaningful force. The creation of institutionally based review procedures was not without conflict, as researchers resisted the notion that they themselves could not be the instruments of their own review.[20] Although the establishment of bodies responsible for the review of the ethics of surveillance need not mirror the already extant IRBs, as has been proposed by some experts on human subjects research,[21] it is clear that some form of explicit, systematic review is necessary.

As we take the first steps in trying to develop the ethics of public health surveillance, we must acknowledge that the guidelines governing research and clinical practice, so focused on protecting individuals, cannot be imported to the public health setting, where the first priority must be the protection of the communal welfare. In the context of public health, individuals may be compelled to do things or desist from doing things to protect or enhance the common good, even in the face of uncertainty.[22] But the invocation to act, especially when the individual rights of privacy and liberty may be impinged, must be subject to limits. Ethical oversight in the public health setting must thus take into account the inevitable tension between the claims of individuals and those of the common good. In developing the ethics of public health surveillance and envisioning a mechanism for oversight, it would be wise to recognize that simple rules will never suffice: ethical sensitivity necessitates an open discussion of how the ethical trade-offs and tensions can be fairly resolved.

Reform may require changes in the legal and regulatory context within which surveillance occurs. That is, states might attempt to legally mandate surveillance practices that ethical oversight would deny, such as linkage of HIV registries with school registries. Alternatively, state law may forbid practices, such as the linkage of HIV and TB registries, which might be ethically demanded if they enhanced the capacity of a health department to fulfill its central mission. The public context of those changes should provide an occasion for a full and transparent airing of the procedural and substantive issues involved. This will be especially challenging given the fact that the proposed changes are not driven by the kinds of scandal and abuse that spurred the ethical oversight of clinical research.

It is inappropriate to regard ethical oversight strictly as an impediment. In the context of public health surveillance, it can serve as a means of avoiding inadvertent breaches in confidentiality and stigma; it can help to ensure that the public understands that surveillance will occur and what purposes it serves; it can protect politically sensitive surveillance efforts. There is, after all, an ethical mandate to undertake surveillance that enhances the well-being of populations.

Notes

1 World Bank Group, "Public health surveillance toolkit," http://survtoolkit.worldbank.org (2002); accessed 12 November 2003.

2 J. Brainard, *Chron. High. Educ.* 50, A25 (31 October 2003).

3 A. L. Fairchild, R. Bayer, "Ethical issues in second-generation surveillance: Guidelines" (WHO, Geneva, in press).

4 CDC, Agendas: HIV/AIDS Public Health Data Uses Project, second preconsultation meeting, 16 June 2003, Atlanta, GA, and 19 September 2003, New York, NY.

5 C. Obermeyer, WHO, personal communication, 12 January 2004.

6 L. Gordis, E. Gold, R. Seltzer, *Am. J. Epidemiol.* 105, 163 (1977).

7 A. M. Capron, *Law Med. Health Care* 19 (fall-winter), 184 (1991).

8 C. I. Cann, K. J. Rothman, *IRB: Rev. Hum. Subj. Res.* 6(4), 5 (1984).

9 Code of Federal Regulations (45 C.F.R. 46).

10 CIOMS, *International Guidelines for Ethical Review of Epidemiological Studies* (Council for International Organizations of Medical Sciences, Geneva, 1991); reprinted in *Law Med. Health Care.* 19 (fall-winter), 247 (1991).

11 Office for Protection from Research Risks, Division of Human Subject Protections, "Evaluation of human subject protections in research conducted by the Centers for Disease Control and Prevention and the Agency for Toxic Substances and Disease Registry," (Centers for Disease Control and Prevention, Atlanta, GA. July 1995), p. 14.

12 CDC meeting on Protection of Human Research Subjects in Public Health, Atlanta, GA, 18 and 19 March 1996, minutes, p. 2.

13 D. E. Snider, D. F. Stroup, *Public Health Rep.* 112, 30 (1997).

14 Ref. (13), p. 32.

15 CDC and the Council of State and Territorial Epidemiologists meeting, Atlanta, GA, 4 and 5 March 1999, minutes, p. 22.

16 J. Middaugh, Council of State and Territorial Epidemiologists, letter to M. Speers, Centers for Disease Control and Prevention, 7 June 1999, p. 2.

17 Ref. (15), p. 27.

18 Reported to authors on condition of anonymity.

19 American Public Health Association, Code of Ethics; available at www.apha.org/codeofethics/.

20 D. J. Rothman, *Strangers at the Bedside: A History of How Law and Bioethics Transformed Medical Decision Making* (Basic Books, New York, 1991), p. 63.

21 R. Levine, interview with A. Fairchild, 2 December 2002.

22 T. H. Ackerman, in *The Ethics of Research Involving Human Subjects: Facing the 21st Century*, H. Y. Vanderpool, Ed. (University Publications Group, Frederick, MD, 1996), pp. 83, 85, 86.

How are scarce medical resources to be justly allocated?

INTRODUCTION TO PART EIGHT

HOW SHOULD HEALTH care resources be allocated since they are insufficient to meet all health care needs? A natural answer is that resources should be used to achieve maximum benefit. The first selection describes one of the most influential versions of this answer, based on QALYs (quality adjusted life years). As Williams explains, the 'essence of a QALY' is that it combines two goods at which health care interventions aim, namely longevity and quality of life. By convention, a year of healthy life expectancy has the benchmark figure of 1; a year of unhealthy life expectancy is valued as less than 1, its precise value being lower the worse the unhealthy person's quality of life. Given this, it is in principle possible to determine the number of QALYs a health care intervention will generate. In turn, the cost-per-QALY of the intervention can be calculated. The upshot is a method of priority setting: 'A high priority health care activity is one where cost-per-QALY is low, and a low priority activity is one where cost-per-QALY is high'.

Despite methodological difficulties, Williams is happy with 'the ethical judgement' the QALYs approach 'imposes'. By contrast, Harris advocates using QALYs to decide between different treatment options for one particular patient, but not to decide which of different patients to treat. At the heart of Harris' objection is his view that the value of a life is the value it has to its bearer; and the value of a life does not vary with contingencies, such as resource scarcity or quality of life. Hence, there is a moral obligation to save the life of each and every person who wants to be saved. By contrast, the QALYs approach assumes that the value of a life and, correlatively, the strength of the obligation to save it, vary according to the bearer's circumstances. For example, to cite Harris' case study, Andrew's circumstances are such that, given the QALYs approach, his life is effectively rendered less valuable than George's.

Harris extrapolates from this core objection to suggest that the QALYs approach perpetrates systematic injustices against groups such as the elderly, lower social classes, and ethnic minorities. The exchange of views between Williams and Grimley Evans pursues the theme by focusing on age based rationing. Part of Williams' case for taking age into account when making allocation decisions is that one's capacity to benefit from health care diminishes with age. Another part of his case comprises the 'fair innings' argument, which trades on the intuition that someone who has led a full life has less of a claim on scarce health resources than someone who has not. Williams' position on age and rationing is in keeping with the original QALYs approach because what justifies treating the young rather than the old is that doing so maximises benefit from scarce resources. Hence, Williams' concession that age matters would do nothing to placate Harris, who decries this maximising agenda and the resulting age discrimination. Likewise, Grimley Evans objects to 'ageism', provocatively likening 'the exclusion from treatment on the basis of a patient's age without reference to his or her physiological condition' to classism, racism and sexism. Echoing Harris, Grimley Evans insists that 'the only person who can put a value on a life is the person living it ... [lives] are therefore formally incommensurable'. Additionally, he distinguishes two ideas for which fair innings codes and questions the usefulness of each in pursuing equity.

The dispute over rationing and the elderly is generated by common intuitions about age and health. Hope focuses on another set of relevant moral intuitions. People are moved to use a disproportionate number of resources to try to save the identifiable victim of a catastrophe in immediate danger of losing their life, thereby incurring the opportunity cost of failing to save numerous unidentifiable people. This intuition, referred to as the 'rule of rescue', opposes the standard QALYs approach, which is impersonal in requiring us to commit resources in whichever way will maximise benefit, irrespective of the identity of beneficiaries. Hope argues that the rule of rescue should not inform resource allocation decisions because even its most plausible justification – namely, a contractualist argument based on people's willingness to incur a slightly higher risk to oneself in order to help identifiable people – is weak.

Maximising benefit seems a natural goal when distributing scarce health care resources, but an equally natural thought is that health care resources should be deployed in order to meet needs. This is represented by the final selection, taken from an early statement of Daniels' influential views, the main components of which are as follows. First, 'the needs which interest us are those things we need in order to achieve or maintain species-typical normal functioning'. Such needs are important because they connect with our interest in 'the range of opportunity we have within which to construct life-plans and conceptions of the good'. In turn, health care needs are things we need in order to achieve normal species functioning. This provides 'a fairly crude measure of the relative importance of health-care needs': the more a condition curtails the normal opportunity range, the more important it is to treat it. Daniels extends Rawls' theory of justice to health care by including health care institutions amongst others that provide for fair equality of opportunity. In ending his piece by sketching some applications of his account, Daniels implicitly clarifies the contrast between a needs based approach, and Williams' QALYs approach, to resource allocation.

Selections

Williams, A. 1985. 'The Value of QALYs', *Health and Social Service Journal* (Supplement), July 18th, 3–5.

Harris, J. 1987. 'QALYfying the Value of Life', *Journal of Medical Ethics*, 13 (3), 117–23 [plus Williams' 'Response'].

Williams, A. and Grimley Evans, J. 1997. 'The Rationing Debate. Rationing Health Care by Age: The Case for, and the Case Against', *British Medical Journal*, 314 (7083), 820–5.

Hope, T. 2001. 'Rationing and Life-saving Treatments: Should Identifiable Patients Have Higher Priority?' *Journal of Medical Ethics*, 27 (3), 179–85.

Daniels, N. 1981. 'Health-care Needs and Distributive Justice', *Philosophy and Public Affairs*, 10 (2), 146–79.

Chapter 43

Alan Williams

THE VALUE OF QALYs

The big strategic issues faced by most health care systems nowadays concern priority setting and the quest for efficiency. Priority setting is essentially a matter of deciding which are the most valuable things that can be done with the available resources. The quest for efficiency has two prongs. At a low level it is essentially a matter of ensuring that no activity costs more than is necessary. At a high level it is a matter of ensuring that no activity is pursued beyond the point where the value of the extra benefits it generates outweighs the value of the extra resources it uses.

Economics is often seen as being about costs, and in that perception economists are virtually indistinguishable from cost accountants. But economics is really about efficiency, and it will be clear from my definition of high-level efficiency that to test it you must be equally concerned with both costs and benefits. Hence the increasing interest shown by economists in the measurement and valuation of the benefits (as well as the costs) of health care.

Quality of life

For the strategic purposes set out above we need a simple, versatile, measure of success which incorporates both life expectancy and quality of life, and which reflects the values and ethics of the community served. The 'quality adjusted life year' (QALY) measure fulfills such a role.

The essence of a QALY is that it takes a year of healthy life expectancy to be worth 1, but regards a year of unhealthy life expectancy as worth less than 1. Its precise value is lower the worse the quality of life of the unhealthy person (which is what the 'quality adjusted' bit is all about). If being dead is worth zero, it is, in principle, possible for a QALY to be negative, i.e. for the quality of someone's life to be judged as worse than being dead.

The general idea is that a beneficial health care activity is one that generates a positive amount of QALYs, and an efficient health care activity is one where the cost-per-QALY is as low as it can be. A high priority health care activity is one where cost-per-QALY is low, and a low priority activity is one where cost-per-QALY is high. Thus an activity which generates only two QALYs but costs only £200 (so that each QALY costs £100 to 'produce') is more efficient than one which generates five QALYs but costs £2,000 (so that each QALY costs £400 to 'produce').

Resource allocation

In this context, no explicit money value is attached to a QALY, one simply calculates what it costs. If one wants to go further and decide which activities can and can't be 'afforded', decision makers will need (implicitly or explicitly) to value QALYs, but that is not a central issue here. I shall concentrate on the processes that lie behind the apparently simple statements made above about the essence of the QALY. Superficially it is an attractive idea and intuitively acceptable in outline but, it contains some very important issues which need to be faced and resolved. It is worth some effort to do this, however, because if we knew the extra cost of an extra QALY for every health care activity this would considerably help anyone involved in resource allocation decisions at all levels in the health care system.

But we must accept the assumptions on which the derivation of a QALY is based.

The first such assumption is that the objective of health care systems is to improve health. The second is that health can be defined in terms that make sense to all sorts of different people. The third is that it is possible to elicit meaningful valuation statements from people about differing degrees of ill-health. The fourth is that it is possible to aggregate these individual valuations in a manner which is consistent with the ethical and political bases of the health-care system.

On the first assumption, cynics say that the welfare of those whose livelihood depends on the supply of health care often seems as important in explaining policy decisions as the welfare of patients or potential patients. So it should not be taken for granted.

The second assumption is much more interesting. It immediately creates potential tensions between the technical clinical data and the layperson's view about health. A clinician may need to know about abstruse parameters (such as gamma globulin count) which a patient cannot readily sense or even understand. Ordinary people do, however, know what it means to be unable to rise from a chair, or to be too confused to be able to look after yourself, or to be in pain, or excessively anxious, or to be tired, irritable, nauseous, feverish, etc. These are the kinds of experiences that lead people to suspect that they may be ill, and on which they base judgements as to how ill they are.

Patients will not usually be aware of the significance of these 'symptoms' for their future health states. That is where the special knowledge of the doctors comes in, especially if there is something that can be done to improve that prognosis. But the very notion of 'improving the prognosis' has to be cast in these same general terms if it is to mean anything to the patient. To offer some treatment that would get my gamma globulin count back to 'normal' doesn't mean anything to me but if I were told that it would completely relieve me of this pain that has been troubling me, and increase my life expectancy by five years, I would be distinctly interested and have some idea of its value.

Basic capabilities

It is perhaps a bit pretentious to label these statements about pain and distress, physical mobility, capacity for self-care, and ability to play normal social roles, as statements about 'quality of life', when some people might wish to reserve that phrase for rather higher things (like an appreciation of Beethoven's late string quartets, the joy of gardening, a talent for water colour painting, or a taste for good wine). Some of these higher things are likely to be adversely affected by illness, but they may not be. Health care is mostly about much more basic capabilities which, when we are healthy we take for granted but when they fail us, tend to be rather pervasively disruptive even though they may be quite 'minor' and not at all life threatening (just think of severe toothache!).

So the 'quality adjustment' initially involves describing different states of ill-health as combinations of these different characteristics. For example, the Rosser classification (Table 43.1) uses eight different degrees of disability, all but one of which may be associated with four different degrees of distress (the exception is 'unconscious', which is assumed to entail no distress). These descriptions are not specific to any particular kind of illness or injury, or to any particular kind of person, and they do not use terms which ordinary people are incapable of understanding.

The 'art' lies in picking on a limited number of states which collectively cover a wide range

Table 43.1 Rosser's Classification of illness states

	Disability	Distress
I	No disability	A No distress
II	Slight social disability	B Mild
III	Severe social disability and/or slight impairment of performance of work. Able to do all housework except very heavy tasks	C Moderate
		D Severe
IV	Choice of work or performance at work very severely limited Housewives and old people able to do light housework only but able to go out shopping	
V	Unable to undertake any paid employment Unable to continue any education Old people confined to home except for escorted outings and short walks and unable to do shopping Housewives able only to perform a few single tasks	
VI	Confined to chair or to wheelchair or able to move around in the house only with support from an assistant	
VII	Confined to bed	
VIII	Unconscious	

See: Kind, Rosser and Williams: 'Valuation of Quality of Life: Some Psychometric Evidence' in Jones-Lee, M.W. (editor) *The Value of Life and Safety*, North Holland, 1982.

of experiences and which provide an array of pigeon-holes into which most experiences of ill-health can be fitted without much distortion or sacrifice of significant details. If the classification system becomes too detailed and complex, people cannot understand and handle it. If it is too crude and simple, it lacks credibility and people will not use it.

This brings us to the third assumption listed concerning the eliciting of relative valuations. There are a variety of techniques for doing this. For example, suppose that each of the 29 Rosser states were printed on a card, and respondents were asked to rank them in order from best to worst. 'No disability and no distress' will surely be best, but it is not so obvious which will be worst.

Suppose now that respondents are asked to say whether they regard any of the bad states as being as bad as (or worse than) death, and if so which ones. This establishes states which have a value of 0 (if as bad as death) or negative values (if worse than death). If the healthy state is said to be equal to 1, and dead is equal to 0, then respondents can then be asked to attach a number (less than 1) to each of the other states, including negative numbers (unconstrained) to each state which is believed to be worse than death. Thus a state valued at 0.5 is felt to be only of half the value as being healthy would be, with the implication that two years life expectancy in that state is felt to be of equal value to one year of healthy life expectancy.

Making judgements

A treatment which offered the prospect of four years of healthy life expectancy in place of four years in a state valued at 0.75, would be 'worth' 1 QALY (0.25 of a QALY is gained each year for four years), whereas a treatment which moved someone from a state valued at 0.25 to one valued at 0.75 for four years would be worth two QALYs. An actual set of such values, (being the median values for each state as elicited from 70 respondents) is set out in Table 43.2.

There are, of course, much easier ways of getting this kind of information, such as asking the professionals to make such judgements, or getting policy makers to express their views in this way, as rather explicit statements of 'policy'. Professionals (e.g., doctors and nurses) will see a great many of people in these states of ill-health, so they may be better able than other people to imagine what it is like to be in them.

But in a democracy no one can really set themselves up as experts in what other people's values ought to be, though the people's representatives do have a legitimate right to express a set of values on behalf of their electorate, and if they get it wrong, they may be held to account.

The issue as to whose values shall count is not a scientific one but a political one. A QALY may be derived by any one of several different valuation processes and that choice is essentially a socio-political judgement requiring socio-political justification. This is equally true of the decisions to use survival rates or life expectancy as a criterion of success since they assume that everyone's survival or life expectancy is of equal value, no matter what the quality of their life is (i.e. only survival is of any value).

If it is decided that the appropriate valuation process is one which involves the aggregation of the valuations of many different individuals and the eliciting of some 'average' value of each health state for that community, then a further issue has to be faced and resolved, as indicated earlier in the fourth assumption namely, what are the ethical implications of this?

Suppose we added together each person's valuations of a particular health state and used the arithmetic mean as the group value of that state. This would be repeated for all states. Since 'healthy' was postulated to be worth 1 for each individual, it will come out as 1 in the averaging process. But there is no way of knowing how any other particular state will come out.

Group norm

What we have done is to impose the ethical judgement that one year of healthy life expectancy is of equal value no matter who gets it, and other states are properly valued relatively to this 'standard' unit of value by each individual, whose valuations have equal weight. If we can accept that ethical judgement as an appropriate basis for health service policy, then we can accept that method of arriving at a group norm. If not, we can't.

But if not, what ethical judgement would we put in its place? The most commonly advocated one is that the value of an additional year of healthy life expectancy is highest somewhere between the ages of 25 and 45, and tails off progressively on either side of that age band. It would be interesting to know whether that is a majority view, and, if so, precisely what the profile of age-adjusted QALYs looks like (assuming that across the whole population the average value is still 1). If we knew that it would be perfectly feasible to use this more complex

Table 43.2 Rosser's valuation matrix: all 70 respondents.

Disability Rating	Distress Rating			
	A	B	C	D
I	1.000	0.995	0.990	0.967
II	0.990	0.986	0.973	0.932
III	0.980	0.972	0.956	0.912
IV	0.964	0.956	0.942	0.870
V	0.946	0.935	0.900	0.700
VI	0.875	0.845	0.680	0.000
VII	0.677	0.564	0.000	−1.486
VIII	−1.028	NOT	APPLICABLE	

Fixed points: Healthy = 1 Dead = 0
Ibid

age adjusted QALY instead of the one based on the assumption that 1 QALY is of equal value to everybody.

Finally, there are a few complications. The first relates to the discounting of QALYs as they become more and more remote in the future as people generally prefer good things sooner rather than later. Since the future stream of costs with which QALYs are to be compared will be discounted to an equivalent present value to reflect this sense of 'time-preference', then the future stream of benefits must also be discounted.

The second complication concerns the fact that prognoses are uncertain, so no treatment actually guarantees a particular amount of QALYs. A treatment may offer a 0.8 probability of 10 QALYs, a 0.1 probability of none, and a 0.1 probability of -5 QALY's (i.e. the loss of the equivalent of five years' healthy life expectancy). Seen in group terms, these probabilities may be interpreted as the fractions of the treated population which will experience each of the outcomes. Either way, we could say that the 'expected' value of the benefits from treatment is $(0.8 \times 10) + (0.1 \times 0) + (0.1 \times -5)$ which is $8 + 0 - 0.5$ or 7.5 QALYs. Thirdly, a treatment may yield QALYs to people other than the person actually being treated (e.g. making person A better may reduce the distress of persons B and C) and in principle all these benefits should be added together to arrive at the proper benefit level for any particular treatment.

To sum up, a QALY is a flexible framework within which a variety of valuation processes can be accommodated. It encompasses both quality of life and life expectancy, and allows individuals to express their own willingness to sacrifice quality for quantity or vice versa, thus enabling one to fuse into a single index the often disjointed criteria of 'survival prospects' and 'symptomatic relief'. It depends on the possibility of finding fairly simple, yet widely applicable, descriptions of different states of ill-health, which are then to be used in two distinct ways.

On the one hand, it must be possible for ordinary individuals to value them relatively to each other, and on the other it must be possible for the practitioners to offer prognoses in terms of them. The former requirement is a challenge to economics and psychometry. The latter is a challenge to clinical research in general (be it medical, nursing or remedial).

The reward for our combined efforts, if successful, would be a great improvement in the usefulness of research on the evaluation of our various activities, and in policy analysis and priority setting in the health care system in general. It might even catch on in the personal social services where it seems equally applicable. It is surely a prize worth striving for.

Chapter 44

John Harris

QALYFYING THE VALUE OF LIFE

Against a background of permanently scarce resources it is clearly crucial that such health care resources as are available be not used wastefully. This point is often made in terms of 'efficiency' and it is argued, not implausibly, that to talk of efficiency implies that we are able to distinguish between efficient and inefficient use of health care resources, and hence that we are in some sense able to measure the results of treatment. To do so of course we need a standard of measurement. Traditionally, in life-endangering conditions, that standard has been easy to find. Successful treatment removes the danger to life, or at least postpones it, and so the survival rates of treatment have been regarded as a good indicator of success.[1] However, equally clearly, it is also of crucial importance to those treated that the help offered them not only removes the threat to life, but leaves them able to enjoy the remission granted. In short, gives them reasonable quality, as well as extended quantity of life.

A new measure of quality of life which combines length of survival with an attempt to measure the quality of that survival has recently[2] been suggested and is becoming influential. The need for such a measure has been thus described by one of its chief architects: 'We need a simple, versatile, measure of success which incorporates both life expectancy and quality of life, and which reflects the values and ethics of the community served. The "Quality Adjusted Life Year" (QALY) measure fulfils such a role'.[3] This is a large claim and an important one, if it can be sustained its consequences for health care will be profound indeed.

There are, however, substantial theoretical problems in the development of such a measure, and more important by far, grave dangers of its misuse. I shall argue that the dangers of misuse, which partly derive from inadequacies in the theory which generates them, make this measure itself a life-threatening device. In showing why this is so I shall attempt to say something positive about just what is involved in making scrupulous choices between people in situations of scarce resources, and I will end by saying something about the entitlement to claim in particular circumstances, that resources are indeed scarce.

We must first turn to the task of examining the QALY and the possible consequences of its use in resource allocation. A task incidentally which, because it aims at the identification and eradication of a life-threatening condition, itself (surprisingly perhaps for a philosophical paper) counts also as a piece of medical research,[4] which if successful will prove genuinely therapeutic.

The QALY

i. *What are QALYs?*

It is important to be as clear as possible as to just what a QALY is and what it might be used for. I cannot do better than let Alan Williams, the architect of QALYs referred to above, tell you in his own words:

> The essence of a QALY is that it takes a year of healthy life expectancy to be worth one, but regards a year of unhealthy life expectancy as worth less than 1. Its precise value is lower the worse the quality of life of the unhealthy person (which is what the "quality adjusted" bit is all about). If being dead is worth zero, it is, in principle, possible for a QALY to be negative, i.e. for the quality of someone's life to be judged worse than being dead.
>
> The general idea is that a beneficial health care activity is one that generates a positive amount of QALYs, and that an efficient health care activity is one where the cost per QALY is as low as it can be. A high priority health care activity is one where the cost-per-QALY is low, and a low priority activity is one where cost-per-QALY is high.[5]

The plausibility of the QALY derives from the idea that 'given the choice, a person would prefer a shorter healthier life to a longer period of survival in a state of severe discomfort and disability'.[6] The idea that any rational person would endorse this preference provides the moral and political force behind the QALY. Its acceptability as a measurement of health then depends upon its doing all the theoretical tasks assigned to it, and on its being what people want, or would want, for themselves.

ii. *How will QALYs be used?*

There are two ways in which QALYs might be used. One is unexceptionable and useful, and fully in line with the assumptions which give QALYs their plausibility. The other is none of these.

QALYs might be used to determine which of rival therapies to give to a particular patient or which procedure to use to treat a particular condition. Clearly the one generating the most QALYs will be the better bet, both for the patient and for a society with scarce resources. However, QALYs might also be used to determine not what treatment to give *these* patients, but which group of patients to treat, or which conditions to give priority in the allocation of health care resources. It is clear that it is this latter use which Williams has in mind, for he specifically cites as one of the rewards of the development of QALYs, their use in 'priority setting in the health care system in general'.[7] It is this use which is likely to be of greatest interest to all those concerned with efficiency in the health service. And it is for this reason that it is likely to be both the most influential and to have the most far-reaching effects. It is this use which is I believe positively dangerous and morally indefensible. Why?

iii. *What's wrong with QALYs?*

It is crucial to realise that the whole plausibility of QALYs depends upon our accepting that they simply involve the generalisation of the 'truth'[8] that 'given the choice a person would prefer a shorter healthier life to a longer period of survival in a state of severe discomfort'. On this view giving priority to treatments which produce more QALYs or for which the cost-per-QALY is low, is both efficient and is also what the community as a whole, and those at risk in particular, actually want. But whereas it follows from the fact that given the choice a person would prefer a shorter healthier life to a longer one of severe discomfort, that the best treatment *for that person* is the one yielding the most QALYs, it does not follow that treatments yielding more

QALYs are preferable to treatments yielding fewer where *different people* are to receive the treatments. That is to say, while it follows from the fact (if it is a fact) that I and everyone else would prefer to have, say one year of healthy life rather than three years of severe discomfort, that we value healthy existence more than uncomfortable existence for ourselves, it does not follow that where the choice is between three years of discomfort for *me* or immediate death on the one hand, and one year of health for *you*, or immediate death on the other, that I am somehow committed to the judgement that you ought to be saved rather than me.

Suppose that Andrew, Brian, Charles, Dorothy, Elizabeth, Fiona and George all have zero life-expectancy without treatment, but with medical care, all but George will get one year complete remission and George will get seven years' remission. The costs of treating each of the six are equal but George's operation costs five times as much as the cost of each of the other operations. It does not follow that even if each person, if asked, would prefer seven years' remission to one for themselves, that they are all committed to the view that George should be treated rather than that they should. Nor does it follow that this is a preference that society should endorse. But it is the preference that QALYs dictate.

Such a policy does not value life or lives at all, for it is individuals who are alive, and individuals who lose their lives. And when they do the loss is principally their loss. The value of someone's life is, primarily and overwhelmingly, its value to him or her; the wrong done when an individual's life is cut short is a wrong to that individual. The victim of a murder or a fatal accident is the person who loses his life. A disaster is the greater the more victims there are, the more lives that are lost. A society which values the lives of its citizens is one which tries to ensure that as few of them die prematurely (that is when their lives could continue) as possible. Giving value to

life-years or QALYs, has the effect in this case of sacrificing six lives for one. If each of the seven *wants* to go on living for as long as he or she can, if each values the prospective term of remission available, then to choose between them on the basis of life-years (quality adjusted or not), is in this case to give no value to the lives of six people.

iv. The ethics of QALYs

Although we might be right to claim that people are not committed to QALYs as a measurement of health simply in virtue of their acceptance of the idea that each would prefer to have more QALYs rather than fewer for themselves, are there good moral reasons why QALYs should none the less be accepted?

The idea, which is at the root of both democratic theory and of most conceptions of justice, that each person is as morally important as any other and hence, that the life and interests of each is to be given equal weight, while apparently referred to and employed by Williams plays no part at all in the theory of QALYs. That which is to be given equal weight is not persons and their interests and preferences, but quality-adjusted life-years. And giving priority to the manufacture of QALYs can mean them all going to a few at the expense of the interests and wishes of the many. It will also mean that all available resources will tend to be deployed to assist those who will thereby gain the maximum QALYs – the young.

v. The fallacy of valuing time

There is a general problem for any position which holds that time-spans are of equal value no matter who gets them, and it stems from the practice of valuing life-units (life-years) rather than people's lives.

If what matters most is the number of life-years the world contains, then the best thing we can do is devote our resources to increasing the population. Birth control, abortion and sex education come out very badly on the QALY scale of priorities.

In the face of a problem like this, the QALY advocate must insist that what he wants is to select the therapy that generates the most QALYs for those people who already exist, and not simply to create the maximum number of QALYs. But if it is people and not units of life-span that matter, if the QALY is advocated because it is seen as a moral and efficient way to fulfil our obligation to provide care for our fellows, then it does matter who gets the QALYs – because it matters how people are treated. And this is where the ageism of QALYs and their other discriminatory features become important.

vi. QALYs are ageist

Maximising QALYs involves an implicit and comprehensive ageist bias. For saving the lives of younger people is, other things being equal, always likely to be productive of more QALYs than saving older people. Thus on the QALY arithmetic we always have a reason to prefer, for example, neonatal or paediatric care to all 'later' branches of medicine. This is because any calculation of the life-years generated for a particular patient by a particular therapy, must be based on the life expectancy of that patient. The older a patient is when treated, the fewer the life-years that can be achieved by the therapy.

It is true that QALYs dictate that we prefer people, not simply who have *more life expectancy*, but rather people who have *more life expectancy to be gained from treatment*. But wherever treatment saves a life, and this will be frequently, for quite simple treatments, like a timely antibiotic, can be life-saving, it will, other things being equal, be the case that younger people have more life expectancy to gain from the treatment than do older people.

vii. Ageism and aid

Another problem with such a view is that it seems to imply, for example, that when looking at societies from the outside, those with a lower average age have somehow a greater claim on our aid. This might have important consequences in looking at questions concerning aid policy on a global scale. Of course it is true that a society's having a low average age might be a good indicator of its need for help, in that it would imply that people were dying prematurely. However, we can imagine a society suffering a disaster which killed off many of its young people (war perhaps) and which was consequently left with a high average age but was equally deserving of aid despite the fact that such aid would inevitably benefit the old. If QALYs were applied to the decision as to whether to provide aid to this society or another much less populous and perhaps with less pressing problems, but with a more normal age distribution, the 'older' society might well be judged 'not worth' helping.

viii. QALYs can be racist and sexist

If a 'high priority health care activity is one where the cost-per-QALY is low, and a low priority activity is one where cost-per-QALY is high' then people who just happen to have conditions which are relatively cheap to treat are always going to be given priority over those who happen to have conditions which are relatively expensive to treat. This will inevitably involve not only a systematic pattern of disadvantage to particular groups of patients, or to people afflicted with particular diseases or conditions, but perhaps also a systematic preference for the survival of some kinds of patients at the expense of others. We usually think that justice requires that we do not allow certain sections of the community or certain types of individual to become the victims of systematic disadvantage and that there are good moral reasons for doing justice, not just when it costs us nothing or when it is convenient or efficient, but also and particularly, when there is a price to be paid. We'll return shortly to this crucial issue of justice, but it is important to be clear about the possible social consequences of adopting QALYs.

Adoption of QALYs as the rationale for the distribution of health care resources may, for the above reasons, involve the creation of a systematic pattern of preference for certain racial groups or for a particular gender or, what is the same thing, a certain pattern of discrimination against such groups. Suppose that medical statistics reveal that say women, or Asian males, do better than others after a particular operation or course of treatment, or, that a particular condition that has a very poor prognosis in terms of QALYs afflicts only Jews, or gay men. Such statistics abound and the adoption of QALYs may well dictate very severe and systematic discrimination against groups identified primarily by race, gender or colour, in the allocation of health resources, where it turns out that such groups are vulnerable to conditions that are not QALY-efficient.[9]

Of course it is just a fact of life and far from sinister that different races and genders are subject to different conditions, but the problem is that QALYs may tend to reinforce and perpetuate these 'structural' disadvantages.

ix. Double jeopardy

Relatedly, suppose a particular terminal condition was treatable, and would, with treatment, give indefinite remission but with a very poor quality of life. Suppose for example that if an accident victim were treated, he would survive, but with paraplegia. This might always cash out at fewer QALYs than a condition which with

treatment would give a patient perfect remission for about five years after which the patient would die. Suppose that both candidates wanted to go on living as long as they could and so both wanted, equally fervently, to be given the treatment that would save their lives. Is it clear that the candidate with most QALYs on offer should always and inevitably be the one to have priority? To judge so would be to count the paraplegic's desire to live the life that was available to him as of less value than his rival's — what price equal weight to the preferences of each individual?

This feature of QALYs involves a sort of double jeopardy. QALYs dictate that because an individual is unfortunate, because she has once become a victim of disaster, we are required to visit upon her a second and perhaps graver misfortune. The first disaster leaves her with a poor quality of life and QALYs then require that in virtue of this she be ruled out as a candidate for life-saving treatment, or at best, that she be given little or no chance of benefiting from what little amelioration her condition admits of. Her first disaster leaves her with a poor quality of life and when she presents herself for help, along come QALYs and finish her off!

x. Life-saving and life-enhancing

A distinction, consideration of which is long overdue, is that between treatments which are life-saving (or death-postponing) and those which are simply life-enhancing, in the sense that they improve the quality of life without improving life-expectancy. Most people think, and for good as well as for prudential reasons, that life-saving has priority over life-enhancement and that we should first allocate resources to those areas where they are immediately needed to save life and only when this has been done should the remainder be allocated to alleviating non-fatal conditions. Of course there are exceptions even here and some conditions, while not life-threatening, are so painful that to leave someone in a state of suffering while we attend even to the saving of life, would constitute unjustifiable cruelty. But these situations are rare and for the vast majority of cases we judge that life-saving should have priority.

It is important to notice that QALYs make no such distinction between types of treatment. Defenders of QALYs often cite with pride the example of hip-replacement operations which are more QALY-efficient than say kidney dialysis.[10] While the difficulty of choosing between treating very different groups of patients, some of whom need treatment simply to stay alive, while others need it to relieve pain and distress, is clearly very acute, and while it may be that life-saving should not *always* have priority over life-enhancement, the dangers of adopting QALYs which regard only one dimension of the rival claims, and a dubious one at that, as morally relevant, should be clear enough.

There is surely something fishy about QALYs. They can hardly form 'an appropriate basis for health service policy'. Can we give an account of just where they are deficient from the point of view of morality? We can, and indeed we have already started to do so. In addition to their other problems, QALYs and their use for priority setting in health care or for choosing not which treatment to give these patients, but for selecting which patients or conditions to treat, involve profound injustice, and if implemented would constitute a denial of the most basic civil rights. Why is this?

Moral constraints

One general constraint that is widely accepted and that I think most people would judge should govern life and death decisions, is the idea that many people believe expresses the values animating the health service as a whole. These are the belief that the life and health of each person matters, and matters as much as that of any other and that each person is entitled to be treated with equal concern and respect both in the way health resources are distributed and in the way they are treated generally by health care professionals, however much their personal circumstances may differ from that of others.

This popular belief about the values which animate the health service depends on a more abstract view about the source and structure of such values and it is worth saying just a bit about this now.

i. The value of life

One such value is the value of life itself. Our own continued existence as individuals is the *sine qua non* of almost everything. So long as we want to go on living, practically everything we value or want depends upon our continued existence. This is one reason why we generally give priority to life-saving over life-enhancing.

To think that *life is valuable*, that in most circumstances, the worst thing that can happen to an individual is that she lose her life when this need not happen, and that the worst thing we can do is make decisions, a consequence of which, is that others die prematurely, we must think that *each life is valuable*. Each life counts for one and that is why more count for more. For this reason we should give priority to saving as many lives as we can, not as many life-years.[11]

One important point must be emphasised at this stage. We talk of 'life-saving' but of course this must always be understood as 'death-postponing'. Normally we want to have our death postponed for as long as possible but where what's possible is the gaining of only very short periods of remission, hours or days, these may not be worth having. Even those who are moribund in this sense can usually recognise this fact, particularly if they are aware that the cost of postponing their death for a few hours or days at the most will mean suffering or death for others. However, even brief remission can be valuable in enabling the individual to put her affairs in order, make farewells and so on, and this can be important. It is for the individual to decide whether the remission that she can be granted is worth having. This is a delicate point that needs more discussion than I can give it here. However, inasmuch as QALYs do not help us to understand the features of a short and painful remission that might none the less make that period of vital importance to the individual, perhaps in terms of making something worth-while out of her life as a whole, the difficulties of these sorts of circumstances, while real enough, do not undermine the case against QALYs.[12]

ii. Treating people as equals

If each life counts for one, then the life of each has the same value as that of any. This is why accepting the value of life generates a principle of equality. This principle does not of course entail that we treat each person equally in the sense of treating each person *the same*. This would be absurd and self-defeating. What it does involve is the idea that we treat each person with the same concern and respect. An illustration provided by Ronald Dworkin, whose work on equality informs this entire discussion, best illustrates this point: 'If I have two children, and one is dying from a disease that is making the other uncomfortable, I do not show equal concern if I flip a coin to decide which should have the remaining dose of a drug'.[13]

It is not surprising then that the pattern of protections for individuals that we think of in terms of civil rights[14] centres on the physical protection of the individual and of her most fundamental interests. One of the prime func- tions of the State is to protect the lives and

fundamental interests of its citizens and to treat each citizen as the equal of any other. This is why the State has a basic obligation, *inter alia*, to treat all citizens as equals in the distribution of benefits and opportunities which affect their civil rights. The State must, in short, treat each citizen with equal concern and respect. The civil rights generated by this principle will of course include rights to the allocation of such things as legal protections and educational and health care resources. And this requirement that the State uphold the civil rights of citizens and deal justly between them, means that it must not choose between individuals, or permit choices to be made between individuals, that abridge their civil rights or in ways that attack their right to treatment as equals.

Whatever else this means, it certainly means that a society, through its public institutions, is not entitled to discriminate between individuals in ways that mean life or death for them on grounds which count the lives or fundamental interests of some as worth less than those of others. If for example some people were given life-saving treatment in preference to others because they had a better quality of life than those others, or more dependants and friends, or because they were considered more useful, this would amount to regarding such people as more valuable than others on that account. Indeed it would be tantamount, literally, to sacrificing the lives of others so that they might continue to live.[15]

Because my own life would be better and even of more value to me if I were healthier, fitter, had more money, more friends, more lovers, more children, more life expectancy, more everything I want, it does not follow that others are entitled to decide that because I lack some or all of these things I am less entitled to health care resources, or less worthy to receive those resources, than are others, or that those resources would somehow be wasted on me.

iii. Civil rights

I have spoken in terms of civil rights advisedly. If we think of the parallel with our attitude to the system of criminal justice the reasons will be obvious. We think that the liberty of the subject is of fundamental importance and that no one should be wrongfully detained. This is why there are no financial constraints on society's obligation to attempt to ensure equality before the law. An individual is entitled to a fair trial no matter what the financial costs to society (and they can be substantial). We don't adopt rubrics for the allocation of justice which dictate that only those for whom justice can be cheaply provided will receive it. And the reason is that something of fundamental importance is at stake – the liberty of the individual.

In health care something of arguably greater importance is often at stake – the very life of the individual. Indeed, since the abolition of capital punishment, the importance of seeing that individuals' civil rights are respected in health care is pre-eminent.

iv. Discrimination

The only way to deal between individuals in a way which treats them as equals when resources are scarce, is to allocate those resources in a way which exhibits no preference. To discriminate between people on the grounds of quality of life, or QALY, or life-expectancy, is as unwarranted as it would be to discriminate on the grounds of race or gender.

So, the problem of choosing how to allocate scarce resources is simple. And by that of course I mean 'theoretically simple', not that the decisions will be easy to make or that it will be anything but agonisingly difficult actually to determine, however justly, who should live and who should die. Life-saving resources should simply be allocated in ways which do not violate the individual's

entitlement to be treated as the equal of any other individual in the society: and that means the individual's entitlement to have his interests and desires weighed at the same value as those of anyone else. The QALY and the other bases of preference we have considered are irrelevant.

If health professionals are forced by the scarcity of resources, to choose, they should avoid unjust discrimination. But how are they to do this?

Just distribution

If there were a satisfactory principle or theory of just distribution now would be the time to recommend its use. Unfortunately there is not a satisfactory principle available. The task is to allocate resources between competing claimants in a way that does not violate the individual's entitlement to be treated as the equal of any other individual – and that means her entitlement to have her fundamental interests and desires weighed at the same value as those of anyone else. The QALY and other quality-of-life criteria are, as we have seen, both dangerous and irrelevant as are considerations based on life-expectancy or on 'life-years' generated by the proposed treatment. If health professionals are forced by the scarcity of resources to choose, not whether to treat but who to treat, they must avoid any method that amounts to unjust discrimination.

I do not pretend that the task of achieving this will be an easy one, nor that I have any satisfactory solution. I do have views on how to approach a solution, but the development of those ideas is a task for another occasion[12]. I will be content for the moment if I have shown that QALYs are not the answer and that efforts to find one will have to take a different direction.

i. Defensive medicine

While it is true that resources will always be limited it is far from clear that resources for health care are justifiably as limited as they are sometimes made to appear. People within health care are too often forced to consider simply the question of the best way of allocating the *health care budget*, and consequently are forced to compete with each other for resources. Where lives are at stake however, the issue is a moral issue which faces the whole community, and in such circumstances, is one which calls for a fundamental reappraisal of priorities. The question should therefore be posed in terms, not of the health care budget alone, but of the *national budget*.[16] If this is done it will be clearer that it is simply not true that the resources necessary to save the lives of citizens are not available. Since the citizens in question are in real and present danger of death, the issue of the allocation of resources to life-saving is naturally one of, among other things, national defence. Clearly then health professionals who require additional resources simply to save the lives of citizens, have a prior and priority claim on the defence budget.

QALYs encourage the idea that the task for health economics is to find more efficient ways of doing the wrong thing – in this case sacrificing the lives of patients who could be saved. All people concerned with health care should have as their priority defensive medicine: defending their patients against unjust and lethal policies, and guarding themselves against devices that tend to disguise the immorality of what they are asked to do.

ii. Priority in life-saving

It is implausible to suppose that we cannot deploy vastly greater resources than we do at present to save the lives of all those in immediate mortal danger. It should be only in exceptional circumstances – unforeseen and massive disasters for example – that we cannot achieve this. However, in such circumstances our first duty is to try to save the maximum number of lives possible. This is because, since each person's life is valuable, and since we are committed to treating each person with the same concern and respect that we show to any, we must preserve the lives of as many individuals as we can. To fail to do so would be to value at zero the lives and fundamental interests of those extra people we could, but do not, save. Where we cannot save all, we should select those who are not to be saved in a way that shows no unjust preference.

We should be very clear that the obligation to save as many lives as possible is *not the obligation to save as many lives as we can cheaply or economically save*. Among the sorts of disasters that force us to choose between lives, is not the disaster of overspending a limited health care budget!

There are multifarious examples of what I have in mind here and just a couple must suffice to illustrate the point. Suppose, as is often the case, providing health care in one region of a country[17] is more expensive than doing so in another, or where saving the lives of people with particular conditions, is radically more expensive than other life-saving procedures, and a given health care budget won't run to the saving of all. Then any formula employed to choose priorities should do just that. Instead of attempting to measure the value of people's lives and select which are worth saving, any rubric for resource allocation should *examine the national budget afresh* to see whether there are any headings of expenditure that are more important to the community than rescuing citizens in mortal danger. For only if all other claims on funding are plausibly more important than that, is it true that resources for life-saving are limited.

Conclusion

The principle of equal access to health care is sustained by the very same reasons that sustain both the principle of equality before the law and the civil rights required to defend the freedom of the individual. These are rightly considered so important that no limit is set on the cost of sustaining them. Equal access to health care is of equal importance and should be accorded the same priority for analogous reasons. Indeed, since the abolition of capital punishment, due process of law is arguably of less vital importance than is access to health care. We have seen that QALYs involve denying that the life and health of each citizen is as important as that of any. If, for example, we applied the QALY principle to the administration of criminal justice we might find that those with little life expectancy would have less to gain from securing their freedom and therefore should not be defended at all, or perhaps given a jury trial only if not in competition for such things with younger or fitter fellow citizens.

A recent BBC television programme calculated[18] that if a health authority had £200,000 to spend it would get 10 QALYs from dialysis of kidney patients, 266 QALYs from hip-replacement operations or 1197 QALYs from anti-smoking propaganda. While this information is undoubtedly useful and while advice to stop smoking is an important part of health care, we should be wary of a formula which seems to dictate that such a health authority would use its resources most efficiently if it abandoned hip replacements and dialysis in favour of advice to stop smoking.

Acknowledgement

This is a revised version of a paper presented to the British Medical Association Annual Scientific Meeting, Oxford, April 1986.

As so often, I must thank my colleague Dr Mary Lobjoit for her generous medical advice. The fact that, like certain patients, I am apt to misunderstand this advice is of course my own fault. Thanks are also due to Don Evans, Alan Williams and the editors of the *Journal of Medical Ethics* for helpful comments.

Notes

1 See the excellent discussion of the recent history of this line of thought in the Office of Health Economics publication *The Measurement of Health* London, 1985.
2 Williams A. 'Economics of Coronary Artery Bypass Grafting.' *British Medical Journal* 1985; 291; and his contribution to the article, Centre eight – in search of efficiency. *Health and Social Service Journal* 1985. These are by no means the first such attempts. See reference (1).
3 Williams A. "The Value of QALYs." *Health and Social Service Journal* 1985.
4 I mention this in case anyone should think that it is only medical scientists who do medical research.
5 See reference (3): 3.
6 See reference (1): 16.
7 See reference (3): 5, and reference (3).
8 I'll assume this can be described as 'true' for the sake of argument.
9 I am indebted to Dr S G Potts for pointing out to me some of these statistics and for other helpful comments.
10 For examples see reference (1) and reference (2).
11 See Parfit D. 'Innumerate Ethics.' *Philosophy and Public Affairs* 1978; 7, 4. Parfit's arguments provide a detailed defence of the principle that each is to count for one.
12 I consider these problems in more detail in my: eQALYty. In: Byrne P, ed. *King's College Studies.* London: King's Fund Press, 1987/8. Forthcoming.
13 Dworkin R. *Taking Rights Seriously.* London: Duckworth, 1977: 227.
14 I do not of course mean to imply that there are such things as rights, merely that our use of the language of rights captures the special importance we attach to certain freedoms and protections. The term 'civil rights' is used here as a 'term of art' referring to those freedoms and protections that are customarily classed as 'civil rights'.

15 For an interesting attempt to fill this gap see Dworkin R. "What is Equality?" *Philosophy and Public Affairs* 1981; 4, 5.

16 And of course the international budget; see my *The Value of Life*. London: Routledge & Kegan Paul 1985: chapter 3.

17 See Townsend P, Davidson N, eds. *Inequalities in Health: The Black Report*. Harmondsworth, Penguin: 1982.

18 BBC 1. *The Heart of the Matter* 1986, Oct.

Response: QALYfying the value of life

Alan Williams

The essence of Harris's position can be encapsulated in the following three propositions:

(1) Health care priorities should not be influenced by any other consideration than keeping people alive;

(2) Everyone has an equal right to be kept alive if that is what they wish, irrespective of how poor their prognosis is, and no matter what sacrifices others have to bear as a consequence;

(3) When allocating health care resources, we must not discriminate between people, not even according to their differential capacity to benefit from treatment.

My position, which he attacks, can be encapsulated in the following three propositions:

(1) Health care priorities should be influenced by our capacity both to increase life expectation and to improve people's quality of life.

(2) A particular improvement in health should be regarded as of equal value, no matter who gets it, and should be provided unless it prevents a greater improvement being offered to someone else.

(3) It is the responsibility of everyone to discriminate wherever necessary to ensure that our limited resources go where they will do the most good.

At the end of the day we simply have to stand up and be counted as to which set of principles we wish to have underpin the way the health care system works.

The rest of Harris's points are really detail and I will deal with them on a subsequent occasion when I have had a chance to study his promised way forward, for that may help to dispel the very serious doubts I hold at present as to whether he realises the grave implications of the position he has adopted.

Chapter 45

Alan Williams and J. Grimley Evans

THE RATIONING DEBATE: RATIONING HEALTH CARE BY AGE

The case for

Alan Williams

As we grow older our recuperative powers diminish. Thus we accumulate a distressing collection of chronic incurable conditions. Some of these are no more than a minor nuisance, and we adapt as best we can; and when adaptation is not possible we learn to tolerate them. Some are more serious, involving severe disability and persistent pain, and may eventually become life threatening.

We are also at risk of various acute conditions (like influenza or pneumonia) which are more serious threats to the health of elderly people than to younger people. We also have more difficulty recovering from what younger people would regard as minor injuries (such as falls). When you add to all this the increased likelihood that illness (and other disruptions of our normal lifestyle) will leave us rather confused and in need of more rehabilitative and social support than a young person it is hardly surprising that NHS expenditure per person rises sharply after about age 65.

The vain pursuit of immortality

People are also living longer, and people aged over 65 now form a much bigger proportion of the population than they used to. From the viewpoint of NHS expenditure this would not matter if the extra years of life were predominantly healthy years but it would if the extra years were ones of disability, pain, and increasing dependence on others.

The evidence on this is ambiguous. Many people remain fit and independent well into their 80s. Others enter their 60s already afflicted with the aftermath of stroke, heart disease, arthritis, or bronchitis. It is not clear whether things are getting worse at each year of age, or whether expectations are rising and people are now more likely to report disabilities once shrugged off as the inevitable consequence of getting old. That many of these conditions are incurable does not mean they are untreatable. Much can be done to reduce their adverse consequences, including many remedial activities which lie outside the NHS (such as home adaptations, domestic support, and special accommodation).

It is important to get away from the notion of "cure" as the criterion of benefit and adopt instead measures of effectiveness that turn on the impact of treatments on people's health related quality of life. Such an approach concentrates on the features that people themselves value, such as mobility, self care, being able to pursue usual activities (whatever they are), and being free of pain and discomfort and anxiety and depression.

Improving the quality of life of elderly people in these ways may not be very costly, but these unglamorous down to earth activities tend to lose out to high tech interventions which gain their emotional hold by claiming that life threatening conditions should always take priority. This vain pursuit of immortality is dangerous for elderly people: taken to its logical conclusion it implies that no one should be allowed to die until everything possible has been done. That means not simply that we shall all die in hospital but that we shall die in intensive care.

Reasonable limits

This attempt to wring the last drop of medical benefit out of the system, no matter what the human and material costs, is not the hallmark of a humane society. In each of our lives there has to come a time when we accept the inevitability of death, and when we also accept that a reasonable limit has to be set on the demands we can properly make on our fellow citizens in order to keep us going a bit longer.

It would be better for that limit to be set, with fairly general consent, before we as individuals get into that potentially harrowing situation. When the time comes we shall probably each want an exception made in our case, because few of us are strong willed enough to act cheerfully in the general public interest when our own welfare is at stake. But if a limit is to be set, on what principles should it be determined? And what is their justification? And what role does age have?

In arguing for this article's proposition I have sought to make two contextual points clear: firstly, that ability to benefit should be measured in rather broader terms than cure or survival, and, secondly, that although chronological age is the best single predictor of increasing health problems, it is only a predictor, not a mechanistic determinant.

But age as an indicator of declining recuperative powers, of future health problems, of increasing need for health care, and of declining capacity to benefit from health care (because of shorter life expectancy) is only half the story. It addresses the issue of whether age is a good indicator of the extent to which people could benefit from health care but not in itself of whether they should be offered it. This more crucial step depends on what the objectives of the NHS are to be.

The NHS's objectives

If we start with the proposition that the objective should be to improve as much as possible the health of the nation as a whole then the people who should get priority are those who will benefit most from the resources available. In some cases the old will benefit most, in others the young. But for treatments which yield benefits that last for the rest of a person's life (or for a long time) the young will generally benefit more, because the rest of a young person's life is usually longer than the rest of an old person's life. And even among old people themselves the life expectancy of a 70 year old is usually greater than that of an 80 year old. Where a treatment offers only modest benefits a person may have to live a long time to make treatment worthwhile – that is, to make the benefit to that person larger than the sacrifices of rival candidates who failed to get treated. So improving the health of the nation as a whole is likely, in some circumstances, to discriminate indirectly against older people.

Is this morally defensible? Well, if we behaved otherwise we would by implication be asserting that in order to provide small benefits for the elderly, young people should sacrifice large benefits. What makes old people more deserving of health benefits than young people? One argument might be that all their lives they have been paying their taxes to finance the health care system (among other things), and just when they need health care most the government lets them down. But the government – that is, their fellow citizens – did not promise to do everything possible no matter what the costs.

The NHS is part of a social insurance system, not a savings club for each individual's health care expenditures. It is the lucky ones who do not get their money's worth out of the system, and the unlucky ones who need heavy NHS expenditures all their lives. The NHS is there to meet certain contingencies but not others. And many of the treatments which the NHS now offers to old people in certain contingencies were not even invented when they started contributing 40 or 50 years ago. So to argue, from a historical viewpoint, about an entitlement to get your money's worth seems inappropriate to any insurance scheme, and in particular to a social insurance scheme such as the NHS.

A different line of argument might be that as the number of years left becomes smaller and smaller, each is more precious. The implication of this argument is that elderly people value their small improvements more highly than young people do their much larger improvements. This raises a fundamental problem about whose values should count in a social insurance setting. Suppose that it were true that older people would spend relatively more on health care to get health improvements rather than other things, whereas younger people would spend relatively more on (say) education for their children and rather less on health benefits for themselves. Rational self interest drives individual citizens operating in private markets precisely in that direction.

But did we not take the NHS out of that context precisely because as citizens (rather than as consumers of health care) we were pursuing a rather different ideal – namely, that health care should be provided according to people's needs, not according to what they were each willing and able to pay. A person's needs (constituting claims on social resources) have to be arbitrated by a third party, whose unenviable task it is to weigh different needs (and different people's needs) one against another. This is precisely what priority setting in health care is all about. So the values of the citizenry as a whole must override the values of a particular interest group within it.

A fair innings

So I can find no compelling argument to justify the view that the young should sacrifice large benefits so that the old can enjoy small ones. But I can find an argument which goes in the opposite direction. It is that one of the objectives of the health care system should be to reduce inequalities in people's lifetime experience of health. The popular folklore is rich in phrases indicating that we all have some vague notion of a "fair innings" in health terms. Put at its crudest, it reflects the biblical idea that the years of our life are three score and ten. Anyone who achieves or exceeds this is reckoned to have had a fair innings, whereas anyone who dies at an earlier age "was cut off in their prime" or "died tragically young." As has been observed, while it is always a misfortune to die if you wish to go on living, it is both a misfortune and a tragedy to die young. Why?

Fom my perspective (approaching the age of 70) I see clearly why it is a tragedy, because someone who dies young has been denied the opportunities that we older people have already had. If reducing inequalities in lifetime health is a worthy social objective, it will lead us to be willing to do more to enable young people to survive than we are willing to do to enable old people to survive.

But I do not think that the notion of a "fair innings" should be restricted to matters of survival and life expectancy. Quality of life considerations concerning health may be just as important. Someone who has suffered a lifetime of pain and disability cannot be said to have had a fair innings even if she did live to be 80, and I would therefore extend the concept to embrace something more than just years of life. My preferred concept would be the number of quality adjusted life years a person had enjoyed. On the whole people's earlier years are healthy years, and their later years less healthy years, so this does not affect the general tenor of my argument. What it implies is that we need to consider, alongside age itself, the quality of a person's lifetime experience of health. The worse it has been, the more consideration they deserve, age for age.

Age matters

So my overall conclusion is that age matters in two respects. Firstly, it affects people's capacity to benefit, and therefore places them at a general disadvantage if the objective is to maximise the benefits of health care. Secondly, the older you are the more likely you will have achieved what your fellow citizens would judge to have been a fair innings, and this will place old people at a disadvantage if the objective is to minimise the differences in lifetime experience of health. I would be the first to admit that I personally have had a fair innings and that it would not be equitable to deny a younger person large benefits in order to provide small ones for me. Indeed, I would go further: it would be equitable to provide small benefits for a young person even if by so doing I were denied large benefits, provided that the young person in question had a low probability of ever achieving a fair innings. Note that this argument does not mean that benefits to young people take absolute priority over benefits to old people. It simply means that we give rather more weight to them than to us.

Surveys of public opinion commonly find that most people, if pushed into a tight situation, would give priority to the young over the old when distributing a given amount of health care benefit. There is also little doubt that health care professionals share this general attitude. It does not, of course, stop them from being kind, considerate, and caring when old people need health care, but it manifests itself at the level of clinical policymaking, when different needs have to be prioritised. For the professionals what may be in their minds may be mostly old people's impaired capacity to benefit from health care. But I strongly suspect that some variant of the fair innings argument also underlies such views, and this is especially likely to be the case among the general public. When the views of older respondents in such surveys have been reported separately, they too give priority to the young over themselves.

So I am encouraged to hope that, in the interests of fairness between the generations, the members of my generation will exercise restraint in the demands we make on the health care system. We should not object to age being one of the criteria (though not the sole criterion) used in the prioritisation of health care, even though it will disadvantage us. The alternative is too outrageous to contemplate – namely, that we expect the young to make large sacrifices so that we can enjoy small benefits. That would not be fair.

The case against

J Grimley Evans

Older people are discriminated against in the NHS. This is best documented in substandard treatment of acute myocardial infarction and other forms of heart disease, where it leads to premature deaths and unnecessary disability. The care for older people with cancer is also poorer than that provided for younger patients.

Age discrimination in the NHS occurs despite explicit statements from the government that withholding treatment on the basis of age is not acceptable. Ageism is mostly instigated by clinicians but condoned by managers. Fundholding general practitioners have a financial incentive to deprive older patients of expensive health care, but there is no ready way to find out whether they do so. Whatever its full extent, the documented instances of age discrimination, together with the occasional published apologia for ageism, show that the morality of age based rationing should be a matter of public concern.

Need to assess individual risk

It is important to be clear what we are talking about. It is proper for a doctor to withhold treatment or investigation that is likely to do more harm than good to a patient. In an individual case actual outcome depends on the patient's physiological condition. The prevalence of impairments that shift the risk:benefit ratio adversely increases with age, so where individual physiological condition is used as the basis for allocating treatment older people are more likely on average to be excluded than are younger people. Nevertheless, wide individual variation exists in aging, and many people in later life function physiologically within the normal range for people much younger. The key issue, therefore, is that each decision should be made on a competent assessment of individual risk

What I am objecting to is the exclusion from treatment on the basis of a patient's age without reference to his or her physiological condition. The patient is being treated as though he or she necessarily had properties identical with those corresponding to the average of the age group. We can draw a contrast with social class and skin colour. Should we withhold health care from members of lower social classes or from black people because of the poorer average outcome of their groups? Rather, most of us would suggest that extra attention should be paid to vulnerable members of such groups to try to compensate for their disadvantage. Why should old people not be viewed similarly?

Ethics, ideology, and the law

I am convinced that in the United Kingdom at present it is unethical to use age as a criterion for depriving people of health care from which they could benefit. The fundamental issue is ideological; and ideologies – and the ethical systems derived from them – can change with circumstances. The notion, implicit in the writings of many ethicists, that there is an objective basis for a universal ethical system is a dangerous illusion. Ethics are no more than logical deductions from primary ideologies. Ideologies are primary in the sense that they cannot be validated by any objective means. They can arise in various ways, and in England they arose by a long process of mutual adaptation of heterogenous people developing efficient ways of living together. Not having a written constitution, we have in Britain to deduce the ideological principles of our society from our history and from the shared rhetoric of our major political parties.

From these I conclude that in times of peace British national values include the equality of citizens in their relation to the institutions of the state and acknowledgement of, and respect for, the uniqueness of individuals regardless of their physical or mental attributes. From the latter follows the equal right of all citizens to live as they wish so long as they do not impede the like rights of others. If these ideas are indeed embodied in the ideology of British society, ageism, as well as racism and sexism, will be unethical.

The founts of ageism

Exploitation of the weak

Several factors generate or are invoked to justify ageism in health care. The first is an issue of real-politik. When health care managers aim to control costs older people are natural victims. They do not riot; they are uncomplaining and politically inactive. The threat of tactical voting by the militant elderly people of the United States caused a major shift in health and social care resources to their benefit. Although comprising more than a quarter of the electorate in Britain, old people are not yet seen by politicians as potential tactical voters. Inevitably they suffer, and inevitably ageism remains legal.

Professional ignorance

Ageism may arise from well intentioned ignorance, where health professionals assume incorrectly that older patients will be harmed rather than benefited by treatment. In reality the absolute benefit of some treatments – in terms, say, of deaths prevented – increases with prior risk while the probability of side effects remains constant. Where prior risk rises with age such treatments may be more effective given to older people than to younger. Moreover, except in the limited area of intensive care medicine, we still know little about the physiological variables that determine individual risks of benefit and harm from medical interventions. We need more research to enable meaningful negotiation over options for care with patients of all ages and to underpin more efficient targeting of resources.

Prejudice

The most important source of ageism is prejudice. Surveys in Britain show that older people are widely seen as of lower social worth than younger, but little has been done to explore the origins and dynamics of this prejudice. Some researchers suggest that public attitudes displayed by such surveys are a valid basis for rationing in the health services. There are several problems with this facile suggestion. People answering questions in a way that indicates low valuation of older people may do so not because of what they really feel but because of what they think the interviewer will regard as the "right" answer.

Typically, questions are in "doctor's dilemma" format in which there is treatment available for only one of two people who differ in age. The possibility of generating equity by allocating the treatment on the toss of a coin is not usually offered and is unlikely to be thought of spontaneously by the average citizen. It is also naive to assume that attitudes exposed by the desperate situation simulated in a doctor's dilemma would also emerge in decisions on real life issues such as the relative lengths of waiting lists for hip replacements and hernia repairs.

Survey interviews are rarely confidential and do not contain control questions in which the two potential patients differ, say, in skin colour. Would researchers suggest that racial prejudice revealed by their questionnaires should be a basis for health service rationing? We may presume not; it would be recognised, as it should be for ageism, that the respondents were failing to conform to the principles of British society. To imply, as some have found it convenient to claim in the ageism debate, that it is paternalistic to esteem the values of society above the ignorant prejudices of some of its members is to confuse demagoguery with democracy.

The power of economics

Economists sometimes claim that their discipline is so fundamental that it can provide a sufficient basis for allocating society's resources in health care. Whether this assertion is acceptable or not is an ethical issue. It can be argued that economists should be restricted to identifying the most cost effective way of achieving a pattern of allocation that has been defined on ideological grounds. We have lived so long under a theocracy of markets, competition, and cost containment that people may forget that these are driven by an ideology of no more validity than the ideology behind common cause, collaboration, and social purpose that it supplanted.

Alan Williams has suggested that if allocations of resources based on quality adjusted life years (QALYs) are thought to bear too heavily on older people, their needs can be weighted to conform more closely with externally derived principles of equity. This approach has the advantage of making the ethical input both explicit and manifestly the responsibility of those who provide it. Virtue still emerges wearing what many will see as the indecency of a price tag. Williams's dialectic derives from what he sees as a necessary trade off between equity and efficiency. In my view his notion of what should be regarded as efficiency in the NHS is questionable. We can find common ground in the assertion that health care resources should be allocated so as to do the most good. The ethical argument crystallises round what view of good should prevail.

There are two perspectives on a health service. On the one side are the purveyors who, like shareholders in a chain of grocery shops, look for the best return on their investment. They may well think it appropriate to measure this return in terms of some measure such as QALYs gained. On the other are the users of the service. Although the NHS has in recent years been forced into a Procrustean bed of market imagery, the average British citizen sees it not as a chain of grocery shops but as something more akin to a motoring organisation to which he pays a subscription so that it will be there to do what he wants when he wants it. He will judge the service on the extent to which it meets his informed desires. There is no reason to expect that maximising the production of QALYs will lead to the same recipe for distributing limited resources as maximising the achievement of users' informed wishes.

British citizens as taxpayers might see themselves alongside Williams with the purveyors but as potential patients would, I suspect, ally themselves more consistently with the users. My assessment is that the users' perspective also provides a rationale more consonant with national values and with the explicit intentions for the NHS at its foundation. There are also unacceptable implications in the purveyors' approach.

Firstly, measurement of output in units based on life years directly or indirectly puts different values on individuals according to their life expectancy. Thus citizens are no longer equal and older people in particular are disadvantaged. Secondly, it assumes that the value of life, at any given level of objectively assessed disability, is determined by its length. But if we assert the unique individuality of citizens, the only person who can put a value on a life is the person living it. Lives of individuals are therefore formally incommensurable and it is mathematically as well as ethically improper to pile weighted valuations of them together as an aggregable commodity like tonnes of coal. There have been nations whose ideologies value citizens only for their potential collective usefulness to the state as soldiers, workers, or breeding females. In the United Kingdom, at least for the time being, are we not spirits of another sort?

The "fair innings" argument

This argument asserts that we have a right only to a certain number of years of life and after then only palliative as distinct from therapeutic care should be provided. Although sometimes mistaken for an economic argument, the fair innings approach will not necessarily save money unless we apply its corollary of compulsory euthanasia at the end of the innings. Palliative care can be more expensive than therapeutic care; the money saved by not providing coronary artery surgery for an elderly woman may be spent several times over if she has to live for months in a nursing home because of her angina.

The fair innings argument has historical roots in Christian theology and its requirement for time to earn one's place in heaven by purging the sins of youth with the good works of later life. For secular man fair innings now codes for two crucially different ideas which commentators sometimes confuse. The first is that as individuals we commonly come to a time when we conclude that we have done all that we wished and were able to do and that life no longer offers the potential of interest or pleasure that might make it preferable to oblivion. For some others of us death may at a particular time offer personal meaning, climactic consummation, or a perfected symbolism to our lives. Dying for a worthy cause may seem better than survival in servitude, failure, or dishonour. Such ideas underlie the existential concept of a fair innings or natural lifespan. Only the person living a life can say when it is complete in this sense, and its length for different individuals might range from 18 to 120 years.

The other version of a fair innings is that owing to overpopulation space on earth has to be rationed and after a time one should make way for someone else to enjoy life. (We could, of course, solve the underlying problem by controlling birth rates rather than limiting lifespan, but let us follow the logical trail.) This form of the fair innings is identified with a fixed number of years, usually assigned by Western authors to the high 70s. The assumption is that life confers some kind of intrinsic good that we can perhaps code as "happiness."

In its simplest form the argument requires that everyone has the same chance of happiness so that the fairness of the innings can be assessed by its length. Clearly this is not true. If the fairness of the innings is actually the area under a happiness/duration curve, the notion should lead to the early turning off of the rich and fortunate in favour of the poor and deprived. It would be theoretically possible to calculate an individual's fair innings allowance on the basis of some form of "happy life expectancy" adjusted for relevant variables such as social class and sex. Whether one should regard this as a serious possibility or an intellectually charming reductio ad absurdum depends on one's estimate of its potential utility. Given their longer life expectancy, women would probably have to take second place to men in access to health care. Rich older people would still, presumably, be able to purchase, in the private sector or abroad, treatments denied to them by the NHS. The fair innings concept is unlikely to provide an acceptable solution to problems of inequity.

Conclusion

Health care resources in Britain are limited, but only because the government limits them. If we continue with the healthcare budget restricted to some seven percent of gross national product rationing is likely also to continue. In a democratic society rationing should be explicit and transparently the responsibility of government. For several reasons it would be timely for Britain to define what its national values and the rights and duties of its citizens are. I should be disturbed if these turned out to differ essentially from those deduced above. If these values are to be translated into the NHS primary rationing has to focus on equitable limits to the type and volume of services. We should not create, on the basis of age or any other characteristic over which individuals have no control, classes of Untermenschen whose lives and well being are deemed not worth spending money on.

Chapter 46

Tony Hope

RATIONING AND LIFE-SAVING TREATMENTS: SHOULD IDENTIFIABLE PATIENTS HAVE HIGHER PRIORITY?

Rationing health care

Rationing of health care is a reality in the UK. The National Health Service (NHS) does not have sufficient funds to ensure the very best treatment for all patients in all situations. Choices have to be made. Managed care systems in the US will have to face similar choices. One level within the British NHS at which decisions are made about which treatments to provide is the level of the health authority.[1] Perhaps as primary care groups and trusts (PCGs and PCTs) develop they will take on an important part of this role.

Daniels and Sabin[2] have emphasised the importance of the process by which resource allocation decisions are made. They argue that for decisions to be just the process by which the decision is made must be a just one. One feature of such a process is what they call a "relevance condition" – that is, that the rationale for a decision must rest on evidence, reasons and principles that all fair-minded people can agree are relevant. The ethical framework which informs the decision making process is part of such a relevance condition.

It is likely that the courts, at least in the UK, will focus on the procedure by which decisions about health care funding are made.[3] If a decision is challenged, the health authority, or whichever body is responsible for the decision, may need to justify the decision in two ways:

1   That the process by which the decision was made was appropriate; and,
2   That the reasons for making that particular decision are justifiable.

It remains unclear, however, how courts will address the conflict between treating each patient in her best interests and the scarcity of resources.[4]

Priorities forum – a decision making process

Oxfordshire Health Authority has developed a procedure for making some resource allocation decisions which meets the criteria laid down by Daniels and Sabin.[2] A "priorities forum" has been set up to advise the health authority on such decisions. The membership includes general practitioners, medical directors of local hospitals, health authority staff, hospital doctors, nurses and lay members. There are about 30 members altogether. Five are lay members, eleven are non-medical managers, two are university academics and the remainder are health professionals, many of whom have managerial responsibilities. Early in the life of this committee the members felt the need for an ethical framework in order to help it to make decisions in a consistent and well thought out manner. The committee decisions also form a kind of case law – as guidance both for the committee in future decisions, and indeed to enable health authority staff to make decisions without bringing them to the committee. The ethical framework is built around three main considerations: cost effectiveness, equity and patient choice. The most difficult of these for the committee is the interpretation of equity in practical situations.

New and expensive treatments

Many of the issues brought to the committee arise from new and expensive treatments, especially drug treatments. The setting in which the committee works, and the political realities of health care delivery mean that wholesale changes in what is funded cannot be made. New treatments can be examined, however, and decisions made as to whether, in the context of what can be afforded, the new treatment should be funded at all, and if so, in which groups of patients. Three examples of issues which have been considered by the forum follow.

Statins

These drugs reduce blood cholesterol with a resulting reduction in the probability of a heart attack or sudden death in people with even mildly raised blood cholesterol. The cost per life-year saved depends on what risk of death or of heart attack the person faces in the first place. This risk depends on blood cholesterol level, gender, age, weight and whether the person smokes. For those at an annual risk of heart attack or death of 1.5 percent, the cost per life-year saved is about £18,000; for those at 3 percent risk it is about £7,000; and for those at 4.5 percent risk it is about £5,000.

Alglucerase

Gaucher's disease is a rare inherited disease. There is only one effective treatment: alglucerase. This drug costs about £80,000 per year per patient.

Beta-interferon

This drug reduces the relapse frequency in some patients with multiple sclerosis. Its effect (at best) is to reduce the number of relapses from about three to two per year. The cost of the drug has been estimated at between £70,000 and £500,000 per quality adjusted life-year.

An approach to decision making

The priorities forum approaches the question of whether, or to what extent, to fund new treatments in the following way. The starting point for the forum's considerations is the evidence with regard to cost-effectiveness. The forum works best if the members have a clear idea of the situation of the patients who stand to benefit from the treatment and a clear idea of the nature of the benefit using current treatment and the nature of the benefit with the proposed treatment. The beneficial effects of most treatments are not certain to occur in all patients and so the forum is usually dealing with probabilities of these benefits.

The forum, in assessing benefits, considers that length of life and quality of life are of major importance. It therefore finds that the approach using "quality adjusted life-years" (QALYs) is a useful starting point (see for example Williams and Edgar et al.[5,6]). This approach allows a consideration of the length and quality of life amalgamated with both cost and probability of effect.

The approach to reasoning about priority decisions which the forum ideally takes can be outlined as follows:

1 It starts with evidence as to how much the treatment costs per year of life saved (in the case of life-saving or life-extending treatment) or per QALY in the case of other treatments.

2 It compares this with a guide cost – that is the cost that the authority normally pays. Roughly speaking the authority can afford to pay about £10,000 to £15,000 per life-year saved.

3 If the proposed treatment is less than this then the forum would normally recommend paying for the treatment. If the proposed treatment is more than this then the forum asks two questions:

  a) are there grounds for paying more than our usual amount (per life-year saved)?

  b) if the answer to this is yes, then the forum asks: do those grounds justify paying that much more?

However, the members of the forum do not, in general, believe that the QALY approach always results in a just or fair result. Some examples of where many members of the forum believe that some deviation from the QALY approach is desirable are as follows:

1 There are situations when, in order to allow an equal quality of service to different groups of people, it may be right to pay more for one group than another. Dental care for people with severe learning disability may be more expensive than for the normal population. However, it may not be right to give a lower priority to such care on the grounds that it is more expensive per QALY.

2 Many believe that the care of the terminally ill should have a higher priority than the QALY calculation suggests.

3 Those who are particularly badly off in terms of their health (for example those suffering from multiple sclerosis) may for this reason merit greater priority than the QALY calculation would consider right.

4 Many people would be prepared to pay more to save the life of identifiable people who are currently ill than to prevent the future death of unidentifiable individuals.

The approach, outlined above, which the forum takes, is useful in practice because it starts with cost-effectiveness data which are both highly relevant and generally available for new drug treatments, at least to some extent. It is an approach which also allows a consideration of many factors other than quantity and quality of life, which may be ethically important in considerations of equity. From the ethical point of view it helps to highlight issues of equality which arise in practice. In this paper I want to consider one such issue: what has been called "the rule of rescue" and what Daniels and Sabin have referred to as "the buried miner". This "rule" is intuitively attractive and operates in practice, for example in forum decisions.

The rule of rescue

The typical situation is when there is an identified person whose life is at high risk and where there is some intervention ("rescue") which has a chance of saving the person's life. The essence of the value I want to consider is that it is normally justified to spend more on saving life in this situation than in situations where we cannot identify who has been helped. For the sake of clarity I will consider two hypothetical, but realistic interventions.

Intervention a (anonymous prevention)

A is a drug which will change the chance of death by a small amount in a large number of people. For example, out of every 2,000 people in the group, if A is not given then 100 people will die over the next few years. If A is given then only 99 will die. Drug A is cheap – the cost per life-year saved is £20,000. Statins provide an example of an intervention of this type.

Intervention b (rescue of identified person)

B is the only effective treatment for an otherwise life-threatening condition. Those with the condition face a greater than 90 percent chance of death over the next year if not given B. If given B then there is a good chance of cure – say greater than 90 percent. B is expensive. The cost per life-year saved is £50,000. Renal replacement therapy is an example of an intervention of this type.

The key difference between these two interventions is that intervention B benefits an identifiable person, whereas intervention A benefits a proportion of patients within a group, but we cannot know who has been benefited.

Let us suppose that the amount of money available to the funding organisation (for example health authority or management care company) allows roughly £15,000 per life-year saved, on average. Therefore both A and B are above the average amount. However, the pressures on a health authority, or other funder, to fund intervention B, can be very great. A particular person whose life can be saved will die if the intervention is not funded. Such a person can take his case to the press or to court.

The question I want to consider is whether it could be right for the funding organisation to choose to fund intervention B but not intervention A?

For shorthand I will call intervention A the "preventive" intervention, and intervention B the "rescue" intervention. I will consider a number of arguments in favour of paying more, per life-year saved, for the rescue intervention

than for the preventive intervention. I will focus on what I think is the strongest of these arguments. I will, however, conclude that this strongest argument is wrong and therefore that it is generally wrong for a funding organisation to pay more for a rescue intervention per life-year saved than for a preventive intervention.

There is one issue which I will not consider in this discussion. No treatment is strictly speaking "life-saving" since none of us is immortal. Treatments at best are death postponing. Some may regard the value of a life-year saved as affected by the age of the person when life is extended. For the sake of the arguments I will consider the age at which interventions A and B have their effects as the same.

I will consider six reasons for paying more, per life-year saved, for rescue interventions than for preventive interventions.

Reason 1: scepticism about the effectiveness of preventive treatment

We can normally be fairly certain of saving the identifiable life (rescue intervention). There may be considerably more scepticism that the preventive treatment really will save lives in the future. This argument, however, is not relevant to most of the situations about funding which are faced by the health authority. Typically the evidence about the effectiveness of the preventive treatment is good. We can be almost certain, for example, that statins will reduce the number of future deaths even though we will never know which individual lives have been saved.

Reason 2: a life in the hand is worth two in the bush

The treatment of the identified person will typically save a life now, or in the very near future, whereas with prevention we are dealing with saving lives further in the future.

It may be psychologically motivating to consider, as it were, that "a life in the hand is worth two in the bush". However, even if there is justification for some discounting of future lives,[7] as some economic models hold, this would only justify a relatively small extra cost per identifiable life saved. Furthermore the difference between saving the life of an identifiable person and preventing the death of an unidentifiable person is not fundamentally about whether one is in the present and one in the future. Some preventive treatments (for example drugs reducing the risk of death following a heart attack) prevent (unidentifiable) deaths over the next few days and months; whereas rescue treatments may save life some time in the future.

Reason 3: rescue is rare, so we can always afford it

Saving the life of an identifiable person is typically rare. We can afford to rescue the occasional person at very high cost – the round-the-world yachtsman who gets into serious trouble, for example; whereas we cannot afford a lower, but still high cost of prevention as that involves many people.

This argument gains its appeal from the fact that when we feel prepared to spend enormous amounts on saving the round-the-world yachtsman we do not imagine that this is at the expense of other people's lives. What we imagine is that the £500,000 (or whatever) is at the expense of things that others might buy. For example, to pay for saving the yachtsman many people may have to go without some luxury (a good meal in a restaurant, for example). However, in the setting of health care funding, where the money available at any one time is fixed, in paying to save the lives of identifiable people we are using money which might otherwise have been spent on saving more lives by funding a prophylactic intervention.

Reason 4: rescue has more effect on quality of life than prevention

When we are moved to want to rescue someone, we may be motivated not only by the wish to save life but also by the wish to help people whose death is likely to be particularly terrible. The kind of situation we have in mind might be that of a buried miner. Thus the view that we should pay large amounts in these circumstances may be motivated by trying to prevent a particularly appalling quality of life rather than simply preventing death.

In the setting of health care it may sometimes be the case that the rescue treatment has more of an impact on quality of life than preventive treatment, but this will not generally be true. In any case, the funding issue is then about the amount to be spent on improvements in quality of life rather than about extending (saving) life.

Reason 5: it is good to care about identifiable individuals

There is a considerably different emotional response to a situation where we can identify an individual who may (or is likely to) benefit from a treatment compared with a situation where the person (or people) who benefit cannot be identified.

We are more moved by, and more moved to help in, the former situation than the latter. Many of our moral intuitions are strongly oriented towards individual relationships. These facts have a number of implications for the current discussion.

(a) In our personal morality many would argue that we have different (and greater) obligations to individuals we know (or have some relationship with) than towards people we don't know, or people whom, as individuals, we could not identify. At one extreme, I have much stronger obligations to my children than to other children. In the setting of health care, a doctor, for example, may have a greater obligation towards her own patients than to other patients – the legal concept of "duty of care" underlines this.

It seems unlikely, however, that this is relevant to funding and the health authority (or other large health care funders). A health authority's "duty of care", or that of a managed care company, is presumably to the population which it serves. This includes both those who stand to gain from the preventive intervention and those who benefit from the rescue intervention. Furthermore, the members of the forum, or the health authority more broadly, will not usually know the people who can benefit from "rescue".

(b) There may be broadly consequentialist grounds for a health service spending more (per life-year saved) on rescue than on prevention. If people in general are moved more by rescue than prevention, then a health service which was seen to "rescue" people may attract more support, and therefore more funding, than one which spent no more on rescue than prevention. In the long run, such a health service might be able to carry out more preventive interventions (and more rescue) because it has more funds overall.

Such an argument depends on empirical assumptions which are quite uncertain, and which are likely to vary depending on how the health service is funded and structured. Even if the empirical facts were to support this view, it remains questionable whether a health authority should be affected by them. The role of the health authority, it might be argued, is to spend its money justly and wisely and in the best interests of the population it serves, and not to fund popular causes where to do so is inequitable. On this view the question of what is equitable must be answered independently of what is popular.

A more convincing argument is that a health authority (or managed care system) should respond to the values of its appropriate constituency. If those constituents believe it is right to spend more on rescue than on prevention then it is right (as an element of democracy) for it to do so. I will consider this argument in more detail when I consider the contractualist argument below.

(c) What we do in health care says something important about the kind of society we are. A society which did not rescue identifiable people, but left such people to die would lack solidarity, and would appear cruel and uncaring. It is right therefore for society to be prepared to put much greater resources into rescuing identifiable people, than into prevention – to do so says something important about the kind of society we are.

This argument combines a number of elements, some of which have been considered already. First, it appeals to the "round-the-world yachtsman" view. Surely we ought to be prepared to forego some luxuries to save a person's life. Second, it raises the whole question of how much overall we ought, as a society, to be prepared to spend to save life. The more caring society would give up many luxuries to help those whose lives are at risk. But, in my view, this argument essentially begs the question at issue. Is it a more caring society which sacrifices several (anonymous) lives which might have been saved (by a preventive intervention) in order to save one identifiable life (through "rescue")?

In the case of rescue we can identify and imagine the particular individual, and the relatives and friends who will be bereaved. This engages our sympathy. But in the case of prevention, although we know that our choice will lead to several deaths – and many more relatives and friends being bereaved – because we do not know which individuals, our sympathy is not engaged. Those who choose to save the more lives through preventive intervention may be regarded as cold-hearted on the grounds that they fail to be properly moved by the individuals who thereby do not receive rescue intervention. But the argument could be reversed. It is not that the first group are not moved by the plight of those needing rescue, it is that the second group are not sufficiently moved by the fact that there are a larger number of people who die, and who are bereaved, as a result of paying more, per life-year saved for rescue than for prevention. The callousness lies in the relative indifference to death and suffering simply because the specific individuals concerned cannot be identified. There is a lack of imagination in those preferring rescue rather than a lack of sympathy in those preferring prophylaxis.

In order to decide which side is right in this debate, I think that first the question of whether the "rule of rescue" is right has to be answered. Thus, the "argument from callousness" stated above cannot be used to support the rule of rescue. The argument depends on having established that it is a good society, or a fair society, which would follow the rule of rescue.

Reason 6: a very small decrease in the chance of death is of only small benefit

The most powerful reason in support of paying more to save the identified life, I believe, is that in the typical cases of prevention the intervention makes only a small difference to the probability of death of any one individual whereas in the typical example of rescue (but by no means in all examples) the intervention makes a large difference to one or more individuals. This enables a broadly "contractualist" argument to be mounted in favour of the rule of rescue. I will try and put this argument in as strong a form as possible.

Premature death is, normally, a very significant harm indeed. It is partly because of this that

good health care is so important. Preventing premature death is a paradigm of what Daniels calls a need[8] – more than almost anything else it interferes with our life goals and opportunities.

But a very small chance of premature death is by no means a great harm – and we cannot claim that we need something which reduces by a very small amount the chance of premature death. All of us in our lives trade small increases in the chance of premature death against really quite small benefits. Consider Peter, the Sunday morning cyclist.

The Sunday morning cyclist

On Sunday mornings Peter cycles along the busy Banbury Road, in Oxford, to buy a newspaper. In doing this he is putting himself at a small but real extra risk of premature death. He is trading this extra risk against the pleasure and value of reading the Sunday morning paper. In balancing these two he finds that the pleasure of the paper – a really rather small pleasure in his life – outweighs the extra risk of premature death. There seems nothing irrational about this. A very small chance of a terrible harm is itself only a small negative weight easily outweighed by other benefits. Most of us will take these small risks not only for our own benefit but for the benefit of others. Consider the friend's job application.

The friend's job application

Suppose that David's friend Sarah is applying for a job which she is keen to get. To meet the deadline the application has to be in the postbox today. Owing to a severe bout of 'flu, Sarah cannot post it herself. To help her, David cycles to Sarah's house to collect the application and post it. This action increases by a very small amount David's chance of premature death. This is easily outweighed by the value of helping Sarah.

With these considerations in mind I will propose a three-stage argument in favour of the health authority's paying for intervention B (at a cost of £50,000 per life-year saved) but not for intervention A (at a cost of £20,000 per life-year saved):

1   A very small increase in the risk of premature death represents only a small consideration to weigh against the benefits.
2   It is rational to trade that small extra risk in order to make a small but definite contribution towards saving someone else's life.
3   In practice this is likely to be what most people would choose and this therefore forms contractualist grounds for a health authority (or other health care purchaser) to operate a rule of rescue.

I will now consider a counterexample to this conclusion: the case of the trapped miner.

The case of the trapped miner

A miner lies trapped following an accident. Without rescue he will die. Given a sufficiently large rescue party the miner can be rescued. Would you, the reader, join the rescue party if you faced a 1:10,000 risk of death in so doing? Would your answer depend on the size of the rescue party needed?

Suppose the facts (perhaps not entirely realistic) are these: there is a small risk of death to those in the rescue party, and this risk varies according to the size of the rescue party. If there were 1,000 rescuers there would be a 1:1,000 chance for each rescuer of death. If there were 10,000 rescuers each would face a 1:2,000 chance of death. If 100,000 rescuers then each would face a 1:10,000 chance of death. If 1,000,000 then each would face a 1:20,000 risk.

Thus the larger the size of the rescue party, the smaller the risk of death faced by each individual rescuer. It is also the case, however, that the larger the size of the rescue party, the more people are likely to die in the rescue attempt. With a rescue party of 1,000,000 each member of the rescue party faces a very small risk of death – well within the risks we normally take for much less important gains than saving a life. However, with such a rescue party about 50 people will die in order to save the life of one person.

If we assume that most people are altruistic at least to a small extent, and most people will accept a very small level of risk of personal death in order to save another's life; and if we assume further that most people, given the choice, would like to face as low a personal risk of death as possible, then respecting the wishes of each potential member of the rescue party would have the following result. The wishes of potential members of the rescue party would be most respected by putting together an enormous rescue party in order to save the trapped miner – at the expense of many lives.

Thus, if the issue of rescue is seen simply as a question of balancing individual risks for each rescuer against the benefit to the individual of being rescued, then it would seem right to pursue a policy which overall was very costly in terms of lives lost. Suppose, for example, that the rescue was being led by a senior army officer. If that army officer were to coordinate the rescue, with the foreseeable result that more people would die in the attempt to rescue than would be saved by the rescue, then the army officer might reasonably be criticised. One reason for such criticism might be that those in the rescue party had no choice but to join in because they had to obey their senior officer. However, if the rescue party were made up of volunteers who knew and accepted the risk to themselves, then might it not be justified to organise a rescue which will foreseeably lead to more deaths overall? The justification would be that fully competent adults in possession of all relevant information have made an individual choice to take part in the rescue party.

The case of the trapped miner, as outlined above, demonstrates, I think, that at the very least the volunteers would need to know not only the individual risk of death which each faces, but also the overall size of the rescue party. It would need to be made clear to the volunteers that the rescue itself will result in more deaths amongst the rescuers than the number of people who will be saved. This information is likely to affect potential volunteers' views on whether or not to join the rescue party.

In any case, returning to the situation which faces the health authority, it is not clear that those who could benefit from the preventive treatment have voluntarily agreed to forego their treatment in order for identifiable patients to receive expensive life-saving treatment. In other words, if the health authority spends more per year of life saved on rescue treatments than prophylactic treatments, the health authority is effectively volunteering those who would benefit from the preventive treatment to take part in the rescue party. Because of limited resources, the health authority in making any of its decisions about treatments which save lives, has to save some people's lives at the expense of other people's lives. In the absence of a clear mandate from the group of people who stand to lose by a particular decision, it seems to me that the core principle must be that those decisions should be taken which overall save more lives.

Why the contractualist theory in favour of rescue fails

A contractualist theory might lead to the view that although there is no explicit mandate (i.e. there has been no explicit volunteering) from those who would benefit from preventive treatment, this is nevertheless what the population would want. In other words, the argument is that the population, say of Oxfordshire, would favour the health authority spending more money to save the lives of identifiable people, thus leading overall to more lives lost through a failure to fund preventive treatments. On this argument it is supposed that most of the population, including those who might benefit from the preventive treatments, would trade the very small increased risk of death (from failure to fund the preventive treatment) for the benefit of rescuing identifiable people. However, if this argument is to be put forward I think that a number of points would need to be clarified.

1   In collecting the evidence, it would need to be made clear that overall more lives would be lost.

2   Even if there were good evidence that this is what people would want there is still the question of whether it is right for the health authority to sacrifice more lives for fewer lives. In the case of the trapped miner, even if it were possible to find a large number of volunteers to join the rescue party in full knowledge both of their personal risk and that overall more lives would be lost, it remains questionable whether an authority should organise such a rescue party.

3   A contractualist defence may make use of Rawls's "veil of ignorance" approach.[9] On this approach Rawls asks us to imagine a whole range of different societies. We can choose which society to join, but we do not know who we will be in the society. In the present context, the two societies under consideration would be one where a health authority (or rescue party) spends more per life-year saved on rescue than on prevention; and the other would be a

society which gives equal priority to prevention. Behind the veil of ignorance we are asked which we would choose. If the veil of ignorance approach is taken to be a genuine empirical question – that is a question as to what people as a matter of fact would choose, given this hypothetical situation, then we simply do not have the empirical facts to know what most people would choose. I imagine there would be considerable variation. If, on the other hand, the veil of ignorance approach is taken to be a method of clarifying the conceptual issues around justice, then it would seem to favour giving no priority to rescue. That is, it would seem most just to choose the society where, overall, our chance of death is least.

I conclude that in the absence of a clear mandate from society, and possibly even if there were such a mandate, it is wrong for a health authority to pay more per life-year saved by rescue than by prevention.

Acknowledgements

I would like to thank Dr John Reynolds for information on new drugs, Professor Siân Griffiths for discussions of rationing procedures and Professor Julian Savulescu and Dr Roger Crisp for detailed discussion of the principles and arguments.

Notes

1   Hope T, Hicks N, Reynolds DJM, Crisp R, Griffiths S. "Rationing and the health authority." British Medical Journal 1998; 317: 1067–9.
2   Daniels N, Sabin J. "Limits to health care: fair procedures, democratic deliberation and the legitimacy problem for insurers." Philosophy and Public Affairs 1997; 26: 303–50.
3   Montgomery J. Health care law. Oxford: Oxford University Press, 1997.
4   Savulescu J. "The cost of refusing treatment and equality of outcome." Journal of Medical Ethics 1998; 24: 231–6.
5   Williams A. "QALYs and ethics – a health economist's view." Society for the Science of Medicine 1996; 43: 1795–804.
6   Edgar A, Salek S, Shickle D, Cohen D. The Ethical QALY – Ethical Issues in Healthcare Resource Allocations. Haslemere, UK: Euromed Communications Ltd, 1998.
7   Savulescu J. "The present-aim theory: a submaximizing theory of rationality?" Australasian Journal of Philosophy. 1998; 76: 229–43.
8   Daniels N. Just Health Care – Studies in Philosophy and Health Policy. Cambridge: Cambridge University Press, 1985.
9   Rawls J. A Theory of Justice. Oxford: Oxford University Press, 1972.

Chapter 47

Norman Daniels

HEALTH-CARE NEEDS AND DISTRIBUTIVE JUSTICE

[. . .]

III. Needs and preferences

Not all preferences are created equal

Before turning to health-care needs in particular, it is worth noting that the concept of needs has been in philosophical disrepute, and with some good reason. The concept seems both too weak and too strong to get us very far toward a theory of distributive justice. Too many things become needs, and too few. And finding a middle ground seems to involve many of the issues of distributive justice one might hope to resolve by appeal to a clear notion of needs.

It is easy to see why too many things appear to be needs. Without abuse of language, we refer to the means necessary to reach any of our goals as needs. To reawaken memories of Miller's, the neighborhood delicatessen of my childhood, I need only the smell of sour pickles in a barrel. To paint my son's swing set, I need a clean brush.[1] The problem of the importance of needs seems to reduce to the problem of the importance or urgency of preferences or wants in general (leaving aside the fact that not all the things we need are expressed as preferences).

But just as not all preferences are on a par – some are more important than others – so too not all the things we say we need are. It is

possible to pick out various things we say we need, including needs for health care, which play a special role in a variety of moral contexts. Taking a cue from T. M. Scanlon's discussion in "Preference and Urgency," we should distinguish *subjective* and *objective* criteria of well-being.[2] We need *some* such criterion to assess the importance of competing claims on resources in a variety of moral contexts. A *subjective* criterion uses the relevant individual's own assessment of how well-off he is with and without the claimed benefit to determine the importance of his preference or claim. An *objective* criterion invokes a measure of importance independent of the individual's own assessment, for example, independent of the *strength* of his preference.

In contexts of distributive justice and other moral contexts, we do *in fact* appeal to some *objective* criteria of well-being. We refuse to rely solely on subjective ones. If I appeal to my friend's duty of beneficence in requesting $100, I will most likely get a quite different reaction if I tell him I need the money to get a root-canal than if I tell him I need the money to go to the Brooklyn neighborhood of my childhood to smell pickles in a barrel. Indeed, it is not likely to matter in his assessment of *obligations* that I strongly *prefer* to go to Brooklyn. Nor is it likely to matter if I insist I feel a great *need* to reawaken memories of my childhood – I am overcome by nostalgia. (He might give me the money for either purpose, but if he gives it so I can smell pickles, we would probably say he is not doing it out of any duty at all, that he feels no obligation.) Similarly, if my appeal was directed to some (even Utopian) social welfare agency rather than my friend, it would adopt objective criteria in assessing the importance of the request independent of my own strength of preference.

The issue as Scanlon has drawn it, between subjective and objective standards of well-being, is not just a claim about the *epistemic* status of our criteria of well-being. He is surely right that we do not rely on subjective standards of well-being: we do not just accept an individual's assessment of his well-being as the *relevant* measure of his well-being in important moral contexts. But the issue here is not just that such a measure is *subjective* and we use an *objective* measure. Nor is the issue that we may be skeptical about the feasibility of developing an objective interpersonal measure of satisfaction, and so we use another measure. Suppose we had an intersubjectively acceptable way of determining individual levels of well-being, where well-being is viewed as the level of satisfaction of the individual's *full range of preferences*. That is, suppose we had some deep social-utility function that enabled us to compare different persons' levels of satisfaction, given the full-range of their preferences and the social goods they have available. Such a scale would be the wrong scale to use in a broad range of moral contexts involving justice and the design of social institutions – at least it is not just an improvement on the scale we do in fact use. We would continue to use a far narrower scale of well-being, one that *does not include the full range of kinds of preferences* people have. So the real issue behind Scanlon's insightful discussion is the choice between objective *truncated* or selective scales of well-being and either objective or subjective *full-range* or "satisfaction" scales of well-being.[3] I shall return shortly to consider

why the truncated scale *ought to be* (and not just *is*) the measure used in issues of social justice.

One indication that we appeal to an objective, truncated standard is that I might say the root-canal, but not the smell of pickles in a barrel, is something I *really* need (assuming the dentist is right). It is a *need* and not just a desire. The implication is that some of the things we claim to need fall into special categories which give them a weightier moral claim in contexts involving the distribution of resources (depending, of course, on how well-off we already are within those categories of need).[4] Our task is to characterize the relevant categories of needs in a way that *explains* two central properties these special needs have. First, these needs are *objectively ascribable*: we can ascribe them to a person even if he does not realize he has them and even if he denies he has them because his preferences run contrary to the ascribed needs. Second, and of greater interest to us, these needs are *objectively important*: we attach a special weight to claims based on them in a variety of moral contexts, and we do so independently of the weight attached to these and competing claims by the relevant individuals. So our philosophical task is to characterize the class of things we need which has these properties and to do so in such a way that we explain why such importance is attached to them.

Needs and species-typical functioning

One plausible suggestion for distinguishing the relevant needs from all the things we can come to need is David Braybrooke's distinction between "course-of-life needs" and "adventitious needs." *Course-of-life needs* are those needs which people "have all through their lives or at certain stages of life through which all must pass." *Adventitious needs* are the things we need because of the particular contingent projects (which may be long-term ones) on which we embark. Human course-of-life needs would include food, shelter, clothing, exercise, rest, companionship, a mate (in one's prime), and so on. Such needs are not themselves deficiencies, for example, when they are anticipated. But a deficiency with respect to them "endangers the normal functioning of the subject of need *considered as a member of a natural species.*"[5] A related suggestion can be found in McCloskey's discussion of the human and personal needs we appeal to in political argument. He argues that needs "relate to what it would be detrimental to us to lack, *where the detrimental is explained by reference to our natures as men and specific persons.*"[6]

The suggestion here is that the needs which interest us are those things we need in order to achieve or maintain species-typical normal functioning. Do such needs have the two properties noted earlier? Clearly they are objectively ascribable, assuming we can come up with the appropriate notion of species-typical functioning. (So, incidentally, are adventitious needs, assuming we can determine the relevant goals by reference to which the adventitious needs become determinate.) Are these needs objectively important in the appropriate way? In a broad range of contexts we do treat them as such – a claim I shall not trouble to argue. What is of interest is to see *why* being in such a need category gives them their special importance.

A tempting first answer might be this: whatever our specific chosen goals or tasks, our ability to achieve them (and consequently our happiness) will be diminished if we fall short of normal species functioning. So, whatever our specific goals, we need these course-of-life needs, and therein lies their objective importance. We need them whatever else we need. For example, it is sometimes said that whatever our chosen goals or tasks, we need our health, and so appropriate health care. But this claim is not strictly speaking true. For many of us, some of our goals, perhaps even those we feel most important to us, are not necessarily undermined by failing health

or disability. Moreover, we can often adjust our goals – and presumably our levels of satisfaction – to fit better with our dysfunction or disability. Coping in this way does not necessarily diminish happiness or satisfaction in life.

Still, there is a clue here to a more plausible account: impairments of normal species functioning reduce the range of opportunity we have within which to construct life-plans and conceptions of the good we have a reasonable expectation of finding satisfying or happiness-producing. Moreover, if persons have a high-order interest in preserving the opportunity to revise their conceptions of the good through time, then they will have a pressing interest in maintaining normal species functioning by establishing institutions – such as health-care systems – which do just that. So the kinds of needs Braybrooke and McCloskey pick out by reference to normal species functioning are objectively important because they meet this high-order interest persons have in maintaining a normal range of opportunities. I shall try to refine this admittedly vague answer, but first I want to characterize health-care needs more specifically and show that they fit within this more general framework.

IV. Health-care needs

Disease and health

To specify a notion of health-care needs, we need clear notions of health and disease. I shall begin with a narrow, if not uncontroversial, "biomedical" model of disease and health. The basic idea is that health is the absence of disease and diseases (I here include deformities and disabilities that result from trauma) are *deviations from the natural functional organization of a typical member of a species.*[7] The task of characterizing this natural functional organization falls to the biomedical sciences, which must include evolutionary theory since claims about the design of the species and its fitness to meeting biological goals underlie at least some of the relevant functional ascriptions. The task is the same for man and beast with two complications. For humans we require an account of the species-typical functions that permit us to pursue biological goals as social animals. So there must be a way of characterizing the species-typical apparatus underlying such functions as the acquisition of knowledge, linguistic communication, and social cooperation. Moreover, adding mental disease and health into the picture complicates the issue further, most particularly because we have a less well-developed theory of species typical mental functions and functional organization. The "biomedical" model clearly presupposes we can, in theory, supply the missing account and that a reasonable part of what we now take to be psychopathology would show up as diseases.[8]

The biomedical model has two controversial features. First, the deviations that play a role in the definition of disease are from species-typical functional organization. In contrast, some treat health as an idealized level of fully developed functioning, as in the WHO definition.[9] Others insist that the notion of disease is strictly normative and that diseases are deviations from socially preferred functional norms.[10] Still, the WHO definition seems to conflate notions of health with those of general well-being, satisfaction, or happiness, over-medicalizing the domain of social philosophy. And, historically, arguments which show that "deviant" functioning – for example, "Drapetomania" (the running-away disease of slaves) or masturbation – have been medicalized and viewed as diseases do not establish the strongly normative thesis that deviance from social norms of functioning constitutes disease. So I shall accept the first feature of the model, noting, of course, that the model does not exclude normative judgments about diseases, for example, about which are undesirable or which excuse us from normally criticizable behavior and justify our entering a "sick role." These judgments circumscribe the normative notion of illness or sickness, not the theoretically more basic notion of disease (which thus admittedly departs from looser ordinary usage).[11]

Second, pure forms of the biomedical model also involve a deeper claim, namely that species-normal functional organization can itself be characterized without invoking normative or value judgments. Here the debate turns on hard issues in the philosophy of biology.[12] Fortunately, these need not detain us since my discussion does not turn on so strong a claim. It is enough for my purposes if the line between disease and the absence of disease is, for the general run of cases, un-controversial and ascertainable through publicly acceptable methods, for example, primarily those of the biomedical sciences. It will not matter if there is some relativization of what counts as a disease category to some features of social roles in a given society, and thus to some normative judgments, provided the core of the notion of species normal functioning is left intact. The model would still, I presume, count infertility as a disease, even though some or many individuals might prefer to be infertile and seek medical treatment to render themselves so. Similarly, unwanted pregnancy is not a disease. Again, dysfunctional noses are diseases, since noses have normal species functions and anatomy. If the dysfunction or deformity is serious, it might warrant treatment as an illness. But deviation of nasal anatomy from individual or social conceptions of beauty does not constitute disease.[13]

Thus the modified biomedical model still allows me to draw a fairly sharp line between uses of health-care services to prevent and treat diseases and uses to meet other social goals. The importance of such other goals may be different and may rest on other bases, for example, in the induced infertility or unwanted pregnancy cases. My intention is to show which principles of justice are relevant to distributing health-care services where we can take as fixed, primarily by nature, a generally uncontroversial baseline of species-normal functional organization. If important moral considerations enter at yet another level, to determine what counts as health and what disease, then the principles I discuss and these others must be reconciled, a task the biomedical model makes unnecessary at this stage and which I want to avoid here in any case. Of course, a complete theory, which I do not pursue, would presumably have to establish priorities among principles governing the meeting of health-care needs and principles for using health-care services to meet other social or individual goals, for example the termination of unwanted pregnancy or the upgrading of the beauty of the population.[14]

Though I have deliberately selected a rather narrow model of disease and health, at least by comparison to some fashionable construals, *health-care needs* emerge as a broad and diverse set. Healthcare needs will be those things we need in order to maintain, restore, or provide functional equivalents (where possible) to normal species functioning. They can be divided into:

(1) adequate nutrition, shelter
(2) sanitary, safe, unpolluted living and working conditions
(3) exercise, rest, and other features of healthy life-styles
(4) preventive, curative, and rehabilitative personal medical services
(5) non-medical personal (and social) support services.

Of course, we do not tend to think of all these things as included among health-care needs, partly because we tend to think narrowly about personal medical services when we think about health care. But the list is not constructed to conform to our ordinary notion of health care but to point out a functional relation between quite diverse goods and services, and the various institutions responsible for delivering them.

Disease and opportunity

The *normal opportunity range* for a given society will be the array of "life-plans" reasonable persons in it are likely to construct for themselves. The range is thus relative to key features of the society – its stage of historical development, its level of material wealth and technological development, and even important cultural facts about it. Facts about social organization, including the conception of justice regulating its basic institutions, will of course determine how that total normal range is distributed in the population. Nevertheless, that issue of distribution aside, normal species-typical functioning provides us with one clear parameter relevant to defining the normal opportunity range. Consequently, impairment of normal functioning through disease constitutes a fundamental restriction on individual opportunity relative to the normal opportunity range.

There are two important points to note about the normal opportunity range. Obviously some diseases constitute more serious curtailments of opportunity than others relative to a given range. But because normal ranges are society relative, the same disease in two societies may impair opportunity differently and so have their importance assessed differently. Thus the social importance of particular diseases is a notion we plausibly ought to relativize between societies, assuming for the moment that impairment of opportunity is a relevant consideration. Within a society, however, the normal opportunity range abstracts from important individual differences in what might be called *effective opportunity*. From the perspective of an individual with a particular conception of the good (life plan or utility function), one who has developed certain skills and capacities needed to carry out chosen projects, *effective* opportunity range will be a sub-space of the normal range. A college teacher whose career and recreational skills rely little on certain kinds of manual dexterity might find his effective opportunity diminished little compared to what a skilled laborer might find if disease impaired that dexterity. By appealing to the normal range I abstract from these differences in effective range, just as I avoid appeals directly to a person's conception of the good when I seek a measure for the social importance (for claims of justice) of health care needs.[15]

What emerges here is the suggestion that we use impairment of the normal opportunity range as a fairly crude measure of the relative importance of health-care needs at the macro level. In general, it will be more important to prevent, cure, or compensate for those disease conditions which involve a greater curtailment of normal opportunity range. Of course, impairment of normal species-functioning has another distinct effect. It can diminish satisfaction or happiness for an individual, as judged by that individual's conception of the good. Such effects are important at the micro level – for example, to individual decision-making about health-care utilization. But I am here seeking the appropriate framework within which to apply principles of justice to health care at the macro level. So we shall have to look further at considerations that weigh against appeals to satisfaction at the macro level.

V. Toward a distributive theory

Satisfaction and narrower measures of well-being

So far my discussion has been primarily descriptive, not normative. As Scanlon suggests, we do not in fact use a full-range satisfaction criterion of well-being when we assess the importance or urgency of individual claims on our resources. Rather, we treat as important only a narrow range of kinds of preferences. More specifically, preferences that bear on the fulfilment of certain kinds of needs are important components of this truncated scale of well-being. In a broad range of moral contexts, we give precedence to claims based on such needs, including health-care needs, over claims based on other kinds of preferences. The Braybrooke and McCloskey suggestion gives us a general characterization of this class of needs: deficiency with regard to them threatens normal species-functioning. More specifically, we can characterize health-care needs as things we need to maintain, restore, or compensate for the loss of, normal species-functioning. Since serious impairments of normal functioning diminish our capacities and abilities, they impair individual opportunity range relative to the range normal for our society. If we suppose people have an interest in maintaining a fair and roughly equal opportunity range, we can give at least a plausible *explanation* why they think healthcare needs are special and important (which is not to say we actually do distribute them accordingly).

In what follows, I shall urge a normative claim: we ought to subsume health care under a principle of justice guaranteeing fair equality of opportunity. Actually, since I cannot here defend such a general principle without going too deeply into the general theory of distributive justice, I shall urge a weaker claim: if an acceptable theory of justice includes a principle providing for fair equality of opportunity, then health-care institutions should be among those governed by it. Indeed, I shall sketch briefly how one general theory, Rawls' theory of justice as fairness, might be extended in this way to provide a distributive theory for health care. But *my account does not presuppose the acceptability of Rawls' theory*. If a rule or ideal code-utilitarianism, or some other theory, establishes a fair equality of opportunity principle, my account will probably be compatible with it (though some of the argument that follows may not be).

In order to introduce some issues relevant to extending Rawls' theory, I want to consider an issue we have thus far left hanging. Should we, for purposes of justice, use the objective, truncated scale of well-being we happen to use rather than a full-range satisfaction scale? Clearly, this too is a general question that takes us beyond the scope of this essay. Moreover, it is unlikely that we could establish conclusively a case against the satisfaction scale by considering the health care context alone. For example, a utilitarian proponent of a satisfaction of enjoyment scale might claim that the general tendencies of different diseases to diminish satisfaction provides, at worst, a rough equivalent to the "impairment of opportunity" criterion I am proposing.[16] Still, it is worth suggesting some of the considerations that weigh against the use of a satisfaction scale.

We can begin by pointing to a special case where our moral judgment would incline us against using a satisfaction scale, namely the case of "social hijacking" by persons with expensive tastes.[17] Suppose we judge how well-off someone is by reference to the full range of individual preferences in a satisfaction scale. Suppose further that moderate people adjust their tastes and preferences so that they have a reasonable chance of being satisfied with their share of social goods. Other more extravagant people form exotic and expensive tastes, even though they have comparable shares to the moderates, and, because their preferences are very strong, they are desperately unhappy when these tastes are not satisfied. Assume we can agree intersubjectively that the extravagants are less satisfied. Then if we are interested in maximizing – or even equalizing – satisfaction, extravagants seem to have a greater claim on further distributions of social resources than

moderates. But something seems clearly unjust if we deny the moderates equal claims on further distributions just because they have been modest in forming their tastes. With regard to tastes and preferences that *could have been otherwise* had the extravagants chosen differently, it seems reasonable to hold them *responsible* for their own low level of satisfaction.[18]

A more general division of responsibility is suggested by this hijacking case. Rawls urges that we hold *society* responsible for guaranteeing the individual a fair share of basic liberties, opportunity, and all-purpose means, like income and wealth, needed for pursuing individual conceptions of the good. But the *individual* is responsible for choosing his ends in such a way that he has a reasonable chance of satisfying them under such just arrangements.[19] Consequently, the special features of an individual's conception of the good – here his extravagant tastes and resulting dissatisfaction – do not give rise to any special claims of justice on social resources. This suggestion about a division of responsibility is really a claim about the *scope* of theories of justice: just arrangements are supposed to guarantee individuals a reasonable share of certain basic social goods which constitute the relevant – truncated – scale of well-being for purposes of justice. The immediate object of justice is not, then, happiness or the satisfaction of desires, though just institutions provide individuals with an acceptable framework within which they can seek happiness and pursue their interests. But individuals remain responsible for the choice of their ends, so there is no injustice in not having sufficient means to reach extravagant ends.

Obviously, a full defense of this claim about the scope of justice and the social division of responsibility, and thus about the reasons for using a truncated scale of well-being, cannot rest on isolated intuitions about cases like the hijacking one. In Rawls' case, a full argument involves the claim that adopting a satisfaction scale commits us to an unacceptable view of persons as mere "containers" for satisfaction, one that departs significantly from our moral practice.[20] Because I cannot pursue these issues here, beyond suggesting there are problems with a satisfaction scale, I am content to show there is a systematic, plausible alternative to using a satisfaction scale (and ultimately to utilitarianism) whose acceptability depends on more general issues. Consequently I stick with my weaker, conditional claim above.

Rawls' argument for a truncated scale is, of course, for a specific scale, one composed of his primary social goods. But my talk about a truncated scale has focused on talk about certain basic needs, in particular, things we need to maintain species-typical normal functioning. Health-care needs are paradigmatic among these. The task that remains is to fit the two scales together. My analysis of the relation between disease and normal opportunity range provides the key to doing that.

Extending Rawls' theory to health care

Rawls' index of primary social goods – his truncated scale of well-being used in the contract – includes five types of social goods: (a) a set of basic liberties; (b) freedom of movement and choice of occupations against a background of diverse opportunities; (c) powers and prerogatives of office; (d) income and wealth; (e) the social bases of self-respect. Actually, Rawls uses two simplifying assumptions when using the index to assess how well-off (representative) individuals are. First, income and wealth are used as approximations to the whole index. Thus the two principles of justice[21] require basic structures to maximize the long term expectations of the least advantaged, estimated by their income and wealth, given fixed background institutions that guarantee equal basic liberties and fair equality of opportunity. More importantly for our

purposes, the theory is idealized to apply to individuals who are "normal, active and fully cooperating members of society over the course of a complete life."[22] There is no distributive theory for health care because no one is sick.

This simplification seems to put Rawls' index at odds with the thrust of my earlier discussion, for the truncated scale of well-being we in fact use includes needs for health care. The primary goods seem to be too truncated a scale, once we drop the idealizing assumption. People with equal indices will not be equally well-off once we allow them to differ in health-care needs. Moreover, we cannot simply dismiss these needs as irrelevant to questions of justice, as we did certain tastes and preferences. But if we simply build another entry into the index, we raise special issues about how to arrive at an approximate weighting of the index items.[23] Similarly, if we treat healthcare services as a specially important primary social good, we abandon the useful generality of the notion of a primary social good. Moreover, we risk generating a long list of such goods, one to meet each important need.[24] Finally, we cannot just finesse the question whether there are special issues of justice in the distribution of health care by assuming fair shares of primary goods will be used in part to buy decent health-care insurance. A constraint on the adequacy of those shares is that they permit one to buy reasonable protection – so we must already know what justice requires by way of reasonable health care.

The most promising strategy for extending Rawls' theory without tampering with useful assumptions about the index of primary goods simply includes health-care institutions among the background institutions involved in providing for fair equality of opportunity.[25] Once we note the special connection of normal species functioning to the opportunity range open to an individual, this strategy seems the natural way to extend Rawls' view that the subject of theories of social justice are the *basic institutions* which provide a framework of liberties and opportunities within which individuals can use fair income-shares to pursue their own conceptions of the good. Insofar as meeting health-care needs has an important effect on the distribution of health, and more to the point, on the distribution of opportunity, the health-care institutions are plausibly included on the list of basic institutions a fair equality of opportunity principle should regulate.[26]

Including health-care institutions among those which are to protect fair equality of opportunity is compatible with the central intuitions behind wanting to guarantee such opportunity in the first place. Rawls is primarily concerned with *the opportunity to pursue careers* – jobs and offices – that have various benefits attached to them. So equality of opportunity is *strategically* important: a person's well being will be measured for the most part by the primary goods that accompany placement in such jobs and offices.[27] Rawls argues it is not enough simply to eliminate formal or legal barriers to persons seeking such jobs – for example, race, class, ethnic, or sex barriers. Rather, positive steps should be taken to enhance the opportunity of those disadvantaged by such social factors as family background.[28] The point is that none of us *deserves* the advantages conferred by accidents of birth – either the genetic or social advantages. These advantages from the "natural lottery" are morally arbitrary, and to let them determine individual opportunity – and reward and success in life – is to confer arbitrariness on the outcomes. So positive steps, for example, through the educational system, are to be taken to provide fair equality of opportunity.[29]

But if it is important to use resources to counter the advantages in opportunity some get in the natural lottery, it is equally important to use resources to counter the natural disadvantages induced by disease (and since class-differentiated social conditions contribute significantly to the etiology of disease, we are reminded disease is not just a product of the natural component of the lottery). But this does not mean we are committed to the futile goal of eliminating all natural differences between persons. Health care has as its goal normal functioning and so concentrates on a specific class of obvious disadvantages and tries to eliminate them. That is its limited contribution to guaranteeing fair equality of opportunity.

The approach taken here allows us to draw some interesting parallels between education and health care, for both are strategically important contributors to fair equality of opportunity. Both address needs which are not equally distributed between individuals. Various social factors, such as race, class, and family background, may produce special learning needs; so too may natural factors, such as the broad class of learning disabilities. To the extent that education is aimed at providing fair equality of opportunity, special provision must be made to meet these special needs. Here educational needs, like health care needs, differ from other basic needs, such as the need for food and clothing, which are more equally distributed between persons. The combination of unequal distribution and the great strategic importance of the opportunity to have health care and education puts these needs in a separate category from those basic needs we can expect people to purchase from their fair-income shares.

It is worth noting another point of fit between my analysis and Rawls' theory. In Rawls' contract situation, a "thick" veil of ignorance is imposed on contractors choosing basic principles of justice: they do not know their abilities, talents, place in society, or historical period. In selecting principles to govern health-care resource-allocation decisions, we need a thinner veil, for we must know about some features of the society, for example, its resource limitations.

Still, using the normal opportunity range and not just the effective range as the baseline has the effect of imposing a plausibly thinned veil. It reflects basic facts about the society but keeps facts about individuals' particular ends from unduly influencing social decisions. Ultimately, defense of a veil depends on the theory of the person underlying the account. The intuition here is that persons are not defined by a particular set of interests but are free to revise their life plans. Consequently, they have an interest in maintaining conditions under which they can revise such plans, which makes the normal range a plausible reference point.

Subsuming health-care institutions under the opportunity principle can be viewed as a way of keeping the system as close as possible to the original idealization under which Rawls' theory was constructed, namely, that we are concerned with normal, fully functioning persons with a complete life span. An important set of institutions can thus be viewed as a first defense of the idealization: they act to minimize the likelihood of departures from the normality assumption. Included here are institutions which provide for public health, environmental cleanliness, preventive personal medical services, occupational health and safety, food and drug protection, nutritional education, and educational and incentive measures to promote individual responsibility for healthy life styles. A second layer of institutions corrects departures from the idealization. It includes those which deliver personal medical and rehabilitative services that restore normal functioning. A third layer attempts, where feasible, to maintain persons in a way that is as close as possible to the idealization. Institutions involved with more extended medical and social support services for the (moderately) chronically ill and disabled and the frail elderly would fit here. Finally, a fourth layer involves health care and related social services for those who can in no way be brought closer

to the idealization. Terminal care and care for the seriously mentally and physically disabled fit here, but they raise serious issues which may not just be issues of justice. Indeed, by the time we get to the fourth layer moral virtues other than justice become prominent.

[. . .]

VII. Applications

The account of health-care needs sketched here has a number of implications of interest to health planners. Here I can only note some of them and set aside the many difficulties that face drawing implications from ideal theory for non-ideal settings.[30]

Access

My account is compatible with (but does not imply) a multi-tiered health-care system. The basic tier would include health-care services that meet important health-care needs, defined by reference to their effects on opportunity. Other tiers would include services that meet less important health-care needs or other preferences. However the upper tiers are to be financed – through cost-sharing, at full-price, at "zero" price[31] – there should be no obstacles, financial, racial, sexual, or geographical to initial access to the system as a whole.

The equality of initial access derives from basic facts about the sociology and epistemology of the determination of health-care needs.[32] The "felt needs" of patients are (unreliable) initial indicators of real health care needs. Financial and geographical barriers to initial access – say to primary care – compel people to make their own determinations of the importance of their symptoms. Of course, every system requires some patient self-assessment, but financial and geographical barriers impose different burdens in such assessment on particular groups. Indeed, where sociological barriers exist to people utilizing services, positive steps are needed (in the schools, at work, in neighborhoods) to make sure unmet needs are detected.

It is sometimes argued that the difficult access problems are ones deriving from geographical barriers and the maldistribution of physicians within specialties. In the United States, it is often argued that achieving more equitable distribution of health care providers would unduly constrain physician liberties. It is important to see that no fundamental liberties need be violated. Suppose that the basic tier of a health-care system is redistributively financed through a national health insurance scheme that eliminates financial barriers, that no alternative insurance for the basic tier is allowed, and that there is central planning of resource allocation to guarantee needs are met. To achieve a more equitable distribution of physicians, planners *license those eligible for reimbursement* in a given health-planning region according to some reasonable formula involving physician-patient ratios.[33] Additional providers might practice in an area, but they would be without benefit of third-party payments for all services in the basic tier (or for other tiers if the national insurance scheme is more comprehensive). Most providers would follow the reimbursement dollar and practice where they are most needed.

Far from violating basic liberties, the scheme merely puts physicians in the same relation to market constraints on job availability that face most other workers and professionals. A college professor cannot simply decide there are people to be taught in Scarsdale or Chevy Chase or Shaker Heights; he must accept what jobs are available within universities, wherever they are. Of course, he is "free" to ignore the market, but then he may not be able to teach. Similarly, managers and many types of workers face the need to locate themselves where there is need for their skills. So the physician's sacrifice of liberty under the scheme (or variants on it, including a National Health Service) is merely the imposition of a burden already faced by much of the working population. Indeed, the scheme does not change in principle the forces that already motivate physicians; it merely shifts where it is profitable for some physicians to practice. The appearance that there is an enshrined liberty under attack is the legacy of an historical accident, one more visible in the United States than elsewhere, namely, that physicians have been more independent of institutional settings for the delivery of their skills than many other workers, and even than physicians in other countries. But this too shall pass.

Resource allocation

My account of health-care needs and their connection to fair equality of opportunity has a number of implications for resource-allocation issues. I have already noted that we get an important distinction between the use of health-care services to meet health-care needs and their use to meet other wants and preferences. The tie of health-care needs to opportunity makes the former use special and important in a way not true of the latter. Moreover, we get a crude criterion – impact on normal opportunity range – for distinguishing the importance of different health-care needs, though I have also noted how far short this falls of being a solution to many hard allocation questions. Three further implications are worth noting here.

There has been much debate about whether the United States' health-care system overemphasizes acute therapeutic services as opposed to preventive and public health measures. Sometimes the argument focuses on the relative efficacy and cost of preventive, as opposed to acute, services. My account suggests there is also an important issue of distributive justice here. Suppose a system is heavily weighted toward acute interventions, yet it provides equal access to its services. Thus anyone with severe respiratory ailments – black lung, brown lung, asbestosis, emphysema, and so on – is given adequate and comprehensive services as needed. Does the system meet the demands of equity? Not if they are determined by the approach of fair equality of opportunity. The point is that people are differentially at risk of contracting such diseases because of work and living conditions. Efficacy aside, preventive measures have distinct distributive implications from acute measures. The opportunity approach requires we attend to both.

My account points to another allocational inequity. One important function of health-care services, here personal medical services, is to restore handicapping dysfunctions, for example, of vision, mobility, and so on. The medical goal is to cure the diseased organ or limb where possible. Where cure is impossible, we try to make function as normal as possible, through corrective lenses or prosthesis and rehabilitative therapy. But where restoration of function is beyond the ability of medicine per se, we begin to enter another area of services, non-medical social support (we move from (4) to (5) on the list of healthcare needs in Section IV). Such support services provide the blind person with the closest he can get to the functional equivalent of vision – for example, he is taught how to navigate, provided with a seeing-eye dog, taught braille, and so on. From the point of view of their impact on opportunity, medical services and social support services that meet health-care needs have the same rationale and are equally important. Yet, for various reasons, probably having to do with the profitability and glamor of personal medical service and careers in them as compared to services for the handicapped, our society has taken only slow and halting steps to meet the health-care needs of those with permanent disabilities. These are matters of justice, not charity; we are not facing conditions of scarcity so severe that these steps to provide equality of opportunity must be forgone in favor of more pressing needs. The point also has implications for the problem of long term care for the frail elderly, but I cannot develop them here.

A final implication of the account raises a different set of issues, namely, how to reconcile the demands of justice with certain traditional views of a physician's obligation to his patients. The traditional view is that the physician's direct responsibility is to the well-being of his patients, that (with their consent) he is to do everything in his power to preserve their lives and well-being. One effect of leaving all resource-allocation decisions in this way to the micro-level decisions of physicians and patients, especially where third-party payment schemes mean little or no rationing by price, is that cost-ineffective utilization results. In the current cost-conscious climate, there is pressure to make physicians see themselves as responsible for introducing economic considerations into their utilization decisions. But the issue raised here goes beyond cost-effectiveness. My account suggests that there are important resource-allocation priorities that derive from considerations of justice. In a context of moderate scarcity, this suggests it is not possible for physicians to see as their ideal the maximization of the quality of care they deliver regardless of cost: pursuing that ideal upsets resource-allocation priorities determined by the opportunity principle. Considerations of justice challenge the traditional (perhaps mythical) view that physicians can act as the unrestrained agents of their patients. The remaining task, which I pursue elsewhere, is to show at what level the constraints should be imposed so as to disturb as little as possible of what is valuable about the traditional view of physician responsibility.[34]

These remarks on applications are frustratingly brief, and fuller development of them is required if we are to assess the practical import of the account I offer. Nevertheless, I think the account offers enough that it is attractive at the theoretical level to warrant further development of its practical implications.

Notes

1 For emphasis, we often refer to things we simply desire or want as things we need. Sometimes we invoke a distinction between noun and verb uses of "need," so that not everything we say we need counts as *a* need. Any distinction we might draw between noun and verb uses depends on our purposes and the context and would still have to be explained by the kind of analysis I undertake above.

2 T. M. Scanlon, "Preference and Urgency," *Journal of Philosophy* 77, no. 19 (November 1975): 655–669.

3 The difference might not be in the *extent* but in the *content* of the scale. An objective full-range satisfaction scale might be constructed so that some categories of (key) preferences are lexically primary to others; preferences not included on a truncated scale never enter the full-range scale except to break ties among those equally well-off on key preferences. Such a scale may avoid my worries, but it needs a rationale for its ranking. The objection raised here to full-range satisfaction measures applies, I believe, with equal force to happiness or enjoyment measures of the sort Richard Brandt defends in *A Theory of the Good and the Right* (Oxford: Oxford University Press, 1979), chap. 14.

4 See Scanlon, "Preference and Urgency," p. 660.

5 David Braybrooke, "Let Needs Diminish That Preferences May Prosper," in *Studies in Moral Philosophy*, American Philosophical Quarterly Monograph Series, No. 1 (Blackwells: Oxford, 1968), p. 90 (my emphasis). Personal medical services do not count as course-of-life needs on the criterion that we need them all through our lives or at certain (developmental) stages, but they do count as course-of-life needs in that deficiency with respect to them may endanger normal functioning.

6 McCloskey, unlike Braybrooke, is committed to distinguishing a narrower noun use of "need" from the verb use. See J. H. McCloskey, "Human Needs,

Rights, and Political Values," *American Philosophical Quarterly* 13, no. 1 (January 1976): 2f (my emphasis). McCloskey's proposal is less clear to me than Braybrooke's: presumably our natures include species-typical functioning but something more as well. Moreover, McCloskey is more insistent than Braybrooke in leaving room for *individual natures*, though Braybrooke at least leaves room for something like this when he refers to the needs that we may have by virtue of individual temperament. The hard problem that faces McCloskey is distinguishing between things we need *to develop our individual natures* and things we come to need in the process of what he calls "self-making," the carrying out of projects one chooses, perhaps in accordance with one's nature but not just by way of developing it.

7 The account here draws on a fine series of articles by Christopher Boorse; see "On the Distinction Between Disease and Illness," *Philosophy & Public Affairs* 5, no. 1 (Fall 1975): 49–68; "What a Theory of Mental Health Should Be," *Journal of the Theory of Social Behavior* 6, no. 1: 61–84; "Health as a Theoretical Concept," *Philosophy of Science* 44 (1977): 542–73. See also Ruth Macklin, "Mental Health and Mental Illness: Some Problems of Definition and Concept Formation," *Philosophy of Science* 39, no. 3 (September 1972): 341–65.

8 Boorse, "What a Theory of Mental Health Should Be," p. 77.

9 "Health is a state of complete physical, mental, and social well-being, and not merely the absence of disease or infirmity." From the Preamble to the Constitution of the World Health Organization. Adopted by the International Health Conference held in New York, 19 June–22 July 1946, and signed on 22 July 1946. *Off. Rec. Wld. Health Org.* 2, no. 100. See Daniel Callahan, "The WHO Definition of 'Health,'" *The Hastings Center Studies* 1, no. 3 (1973): 77–88.

10 See H. Tristram Engelhardt, Jr., "The Disease of Masturbation: Values and the Concept of Disease," *Bulletin of the History of Medicine* 48, no. 2 (Summer 1974): 234–48.

11 Boorse's critique of strongly normative views of disease is persuasive independently of some problematic features of his own account.

12 For example, we need an account of functional ascriptions in biology (see Boorse, "Wright on Functions," *Philosophical Review* 85, no. 1 [January 1976]: 70–86). More specifically, we need to be able to distinguish genetic variations from disease, and we must specify the range of environments taken as "natural" for the purpose of revealing dysfunction. The latter is critical to the second feature of the biomedical model: for example, what range of social roles and environments is included in the natural range? If we allow too much of the social environment, then racially discriminatory environments might make being of the wrong race a disease; if we disallow all socially created environments, then we seem not to be able to call dyslexia a disease (disability).

13 Anyone who doubts the appropriateness of treating some physiognomic deformities as serious diseases with strong claims on surgical resources should look at Frances C. MacGreggor's *After Plastic Surgery: Adaptation and Adjustment* (New York: Praeger, 1979). Even where there is no disease or deformity, there is nothing in the analysis I offer that precludes individuals or society from deciding to use health-care technology to make physiognomy conform to some standard of beauty. But such uses of health technology will not be justifiable as the fulfillment of health-care *needs*.

14 My account has the following bearing on the debate about Medicaid-funded abortions. Non-therapeutic abortions do not count as health-care needs, so if Medicaid has as its only function the meeting of the health-care needs of the poor, then we cannot argue for funding the abortions just like any other procedure. Their justifications will be different. But if Medicaid should serve other important goals, like ensuring that poor and well-off women can equally well control their bodies, then there is justification for funding abortions. There is also the worry that not funding them will contribute to other health problems induced by illegal abortions.

15 One issue here is to avoid "hijacking" by past preferences which themselves define the effective range. Of course, effective range may be important in micro allocation decisions.

16 Presumably, he must also claim that we improve satisfaction more by treating and preventing disease than by finding ways to encourage people to adjust to their conditions by reordering their preference curves.

17 I draw on Rawls' unpublished lecture, "Responsibility for Ends," in the following three paragraphs.

18 Here again the utilitarian proponent of the satisfaction scale may issue a typical promissory note, assuring us that maximizing satisfaction overall requires institutional arrangements that act to minimize social hijacking.

19 The division presupposes, as Rawls points out in response to Scanlon, that people have the ability and know they have the responsibility to adjust their desires in view of their fair shares of (primary) social goods. See Scanlon, "Preference and Urgency," pp. 665–66.

20 Satisfaction scales leave us no basis for not wanting to be whatever person, construed as a set of preferences, has higher satisfaction. To borrow Bernard Williams' term, they leave us with no basis for insisting on the *integrity* of persons. See Rawls, "Responsibility for Ends." The view that issues here turn in a fundamental way on the nature of persons is pursued in Derek Parfit, "Later Selves and Moral Principles," *Philosophy and Personal Relations,* ed. Alan Montefiore (London: Routledge & Kegan Paul, 1973): 137–69; Rawls, "Independence of Moral Theory," *Proceedings and Addresses of the American Philosophical Association,* 48 (1974–1975): 5–22; and Daniels, "Moral Theory and the Plasticity of Persons," *Monist* 62, no. 3 (July 1979): 265–87.

21 See *A Theory of Justice* (Cambridge: Harvard, 1971), p. 302.

22 Rawls, "Responsibility for Ends."

23 Some weighting problems will have to be faced anyway; see my "Rights to Health Care" for further discussion. Also see Kenneth Arrow, "Some Ordinalist Utilitarian Notes on Rawls's Theory of Justice," *Journal of Philosophy* 70, no. 9 (1973): 245–63. Also see Joshua Cohen, "Studies in Political Philosophy," Ph.D. diss. (Harvard University, 1978), Part III and Appendices.

24 See Ronald Greene, "Health Care and Justice in Contract Theory Perspective," in *Ethics & Health Policy*, ed. Robert Veatch and Roy Branson (Cambridge, MA: Ballinger, 1976), pp. 111–26.

25 The primary social goods themselves remain general and abstract properties of social arrangements – basic liberties, opportunities, and certain all-purpose exchangeable means (income and wealth). We can still simplify matters in using the index by looking solely at income and wealth – assuming a background of equal basic liberties and fair equality of opportunity. Health care is not a primary social good – neither are food, clothing, shelter, or other basic needs. The presumption is that the latter will be adequately provided for from fair shares of income and wealth. The special importance and unequal distribution of health-care needs, like educational needs, are acknowledged by their connection to other institutions that provide for fair equality of opportunity. But opportunity, not health care or education, is the primary social good.

26 Here I shift emphasis from Rawls when he remarks that health is a *natural* as opposed to *social* primary good because its possession is less influenced by basic institutions. See *A Theory of Justice*, p. 62. Moreover, it seems to follow that where health care is generally inefficacious – say, in earlier centuries – it loses its status as a special concern of justice and the "caring" it offers may more properly be viewed as a concern of charity.

27 The ways in which disease affects normal opportunity range are more extensive than the ways in which it affects opportunity to pursue careers, a point I return to later.

28 Of course, the effects of family background cannot all be eliminated. See *A Theory of Justice*, p. 74.

29 Rawls allows individual differences in talents and abilities to remain relevant to issues of job placement, for example, through their effects on productivity. So fair equality of opportunity does not mean that individual differences no longer confer advantages. Advantages are constrained by the difference principle. See my "Merit and Meritocracy," *Philosophy & Public Affairs* 7, no. 3 (Spring 1978): 206–23.

30 I discuss these difficulties in "Conflicting Objectives and the Priorities Problem," to appear in Peter Brown, Conrad Johnson, and Paul Vernier, eds., *Income Support: Conceptual and Policy Issues* (Rowman and Littlefield, forthcoming). My *Justice and Health Care Delivery* develops some applications in detail.

31 The strongest objections to such mixed systems is that the upper tier competes for resources with the lower tiers. See Claudine McCreadie, "Rawlsian Justice and the Financing of the National Health Service," *Journal of Social Policy* 5, no. 2 (1976): 113–31.

32 See Avedis Donabedian, *Aspects of Medical Care Administration* (Cambridge: Harvard, 1973).

33 I ignore the crudeness of such measures. For fuller discussion of these manpower distribution issues see my "What is the Obligation of the Medical Profession in the Distribution of Health Care?" presented to the Conference on Health Care and Human Rights, University of Cincinnati Medical Center, 6 March 1980.

34 See Avedis Donabedian "The Quality of Medical Care: A Concept in Search of a Definition," *Journal of Family Practice* 9, no. 2 (1979), pp. 277–84; and Daniels, "Cost-Effectiveness and Patient Welfare," in Marc Basson, ed., *Ethics, Humanism and Medicine* (Aldon Liss, forthcoming).

Research for this paper was supported by Grant Number HS03097 from the National Center for Health Services Research, OASH, and by a Tufts Sabbatical Leave. I am also indebted to the Commonwealth Fund, which sponsored a seminar on this material at Brown University. Earlier drafts benefited from presentations to the Hastings Center Institute project on Ethics and Health Policy (funded by the Kaiser Foundation), a NCHSR staff seminar, and colloquia at Tufts, NYU Medical Center, University of Michigan, and University of Georgia. Helpful comments were provided by Ronald Bayer, Hugo Bedau, Richard Brandt, Dan Brock, Arthur Caplan, Josh Cohen, Allen Gibbard, Ruth Macklin, Carola Mone, John Rawls, Daniel Wikler, and the Editors of *Philosophy & Public Affairs*. This essay is excerpted from my *Justice and Health Care Delivery*, Cambridge University Press, in preparation.

Do Western principles of research ethics apply in the developing world?

INTRODUCTION TO PART NINE

B IOETHICS IS A global discipline because medicine, health care and research are international undertakings. This brings into focus the health status of people in the developing world. It also creates methodological challenges, such as how to resolve bioethical disputes given the plurality of cultural and ethical viewpoints. And it raises political questions about which institutions have authority to regulate global activities. Such concerns have coalesced into a sub-discipline, 'developing world bioethics'. A major topic in developing world bioethics concerns biomedical research using human participants and, in particular, the dilemma between benefiting from research data and protecting research subjects. Principles governing biomedical research, including standards for the design of clinical trials, are established in guide-lines; but these were devised in and for industrialised nations, so the question arises as to how well they apply to biomedical research taking place in the developing world.

The question was given impetus by the controversy generated by clinical trials described by Lurie and Wolfe, and commented on by Angell. The vertical (or perinatal) transmission of HIV occurs when a mother infects her child around the time of birth. The AIDS Clinical Trials Group (ACTG) 076 regimen, involving the antiretroviral drug zidovudine, was proven to be effective in avoiding vertical transmission. Infection rates for HIV in the developing world are staggering but the 076 regimen is compli-cated and expensive so a simpler and cheaper alternative treatment is desirable. Lurie and Wolfe report relevant randomised control trials (RCTs) that allocated some participants to a placebo group. These participants did not receive any antiretroviral drugs, despite it being known that zidovudine is effective. This flouts important prin-ciples of research ethics: that subjects should not be denied a known best treatment; that there must be genuine uncertainty as to whether a proposed treatment is more or less effective than either an existing, or no, treatment (equipoise); and that subjects should not be used as means to ends. Angell provocatively describes the research as analogous to the infamous Tuskegee syphilis study, a 'textbook example of unethical research'.

Arguing in favour of the controversial trials, Varmus and Satcher dispute the analogy with Tuskegee. For one thing, the trials enjoy considerable 'local support and approval'. For another, the trials are well-suited to the health needs of the local population; for example, the 076 regimen is prohibitively expensive for developing countries so the important question is whether the proposed alternative is better than nothing, and placebo-controlled trials are well designed to answer this. Levine concurs, focusing on the principle that 'every patient . . . should be assured of the best proven . . . therapeutic method'. Levine points out that, although the controversial trials seem to flout this principle because subjects allocated to the placebo group were denied antiretroviral drugs, the phrase is ambiguous: 'does it mean the best therapy available anywhere in the world? Or does it mean the standard that prevails in the country in which the trial is conducted?' He argues that, on its most plausible interpretation, the standard relevant to the controversial trials is the prevailing in-country medical practice, i.e. 'no antiretroviral therapy'. In keeping with the view of Varmus and Satcher, Levine concludes that the trials were best suited to answering the crucial question of whether a short-term, low-dose regimen is better than nothing.

Orentlicher's discussion clarifies that the implications of the debate reach further than the controversy over the 076 regimen. What is the status and authority of principles of biomedical research: are they absolute, or can they be overridden given local circumstances? For example, is the principle that no subject is to be denied a known best treatment never to be flouted, or can the circumstances of the research justify withholding a known best treatment? Orentlicher presents three scenarios in which 'U.S. physicians treat their Kenyan patients or employees differently than they would patients or employees in the U.S. because of differences in local circumstances'. He points out the irony that only the research scenario would be intuited as unethical yet that is the superior one 'from the perspective of a pregnant Kenyan worried about transmitting HIV to her child'. Orentlicher concludes that there are universal standards of human research ethics, but these apply differently according to local circumstances, so the controversial trials do not entail a double standard.

One of the arguments against the trials Orentlicher discusses in the latter half of his paper is that the treatments studied, whilst cheaper than the 076 regimen, would still be unaffordable and therefore unavailable in the host countries. This segues into a discussion of exploitation because the point of the availability condition is to avoid the exploitation of Third World subjects by researchers intent on producing lucrative drugs for First World markets. Participants in the 2001 Conference on Ethical Aspects of Research in Developing Countries argue against the 'general agreement that "reasonable availability" is necessary in order to ensure that the subject population is not exploited', and for a 'fair benefit' framework. They suggest that the 'reasonable availability' condition says both too much and too little: it is overly prescriptive about what benefit is required to avoid exploitation, and fails to recognise the multifaceted nature of suitable benefits to host countries. Arras' brief comment on the participants' paper widens the focus from the details of what constitutes an appropriate benefit accruing from a study to the broader context of all such research, namely global injustice.

The relationship between bioethical arguments of the kind contained in these selections, and published guidelines, is symbiotic: the guidelines inform current practice; the arguments require revisions of the guidelines. The Declaration of Helsinki has been a particularly influential set of guidelines, so it is instructive to ask how the bioethical dispute has impacted on it. Macklin presents revisions to the Declaration, insightfully distinguishing between the more and less successful.

Selections

Lurie, P. and Wolfe, S.M. 1997. 'Unethical Trials of Interventions to Reduce Perinatal Transmission of the Human Immunodeficiency Virus in Developing Countries', *The New England Journal of Medicine*, 337 (12), 853–6.

Angell, M. 1997. 'The Ethics of Clinical Research in the Third World', *The New England Journal of Medicine*, 337 (12), 847–9.

Varmus, H. and Satcher, D. 1997. 'Ethical Complexities of Conducting Research in Developing Countries', *The New England Journal of Medicine*, 337 (14), 1003–5.

Levine, R.J. 1998. 'The "Best Proven Therapeutic Method" Standard in Clinical Trials in Technologically Developing Countries', *IRB: Ethics and Human Research*, 20 (1), 5–9.

Orentlicher, D. 2002. 'Universality and its Limits: When Research Ethics Can Reflect Local Circumstances', *Journal of Law, Medicine and Ethics*, 30 (3), 403–10.

Participants in the 2001 Conference on Ethical Aspects of Research in Developing Countries 2004. 'Moral Standards for Research in Developing Countries: From "Reasonable Availability" to "Fair Benefits"', *Hastings Center Report*, 34 (3), 17–27.

Arras, J.D. 2004. 'Another Voice: Fair Benefits in International Medical Research', *Hastings Center Report*, 34 (3), 3.

Macklin, R. 2009. 'The Declaration of Helsinki: Another Revision', *Indian Journal of Medical Ethics*, 6 (1), 2–4.

Chapter 48

Peter Lurie and Sidney M. Wolfe

UNETHICAL TRIALS OF INTERVENTIONS TO REDUCE PERINATAL TRANSMISSION OF THE HUMAN IMMUNODEFICIENCY VIRUS IN DEVELOPING COUNTRIES

It has been almost three years since the Journal[1] published the results of AIDS Clinical Trials Group (ACTG) Study 076, the first randomized, controlled trial in which an intervention was proved to reduce the incidence of human immunodeficiency virus (HIV) infection. The antiretroviral drug zidovudine, administered orally to HIV-positive pregnant women in the United States and France, administered intravenously during labor, and subsequently administered to the newborn infants, reduced the incidence of HIV infection by two thirds.[2] The regimen can save the life of one of every seven infants born to HIV-infected women.

Because of these findings, the study was terminated at the first interim analysis and within two months after the results had been announced, the Public Health Service had convened a meeting and concluded that the ACTG 076 regimen should be recommended for all HIV-positive pregnant women without substantial prior exposure to zidovudine and should be considered for other HIV-positive pregnant women on a case-by-case basis.[3] The standard of care for HIV-positive pregnant women thus became the ACTG 076 regimen.

In the United States, three recent studies of clinical practice report that the use of the ACTG 076 regimen is associated with decreases of 50 percent or more in perinatal HIV transmission.[4–6]

But in developing countries, especially in Asia and sub-Saharan Africa, where it is projected that by the year 2000, 6 million pregnant women will be infected with HIV,[7] the potential of the ACTG 076 regimen remains unrealized primarily because of the drug's exorbitant cost in most countries.

Clearly, a regimen that is less expensive than ACTG 076 but as effective is desirable, in both developing and industrialized countries. But there has been uncertainty about what research design to use in the search for a less expensive regimen. In June 1994, the World Health Organization (WHO) convened a group in Geneva to assess the agenda for research on perinatal HIV transmission in the wake of ACTG 076. The group, which included no ethicists, concluded, "Placebo-controlled trials offer the best option for a rapid and scientifically valid assessment of alternative antiretroviral drug regimens to prevent [perinatal] transmission of HIV."[8] This unpublished document has been widely cited as justification for subsequent trials in developing countries. In our view, most of these trials are unethical and will lead to hundreds of preventable HIV infections in infants.

Primarily on the basis of documents obtained from the Centers for Disease Control and Prevention (CDC), we have identified 18 randomized, controlled trials of interventions to prevent perinatal HIV transmission that either began to enroll patients after the ACTG 076 study was completed or have not yet begun to enroll patients. The studies are designed to evaluate a variety of interventions: antiretroviral drugs such as zidovudine (usually in regimens that are less expensive or complex than the ACTG 076 regimen), vitamin A and its derivatives, intrapartum vaginal washing, and HIV immune globulin, a form of immunotherapy. These trials involve a total of more than 17,000 women.

In the two studies being performed in the United States, the patients in all the study groups have unrestricted access to zidovudine or other antiretroviral drugs. In 15 of the 16 trials in developing countries, however, some or all of the patients are not provided with antiretroviral drugs. Nine of the 15 studies being conducted outside the United States are funded by the U.S. government through the CDC or the National Institutes of Health (NIH), five are funded by other governments, and one is funded by the United Nations AIDS Program. The studies are being conducted in Cote d'Ivoire, Uganda, Tanzania, South Africa, Malawi, Thailand, Ethiopia, Burkina Faso, Zimbabwe, Kenya, and the Dominican Republic. These 15 studies clearly violate recent guidelines designed specifically to address ethical issues pertaining to studies in developing countries. According to these guidelines, "The ethical standards applied should be no less exacting than they would be in the case of research carried out in [the sponsoring] country."[9] In addition, U.S. regulations governing studies performed with federal funds domestically or abroad specify that research procedures must "not unnecessarily expose subjects to risk."[10]

The 16th study is noteworthy both as a model of an ethically conducted study attempting to identify less expensive antiretroviral regimens and as an indication of how strong the placebo-controlled trial orthodoxy is. In 1994, Marc Lallemant, a researcher at the Harvard School of Public Health, applied for NIH funding for an equivalency study in Thailand in which three shorter zidovudine regimens were to be compared with a regimen similar to that used in the ACTG 076 study. An equivalency study is typically conducted when a particular regimen has already been proved effective and one is interested in determining whether a second regimen is about as effective but less toxic or expensive.[11] The NIH study section repeatedly put pressure on Lallemant and the Harvard School of Public Health to conduct a placebo-controlled trial instead, prompting the director of Harvard's human subjects committee to reply, "The conduct of a placebo-controlled trial for [zidovudine] in pregnant women in Thailand would be unethical and unacceptable, since an active-controlled trial is feasible."[12] The NIH eventually relented, and the study is now under way. Since the nine studies of antiretroviral drugs have attracted the most attention, we focus on them in this article.

Asking the wrong research question

There are numerous areas of agreement between those conducting or defending these placebo-controlled studies in developing countries and those opposing such trials. The two sides agree that perinatal HIV transmission is a grave problem meriting concerted international attention; that the ACTG 076 trial was a major breakthrough in perinatal HIV prevention; that there is a role for research on this topic in developing countries; that identifying less expensive, similarly effective interventions would be of enormous benefit, given the limited resources for medical care in most developing countries; and that randomized studies can help identify such interventions.

The sole point of disagreement is the best comparison group to use in assessing the effectiveness of less-expensive interventions once an effective intervention has been identified. The researchers conducting the placebo-controlled trials assert that such trials represent the only appropriate research design, implying that they answer the question, "Is the shorter regimen better than nothing?" We take the more optimistic view that, given the findings of ACTG 076 and other clinical information, researchers are quite capable of designing a shorter antiretroviral regimen that is approximately as effective as the ACTG 076 regimen. The proposal for the Harvard study in Thailand states the research question clearly: "Can we reduce the duration of prophylactic [zidovudine] treatment without increasing the risk of perinatal transmission of HIV, that is, without compromising the demonstrated efficacy of the standard ACTG 076 [zidovudine] regimen?"[13] We believe that such equivalency studies of alternative antiretroviral regimens will provide even more useful results than placebo-controlled trials, without the deaths of hundreds of newborns that are inevitable if placebo groups are used.

At a recent congressional hearing on research ethics, NIH director Harold Varmus was asked how the Department of Health and Human Services could be funding both a placebo-controlled trial (through the CDC) and a non-placebo-controlled equivalency study (through the NIH) in Thailand. Dr. Varmus conceded that placebo-controlled studies are "not the only way to achieve results."[14] If the research can be satisfactorily conducted in more than one way, why not select the approach that minimizes loss of life?

Inadequate analysis of data from ACTG 076 and other sources

The NIH, CDC, WHO, and the researchers conducting the studies we consider unethical argue that differences in the duration and route

of administration of antiretroviral agents in the shorter regimens, as compared with the ACTG 076 regimen, justify the use of a placebo group.[15–18] Given that ACTG 076 was a well-conducted, randomized, controlled trial, it is disturbing that the rich data available from the study were not adequately used by the group assembled by WHO in June 1994, which recommended placebo-controlled trials after ACTG 076, or by the investigators of the 15 studies we consider unethical.

In fact, the ACTG 076 investigators conducted a subgroup analysis to identify an appropriate period for prepartum administration of zidovudine. The approximate median duration of prepartum treatment was 12 weeks. In a comparison of treatment for 12 weeks or less (average, 7) with treatment for more than 12 weeks (average, 17), there was no univariate association between the duration of treatment and its effect in reducing perinatal HIV transmission ($P = 0.99$) (Gelber R: personal communication). This analysis is somewhat limited by the number of infected infants and its post hoc nature. However, when combined with information such as the fact that in non-breast-feeding populations an estimated 65 percent of cases of perinatal HIV infection are transmitted during delivery and 95 percent of the remaining cases are transmitted within two months of delivery,[19] the analysis suggests that the shorter regimens may be equally effective. This finding should have been explored in later studies by randomly assigning women to longer or shorter treatment regimens.

What about the argument that the use of the oral route for intrapartum administration of zidovudine in the present trials (as opposed to the intravenous route in ACTG 076) justifies the use of a placebo? In its protocols for its two studies in Thailand and Cote d'Ivoire, the CDC acknowledged that previous "pharmacokinetic modelling data suggest that [zidovudine] serum levels obtained with this [oral] dose will be similar to levels obtained with an intravenous infusion."[20]

Thus, on the basis of the ACTG 076 data, knowledge about the timing of perinatal transmission, and pharmacokinetic data, the researchers should have had every reason to believe that well-designed shorter regimens would be more effective than placebo. These findings seriously disturb the equipoise (uncertainty over the likely study result) necessary to justify a placebo-controlled trial on ethical grounds.[21]

Defining placebo as the standard of care in developing countries

Some officials and researchers have defended the use of placebo-controlled studies in developing countries by arguing that the subjects are treated at least according to the standard of care in these countries, which consists of unproven regimens or no treatment at all. This assertion reveals a fundamental misunderstanding of the concept of the standard of care. In developing countries, the standard of care (in this case, not providing zidovudine to HIV-positive pregnant women) is not based on a consideration of alternative treatments or previous clinical data, but is instead an economically determined policy of governments that cannot afford the prices set by drug companies. We agree with the Council for International Organizations of Medical Sciences that researchers working in developing countries have an ethical responsibility to provide treatment that conforms to the standard of care in the sponsoring country, when possible.[9] An exception would be a standard of care that required an exorbitant expenditure, such as the cost of building a coronary care unit. Since zidovudine is usually made available free of charge by the manufacturer for use in clinical trials, excessive cost is not a factor in this case. Acceptance of a standard of care that does not conform to the standard in the sponsoring country results in a double standard in research. Such a double standard, which permits research designs that are unacceptable in the sponsoring country, creates an incentive to use as research subjects those with the least access to health care.

What are the potential implications of accepting such a double standard? Researchers might inject live malaria parasites into HIV-positive subjects in China in order to study the effect on the progression of HIV infection, even though the study protocol had been rejected in the United States and Mexico. Or researchers might randomly assign malnourished San (bushmen) to receive vitamin-fortified or standard bread. One might also justify trials of HIV vaccines in which the subjects were not provided with condoms or state-of-the-art counseling about safe sex by arguing that they are not customarily provided in the developing countries in question. These are not simply hypothetical worst-case scenarios; the first two studies have already been performed,[22,23] and the third has been proposed and criticized.[24]

Annas and Grodin recently commented on the characterization and justification of placebos as a standard of care: " 'Nothing' is a description of what happens; 'standard of care' is a normative standard of effective medical treatment, whether or not it is provided to a particular community."[25]

Justifying placebo-controlled trials by claiming they are more rapid

Researchers have also sought to justify placebo-controlled trials by arguing that they require fewer subjects than equivalency studies and can therefore be completed more rapidly. Because equivalency studies are simply concerned with excluding alternative interventions that fall below some preestablished level of efficacy (as opposed to establishing which intervention is superior), it is customary to use one-sided statistical testing in such studies.[11] The numbers of women needed for a placebo-controlled trial and an equivalency study are similar.[26] In a placebo-controlled trial of a short course of zidovudine, with rates of perinatal HIV transmission of 25 percent in the placebo group and 15 percent in the zidovudine group, an alpha level of 0.05 (two-sided), and a beta level of 0.2, 500 subjects would be needed. An equivalency study with a transmission rate of 10 percent in the group receiving the ACTG 076 regimen, a difference in efficacy of 6 percent (above the 10 percent), an alpha level of 0.05 (one-sided), and a beta level of 0.2 would require 620 subjects (McCarthy W: personal communication).

Toward a single international standard of ethical research

Researchers assume greater ethical responsibilities when they enroll subjects in clinical studies, a precept acknowledged by Varmus recently when he insisted that all subjects in an NIH-sponsored needle-exchange trial be offered hepatitis B vaccine.[27] Residents of impoverished, postcolonial countries, the majority of whom are people of color, must be protected from potential exploitation in research. Otherwise, the abominable state of health care in these countries can be used to justify studies that could never pass ethical muster in the sponsoring country.

With the increasing globalization of trade, government research dollars becoming scarce, and more attention being paid to the hazards posed by "emerging infections" to the residents of industrialized countries, it is likely that studies in developing countries will increase. It is time to develop standards of research that preclude the kinds of double standards evident in these trials. In an editorial published nine years ago in the Journal, Marcia Angell stated, "Human

subjects in any part of the world should be protected by an irreducible set of ethical standards."[28] Tragically, for the hundreds of infants who have needlessly contracted HIV infection in the perinatal-transmission studies that have already been completed, any such protection will have come too late.

Notes

1   Connor EM, Sperling RS, Gelber R, et al. "Reduction of maternal-infant transmission of human immunodeficiency virus type 1 with zidovudine treatment," N Engl J Med 1994; 331: 1173–80.

2   Sperling RS, Shapiro DE, Coombs RW, et al. "Maternal viral load, zidovudine treatment, and the risk of transmission of human immunodeficiency virus type 1 from mother to infant," N Engl J Med 1996; 335: 1621–9.

3   "Recommendations of the U.S. Public Health Service Task Force on the use of zidovudine to reduce perinatal transmission of human immunodeficiency virus." MMWR Morb Mortal Wkly Rep 1994; 43(RR–11): 1–20.

4   Fiscus SA, Adimora AA, Schoenbach VJ, et al. "Perinatal HIV infection and the effect of zidovudine therapy on transmission in rural and urban counties." JAMA 1996; 275: 1483–8.

5   Cooper E, Diaz C, Pitt J, et al. "Impact of ACTG 076: use of zidovudine during pregnancy and changes in the rate of HIV vertical transmission," in Program and Abstracts of the Third Conference on Retroviruses and Opportunistic Infections, Washington, D.C., January 28-February 1, 1996. Washington, D.C.: Infectious Diseases Society of America, 1996: 57.

6   Simonds RJ, Nesheim S, Matheson P, et al. "Declining mother to child HIV transmission following perinatal ZDV recommendations." Presented at the 11th International Conference on AIDS, Vancouver, Canada, July 7–12, 1996. abstract.

7   Scarlatti G. Paediatric HIV infection. Lancet 1996; 348: 863–8.

8   Recommendations from the meeting on mother-to-infant transmission of HIV by use of antiretrovirals, Geneva, World Health Organization, June 23–25, 1994.

9   World Health Organization. "International ethical guidelines for biomedical research involving human subjects." Geneva: Council for International Organizations of Medical Sciences, 1993.

10 45CFR46.111(a)(1).

11 "Testing equivalence of two binomial proportions." in Machin D, Campbell MJ. *Statistical tables for the design of clinical trials.* Oxford, England: Blackwell Scientific, 1987: 35–53.

12 Brennan TA. Letter to Gilbert Meier, NIH Division of Research Ethics, December 28, 1994.

13 Lallemant M, Vithayasai V. *A Short ZDV Course to Prevent Perinatal HIV in Thailand.* Boston: Harvard School of Public Health, April 28, 1995.

14 Varmus H. Testimony before the Subcommittee on Human Resources, Committee on Government Reform and Oversight, U.S. House of Representatives, May 8, 1997.

15 "Draft talking points: responding to Public Citizen press conference." Press release of the National Institutes of Health, April 22, 1997.

16 "Questions and answers: CDC studies of AZT to prevent mother-to-child HIV transmission in developing countries." Press release of the Centers for Disease Control and Prevention, Atlanta. (undated document.)

17 "Questions and answers on the UNAIDS sponsored trials for the prevention of mother-to-child transmission: background brief to assist in responding to issues raised by the public and the media." Press release of the United Nations AIDS Program. (undated document.)

18 Halsey NA, Meinert CL, Ruff AJ, et al. "Letter to Harold Varmus, Director of National Institutes of Health." Baltimore: Johns Hopkins University, May 6, 1997.

19 Wiktor SZ, Ehounou E. "A randomized placebo-controlled intervention study to evaluate the safety and effectiveness of oral zidovudine administered in late pregnancy to reduce the incidence of mother-to-child transmission of HIV-1 in Abidjan, Cote D'Ivoire." Atlanta: Centers for Disease Control and Prevention, (undated document.)

20 Rouzioux C, Costagliola D, Burgard M, et al. "Timing of mother-to-child HIV-1 transmission depends on maternal status." *AIDS* 1993; 7:Suppl 2: 549–552.

21 Freedman B. "Equipoise and the ethics of clinical research." *N Engl J Med* 1987; 317: 141–5.

22 Heimlich HJ, Chen XP, Xiao BQ, et al. "CD4 response in HIV-positive patients treated with malaria therapy." Presented at the 11th International Conference on AIDS, Vancouver, B.C., July 7–12, 1996. abstract.

23 Bishop WB, Laubscher I, Labadarios D, Rehder P, Louw ME, Fellingham SA. "Effect of vitamin-enriched bread on the vitamin status of an isolated rural community – a controlled clinical trial." *S Afr Med J* 1996; 86:Suppl: 458–62.

24 Lurie P, Bishaw M, Chesney MA, et al. "Ethical, behavioral, and social aspects of HIV vaccine trials in developing countries." *JAMA* 1994; 271: 295–301.

25 Annas G, Gradin M. "An apology is not enough." *Boston Globe.* May 18, 1997: C1–C2.

26 Freedman B, Weijer C, Glass KC. "Placebo orthodoxy in clinical research. I. Empirical and methodological myths." *J Law Med Ethics* 1996; 24: 243–51.

27 Varmus H. Comments at the meeting of the Advisory Committee to the Director of the National Institutes of Health, December 12, 1996.

28 Angell M. "Ethical imperialism? Ethics in international collaborative clinical research." *N Eng J Med* 1988; 319: 1081–3.

Chapter 49

Marcia Angell

THE ETHICS OF CLINICAL RESEARCH IN THE THIRD WORLD

An essential ethical condition for a randomized clinical trial comparing two treatments for a disease is that there be no good reason for thinking one is better than the other.[1,2] Usually, investigators hope and even expect that the new treatment will be better, but there should not be solid evidence one way or the other. If there is, not only would the trial be scientifically redundant, but the investigators would be guilty of knowingly giving inferior treatment to some participants in the trial. The necessity for investigators to be in this state of equipoise[2] applies to placebo-controlled trials, as well. Only when there is no known effective treatment is it ethical to compare a potential new treatment with a placebo. When effective treatment exists, a placebo may not be used. Instead, subjects in the control group of the study must receive the best known treatment. Investigators are responsible for all subjects enrolled in a trial, not just some of them, and the goals of the research are always secondary to the well-being of the participants. Those requirements are made clear in the Declaration of Helsinki of the World Health Organization (WHO), which is widely regarded as providing the fundamental guiding principles of research involving human subjects.[3] It states, "In research on man [sic], the interest of science and society should never take precedence over considerations related to the wellbeing of the subject," and "In any medical study, every patient – including those of a control group, if any – should be assured of the best proven diagnostic and therapeutic method."

One reason ethical codes are unequivocal about investigators' primary obligation to care for the human subjects of their research is the strong temptation to subordinate the subjects' welfare to the objectives of the study. That is particularly likely when the research question is extremely important and the answer would probably improve the care of future patients substantially. In those circumstances, it is sometimes argued explicitly that obtaining a rapid, unambiguous answer to the research question is the primary ethical obligation. With the most altruistic of motives, then, researchers may find themselves slipping across a line that prohibits treating human subjects as means to an end. When that line is crossed, there is very little left to protect patients from a callous disregard of their welfare for the sake of research goals. Even informed consent, important though it is, is not protection enough, because of the asymmetry in knowledge and authority between researchers and their subjects. And approval by an institutional review board, though also important, is highly variable in its responsiveness to patients' interests when they conflict with the interests of researchers.

A textbook example of unethical research is the Tuskegee Study of Untreated Syphilis.[4] In that study, which was sponsored by the U.S. Public Health Service and lasted from 1932 to 1972, 412 poor African-American men with untreated syphilis were followed and compared with 204 men free of the disease to determine the natural history of syphilis. Although there was no very good treatment available at the time the study began (heavy metals were the standard treatment), the research continued even after penicillin became widely available and was known to be highly effective against syphilis. The study was not terminated until it came to the attention of a reporter and the outrage provoked by front-page stories in the Washington Star and New York Times embarrassed the Nixon administration into calling a halt to it.[5] The ethical violations were multiple: Subjects did not provide informed consent (indeed, they were deliberately deceived); they were denied the best known treatment; and the study was continued even after highly effective treatment became available. And what were the arguments in favor of the Tuskegee study? That these poor African-American men probably would not have been treated anyway, so the investigators were merely observing what would have happened if there were no study; and that the study was important (a "never-to-be-repeated opportunity," said one physician after penicillin became available).[6] Ethical concern was even stood on its head when it was suggested that not only was the information valuable, but it was especially so for people like the subjects – an impoverished rural population with a very high rate of untreated syphilis. The only lament seemed to be that many of the subjects inadvertently received treatment by other doctors.

Some of these issues are raised by Lurie and Wolfe elsewhere in this issue of the Journal. They discuss the ethics of ongoing trials in the Third World of regimens to prevent the vertical transmission of human immunodeficiency virus (HIV) infection[7]. All except one of the trials employ placebo-treated control groups, despite the fact that zidovudine has already been clearly shown to cut the rate of vertical transmission greatly and is now recommended in the United States for all HIV-infected pregnant women. The justifications are reminiscent of those for the Tuskegee study: Women in the Third World would not receive antiretroviral treatment anyway, so the investigators are simply observing what would happen to the subjects' infants if there were no study. And a placebo-controlled study is the fastest, most efficient way to obtain unambiguous information that will be of greatest value in the Third World. Thus, in response to protests from Wolfe and others to the secretary of Health and Human Services, the directors of the National Institutes of Health (NIH) and the Centers for Disease Control and Prevention (CDC) – the organizations sponsoring the studies – argued, "It is an unfortunate fact that the current standard of perinatal care for the HIV-infected pregnant women in the sites of the studies does not include any HIV prophylactic intervention at all," and the inclusion of placebo controls "will result in the most rapid, accurate, and reliable answer to the question of the value of the intervention being studied compared to the local standard of care."[8]

Also in this issue of the Journal, Whalen et al. report the results of a clinical trial in Uganda of various regimens of prophylaxis against tuberculosis in HIV-infected adults, most of whom had positive tuberculin skin tests.[9] This study, too, employed a placebo-treated control group, and in some ways it is analogous to the studies criticized by Lurie and Wolfe. In the United States it would probably be impossible to carry out such a study, because of long-standing official recommendations that HIV-infected persons with positive tuberculin skin tests receive prophylaxis against tuberculosis. The first was issued in 1990 by the CDC's Advisory Committee for Elimination of Tuberculosis.[10] It stated that tuberculin-test-positive persons with HIV infection "should be considered candidates for preventive therapy." Three years later, the recommendation was reiterated more strongly in a joint statement by the American Thoracic Society and the CDC, in collaboration with the Infectious Diseases Society of America and the American Academy of Pediatrics.[11] According to this statement, ". . . the identification of persons with dual infection and the administration of preventive therapy to these persons is of great importance." However, some believe that these recommendations were premature, since they were based largely on the success of prophylaxis in HIV-negative persons.[12]

Whether the study by Whalen et al. was ethical depends, in my view, entirely on the strength of the preexisting evidence. Only if there was genuine doubt about the benefits of prophylaxis would a placebo group be ethically justified. This is not the place to review the scientific evidence, some of which is discussed in the editorial of Msamanga and Fawzi elsewhere in this issue.[13] Suffice it to say that the case is debatable. Msamanga and Fawzi conclude that "future studies should not include a placebo group, since preventive therapy should be considered the standard of care." I agree. The difficult question is whether there should have been a placebo group in the first place.

Although I believe an argument can be made that a placebo-controlled trial was ethically justifiable because it was still uncertain whether prophylaxis would work, it should not be argued that it was ethical because no prophylaxis is the "local standard of care" in sub-Saharan Africa. For reasons discussed by Lurie and Wolfe, that reasoning is badly flawed[7]. As mentioned earlier, the Declaration of Helsinki requires control groups to receive the "best" current treatment, not the local one. The shift in wording between

"best" and "local" may be slight, but the implications are profound. Acceptance of this ethical relativism could result in widespread exploitation of vulnerable Third World populations for research programs that could not be carried out in the sponsoring country.[14] Furthermore, it directly contradicts the Department of Health and Human Services' own regulations governing U.S.-sponsored research in foreign countries,[15] as well as joint guidelines for research in the Third World issued by WHO and the Council for International Organizations of Medical Sciences,[16] which require that human subjects receive protection at least equivalent to that in the sponsoring country. The fact that Whalen et al. offered isoniazid to the placebo group when it was found superior to placebo indicates that they were aware of their responsibility to all the subjects in the trial.

The Journal has taken the position that it will not publish reports of unethical research, regardless of their scientific merit.[14-17] After deliberating at length about the study by Whalen at al., the editors concluded that publication was ethically justified, although there remain differences among us. The fact that the subjects gave informed consent and the study was approved by the institutional review board at the University Hospitals of Cleveland and Case Western Reserve University and by the Ugandan National AIDS Research Subcommittee certainly supported our decision but did not allay all our misgivings. It is still important to determine whether clinical studies are consistent with preexisting, widely accepted ethical guidelines, such as the Declaration of Helsinki, and with federal regulations, since they cannot be influenced by pressures specific to a particular study.

Quite apart from the merits of the study by Whalen et al., there is a larger issue. There appears to be a general retreat from the clear principles enunciated in the Nuremberg Code and the Declaration of Helsinki as applied to research in the Third World. Why is that? Is it because the "local standard of care" is different? I don't think so. In my view, that is merely a self-serving justification after the fact. Is it because diseases and their treatments are very different in the Third World, so that information gained in the industrialized world has no relevance and we have to start from scratch? That, too, seems an unlikely explanation, although here again it is often offered as a justification. Sometimes there may be relevant differences between populations, but that cannot be assumed. Unless there are specific indications to the contrary, the safest and most reasonable position is that people everywhere are likely to respond similarly to the same treatment.

I think we have to look elsewhere for the real reasons. One of them may be a slavish adherence to the tenets of clinical trials. According to these, all trials should be randomized, double-blind, and placebo-controlled, if at all possible. That rigidity may explain the NIH's pressure on Marc Lallemant to include a placebo group in his study, as described by Lurie and Wolfe[7]. Sometimes journals are blamed for the problem, because they are thought to demand strict conformity to the standard methods. That is not true, at least not at this journal. We do not want a scientifically neat study if it is ethically flawed, but like Lurie and Wolfe we believe that in many cases it is possible, with a little ingenuity, to have both scientific and ethical rigor.

The retreat from ethical principles may also be explained by some of the exigencies of doing clinical research in an increasingly regulated and competitive environment. Research in the Third World looks relatively attractive as it becomes better funded and regulations at home become more restrictive. Despite the existence of codes requiring that human subjects receive at least the same protection abroad as at home, they are still honored partly in the breach. The fact remains that many studies are done in the Third World that simply could not be done in the

countries sponsoring the work. Clinical trials have become a big business, with many of the same imperatives. To survive, it is necessary to get the work done as quickly as possible, with a minimum of obstacles. When these considerations prevail, it seems as if we have not come very far from Tuskegee after all. Those of us in the research community need to redouble our commitment to the highest ethical standards, no matter where the research is conducted, and sponsoring agencies need to enforce those standards, not undercut them.

Notes

1 Angell M. "Patients' preferences in randomized clinical trials," N Engl J Med 1984; 310: 1385–7.

2 Freedman B. "Equipoise and the ethics of clinical research," N Engl J Med 1987; 317: 141–5.

3 "Declaration of Helsinki IV, 41st World Medical Assembly, Hong Kong, September 1989," in Annas GJ, Grodin MA, eds. *The Nazi doctors and the Nuremberg Code: human rights in human experimentation*. New York: Oxford University Press, 1992: 339–42.

4 "Twenty years after: the legacy of the Tuskegee syphilis study," *Hastings Cent Rep* 1992; 22(6): 29–40.

5 Caplan AL. "When evil intrudes." *Hastings Cent Rep* 1992; 22(6): 29–32.

6 "The development of consent requirements in research ethics." In: Faden RR, Beauchamp TL. *A history and theory of informed consent*. New York: Oxford University Press, 1986: 151–99.

7 Lurie P, Wolfe SM. "Unethical trials of interventions to reduce perinatal transmission of the human immunodeficiency virus in developing countries." N Engl J Med 1997; 337: 853–6.

8 The conduct of clinical trials of maternal-infant transmission of HIV supported by the United States Department of Health and Human Services in developing countries. Washington, D.C.: Department of Health and Human Services, July 1997.

9 Whalen CC, Johnson JL, Okwera A, et al. "A trial of three regimens to prevent tuberculosis in Ugandan

adults infected with the human immuno-deficiency virus." *N Engl J Med* 1997; 337: 801–8.

10 "The use of preventive therapy for tuberculous infection in the United States: recommendations of the Advisory Committee for Elimination of Tuberculosis." *MMWR Morb Mortal Wkly Rep* 1990; 39(RR-8): 9–12.

11 Bass JB Jr, Farer LS, Hopewell PC, et al. "Treatment of tuberculosis and tuberculosis infection in adults and children." *Am J Respir Crit Care Med* 1994; 149: 1359–74.

12 De Cock KM, Grant A, Porter JD. "Preventive therapy for tuberculosis in HIV-infected persons: international recommendations, research, and practice." *Lancet* 1995; 345: 833–6.

13 Msamanga GI, Fawzi WW. "The double burden of HIV infection and tuberculosis in sub-Saharan Africa." *N Engl J Med* 1997; 337: 849–51.

14 Angell M. "Ethical imperialism? Ethics in international collaborative clinical research," *N Engl J Med* 1988; 319: 1081–3.

15 Protection of human subjects, 45 CFR (section mark)46 (1996).

16 International ethical guidelines for biomedical research involving human subjects. Geneva: Council for International Organizations of Medical Sciences, 1993.

17 Angell M. "The Nazi hypothermia experiments and unethical research today," *N Engl J Med* 1990; 322: 1462–4.

Chapter 50

Harold Varmus and David Satcher

ETHICAL COMPLEXITIES OF CONDUCTING RESEARCH IN DEVELOPING COUNTRIES

One of the great challenges in medical research is to conduct clinical trials in developing countries that will lead to therapies that benefit the citizens of these countries. Features of many developing countries – poverty, endemic diseases, and a low level of investment in health care systems – affect both the ease of performing trials and the selection of trials that can benefit the populations of the countries. Trials that make use of impoverished populations to test drugs for use solely in developed countries violate our most basic understanding of ethical behavior. Trials that apply scientific knowledge to interventions that can be used to benefit such populations are appropriate but present their own ethical challenges. How do we balance the ethical premises on which our work is based with the calls for public health partnerships from our colleagues in developing countries?

Some commentators have been critical of research performed in developing countries that might not be found ethically acceptable in developed countries. Specifically, questions have been raised about trials of interventions to prevent maternal-infant transmission of the human immunodeficiency virus (HIV) that have been sponsored by the National Institutes of Health (NIH) and the Centers for Disease Control and Prevention (CDC).[1,2] Although these commentators raise important issues, they have

not adequately considered the purpose and complexity of such trials and the needs of the countries involved. They also allude inappropriately to the infamous Tuskegee study, which did not test an intervention. The Tuskegee study ultimately deprived people of a known, effective, affordable intervention. To claim that countries seeking help in stemming the tide of maternal-infant HIV transmission by seeking usable interventions have followed that path trivializes the suffering of the men in the Tuskegee study and shows a serious lack of understanding of today's trials.

After the Tuskegee study was made public, in the 1970s, a national commission was established to develop principles and guidelines for the protection of research subjects. The new system of protection was described in the Belmont report.[3] Although largely compatible with the World Medical Association's Declaration of Helsinki,[4] the Belmont report articulated three principles: respect for persons (the recognition of the right of persons to exercise autonomy), beneficence (the minimization of risk incurred by research subjects and the maximization of benefits to them and to others), and justice (the principle that therapeutic investigations should not unduly involve persons from groups unlikely to benefit from subsequent applications of the research).

There is an inherent tension among these three principles. Over the years, we have seen the focus of debate shift from concern about the burdens of participation in research (beneficence) to equitable access to clinical trials (justice). Furthermore, the right to exercise autonomy was not always fully available to women, who were excluded from participating in clinical trials perceived as jeopardizing their safety; their exclusion clearly limited their ability to benefit from the research. Similarly, persons in developing countries deserve research that addresses their needs.

How should these principles be applied to research conducted in developing countries? How can we – and they – weigh the benefits and risks? Such research must be developed in concert with the developing countries in which it will be conducted. In the case of the NIH and CDC trials, there has been strong and consistent support and involvement of the scientific and public health communities in the host countries, with local as well as United States-based scientific and ethical reviews and the same requirements for informed consent that would exist if the work were performed in the United States. But there is more to this partnership. Interventions that could be expected to be made available in the United States might be well beyond the financial resources of a developing country or exceed the capacity of its health care infrastructure. Might we support a trial in another country that would not be offered in the United States? Yes, because the burden of disease might make such a study more compelling in that country. Even if there were some risks associated with intervention, such a trial might pass the test of beneficence. Might we elect not to support a trial of an intervention that was beyond the reach of the citizens of the other country? Yes, because that trial would not pass the test of justice.

Trials supported by the NIH and the CDC, which are designed to reduce the transmission of HIV from mothers to infants in developing countries, have been held up by some observers as examples of trials that do not meet ethical standards. We disagree. The debate does not hinge on informed consent, which all the trials have obtained. It hinges instead on whether it is ethical to test interventions against a placebo control when an effective intervention is in use elsewhere in the world. A background paper sets forth our views on this matter more fully.[5] The paper is also available on the World Wide Web (at www.nih.gov/news/mathiv/mathiv.htm).

One such effective intervention – known as AIDS Clinical Trials Group protocol 076 – was a major breakthrough in the search for a way to interrupt the transmission of HIV from mother to infant. The regimen tested in the original study, however, was quite intensive for pregnant women and the health care system. Although this regimen has been proved effective, it requires that women undergo HIV testing and receive counseling about their HIV status early in pregnancy, comply with a lengthy oral regimen and with intravenous administration of the relatively expensive antiretroviral drug zidovudine, and refrain from breast-feeding. In addition, the newborn infants must receive six weeks of oral zidovudine, and both mothers and infants must be carefully monitored for adverse effects of the drug. Unfortunately, the burden of maternal-infant transmission of HIV is greatest in countries where women present late for prenatal care, have limited access to HIV testing and counseling, typically deliver their infants in settings not conducive to intravenous drug administration, and depend on breastfeeding to protect their babies from many diseases, only one of which is HIV infection. Furthermore, zidovudine is a powerful drug, and its safety in the populations of developing countries, where the incidences of other diseases, anemia, and malnutrition are higher than in developed countries, is unknown. Therefore, even though the 076 protocol has been shown to be effective in some countries, it is unlikely that it can be successfully exported to many others.

In addition to these hurdles, the wholesale cost of zidovudine in the 076 protocol is estimated to be in excess of $800 per mother and infant, an amount far greater than most developing countries can afford to pay for standard care. For example, in Malawi, the cost of zidovudine alone for the 076 regimen for one HIV-infected woman and her child is more than 600 times the annual per capita allocation for health care.

Various representatives of the ministries of health, communities, and scientists in developing countries have joined with other scientists to call for less complex and less expensive interventions to counteract the staggering impact of maternal-infant transmission of HIV in the developing world. The World Health Organization moved promptly after the release of the results of the 076 protocol, convening a panel of researchers and public health practitioners from around the world. This panel recommended the use of the 076 regimen throughout the industrialized world, where it is feasible, but also called for studies of alternative regimens that could be used in developing countries, observing that the logistical issues and costs precluded the widespread application of the 076 regimen.[6] To this end, the World Health Organization asked UNAIDS, the Joint United Nations Programme on HIV/AIDS, to coordinate international research efforts to develop simpler, less costly interventions.

The scientific community is responding by carrying out trials of several promising regimens that developing countries recognize as candidates for widespread delivery. However, these trials are being criticized by some people because of the use of placebo controls. Why not test these new interventions against the 076 regimen? Why not test them against other interventions that might offer some benefit? These questions were carefully considered in the development of these research projects and in their scientific and ethical review.

An obvious response to the ethical objection to placebo-controlled trials in countries where there is no current intervention is that the assignment to a placebo group does not carry a risk beyond that associated with standard practice, but this response is too simple. An additional response is that a placebo-controlled study usually provides a faster answer with fewer subjects, but the same result might be achieved with more sites or more aggressive enrollment. The most compelling reason to use a placebo-controlled study is that it provides definitive answers to questions about the safety and value of an intervention in the setting in which the study is performed, and these answers are the point of the research. Without clear and firm answers to whether and, if so, how well an intervention works, it is impossible for a country to make a sound judgment about the appropriateness and financial feasibility of providing the intervention.

For example, testing two or more interventions of unknown benefit (as some people have suggested) will not necessarily reveal whether either is better than nothing. Even if one surpasses the other, it may be difficult to judge the extent of the benefit conferred, since the interventions may differ markedly in other ways – for example, cost or toxicity. A placebo-controlled study would supply that answer. Similarly, comparing an intervention of unknown benefit – especially one that is affordable in a developing country – with the only intervention with a known benefit (the 076 regimen) may provide information that is not useful for patients. If the affordable intervention is less effective than the 076 regimen – not an unlikely outcome – this information will be of little use in a country where the more effective regimen is unavailable. Equally important, it will still be unclear whether the affordable intervention is better than nothing and worth the investment of scarce health care dollars. Such studies would fail to meet the goal of determining whether a treatment that could be implemented is worth implementing.

A placebo-controlled trial is not the only way to study a new intervention, but as compared with other approaches, it offers more definitive answers and a clearer view of side effects. This is not a case of treating research subjects as a means to an end, nor does it reflect "a callous disregard of their welfare".[2] Instead, a placebo-controlled trial may be the only way to obtain an answer that is ultimately useful to people in similar circumstances. If we enroll subjects in a study that exposes them to unknown risks and is designed in a way that is unlikely to provide results that are useful to the subjects or others in the population, we have failed the test of beneficence.

Finally, the NIH- and CDC-supported trials have undergone a rigorous process of ethical review, including not only the participation of the public health and scientific communities in the developing countries where the trials are being performed but also the application of the U.S. rules for the protection of human research subjects by relevant institutional review boards in the United States and in the developing countries. Support from local governments has been obtained, and each active study has been and will continue to be reviewed by an independent data and safety monitoring board.

To restate our main points: these studies address an urgent need in the countries in which they are being conducted and have been developed with extensive in-country participation. The studies are being conducted according to widely accepted principles and guidelines in bioethics. And our decisions to support these trials rest heavily on local support and approval.

In a letter to the NIH dated May 8, 1997, Edward K. Mbidde, chairman of the AIDS Research Committee of the Uganda Cancer Institute, wrote:

> These are Ugandan studies conducted by Ugandan investigators on Ugandans. Due to lack of resources we have been sponsored by organizations like yours. We are grateful that you have been able to do so. . . . There is a mix up of issues here which needs to be clarified. It is not NIH conducting the studies in Uganda but Ugandans conducting their study on their people for the good of their people.

The scientific and ethical issues concerning studies in developing countries are complex. It is a healthy sign that we are debating these issues so that we can continue to advance our knowledge and our practice. However, it is essential that the debate take place with a full understanding of the nature of the science, the interventions in question, and the local factors that impede or support research and its benefits.

Notes

1 Lurie P, Wolfe SM. "Unethical trials of interventions to reduce perinatal transmission of the human immunodeficiency virus in developing countries." N Engl J Med 1997; 337: 853–6.

2 Angell M. "The ethics of clinical research in the third world". N Engl J Med 1997; 337:847–9.

3 National Commission for the Protection of Human Subjects of Biomedical and Behavioral Research. Belmont report: ethical principles and guidelines for the protection of human subjects of research. Washington, D.C.: Government Printing Office, 1988. (GPO 887–809.)

4 World Medical Association Declaration of Helsinki. Adopted by the 18th World Medical Assembly, Helsinki, 1964, as revised by the 48th

World Medical Assembly, Republic of South Africa, 1996.

5   The conduct of clinical trials of maternal-infant transmission of HIV supported by the United States Department of Health and Human Services in developing countries. Washington, D.C.: Department of Health and Human Services, July 1997.

6   Recommendations from the meeting on mother-to-infant transmission of HIV by use of antiretrovirals. Geneva, World Health Organization, June 23–25, 1994.

Chapter 51

Robert J. Levine

THE "BEST PROVEN THERAPEUTIC METHOD" STANDARD IN CLINICAL TRIALS IN TECHNOLOGICALLY DEVELOPING COUNTRIES

In September 1997 two articles[1,2] published in the *New England Journal of Medicine* incited intense controversy about the ethical justification of randomized clinical trials.[3] These articles criticized as unethical the placebo-controlled clinical trials of AZT[4] for the prevention of perinatal transmission of HIV infection[5] currently being carried out in several technologically developing countries. These trials were called unethical principally on the ground that they violated Article II.3 of the Declaration of Helsinki:

> In any medical study, every patient – including those of a control group, if any – should be assured of the best proven diagnostic and therapeutic method.[6]

In this essay I will consider the intended meaning of the "best proven therapeutic method" standard and its implications for the choice of a control group for clinical trials in technologically developing countries.

The best proven therapeutic method for the prevention of perinatal transmission of HIV infection is the "076 regimen," so named because it was AIDS Clinical Trials Group protocol number 076 that demonstrated the effectiveness of AZT in reducing perinatal transmission of HIV infection about 67%, from about 25% to about 8%.[7] The 076 regimen, which is now accepted as the "standard of care" in the United States and many other technologically developed countries, entails administration of AZT to pregnant women orally during pregnancy, intravenously during delivery, and orally to the newborn infant.

The clinical trials now in progress in developing countries are designed to evaluate the efficacy of a much lower dose of AZT administered for a shorter term than the 076 regimen. Further, there is no parenteral administration of AZT during delivery nor is AZT given to the newborn infants. I will refer to the regimen being evaluated as the short-term, low-dose AZT regimen.[8]

It is essential to evaluate the short-term, low-dose AZT regimen because large-scale use of the 076 regimen is not a realistic consideration in most developing countries. The reasons for this are largely economic. The cost of the drug alone in the 076 regimen is approximately $800 (US) per pregnant woman; this does not include any of the other costs of providing treatment. The sub-Saharan African countries in which some of these trials are being conducted typically have annual per capita health budgets of less than $10 (US).

Moreover, in the developing countries in which these trials are being carried out, the 076 regimen is impracticable because women usually do not come to clinics for prenatal care early enough to start the 076 regimen. In addition, there are generally no facilities for intravenous administration of AZT during delivery.

Perhaps the most important difference from developed countries is that women in developing countries almost always breast-feed their newborn infants even though it is well known that breast-feeding is an efficient method of transmitting HIV from mother to infant.[9] Women are encouraged by health authorities to breast-feed their new-born children even if they have HIV infection. There are generally no facilities for making sterile infant formula available to the infants: even if formula were available, there is no safe water supply in which to dissolve it. Infants in these countries are even more likely to die of infant diarrhea than of HIV infection.

As a consequence of breast-feeding, the conditions are greatly different from those in which the 076 regimen was proved effective. The children are being exposed to a source of the virus for a much longer period of time than they are in developed countries. For this and other reasons the 076 regimen in developing countries could be less effective in preventing perinatal transmission.

When Helsinki calls for the "best proven therapeutic method" does it mean the best therapy available anywhere in the world? Or does it mean the standard that prevails in the country in which the trial is conducted? Helsinki is not clear about this. But I think that a careful analysis of this document and its history suggests that the best proven therapy standard was intended primarily as a standard of medical practice. A consideration of that conclusion yields a second conclusion: that the best proven therapy standard must necessarily mean the standard that prevails in the country in which the clinical trial is carried out.

The document that came to be known as the Declaration of Helsinki was drafted by the World Medical Association's Committee on Medical Ethics. In 1953, when this committee began its work, it identified:

> a need for professional guidelines designed by physicians for physicians (as opposed to the Nuremberg Code, which was formed by jurists for use in a legal trial). Moreover, it was recognized that experiments must be classified into two groups: "experiments in new diagnostic and therapeutic methods" and "experiments undertaken to serve other purposes than simply to cure an individual."[10]

It is worth emphasizing that in 1953 the Committee on Medical Ethics understood as its mission the development of a document by *physicians for physicians*. The same commitment to the primacy of the values of the medical profession was expressed 11 years later in the final draft of the Declaration:

> The Declaration of Geneva of the World Medical Association *binds the physician* with the words, "The health of my patient will be my first consideration," and the International Code of Medical Ethics declares that, "A physician shall act only in the patient's interest when providing medical care which might have the effect of weakening the physical and mental condition of the patient" (*emphasis added*).

It is further worth emphasizing that a second component of its mission was to provide guidance "for *experiments in new diagnostic and therapeutic methods*" (emphasis added), an activity that can be and should be distinguished from the evaluation of such methods. As the Council of International Organizations of Medical Sciences (CIOMS) noted in 1993:

> the Declaration [of Helsinki] does not provide for controlled clinical trials. Rather, it assures

the freedom of the physician "to use a new diagnostic or therapeutic measure, if in his or her judgment it offers hope of saving life, reestablishing health or alleviating suffering."[11]

It is of further interest that Section II of the Declaration, subtitled "Clinical Research," consistently refers to the person with whom the physician interacts as the "patient." In Sections I (Basic Principles) and III (Nonclinical Biomedical Research) the physician (sometimes called "investigator") interacts with either "subjects," "individuals," or, in only one instance (Article III.2), "patients." According to Article III.2 patients may serve as subjects but only if they have illnesses unrelated to the "experimental design."[12]

In the light of these considerations, I conclude that the most plausible explanation of the "best proven therapeutic method" requirement is that it is a standard of medical practice. Its appearance in Article II.3 of the Declaration of Helsinki is to serve as a reminder to physician-investigators that they are not to allow the needs of science to override the values of medical practice.[13]

If one wishes to fault anyone for not making the 076 regimen available to residents of developing countries, then this should be done on grounds of failure to meet a standard of medical practice. Access to this regimen should be assured to all HIV-infected pregnant women and not just those who are enrolled in a clinical trial. In passing, it is worth noticing that it would be inappropriate to fault the governments of the developing countries for failure to measure up to the requirement to provide the 076 regimen. For reasons elaborated above, it would be virtually impossible for the governments in question to purchase sufficient supplies of AZT to meet the needs of their people. And one cannot hold an agent or an agency ethically responsible for doing anything it lacks the capacity to do.[14]

The initiation of a research program is not the same as establishment of an entitlement. If it were, then no Western drug company would ever develop any products to deal with problems that exist primarily or exclusively in developing countries.

Critics of the clinical trials now in progress in developing countries claim that these trials violate Article II.3 of the Declaration of Helsinki because the control group receives placebo rather than AZT. Some critics insist that women in the control group must receive nothing less than the 076 regimen because this is the only way to measure up to Helsinki's standard. Others would appear to be satisfied if all women in the clinical trial received some AZT (either the short-term, low-dose regimen or some other dose); they call for either a dose-response design[15] or historical controls.

The relevant standard therapy that ought to be made available to the subjects in a clinical trial is the standard that prevails in the country in which the trial is being conducted. And in the case of the clinical trials of the short-term, low-dose AZT regimen the standard in the countries in which the trials are being carried out is no antiretroviral therapy. What they need to find out is whether the short-term, low-dose regimen of AZT is better than their standard therapy – that is, better than nothing. Their critics who argue that the new dosage schedule should be compared with the 076 regimen appear to have missed the point of the clinical trials. The developing countries have no need for an answer to the question: "Is the short-term, low-dose regimen equal to or superior to the 076 regimen?" They need to know whether the short-term, low-dose, regimen is better than the current standard of no antiretroviral therapy.

In the current controversy about the conduct of clinical trials, there is much concern about the possibility that the inhabitants of developing countries might be exploited to get data to support the marketing of products in the more lucrative markets of technologically developed countries. I must acknowledge that in the past this was a serious problem. However, in 1993, CIOMS, in collaboration with the World Health Organization promulgated international ethical guidelines which are responsive to this problem; for example these guidelines require that:

> Externally sponsored research designed to develop a therapeutic, diagnostic or preventive product must be responsive to the health needs [and priorities] of the host country. It should be conducted only in host countries in which the disease ... for which the product is indicated is an important problem. As a general rule, the sponsoring agency should agree in advance of the research that any product developed through such research will be made reasonably available to the inhabitants of the host community or country at the completion of successful testing.[16]

These requirements are very much more protective against exploitation of inhabitants of developing countries than the more familiar ethical justifications such as informed consent that get so much more attention. In the countries in which the short-term, low-dose AZT regimen is being evaluated there can be no doubt that perinatal transmission of HIV is an important problem. With regard to the "reasonably available" standard, the short-term, low-dose regimen of AZT is apparently now affordable in Thailand and probably also in Brazil. If it proves to be safe and effective, international agencies have expressed their willingness to assist those countries that could not afford it (that is, countries in sub-Saharan Africa) in raising the necessary funds.

Ironically, the conduct of a clinical trial designed to compare the short-term, low-dose AZT regimen with the 076 regimen – as called for by some critics of the current placebo-controlled trials – not only would not be responsive to the needs of the developing countries, it

would provide an answer that could be of great interest in the developed countries. If the short-term, low-dose regimen turned out to be equal in effectiveness to the 076 regimen, this finding would have major economic and therapeutic implications for the developed countries. As such, then, it should be considered unethical to conduct such a trial anywhere but in a technologically developed country.[17]

Those critics who have argued that the placebo-control arm of the current clinical trials of the short-term, low-dose AZT regimen should be replaced by historical controls appear to be relying on two erroneous assumptions. The first is that the historical control data already available are sufficient for comparison purposes. This is incorrect. The currently available historical control data are largely derived from epidemiological studies carried out in developed countries as well as the data derived from an analysis of the outcomes in the placebo-control arm of the 076 trial. There are good reasons to predict that these data are not relevant to developing countries. For example, the nearly universal practice of breast-feeding should have an important effect on the incidence of HIV infection in the children of HIV infected mothers.

The second incorrect assumption is that the short-term, low-dose AZT regimen will have at least some good effect. In other words, a little AZT is likely to be better than no AZT at all. By no means is this a foregone conclusion. The worst case scenario is that it could have no effect on the rate of transmission and the virus that is transmitted could be much more likely to be resistant to AZT.[18]

Conclusion

The most plausible explanation of the "best proven therapeutic method" requirement is that it is a standard of medical practice. Its appearance in Article II.3 of the Declaration of Helsinki is to serve as a reminder to physician-investigators that they are not to allow the needs of science to override the values of medical practice. The relevant "standard therapy" that ought to be made available to the subjects in a clinical trial is the standard that prevails in the country in which the trial is being conducted; in a randomized clinical trial, the agent or agents that are to be compared should be in a state of clinical equipoise and none should be known to be inferior to the relevant standard therapy.[19]

On the basis of this understanding of the best proven therapeutic method standard, and in consideration of the sponsors' and investigators' apparent compliance with other standards set forth in the CIOMS/WHO International Ethical Guidelines – particularly those standards designed to prevent exploitation of residents of developing countries – I conclude that the current placebo-controlled clinical trials of the short-term, low-dose AZT regimen in the prevention of perinatal transmission of HIV infection are ethically justified.

Acknowledgment

This work was funded in part by grant number PO1 MH/DA 56 826–01A1 from the National Institute of Mental Health and the National Institute on Drug Abuse.

Notes

1 Lurie P, Wolfe SM: "Unethical trials of interventions to reduce perinatal transmission of the human immunodeficiency virus in developing countries." *New England Journal of Medicine* 1997; 337: 853–6.

2 Angell M: "The ethics of clinical research in the third world." *New England Journal of Medicine* 1997; 337: 847–9.

3 For rebuttals to the critiques by Lurie and Wolfe and Angell see Varmus H, Satcher D: "Ethical complexities of conducting research in developing countries." *New England Journal of Medicine* 1997; 337: 1003–5; and Bloom BR: "The highest attainable standard: ethical issues in AIDS vaccines." *Science* 1998; 279:186–8.

4 The antiretroviral drug, originally known as azidothymidine (AZT), and subsequently renamed zidovudine (ZDV) and Retrovir(R).

5 The passage of the human immunodeficiency virus (HIV), the causative agent of AIDS, from infected pregnant women to their fetuses during pregnancy or delivery is known as perinatal transmission.

6 The Declaration of Helsinki, the code of research ethics of the World Medical Association, was first promulgated in Helsinki in 1964. Quoted above is the fourth and most recent amended version adopted in 1996 in the Republic of South Africa. The second (and last) sentence of Article II.3 is "This does not exclude the use of inert placebo in studies where no proven diagnostic or therapeutic method exists." This final sentence was added because the WMA was distressed that its Helsinki Declaration was being misinterpreted as proscribing all placebo controls.

The Declaration of Helsinki is an illogical document owing to its reliance on the spurious distinction between therapeutic and nontherapeutic research. The WMA's occasional revisions of the Declaration have not succeeded in correcting its conceptual errors. In recognition of this, the American Medical Association has recently proposed a thorough revision.

For further discussion of the conceptual problems presented by the Declaration of Helsinki see Levine RJ: "International codes and guidelines for research ethics: a critical appraisal." In Vanderpool HY, ed.: *The Ethics of Research Involving Human Subjects: Facing the 21st Century.* Frederick, Md.: University Publishing Group, 1996: 235–59.

7 Connor EM, Sperling RS, Gelber R et al. "Reduction of maternal-infant transmission of human immunodeficiency virus type I with zidovudine treatment." *New England Journal of Medicine* 1994; 331: 1173–80; Sperling RS, Shapiro DE, Coombs RW et al.: "Maternal viral load, zidovudine treatment, and the risk of transmission of human

immunodeficiency virus type I from mother to infant." *New England Journal of Medicine* 1996; 335: 1621–9.

8 The daily dose of AZT is the same as that of the 076 regimen. The total dose is lower owing to the shorter duration of administration.

9 Dunn DT, Newell ML, Ades AE, and Peckham CS: "Risk of human immunodeficiency virus type I transmission through breastfeeding." *Lancet* 1992; 340: 585–8.

10 Perley S, Fluss SS, Bankowski Z, Simon F: "The Nuremberg Code: an international overview." In Annas GJ, Grodin MA, eds.: *The Nazi Doctors and the Nuremberg Code: Human Rights in Human Experimentation.* New York: Oxford University Press, 1992: 149–73, at p. 157.

11 CIOMS/WHO. *International Ethical Guidelines for Biomedical Research Involving Human Subjects.* Geneva: CIOMS, 1993: 52. The passage quoted above is followed by: "Also in regard to Phase II and Phase III drug trials there are customary and ethically justified exceptions to the requirements of the Declaration of Helsinki. A placebo given to a control group, for example, cannot be justified by its 'potential diagnostic or therapeutic value for the patient,' as Article II.6 prescribes. Many other interventions and procedures characteristic of late-phase drug development have no possible diagnostic or therapeutic value for the patients and thus must be justified on other grounds; usually such justification consists of a reasonable expectation that they carry little or no risk and that they will contribute materially to the achievement of the goals of the research."

12 This is another example of an illogical requirement in the Declaration of Helsinki. For further discussion of this see my chapter in Vanderpool (*supra*, at note 6).

13 In a personal communication to Barry Bloom (see note 3), Povl Riis, Chairman of the Central Scientific-Ethical Committee of Denmark, a member of the Committee that drafted the 1975 (Tokyo) revision of the Declaration of Helsinki and co-author of the best proven therapeutic method statement in the Declaration, stated: "In the 1970s we focused mainly on clinical-pharmacological

trials in developed countries. During the last 15 years, the Danish Committee System has interpreted the above sentence as implicitly qualified by '... in accordance with national accessibility' and 'with the indispensable condition that the project is of great importance to the developing country and consequently is not intending to test drugs or methods of primary interest to the donating country.' "

14 This idea is often expressed aphoristically: "Ought implies can."

Those who would find in the fact that the 076 regimen is not available in developing countries a breach of ethical duty must also identify the agent or agency that should be held accountable for performing this duty.

15 Those who oppose placebo controls on ethical grounds often recommend dose response designs as "ethically superior." However, the ethical problems presented by such designs are essentially the same as those of placebo controls. See Levine RJ: *Ethics and Regulation of Clinical Research,* 2d ed. Baltimore, Md.: Urban & Schwarzenberg, 1986: 202–7.

16 See note 11, CIOMS/WHO 1993: 44–5.

17 Actually, it would probably be unethical to conduct such a trial even in a developed country unless the design were adapted to the empirical realities of the developed country. There would be no reason, for example, not to provide orally administered AZT to the newborn infant or intravenous AZT to the mother during delivery.

18 Because administration of single antiretroviral agents can be associated with the emergence of resistant strains in those so treated it is preferred, when possible, to use combination therapies. In fact, the only adult population with HIV infection that is routinely treated with a single antiretroviral in the United States is pregnant women. This suggests strongly that the interests of the fetuses are routinely being given a higher priority than those of the pregnant women. This raises another set of ethical considerations that I shall address at another time.

19 It is not customary to insist on a state of clinical equipoise in placebo-controlled trials unless the purpose of the active agent being evaluated is to

mitigate that component of a disease process that leads to lethal or disabling complications. Thus, it is not uncommon to see placebo-controlled trials of new analgesics, anxiolytics, and antihypertensives that are not justified by a state of clinical equipoise. See note 15, Levine 1986:202–7 and Levine RJ: "Uncertainty in clinical research." *Law, Medicine & Health Care* 1988; 16:174–82.

Chapter 52

David Orentlicher

UNIVERSALITY AND ITS LIMITS: WHEN RESEARCH ETHICS CAN REFLECT LOCAL CIRCUMSTANCES

Studies in several developing countries for treatment to prevent HIV-transmission from mother to child generated considerable controversy in 1997. Critics of the studies argued that basic principles of research ethics were violated. According to the critics, researchers subjected women in developing countries to studies that would have been unethical in the United States (and other developed countries) and the researchers were therefore engaged in unethical exploitation of citizens of the developing countries in which the studies were conducted.

While the critics agreed that unethical exploitation had occurred, they differed on the exact nature of the exploitation. Some observers condemned the researchers for employing a double standard – because the researchers were applying a standard of care that would have been unacceptable in their own country. In the view of these critics, researchers should have been comparing the experimental treatment to established therapy rather than to placebo, as would have been required in the United States or other developed countries.[1] Other critics objected on the ground that once the trials demonstrated the efficacy of the experimental therapy, the therapy would not become available in developing countries because of its high cost. In this view, a study in developing countries need not always conform to the standard of care in developed countries, but studies on residents of developing countries cannot be conducted solely for the benefit of residents of developed countries.[2]

The trials and their fallout raise important questions. Can researchers conduct studies in some countries that would be unethical in other countries? In other words, are there universal principles of research ethics that prevent researchers from avoiding moral constraints in one country by moving their studies to another country? And if trials can be conducted in some countries even though unacceptable in other countries, does the fact that a study would be unacceptable elsewhere place special limits on researchers conducting the study where it is acceptable?

I will argue that one can accept the idea of universal ethical standards for research and still permit different trials for different countries. It does not follow that, if a research study is unethical in the United States, it is also unethical in Kenya. Rather, one can accept the same principles of research ethics for Kenya and the United States and still conclude that those universal principles allow for different studies in different countries because of differences in local circumstances. I will also argue that concerns about exploitation should place constraints on researchers conducting studies with patients in developing countries, but I will conclude that these constraints need not be as strict as those suggested by existing guidelines or other commentators. I will use the AZT trials in several developing countries during the mid 1990s to illustrate my arguments.

The controversial AZT trials

In 1994, researchers reported the success of zidovudine therapy (AZT) to reduce HIV-transmission from mother to child.[3] In the studies, which were conducted in France and the United States, half of the pregnant women were given AZT for up to twenty-five weeks during pregnancy[4] as well as during labor and delivery, and their infants were given AZT for six weeks after birth. The other half of the women were given a placebo. The researchers reported a two-thirds reduction in HIV-transmission for the women and children who received AZT therapy.

Although the results were dramatic, the researchers could not affect treatment for all pregnant women. The treatment began when the women initiated prenatal care, and many pregnant women do not receive care from an obstetrician until they are ready to deliver. In addition, the treatment's costs have been estimated at $800 per patient, an amount unaffordable in most developing countries.[5]

In response to the study's limitations, other researchers conducted studies involving a less aggressive, less expensive course of AZT therapy to see if it would also reduce the risk of HIV-transmission from mother to child. The studies were conducted in developing countries, mostly in Africa, and pregnant women in nearly all of those studies were divided into two groups, one of which received the less aggressive therapy, the other of which received placebo.[6]

These studies were criticized initially because pregnant women in the placebo group were denied any therapy, even though a proven therapy to prevent HIV-transmission existed.

According to the critics, an experimental treatment should be compared to established therapy when there is an established therapy. In this view, the less aggressive therapy should have been compared to standard AZT treatment rather than to placebo. If research subjects are instead given placebo, they are put at unnecessary risk for the disease being studied.[7] It is well-recognized that in the United States, a placebo control would not have been allowed. In other words, the critics said, researchers should not conduct studies in developing countries that would have been unethical in the United States.[8]

Defenders of the research replied that the women given placebo were not harmed by their participation in the studies. Because of their countries' poverty, the women would not have had access to AZT treatment outside of the study. By enrolling in the study, they had a 50 percent chance of getting a potentially effective treatment. Moreover, it was argued, a placebo arm to the study was necessary to find out whether the less aggressive course of therapy was effective and how effective it was. If, as expected (and as turned out), the less aggressive therapy was less effective than the more aggressive therapy, and the two forms of therapy were compared only with each other, we would not know if the less aggressive therapy was better than nothing.[9]

Still, even if a placebo control was permissible, other critics objected to the studies on the ground that the studies were validating experimental therapy that would become available primarily in developed countries. Even though the less aggressive course of therapy was much less expensive than the more aggressive course, it would still be unaffordable in most of the countries in which the AZT trials took place. Under such circumstances, the critics said, the research subjects were assuming the risks of research solely for the benefit of people living in developed countries, a situation that clearly constitutes exploitation.[10]

Different research in different countries – A double standard?

Who was correct, the opponents or proponents of the study? To answer this question, I think it is helpful to imagine it is 1995, a year after researchers first reported the success of AZT in reducing HIV-transmission from mother to child. Imagine also that the following scenarios are about to occur.

Clinic physicians

A group of physicians from the U.S. decides to spend a year in a clinic in rural Kenya, delivering medical care to patients at the clinic. Medical students from the U.S. will rotate through the clinic during the year under the supervision of the visiting physicians. The physicians and students have a number of reasons for doing this. They think it is important to do some practice in a severely underserved community, they think the experience will sharpen their clinical skills, and they plan to spend some time in Kenya and neighboring countries as tourists. At the end of the year, the physicians will return to their U.S. practices, and the students will go on to practice in the U.S. The physicians and students will bring some equipment and medicines with them, but for the most part, they will rely on the resources of the clinic. That means that pregnant women receiving prenatal care at the clinic will not receive AZT to prevent the spread of HIV to their fetuses. The cost of the AZT treatment is well beyond what is affordable for the clinic.[11]

Physician entrepreneurs

A second group of physicians from the U.S. will establish a business in rural Kenya to process natural substances from the area into pharmaceuticals. In particular, the physicians expect to manufacture some drugs that will be effective

agents against cancer, including cancers that are common in developing countries but very rare in developed countries. The company will pay appropriate royalties to the Kenyan government if any of the drugs are marketed. The physicians will hire workers from the local population, and they will pay them a good local wage, but one that is below the minimum wage in the U.S. The physicians will also provide health insurance that will cover the standard of care in the local community. That means that there will not be coverage for pregnant women to receive AZT to prevent HIV-transmission, again because such treatment is too expensive for that part of Kenya.

Research physicians

A third group of physicians from the U.S. comes to rural Kenya to conduct a research trial. The physicians are interested in seeing whether a less aggressive, less expensive regimen of AZT treatment can also reduce HIV-infection in newborns born to HIV-infected women. The trial will have two arms: half the subjects will be randomized to receive the less intensive regimen of AZT to prevent transmission of HIV; the other half of the subjects will receive a placebo.

I take it that of the three scenarios I have described – the doctors and medical students practicing in a local clinic, the physicians establishing a drug company, and the doctors conducting the AZT trial – only the research scenario would be condemned as unethical, even though in all three examples, the U.S. physicians treat their Kenyan patients or employees differently than they would patients or employees in the U.S. because of differences in local circumstances. The physicians in the local clinic give different treatment because of differences in affordability between the U.S. and Kenya, and the physician-entrepreneurs offer lower wages and less generous health insurance because of differences in the standard of living between the U.S. and Kenya.[12]

Yet from the perspective of a pregnant Kenyan worried about transmitting HIV to her child, the research scenario is clearly superior to the other scenarios. Pregnant Kenyans are less likely to transmit HIV to their children under the research protocol than they are if they receive care in the clinic or if they work at the drug company.

The question then is: Why would we condemn the research scenario? If U.S. doctors treating patients, and U.S. companies hiring employees can operate according to local ethical standards,[13] why cannot U.S. medical researchers rely on local standards in meeting their ethical obligations?

In fact, we do not need to assume different ethical standards to justify the research study. We can employ a universal set of standards whose application may vary depending on local resources. The idea here is similar to the principle in medical malpractice law that a physician's obligations to patients can take into account the facilities available. A doctor practicing at a small rural hospital cannot be expected to provide the same care as a doctor practicing at a major academic medical center.[14] Thus, we would expect the U.S. researchers to follow the same human research rules overseas that they follow here (e.g., informed consent, reasonable risk-benefit ratio, minimization of risk). But if U.S. doctors treating Kenyans are not obligated to deliver the U.S. level of care but only care that is reasonable with Kenya's resources, researchers also should be able to tailor their protocols to Kenya's resources. In other words, letting researchers conduct studies in other countries that they could not conduct in their own country does not necessarily entail a double standard. It is a double standard if we define the standard in terms of which research trials can be performed. In this view, we would have a double standard if a particular study could be conducted in one country but not the other. However, there is no double standard if we define the standard in terms of which ethical guidelines must be followed in designing a study's procedures. And the important standard is the ethical guidelines that must be employed.

Let me suggest another example that responds to the double standard argument. This example uses the ethical principle that research subjects can be compensated for their participation, but compensation should not be so great that it becomes coercive. Thus, we might allow research subjects to be paid a few hundred dollars but not tens of thousands of dollars. Now if we take a strict view of universal standards, we would have to have the same limits on compensation everywhere. If a certain payment would be unethical in Kenya, we would have to consider it unethical in the U.S. But we do not think that this is the consequence of universal standards. Just because a $50 or $100 payment in Kenya might be coercive, it would not be coercive in the U.S. We apply the same standard of no coercive payments, and that leads to cut-offs at different levels of payment around the world, because the standard of no coercive payments has to take into account local economic conditions.

In short, we need not condemn medical research in developing countries simply because a research protocol takes account of the economic resources of the host country and therefore has elements that would not be permitted in the U.S. or other countries with more wealth.[15]

Other objections to the AZT trials

Objections to the AZT trials went beyond the double standard argument. I will now consider other arguments that have been made against the AZT trials in developing countries in Africa and elsewhere.

The AZT trials in Africa were reprehensible in the way the Tuskegee study and other notorious research was reprehensible

These kinds of comparison are misguided. Tuskegee (the syphilis study of poor African-American men) was bad because the subjects were deceived from the outset of the study and because they were later deprived of a treatment that they should have received.[16] The AZT studies did not entail deception of the subjects, nor did they entail the withholding of care that the subjects were entitled to receive outside of the study. The pregnant women in the AZT trials were not made worse off by virtue of their participation in the trials.[17]

Note in this regard how the three-scenario comparison that I began with helps us with the analysis. I suggested that, if treating physicians do not have an obligation to provide the U.S. standard of care in developing countries, researchers would also not have that obligation. If we look at other studies that have been compared to the AZT trials, we see that the researchers deviated from obligations that treating physicians would have had. The Tuskegee case is a good example; treating doctors would have had an obligation to be truthful and to provide penicillin (once penicillin became widely available).

Peter Lurie and Sidney Wolfe mention some other examples in an article in the New England Journal of Medicine that have the same problem as the example of the Tuskegee study. Lurie and Wolfe claim that, if the AZT trials were acceptable, it would also be permissible for researchers to inject live malaria parasites in HIV-positive subjects in China in order to study the effect on the progression of HIV-infection.[18] This argument is not persuasive because researchers injecting malaria parasites would be doing something that treating physicians could not do. Similarly, it does not follow from the AZT trials that researchers could assign malnourished aboriginals to receive either vitamin-fortified or standard bread.[19] It is true that the aboriginals were not made worse off by the trials, but I think we would say that, if treating physicians offer the aboriginals bread, they should offer vitamin-fortified rather than standard bread.

For the same reason that the Tuskegee and similar analogies are mistaken, so is it mistaken to argue that the AZT trials in developing countries were wrong because we do not let researchers treat poor people in this country differently than wealthy persons. We could not justify the AZT trials on poor persons in the U.S. by saying that the poor persons would not otherwise have received the aggressive regimen of AZT and therefore were not made worse off by their participation in the research trials. But the reason we say that the AZT trials would have been unethical here is because we believe that doctors in the U.S. have an ethical (and legal) obligation to give the same quality treatment to all persons in their community, regardless of their wealth (once treatment is commenced). We do not believe, on the other hand, that U.S. doctors have an obligation to give poor persons overseas the same care as wealthy (or poor) persons in the United States. In other words, while my scenario of the U.S. physicians practicing in a Kenyan clinic is ethically acceptable, it would not be acceptable if the physicians practiced in the same way in an inner city clinic in the U.S.[20]

Researchers have greater obligations to their subjects than do treating physicians to their patients

The short answer to this argument is that the researchers in the AZT trials did more for their subjects than treating physicians in the host countries would have done for the pregnant women as patients (as illustrated by my scenarios of the treating physicians and the research physicians). The researchers gave the pregnant women a 50 percent instead of 0 percent chance at a treatment that might have substantially reduced the risk of HIV-transmission.[21] The hypothetical treating physicians were not in a position to offer AZT treatment to any patients.

Moreover, we hold researchers to higher standards because research often entails the taking of special risks. If we are going to ask research subjects to assume a risk for the benefit of society, we owe them special duties. But in this case, the research subjects were given an opportunity to reduce their risk of HIV-transmission by participating in the study. In other words, there was no heightened risk that required special consideration.

In observing that the pregnant women might have benefited from their participation in the AZT trials, I am not succumbing to the "therapeutic illusion" of medical research. I recognize that research is designed to accumulate knowledge, not to benefit research subjects. Nevertheless, in many research trials, the subjects do have a reasonable expectation of benefit. Pregnant women in the AZT trials genuinely could think that they might reduce the risk of HIV-transmission to their children by participating in the trials.

Although the requirement of special standards for research was satisfied in the context of the AZT trials, it would impose greater limits for other studies in which experimental therapies are being tested. In other words, the fact that the pregnant women were clearly better off for their participation in the AZT trials is a feature of those studies that will not be present in other studies. Accordingly, the analysis will be different for other studies. I will indicate how this is so when I further consider the exploitation argument below.

The AZT trials employed a placebo control rather than a standard therapy control

In this view, the trials were unacceptable because experimental therapies for HIV should be compared to existing therapy, not to placebo.[22] According to The Declaration of Helsinki, new treatments "should be tested against . . . the best current" treatments, with placebos reserved for studies in which no proven treatment exists.[23]

This argument fails as well. The best therapy available varies from country to country. For pregnant women in Kenya at the time of the AZT trials, there was no treatment to prevent transmission of HIV to their children. Thus, even though we also expect treating physicians to assure patients of the best current therapy, treating physicians need only provide those therapies that are reasonably available in their community. The treating physicians in my scenario of the local clinic in Kenya were obligated to provide the best therapy available in Kenya, not the best therapy available anywhere in the world. Similarly, physicians conducting research in Kenya should be obliged only to provide the best therapy available in Kenya.[24]

Moreover, as others have observed, a placebo control was important in the AZT trials.[25] If researchers had tested the standard therapy against the less aggressive treatment, and the less aggressive treatment provided less protection against HIV-transmission, we still would not know if the less aggressive treatment was better than placebo, or whether its advantages over placebo could justify its costs.

Although I would not criticize the AZT trials in Africa for including a placebo arm, I think the trials would have been better studies if they had included a third arm – we would have learned more about AZT treatment of HIV-infected women if the less aggressive treatment had been compared not only with placebo but also with standard therapy. That way we would have known how the less aggressive therapy compared to the more aggressive therapy as well as how the less aggressive therapy compared to the placebo.[26]

Note that there is a tension between an obligation to avoid placebo controls and an obligation to conduct research whose results will benefit people living in the country where the research is conducted (an obligation taken up in the next subsection). As I and others have argued, a placebo control was necessary to answer the question whether the less aggressive course of therapy was better, or sufficiently better, than no treatment, a critical question for people living in countries that could not afford the more expensive course of AZT.

The treatments being studied would not be affordable even at their much lower cost in the host countries and therefore should not have been studied in those countries

This argument accepts the use of a placebo arm in developing countries – even though forbidden in developed countries – as long as the therapy tested will be used in the host country. The requirement that the therapy be available in the host country after the study is over helps ensure that researchers do not exploit vulnerable citizens of developing countries. Some researchers, it is thought, are like the colonialists of old, raping poor countries for the benefit of their compatriots back home. According to the research guidelines of the Council for International Organizations of Medical Sciences (CIOMS), research conducted in a country must be responsive to that country's health needs.[27]

This is a complicated argument because it requires us to define exploitation, and people use the term in different ways. Some people would label a practice as exploitative only when the practice is immoral; other people would include a wider set of practices in their definition of exploitation and distinguish between acceptable and unacceptable exploitation.

Before I consider the definition of exploitation, it is worthwhile observing that most practices that we condemn as exploitative are practices that raise concerns about coercion or

the taking advantage of someone's desperate situation. I do not think the AZT trials raised either of these concerns (assuming proper informed consent). We worry about using prisoners as research subjects because inmates may think participation is required to have a good relationship with prison authorities. We worry about letting people sell their kidneys because indigent persons may feel that they have no choice but to sell a kidney in order to feed or shelter their family. That is, we worry that their desperation will lead them to engage in action that is harmful to them. To be sure, one might argue that the subjects in the AZT trial participated because of their poverty, but the participation did not require them to act against their interests. We cannot condemn practices simply because poor people are more likely to engage in them than wealthy people. To do that, we would have to condemn much of capitalism.

Defining exploitation

Returning now to the meaning of exploitation, I distinguish between fair and unfair exploitation rather than saying exploitation includes only unfair practices. Following Joel Feinberg, I will define exploitation as occurring when one person gains by using a characteristic of another person to his/her own advantage.[28] Exploitation, then, would include my raising money for my political campaign by taking advantage of a donor's generosity, as well as my obtaining my older brother's birthright by offering a bowl of porridge for the birthright when my brother is famished. While both are forms of exploitation, they differ in terms of their morality.

That takes us to the question: When is exploitation unfair? It is unfair if: (1) the other person does not give truly voluntary consent; (2) the other person is harmed; (3) the exploiter is profiting off the desperation of other persons (e.g., selling a worthless drug to the terminally ill who are adequately warned of the drug's uselessness); or (4) the exploiter's gain is disproportionate when compared with the exploited person's gain.[29]

Were the AZT trials exploitative?

If we are going to condemn the AZT trials as exploitative, it would have to be on the ground that the gain for the United States (the exploiter) is disproportionate when compared with the gain of Kenyans (the exploited person) (definition 4 above). In fact, Leonard Glantz, George Annas, Michael Grodin, and Wendy Mariner have argued that there is exploitation if research in a developing country will not be used to benefit residents of the developing country,[30] a principle consistent with the CIOMS guideline requiring that, when a therapy is developed, it must "be made reasonably available for the benefit of [the] population or community" in the host country.[31] If research results will be used only for the benefit of people living in the developed country, we should worry that the developed country's gain is disproportionate to the developing country's gain.

Whether this is so is a hard question. One could say that it is too strong a definition of unfair exploitation. In other settings, we do not consider it unfair when two people enter a contract under which different kinds of gain are realized. It is not unfair exploitation if workers at a Rolls Royce factory cannot afford to purchase a Rolls Royce. Using an example from medicine (the second scenario at the beginning of this article), we do not think it is unethical to develop drugs from substances in developing countries – even if the drugs will be used primarily in developed countries – as long as there is fair compensation to the developing country in the form of royalties or other payments. Nor do we think it unethical if physicians go to Kenya and use the experience there to improve their

clinical skills for patients in the U.S. (the first scenario at the beginning of this article).

Still we might say that research is different from business or clinical practice and that researchers have higher obligations than do employers or treating physicians. But even so, the principle enunciated by Glantz and the CIOMS guideline may go too far. With regard to the AZT trials, it is not clear that there was disproportionate gain. The research subjects got the chance to save their children from HIV-infection, a very important benefit. And all participants received access to good general medical care.

The example of the AZT trials suggests a distinction between trials in which the research subjects will clearly be better off by their participation and trials in which the subjects may end up being harmed by their participation. For an example of the latter kind of study, consider a trial of an experimental drug that may have valuable therapeutic effects but may also have serious toxicity. With this kind of study, the possibility of exploitation is much greater than with the AZT trials or other studies in which the advantages of participation clearly outweigh any risks. Accordingly, it is important to have stronger safeguards against exploitation for studies in which research subjects assume a real risk by virtue of their participation than for studies like the AZT trials in which the research subjects have much to gain at little risk to their health.

Appropriate safeguards to prevent exploitation

The question then is whether the stronger safeguards need be as strict as suggested by Glantz and the CIOMS guideline. Should we require that the intervention be made available after study in the host country? Should we also require, as would Glantz and CIOMS, that researchers establish in advance that the intervention will definitely be used in the host country?

One can object to either part of this strict position. Robert Crouch and John Arras have criticized the second prong of the Glantz/ CIOMS guidelines. They have argued that we might not expect funders of research to commit in advance to making available a treatment with hypothetical benefits and hypothetical costs. Rather, a firm commitment may not be achievable until the study's results begin to come in, and the treatment's success gives more reason, and generates more pressure, to make the therapy available in the host country.[32] Moreover, progress sometimes occurs in multiple steps. The experimental intervention might not make it back to the host country, but the results of the trial could easily lead to other trials with interventions that would make it back to the host country. If research demonstrates that a $50 therapy is almost as good as an $800 therapy, one can more readily justify a study of a $5 therapy.

While this objection to the second prong of the Glantz/CIOMS position has force, it seems insufficient to overcome that part of the position. If the intervention must be made available in the host country after the study, there are good reasons to require that the study's sponsor establish in advance that the intervention will in fact be made available. Given the examples of exploitation by researchers in the past and the serious harm caused to the research enterprise when the public becomes suspicious of the motives of researchers, we may want a strong rule to minimize the possibility of exploitation.

The response to the first part of the strict position of Glantz and CIOMS follows from my earlier point about people engaged in a mutual effort realizing different kinds of gain. Preventing exploitation does not demand that a treatment be made available in the country in which it is studied. Rather, it demands that the benefits to

the host country be proportionate to the benefits realized by the country sponsoring the research. If the host country can benefit in ways other than using the studied treatment – by receiving appropriate royalties from the sale of the drug, for example – then no exploitation would result.

Moreover, it is possible to permit other kinds of benefit without compromising on the requirement that a research sponsor establish in advance that the benefit will definitely be provided (i.e., we can preserve the second prong of the Glantz/CIOMS position when we amend their first prong). Indeed, it may be easier to establish in advance the provision of appropriate royalties than to establish in advance that a drug being studied will be made available in the host country. A research sponsor may not be able to guarantee the proper functioning of all of the channels of drug distribution in a developing country, but it can guarantee the payment of a royalty on sales. In short, one can have a strong safeguard to prevent exploitation without requiring that a studied intervention be made available in the host country after the study.

It is not only feasible to allow different kinds of benefit, it may be desirable to do so. Denying the option of different kinds of benefit might wrongly tie the developing country's values to the values of the developed country. A wealthier country will likely place a higher value on treatments for disease than a poorer country that lacks clean water and other public health necessities. The developing country would do better to receive royalties or other payments that it could allocate to measures for preventing disease than to receive a drug for treating a disease that could have been prevented. Furthermore, by providing fixed payments rather than royalty payments, the study sponsors can ensure that a benefit is gained by the study subjects. If a developing country's benefit is to be realized solely through royalties, the country will receive nothing if the experimental therapy does not make it to market.

Conclusion

In sum, we can accept the important principle that universal standards of ethics for human research (e.g., no exploitation) exist and also recognize that those principles apply differently when local circumstances vary. Studies that might be unethical in the U.S. may nevertheless be ethical in other countries, just as studies that are unethical in other countries may be ethical in the United States (e.g., a study in Kenya with compensation of $500).

When studies in developing countries take into account local circumstances, it is essential to employ strong safeguards that prevent exploitation. However, appropriately strong safeguards can be employed without requiring that a studied intervention be made available in the host country after the study is completed. Other kinds of benefit (e.g., royalties) can be provided to protect the host country from exploitation.

Notes

1 P. Lurie and S.M. Wolfe, "Unethical Trials of Interventions to Reduce Perinatal Transmission of the Human Immunodeficiency Virus in Developing Countries," *N. Engl. J. Med.*, 337 (1997): 853–56, at 854–55.

2 R.A. Crouch and J.D. Arras, "AZT Trials and Tribulations," *Hastings Center Report*, 28, no. 6 (1998): 26–34, at 29; L.H. Glantz et al, "Research in Developing Countries: Taking Benefit Seriously," *Hastings Center Report*, 28, no. 6 (1998): 38–42, at 40–42.

3 E.M. Connor et al., "Reduction of Maternal-Infant Transmission of Human Immunodeficiency Virus Type I with Zidovudine Treatment," *N. Engl. J. Med.*, 331 (1994): 1173–80.

4 AZT treatment began when the women entered the study, which occurred between weeks fourteen and thirty-four of their pregnancies. *Id.* at 1174.

5 C. Levine, "Placebos and HIV: Lessons Learned," *Hastings Center Report*, 28, no. 6 (1998): 43–48, at 44. To be sure, there is some artificiality to the stated cost of drug therapy. Companies may be able to price far above actual costs and exact monopoly profits for drugs under patent protection. Furthermore, public pressure has been successful in lowering the market price of important drugs. Nevertheless, it is clear that citizens of developing countries cannot afford drug treatments to the same extent as citizens of developed countries.

6 Lurie and Wolfe, *supra* note 1, at 853–54.

7 *Id.* at 854–55.

8 K.J. Rothman and K.B. Michels, "Declaration of Helsinki Should Be Strengthened: FOR," *British Medical Journal*, 321 (2000): 442–45, at 443–44.

9 H. Varmus and D. Satcher, "Ethical Complexities of Conducting Research in Developing Countries," *N. Engl. J. Med.*, 337 (1997): 1003–05, at 1004–05; Crouch and Arras, *supra* note 2, at 27.

10 Glantz et al., *supra* note 2, at 40–41.

11 This example is based loosely on the Kenya Program of the Indiana University School of Medicine.

12 Some people might reject the view that a U.S. citizen's obligations to citizens and residents of the U.S. are greater than the citizen's obligations to people living in other countries (who are not citizens of the U.S.). However, as a matter of ethics and law, national borders matter. When Congress funds health care coverage for older persons, it acts ethically when it limits coverage to people living in the United States. Similarly, the constitutional rights of U.S. citizens and residents under U.S. law exceed those of noncitizens living in other countries. This distinction between local residents and people living in other countries reflects a number of considerations. First, ties of kinship matter. We recognize the interest of people in devoting more of their money and time to family members over strangers, to local non-profit organizations over non-profits in other cities or states, to co-religionists over members of other religions, and to countrymen and women over

people who live elsewhere. In addition, when people share citizenship or residency with each other, they share in a collection of rights and responsibilities. Nonresidents do not bear most of the responsibilities of community membership, and they therefore do not bear most of the rights of community membership.

13 Some people might find the drug company example unethical on grounds of exploitation, but I do not think we can condemn businesses simply because they bring a resource from a less developed country to a more developed country. It is not in the interest of developing countries if they are prevented from sending their fruits and vegetables, oil and gas, or precious minerals to other countries in exchange for goods that cannot be produced in their countries. The exploitation concern is discussed at length later in this article.

14 B.R. Furrow et al., *Health Law*, 2d ed. (St. Paul: West Group, 2000): at 265.

15 M. Baum, "Declaration of Helsinki Should Be Strengthened: AGAINST," *British Medical Journal*, 321 (2000): 444–45, at 445.

16 M. Angell, "The Ethics of Clinical Research in the Third World," *N. Engl. J. Med.*, 337 (1997): 847–49, at 847.

17 Some critics have questioned whether in fact the women gave true informed consent, and other critics have observed that many of the women suffered stigmatization when their HIV status became known. These objections are important, but they would apply to any research done in the same communities, regardless of whether a placebo control was used or whether the experimental therapy studied was destined for use in developing or developed countries.

18 Lurie and Wolfe, *supra* note 1, at 855.

19 *Id.* at 855.

20 Medical malpractice standards take into account local resources, but a physician practicing in an inner city clinic could not provide a lower level of care on account of the clinic's resources. Rather, if the patient needed care that could not be adequately provided in the clinic, the physician would transfer the patient to an appropriate facility.

21 In fact, the less aggressive, less expensive course of AZT treatment substantially reduces the risk of HIV-transmission from mother to infant, though not as effectively as the more aggressive, more expensive course of AZT. C. Grady, "Science in the Service of Healing," *Hastings Center Report*, 28, no. 6 (1998): 34–38, at 35–36.

22 H.T. Shapiro and E.M. Meslin, "Ethical Issues in the Design and Conduct of Clinical Trials in Developing Countries," *N. Engl. J. Med.*, 345 (2001): 139–42, at 140. See also National Bioethics Advisory Commission, *Ethical and Policy Issues in International Research: Clinical Trials in Developing Countries*, ISBN 931022-13-5 (April 2001), available at www.georgetown.edu/research/nrcbl/nbac/pubs.html ("[r]esearchers and sponsors should design clinical trials that provide members of any control group with an established effective treatment, whether or not such treatment is available in the host country.").

23 World Medical Association, "Declaration of Helsinki: Ethical Principles for Medical Research Involving Human Subjects," *JAMA*, 284 (2000): 3043–45, ¶ 29, at 3045. In an October 2001 clarification, the World Medical Association indicated that placebo controls can be used even when proven therapy exists if (1) the placebo control is necessary to establish the safety or efficacy of an experimental therapy, or (2) the experimental therapy is designed for treatment of a "minor condition" and receiving the placebo will not increase the risk of "serious or irreversible harm." See World Medical Association, *Note of Clarification on Paragraph 29 of the WMA Declaration of Helsinki* (October 7, 2001), at www.wma.net/e/policy/17-c_e.html#clarification (last visited October 4, 2002).

24 R.J. Levine, "Some Recent Developments in the International Guidelines on the Ethics of Research Involving Human Subjects," *Annals of the New York Academy of Sciences*, 918 (2000): 170–78, at 174–76.

25 Varmus and Satcher, *supra* note 9, at 1004–05; Crouch and Arras, *supra* note 2, at 27.

26 Placebo controls – even when there is an established therapy – also have value for other kinds of studies. When testing an experimental therapy, one can determine whether it is effective with a smaller

number of subjects if the experimental therapy is compared to a placebo rather than to established therapy. E.J. Emanuel and F.G. Miller, "The Ethics of Placebo-Controlled Trials – A Middle Ground," *N. Engl. J. Med*, 345 (2001): 915–19, at 916. For additional discussion of the justifications of placebo-controlled trials, see D. Orentlicher, "Placebo-Controlled Trials of New Drugs: Ethical Considerations," *Diabetes Care*, 24 (2001): 771–72.

27 See Council for International Organizations of Medical Sciences, *International Ethical Guidelines for Biomedical Research Involving Human Subjects*, Guideline 10, *at* http://cioms.ch/frame_guidelines_sept_2002. htm (revised August 2002).

28 J. Feinberg, *The Moral Limits of the Criminal Law: Harmless Wrongdoing* (New York: Oxford University Press, 1988): 176–210.

29 Id. at 204–10.

30 Glantz et al., *supra* note 2, at 40–41.

31 Council for International Organizations of Medical Sciences, *supra* note 27.

32 Crouch and Arras, *supra* note 2, at 30–31.

Chapter 53

The participants in the 2001 conference on ethical aspects of research in developing countries

MORAL STANDARDS FOR RESEARCH IN DEVELOPING COUNTRIES: FROM "REASONABLE AVAILABILITY" TO "FAIR BENEFITS"

Commentators have argued that when research conducted in a developing country shows an intervention to be effective, the intervention must be made "reasonably available" to the host population after the trial. But this standard is sometimes too stringent, and sometimes too lenient. It offers a benefit, but not necessarily a fair benefit.

Over the last decade, clinical research conducted by sponsors and researchers from developed countries in developing countries has grown very controversial.[1] The perinatal HIV transmission studies that were sponsored by the National Institutes of Health and the Centers for Disease Control and conducted in Southeast Asia and Africa inflamed this controversy and focused it on the standard of care – that is, on whether treatments tested in developing countries should be compared to the treatments provided locally or to the best interventions available anywhere.[2] Since then, this debate has expanded to include concerns about informed consent.

A subject that has received less discussion but is potentially even more important is the requirement that any drugs proven effective in the trial be made available to the host population after the trial.[3] There seems to be general agreement that "reasonable availability" is necessary in order to ensure that the subject population is not exploited.

This consensus is mistaken, however. A "fair benefits" framework offers a more reliable and justifiable way to avoid exploitation. In this paper we develop the argument for the fair benefits framework in detail and compare the two approaches in a specific case – the trial of hepatitis A vaccine in Thailand.

Current views on the reasonable availability requirement

The idea of making interventions reasonably available was emphasized in the *International Ethical Guidelines* issued in 1993 by the Council for International Organizations of Medical Sciences (CIOMS), and it was reiterated in the 2002 revision in Guideline 10 and its commentary.

> As a general rule, the sponsoring agency should agree in advance of the research that any product developed through such research will be made reasonably available to the inhabitants of the host community or country at the completion of successful testing. Exceptions to this general requirement should be justified and agreed to by all concerned parties before the research begins.[4]

Four issues have generated disagreement. First, how strong or explicit should the commitment to provide the drug or vaccine be at the initiation of the research trial? CIOMS required an explicit, contract-like mechanism, agreed to before the trial, and it assigns this responsibility to the sponsors of research. The Declaration of Helsinki's 2000 revision endorses a less stringent guarantee that does not require availability of interventions to be "ensured" "in advance."[5] Several other ethical guidelines suggest "discussion in advance" but do not require formal, prior agreements.[6] Conversely, some commentators insist that the CIOMS guarantee is "not strong or specific enough."[7] For instance, the chair and executive director of the U.S. National Bioethics Advisory Commission (NBAC) contended:

> If the intervention being tested is not likely to be affordable in the host country or if the health care infrastructure cannot support its proper distribution and use, it is unethical to ask persons in that country to participate in the research, since they will not enjoy any of its potential benefits.[8]

To address these concerns, others advocate that research in developing countries ethically requires a formal and explicit prior agreement that "includes identified funding" and specifies improvements necessary in the "country's health care delivery capabilities."[9]

The second area of disagreement has concerned who is responsible for ensuring reasonable availability. Are sponsors responsible, as the original CIOMS guideline called for? Does responsibility rest with host country governments? Or international aid organizations? The third area of disagreement focuses on what it means for drugs to be made reasonably available. Does it require that the drug or vaccine be free, subsidized, or at market prices?

Finally, to whom should interventions be made reasonably available? Should they be restricted to participants in the research study? Should they include the village or tribe from which individual participants were enrolled? Or the whole country in which the research was conducted?

The justification of reasonable availability

Why is reasonable availability thought to be a requirement for ethical research in developing countries? Research uses participants to develop generalizable knowledge that can improve health and health care for others.[10] The potential for exploitation of individual participants enrolled in research as well as communities that support and bear the burdens of research is inherent in every research trial. Historically, favorable risk-benefit ratios, informed consent, and respect for enrolled participants have been the primary mechanisms for minimizing the potential exploitation of individual research participants.[11] In developed countries, exploitation of populations has been a less significant concern because there is a process, albeit an imperfect one, for ensuring that interventions proven effective through clinical research are introduced into the health care system and benefit the general population.[12] In contrast, the potential for exploitation is acute in research trials in developing countries. Target populations may lack access to regular health care, political power, and an understanding of research. Hence, they may be exposed to the risks of research with few tangible benefits. The benefits of research – access to new effective drugs and vaccines – may be predominantly for people in developed countries with profits to the pharmaceutical industry. Many consider this scenario the quintessential case of exploitation.[13]

Supporters deem that reasonable availability is necessary to prevent such exploitation of communities. As one group of commentators put it:

[I]n order for research to be ethically conducted [in a developing country] it must offer the potential of actual benefit to the inhabitants of that developing country. . . . [F]or underdeveloped communities to derive potential benefit from research, they must have access to the fruits of such research.[14] (emphasis added)

Or as the commentary to the 2002 CIOMS Guideline 10 put it:

[I]f the knowledge gained from the research in such a country [with limited resources] is used primarily for the benefit of populations that can afford the tested product, the research may rightly be characterized as exploitative and, therefore, unethical.[15]

What is exploitation?

Even though it seems initially plausible, there are a number of problems with making reasonable availability a necessary ethical requirement for multinational research in developing countries. The most important problem is that the reasonable availability requirement embodies a mistaken conception of exploitation and therefore offers wrong solution to the problem of exploitation.

There are numerous ways of harming other individuals, only one of which is exploitation. Oppression, coercion, assault, deception, betrayal, and discrimination are all distinct ways of harming people. They are frequently all conflated and confused with exploitation.[16] One reason for distinguishing these different wrongs is that they require very different remedies. Addressing coercion requires removing threats, and addressing deception requires full disclosure, yet removing threats and requiring full disclosure will not necessarily prevent exploitation.

What is exploitation? In the useful analysis developed by Alan Wertheimer, Party A exploits party B when B receives an unfair level of benefits as a result of B's interactions with A.[17] Whether B's benefits are fair depends upon the burdens that B bears as part of the interaction and the benefits that A and others receive as a result of B's participation in the interaction. If B runs his car into a snow bank and A offers to tow him out but only at the cost of $200 – when the normal and fair price for the tow is $75 – then A exploits B.

Wertheimer's conception of exploitation is distinct from the conventional idea that exploitation entails the "use" of someone else for one's own benefit. There are many problems with this familiar conception. Most importantly, if exploitation is made to depend only on instrumental use of another person, then almost all human interactions are exploitative. We constantly and necessarily use other people.[18] In the example above, not only does A exploit B, but B also exploits A, because B uses A to get his car out of the snow bank. Sometimes the word "exploit" refers to a *neutral* use – as when we say that a person exploited the minerals or his own strength. However, in discussions of research, especially but not exclusively when the research occurs in developing countries, exploitation is never neutral; it is always a moral wrong. Consequently, we do not need to mark out all cases of use. We need only to identify those that are morally problematic.[19]

The Wertheimerian conception of exploitation also departs from the commonly cited Kantian conception. As Allen Buchanan characterizes the Kantian conception, "To exploit a person involves the *harmful, merely instrumental utilization* of him or his capacities, for one's own advantage or for the sake of one's own ends."[20] The Kantian conception of exploitation seems to expand beyond use to include a separate harm. But in the case of exploitation, what is this "other harm"? For a Kantian, *to exploit* must mean to use in a way that the other person could not consent to, a way that undermines their autonomy.[21] However in many cases, people consent – with full knowledge and without threats – and yet we think they are exploited. People in developing countries could consent to being on a research study after full informed consent and still be exploited. Similarly, snow bank-bound B seems exploited even if he consents to being towed out for $200. Thus the Kantian conception seems mistaken in fusing exploitation with inadequate consent.

In any event, the reasonable availability requirement is not grounded in Kantian claims about use and violation of autonomy. Rather, it is aimed at ensuring that people have access to the interventions that they helped to demonstrate were effective. It is related to the benefits people receive from participating in a research study, not to their autonomy in consent. Consequently, whatever the merits of the Kantian conception of exploitation, it seems irrelevant to deciding whether making the trial intervention reasonably available can prevent exploitation. In contrast, the Wertheimerian view, which locates the core moral issue inherent in exploitation in the fair level of benefits each party of an interaction receives, captures the ethical concern underlying the reasonable availability requirement.

In determining whether exploitation has occurred in any case, the Wertheimerian conception gives us at least six important considerations to bear in mind. First, exploitation is a micro-level concern. Exploitation is about harms from discrete interactions, rather than about the larger social justice of the distribution of background rights and resources. Certainly macro-level distributions of resources can influence exploitation, but the actual exploitation is distinct. Furthermore, while past events may lead people to feel and claim that they have been exploited, whether exploitation occurred does not depend either on their feelings or on historical injustices. Exploitation is about the fairness of an individual exchange. Indeed, as we shall note below, exploitation can happen even in a just society, and it can fail to occur even when there is gross inequality between the parties. As Wertheimer argues:

> [W]hile the background conditions shape our existence, the primary experiences occur at the micro level. Exploitation matters to people. People who can accept an unjust set of aggregate resources with considerable equanimity will recoil when they feel exploited in an individual or local transaction. . . . Furthermore, micro-level exploitation is not as closely linked to macro-level injustice as might be thought. Even in a reasonably just society, people will find themselves in situations [that] will give rise to allegations of exploitation.[22]

The reasonable availability requirement recognizes the possibility of exploitation associated with a particular study, and it does not require ensuring the just distribution of all rights and resources or a just international social order. This is more than just a pragmatic point; it reflects the deep experience that exploitation is transactional.

Second, because exploitation is about interactions at a micro level, between researcher and community, it can occur only once an interaction is initiated. In this sense, the obligations to avoid exploitation are obligations that coexist with initiating an interaction.

Third, exploitation is about "how much," not "what," each party receives. The key issue is fairness in the level of benefits. Moreover, exploitation depends upon fairness, not "equalness." An unequal distribution of benefits may be fair if there are differences in the burdens and contributions of each party. Fairness in the distribution of benefits is common to both Wertheimer's theory

of exploitation and Rawls's theory of justice, but
the notion of fairness important for exploitation
is not Rawlsian. They differ in that Rawls addresses
macro- and Wertheimer micro-level distributions
of benefits. The Rawlsian conception of fairness
addresses the distribution of rights, liberties, and
resources for the basic structure of society within
which individual transactions occur.[23] In other
words, Rawlsian fairness is about constitutional
arrangements, taxes, and opportunities. Rawls's
conception has often but wrongly been applied to
micro-level decisions, where it usually issues in
implausible and indefensible recommendations.
Fairness in individual interactions, which is the
concern of exploitation, is based on ideal market
transactions.[24] Thus a fair distribution of benefits
at the micro-level is based on the level of benefits
that would occur in a market transaction devoid
of fraud, deception, or force, in which the parties
have full information. While this is always
idealized – in just the way that economic theory is
idealized – it is the powerful ideal informing the
notion of fairness of micro-level transactions. This
notion of fairness is also relative: just as fair price
in a market is based on comparability, so too is the
determination of fair benefits based on comparisons to the level of benefits received by other
parties interacting in similar circumstances.

Fourth, that one party is vulnerable may make
exploitation more likely, but does not inherently
entail exploitation. Since exploitation involves the
distribution of benefits and burdens, vulnerability
is neither necessary nor sufficient for its occurrence. The status of the parties is irrelevant in
determining whether exploitation has occurred.
If the exchange is fair to both parties, then no one
is exploited, regardless of whether one party is
poor, uneducated, or otherwise vulnerable and
disadvantaged. In the case of snow-bound B, if A
charges B $75 for towing the car out, then B is
not exploited even though B is vulnerable.

Fifth, since exploitation is about the fairness
of micro-level interactions, the key question is

the level of benefits provided to the parties who
interact. Determining whether exploitation has
occurred does not involve weighing the benefits
received by people who do not participate in the
interaction.

Finally, because fairness depends on idealized
market transactions, determining when exploitation occurs – when the level of benefits is
unfair – will require interpretation. As with the
application of legal principles and constitutional
provisions, the inevitability of interpretation
means that reasonable people can and will disagree. But such interpretation and controversy
does not invalidate either judicial or moral
judgments.

Problems with the reasonable availability requirement

The fundamental problem with the reasonable
availability standard is that it guarantees a benefit
– the proven intervention – but not a fair level of
benefits, and therefore it does not necessarily
prevent exploitation. Reasonable availability
focuses on what – the products of research – but
exploitation requires addressing how much – the
level of benefit. For some research in which
either the subjects would be exposed to great
risks or the sponsor stands to gain enormously,
reasonable availability might be inadequate and
unfair. Conversely, for very low- or no-risk
research in which the population would obtain
other benefits, or in which the benefits to the
sponsor are minimal, requiring the sponsor to
make a product reasonably available could be
excessive and unfair.

There are also other problems with the
reasonable availability standard. First, it
embodies a very narrow notion of benefits. It
suggests that only one type of benefit – a proven
intervention – can justify participation in clinical research. But a population in a developing
country could consider a diverse range of other

benefits from research, including the training of health care or research personnel, the construction of health care facilities and other physical infrastructure, and the provision of public health measures and health services beyond those required as part of the research trial. The reasonable availability standard ignores such benefits, and hence cannot reliably determine when exploitation has occurred.

Second, at least as originally formulated by CIOMS, the reasonable availability standard applies to only a narrow range of clinical research – successful Phase III testing of interventions.[25] It does not apply to Phase I and II drug and vaccine testing, or to genetic, epidemiology, and natural history research, which are all necessary and common types of research in developing countries but may be conducted years or decades before any intervention is proven safe and effective. Consequently, either the reasonable availability requirement suggests that Phase I and II studies cannot be ethically conducted in developing countries – a position articulated in the original CIOMS guidelines but widely repudiated – or there is no ethical requirement to provide benefits to the population when conducting such early phase research, or reasonable availability is not the only way to provide benefits from a clinical research study.

To address this gap, CIOMS altered the reasonable availability requirement in 2002:

> Before undertaking research in a population or community with limited resources, the sponsor and the investigator must make every effort to ensure that . . . any intervention or product developed, or *knowledge generated*, will be made reasonably available for the benefit of that population or community.[26] (emphasis added)

According to CIOMS, some knowledge alone may constitute a fair level of benefits for some

non-Phase III studies. But in many non-Phase III studies, it may not match either the risks to subjects or the benefits to others. Indeed, the requirement could permit pharmaceutically sponsored Phase I and II testing of drugs in developing countries while shifting Phase III testing and sales to developed countries as long as data from the early studies are provided to the developing countries. This modification to encompass non-Phase III studies might actually invite *more* exploitation of developing countries.

Third, even in Phase III studies, the reasonable availability requirement provides an *uncertain* benefit to the population, since it makes benefit depend on whether the trial is a "successful testing" of a new product. If there is true clinical equipoise at the beginning of Phase III trials conducted in developing countries, then the new intervention should be proven more effective in only about half of the trials.[27] Consequently, reliance on reasonable availability alone to provide benefits implies that the host country will receive sufficient benefits from half or fewer of all Phase III studies.

Fourth, assuring reasonable availability does not avert the potential for undue inducement of a deprived population. One worry about research in developing countries is that collateral benefits will be escalated to induce the population to enroll in excessively risky research. If the population lacks access to public health measures, routine vaccines, medications for common ailments, and even trained health care personnel, then providing these services as part of a research study might induce them to consent to the project despite its risks, and despite the fact that it disproportionately benefits people in developed countries.[28] Similarly, guaranteeing reasonable availability to a safe and effective drug or vaccine after a study could also function as an undue inducement if the population lacks basic health care.

Fifth, it is beyond the authority of researchers and even of many sponsors of research to guarantee reasonable availability. Clinical researchers and even some sponsors in developed countries, such as the NIH and Medical Research Council, do not control drug approval processes in their own countries, much less in other countries. Similarly they do not control budgets for health ministries or foreign aid to implement research results, and may be, by law, prevented from providing assistance with implementation of research results. At best, they can generate data to inform the deliberations of ministers of health, aid officials, international funding organizations, and relevant others, and then try to persuade those parties to implement effective interventions.

Further, because most Phase III trials take years to conduct, policymakers in developing countries and aid agencies may resist agreements to provide an intervention before they know how beneficial it is, the logistical requirements for implementing and distributing it, and how it compares to other potential interventions. Such cautiousness seems reasonable given the scarce resources available for health delivery.

Sixth, requiring reasonable availability tacitly suggests that the population cannot make its own, autonomous decisions about what benefits are worth the risks of a research trial. In many cases the resources expended on making a drug or vaccine available could be directed to other benefits instead, which the host community might actually prefer. Disregarding the community's view about what constitutes appropriate benefits for them – insisting that a population must benefit in a specific manner – implies a kind of paternalism.

Finally, requiring a prior agreement to supply a proven product at the end of a successful trial can become a "golden handcuff," constraining rather than benefiting the population. If there is a prior agreement to receive a specific drug or vaccine, rather than cash or some other transferable commodity, the prior agreement commits the population to using the specific intervention tested in the trial. (Pharmaceutical companies are likely to provide their own product directly and avoid agreements in which they are required to provide the product of a competitor.) Yet if other, more effective or desirable interventions are developed, the population is unlikely to have the resources to obtain those interventions. Hence prior agreements can actually limit access of the population to appropriate interventions.

Because of these difficulties, the reasonable availability requirement is recognized more in the breech than in its fulfillment; consequently much effort has been devoted to identifying and justifying exceptions.

The fair benefits framework

Certainly, targeted populations in developing countries ought to benefit when clinical research is performed in their communities. Making the results of the research available is one way to provide benefits to a population, but it is not the only way. Hence it is not a necessary condition for ethical research in developing countries, and it should not be imposed unless the developing countries have themselves affirmed it.

This was the consensus of the clinical researchers, bioethicists, and IRB chairs and members from eight African and three Western countries – Egypt, Ghana, Kenya, Malawi, Mali, Nigeria, Tanzania, Uganda, Norway, the United Kingdom, and the United States – who participated in the 2001 Conference on Ethical Aspects of Research in Developing Countries (EARD). As an alternative to reasonable availability, this group proposes the "fair benefits framework."[29]

The fair benefits framework supplements the usual conditions for the ethical conduct of research trials, such as independent review by an institutional review board or research ethics committee and individual informed consent.[30]

In particular, it relies on three background principles that are widely accepted as requirements for ethical research. First, the research should have social value: it should address a health problem of the developing country population. Second, the subjects should be selected fairly: the scientific objectives of the research itself, not poverty or vulnerability, must provide a strong justification for conducting the research in a specific population. The subjects might be selected, for example, because the population has a high incidence of the disease being studied or of the transmission rates of infection necessary to evaluate a vaccine. Third, the research must have a favorable risk-benefit ratio: benefits to participants must outweigh the risks, or the net risks must be acceptably low.

To these widely accepted principles, the fair benefits framework adds three further principles, which are specified by fourteen benchmarks (see Table 53.1):

Principle 1: Fair Benefits. There should be a comprehensive delineation of tangible benefits to the research participants and the population from both the conduct and results of the research. These benefits can be of three types: (1) benefits to research participants during the research; (2) benefits to the population during the research; or (3) benefits to the participants and population after completion of the research. It is not necessary to provide each of these types of benefits; the ethical imperative based on the conception of exploitation is only for a fair level of benefits. It would seem fair that as the burdens and risks of the research increase, the benefits should also increase. Similarly, as the benefits to the sponsors, researchers, and others outside the population increase, the benefits to the host population should also increase.

Because the aim of the fair benefits framework is to avoid exploitation, the population at risk for exploitation is the relevant group to receive benefits and determine their fairness.

Indeed, determination of whether the distribution of benefits is fair depends on the level of benefits received by those members of the community who actually participate in the research, for it is they who bear the burdens of the interaction. However, each benefit does not have to accrue solely to the research participants; a benefit could be directed instead to the entire community. For instance, capacity development or enhanced training in ethics review would be provided to the community, and then benefit the participants indirectly. The important question is how much the participants will benefit from these measures.

In addition, the community will likely bear some burdens and impositions of the research because its health care personnel are recruited to staff the research teams, and its physical facilities and social networks are utilized to conduct the study. Thus, to avoid exploitation, consideration of the benefits for the larger community may also be required. However, since exploitation is a characteristic of micro-level transactions, there is no justification for including everybody in an entire region or country in the distribution of benefits (nor in the decisionmaking that is required by the next principle) unless the whole region or country is involved in bearing the burdens of the research and at risk for exploitation.

Principle 2: Collaborative Partnership. The population being asked to enroll determines whether a particular array of benefits is sufficient and fair. Currently, there is no shared international standard of fairness; reasonable people disagree.[31] More importantly, only the host population can determine the value of the benefits for itself. Outsiders are likely to be poorly informed about the health, social, and economic context in which the research is being conducted, and they are unlikely to fully appreciate the importance of the proposed benefits to the population.

Table 53.1 The Fair Benefits Framework

Principles	Benchmarks for determining whether the principle is honored.
Fair benefits	• Benefits to participants during the research 1) **Health improvement:** Health services that are essential to the conduct of the research will improve the health of the participants. 2) **Collateral health services:** Health services beyond those essential to the conduct of the research are provided to the participants. • Benefits to participants and population during the research 3) **Collateral health services:** Additional health care services are provided to the population. 4) **Public health measures:** There are additional public health measures provided to the population. 5) **Employment and economic activity:** The research project provides jobs for the local population and stimulates the local economy. • Benefits to population after the research 6) **Availability of the intervention:** If proven effective, the intervention should be made available to the population. 7) **Capacity development:** There are improvements in health care physical infrastructure, training of health care and research personnel, or training of health personnel in research ethics. 8) **Public health measures:** Additional public health measures provided to the population will have a lasting benefit. 9) **Long-term collaboration:** The particular research trial is part of a long-term research collaboration with the population. 10) **Financial rewards:** There is a plan to share fairly with the population the financial rewards or intellectual property rights related to the intervention being evaluated.
Collaborative partnership	1) **Free, uncoerced decisionmaking:** The population is capable of making a free, uncoerced decision: it can refuse participation in the research. 2) **Population support:** When it has understood the nature of the research trial, the risks and benefits to individual subjects, and the benefits to the population, the population decides that it wants the research to proceed.
Transparency	1) **Central repository of benefits agreements:** An independent body creates a publicly accessible repository of all formal and informal benefits agreements. 2) **Community consultation:** Forums with populations that may be invited to participate in research, informing them about previous benefits agreements.

Furthermore, the population's choice to participate must be free and uncoerced; refusing to participate in the research study must be a realistic option. While there can be controversy about who speaks for the population being asked to enroll, this is a problem that is not unique to the fair benefits framework. Even – or especially – in democratic processes, unanimity of decisions cannot be the standard; disagreement is inherent. But how consensus is determined in the absence of an electoral process is a complex question in democratic theory beyond the scope of this article.

Principle 3: Transparency. Fairness is relative, since it is determined by comparisons with similar interactions. Therefore transparency – like the full information requirement for ideal market transactions – allows comparisons with similar transactions. A population in a developing country is likely to be at a distinct disadvantage relative to the sponsors from the developed country in determining whether a proposed level of benefits is fair. To address these concerns, a publicly accessible repository of all benefits agreements should be established and operated by an independent body, such as the World Health Organization. A central repository permits independent assessment of the fairness of benefits agreements by populations, researchers, governments, and others, such as nongovernmental organizations. There could also be a series of community consultations to make populations in developing countries aware of the terms of the agreements reached in other research projects. Such information will facilitate the development of "case law" standards of fairness that evolve out of a number of agreements.

Together with the three background conditions, these three new principles of the fair benefits framework ensure that: (1) the population has been selected for good scientific reasons, (2) the research poses few net risks to the research participants, (3) there are sufficient benefits to the participants and population, (4) the population is not subject to a coercive choice, (5) the population freely determines whether to participate and whether the level of benefits is fair given the risks of the research, and (6) there is an opportunity for comparative assessments of the fairness of the benefit agreements.

Application to the hepatitis A vaccine case

We can compare the reasonable availability requirement with the fair benefits framework in the case of Havrix, an inactivated hepatitis A vaccine that was tested in 1990 among school children from Kamphaeng Phet province in northern Thailand.[32] The study was a collaboration of the Walter Reed Army Institute of Research (in the United States), SmithKline Beecham Biologicals, and Thailand's Ministry of Public Health. Initially, there was a randomized,

double-blind Phase II study involving 300 children, primarily family members of physicians and nurses at the Kamphaeng Phet provincial hospital. After a demonstration of safety and of an antibody response that neutralizes hepatitis A, a randomized, double blind Phase III study with a hepatitis B vaccine control involving 40,000 children, one to sixteen years old, was initiated to assess protection against hepatitis A infection.

The study was conducted in Thailand for several reasons. First, there were increasingly common episodes of hepatitis A infection during adolescence and adulthood, including hepatitis A outbreaks, such as at the National Police Academy in 1988. Second, while hepatitis A transmission was focal, there was a sufficiently high transmission rate – 119 per 100,000 population – in rural areas to assess vaccine efficacy. Third, the area had been the site of a prior Japanese encephalitis vaccine study.[33] Ultimately, the Japanese encephalitis vaccine was registered in Thailand in 1988 and included in the Thai mandatory immunization policy in 1992.

Prior to the Phase III study, there was no formal agreement to make Havrix widely available in Thailand. Due to competing vaccination priorities (especially for implementation of hepatitis B vaccine), the cost of a newly developed hepatitis A vaccine, and the available health care budget in Thailand, it was unlikely that Havrix would be included in the foreseeable future in Thailand's national immunization program, in which vaccines are provided to the population at no cost. In addition, SmithKline Beecham Biologicals made no commitment to provide free Havrix to Thailand. However, the company did commit to provide the vaccine to all research participants and to pursue Havrix registration in Thailand, enabling the vaccine to be sold in the private market. While there was no promise about what the prices would be for the private market, SmithKline Beecham Biologicals had previously utilized tiered pricing on vaccines. Registration and distribution would enable the Ministry of Public Health to use Havrix to control hepatitis A outbreaks at schools and other institutions. Nevertheless, at the start of the trial, all collaborators recognized that the largest market for Havrix would be travelers from developed countries.

Was the Havrix study ethical? Although all the study participants ultimately received hepatitis A and B vaccines, the study did not fulfill the reasonable availability requirement. There was no prior agreement to provide the vaccine to everyone in Kamphaeng Phet province, and since most Thais would not be able to afford the vaccine, committing to registering and selling it on the private market does not seem to be "reasonably available." Thus, by this standard, the trial seems to be unethical.

The fair benefits framework, however, requires a more multifaceted assessment. First, the study seemed to fulfill the background requirements of social value, fair subject selection, and favorable risk-benefit ratio. Hepatitis A was a significant health problem in northern Thailand and recognized as such by the Thai Ministry of Public Health. Although the population in Kamphaeng Phet province was poor, the epidemiology of hepatitis A provided an independent scientific rationale for site selection. The preliminary data indicated that the candidate vaccine had an excellent safety profile and probable protective efficacy, suggesting a highly favorable risk-benefit ratio for participants.

The benefits of the Havrix trial were of several sorts. By design, all 40,000 children in the trial received both hepatitis A and B vaccines. In addition, regional medical services were augmented. The research team contracted with the community pubic health workers to examine all enrolled children absent from school at their homes, to provide necessary care, and, if appropriate, to arrange transfer to the district or provincial hospital.

There were also benefits for the provincial population. Public health stations throughout Kamphaeng Phet province that lacked adequate refrigeration to store vaccines, medicines, and blood specimens received new refrigerators. Similarly, rural health stations lacking reliable access to the existing FM wireless network link with the provincial hospital's consultants were joined to the network. In the six schools that had hepatitis A outbreaks during the study, the research team arranged for inspection of the schools and identification of deficiencies in toilet, hand-washing facilities, and water storage contributing to the outbreak. At each school, the researchers contracted and paid to have recommended improvements implemented. In addition, public health workers were provided with unlimited stocks of disposable syringes and needles, as well as training on measures to reduce the incidence of blood-borne diseases. Hepatitis B vaccinations were provided to all interested government personnel working on the trial, including approximately 2,500 teachers, public health workers, nurses, technicians, and physicians. Since deaths of enrolled research participants were tracked and investigated, the research team identified motor vehicle accidents, especially pedestrians struck by cars, as a major cause of mortality in the province and recommended corrective measures.[34] Finally, the training of Thai researchers and experience in conducting the Havrix trial may have facilitated subsequent research trials, including the current HIV vaccine trials in Thailand.

Regarding the principle of collaborative partnership, there were extensive consultations in Kamphaeng Phet province prior to initiating and conducting the trial. The provincial governor, medical officer, education secretary, and hospital director provided comments before granting their approval. In each of the 146 participating communities, researchers made public presentations about the study and held briefings for interested parents and teachers. Each school appointed a teacher to maintain a liaison with the research team. Parental and community support appeared to be related to the provision of hepatitis B vaccine to all participants, since hepatitis was seen as a major health problem and the children lacked access to the vaccine.

Furthermore, the protocol was reviewed by the Thai Ministry of Public Health's National Ethical Review Committee, as well as by two IRBs in the United States. The Ministry of Public Health appointed an independent committee composed of thirteen senior physicians and ministry officials to monitor the safety and efficacy of the trial. And rejecting the trial appeared to be a genuine option; certainly those Thai scientists who tried hard to prevent it, including by lobbying the National Ethics Review Committee, seemed to think so.

At the time of this trial, there was no central repository of benefits agreements to fulfill the transparency principle. However, the measures taken to benefit the population, including provision of the hepatitis A and B vaccines and registration of Havrix in Thailand, were discussed with the Ministry of Public Health and provincial officials and published.

Did the Havrix study provide fair benefits? Clearly some in Thailand thought not. They argued that the trial did not address a pressing health need in a manner appropriate to the country; instead, they held, it addressed a health interest of the U.S. army. Second, some have alleged there was insufficient technology transfer. In particular, no training was provided to Thai researchers to conduct testing for the antibody to hepatitis A or to develop other laboratory skills. Third, it was claimed that inadequate respect was accorded to the Thai researchers, as none were among the study's principal investigators and none were named in the original protocol (they were simply referred to as "Thai researchers"). Only after protests

were they individually identified. The American investigators claim vehemently that this charge is inaccurate. A prominent vaccine researcher summarized the sentiment against Thai participation:

> Journalists in the country have accused the government and medical community of a national betrayal in allowing Thai children to be exploited.... The role of Thailand in rounding up its children for immunization was hardly seen as a meaningful partnership in this research aim. In private, government ministers agreed with this, but the sway of international politics and money was too persuasive.[35]

Many others argued that the benefits to the population of Kamphaeng Phet province were sufficient, especially given the minimal risk of the study. Still others are uncertain. In their view, the level of benefits were not clearly inadequate, but more long-term benefits could have been provided to the community depending on the level of the sponsors' benefits – in this case, SmithKline Beecham's profits from vaccine sales. To address the uncertainty of how much a company might benefit from drug or vaccine sales, some propose profit-sharing agreements that provide benefits to the community related to the actual profits.

Universal agreement is a naïve and unrealistic goal. The goal is only a consensus in the population to be enrolled in the trial. Consensus on the appropriateness of a research study acknowledges that some disagreement is not only possible but likely, and even a sign of a healthy partnership.[36] In this trial, the national ministry, the provincial governmental and health officials, and the Kamphaeng Phet population seemed supportive.

Further, the dissent focused not on whether the vaccine would be made available to the population if it were proven effective, but on the level of a broad range of burdens and benefits, both to the community and to the sponsors. It is precisely this sort of broad, nuanced, and realistic assessment of the community's interests that is permitted and promoted by the fair benefits framework. Rather than making any one type of benefit into a moral litmus test, the fair benefits framework takes into account all of the various ways the community might benefit from the research.

Acknowledgements

We thank Dean Robin Broadhead of Malawi, and Dan Brock and Jack Killen for support and critical reviews of the manuscript. We thank Bruce Innis of SmithKline Beecham and Thai clinical researchers who prefer to remain anonymous for discussing the Havrix case with us and providing many substantive details.

Notes

1 M. Barry, "Ethical Considerations of Human Investigation in Developing Countries: the AIDS Dilemma," NEJM 319 (1988) 1083–86; M. Angell, "Ethical Imperialism? Ethics in International Collaborative Clinical Research," NEJM 319 (1988): 1081–83; N.A. Christakas, "The Ethical Design of an AIDS Vaccine Trial in Africa," Hastings Center Report 18, no. 3 (1988): 31–37.

2 P. Lurie and S.M. Wolfe, "Unethical Trials of Interventions to Reduce Perinatal Transmission of the Human Immunodeficiency Virus in Developing Countries," NEJM 337 (1997): 853–56; M. Angell, "The Ethics of Clinical Research in the Third World," NEJM 337 (1997): 847–49; H. Varmus and D. Satcher, "Ethical Complexities of Conducting Research in Developing Countries," NEJM 337 (1997): 1003–1005; R. Crouch and J. Arras, "AZT Trials and Tribulations," Hastings Center Report 28, no. 6 (1998): 26–34; C. Grady, "Science in the Service of Healing," Hastings Center Report 28, no. 6 (1998):

34–38; R.J. Levine, "The 'Best Proven Therapeutic Method' Standard in Clinical Trials in Technologically Developing Countries," IRB 20 (1998): 5–9; B.R. Bloom, "The Highest Attainable Standard: Ethical Issues in AIDS Vaccines," Science 279 (1998): 186–88.

3 World Medical Association. Declaration of Helsinki, 2000 at www.wma.net/e/policy12-c_e.html; Council for International Organizations of Medical Science, International Ethical Guidelines for Biomedical Research Involving Human Subjects (Geneva: CIOMS, 1993); P. Wilmshurst, "Scientific Imperialism: If They Won't Benefit from the Findings, Poor People in the Developing World Shouldn't be Used in Research," BMJ 314 (1997): 840–41; P.E. Cleaton-Jones, "An Ethical Dilemma: Availability of Anti-retroviral Therapy after Clinical Trials with HIV Infected Patients are Ended," BMJ 314 (1997): 887–88.

4 CIOMS, International Ethical Guidelines, 1993.

5 World Medical Association Declaration of Helsinki.

6 Medical Research Council of the United Kingdom, Interim Guidelines – Research Involving Human Participants in Developing Societies: Ethical Guidelines for MRC-sponsored Studies (London: MRC, 1999); Joint United National Programme on HIV/AIDS (UNAIDS), Ethical Considerations in HIV Preventive Vaccine Research (Geneva: UN-AIDS, 2000); National Consensus Conference, Guidelines for the Conduct of Health Research Involving Human Subjects in Uganda (Kampala, Uganda: National Consensus Conference, 1997); Medical Research Council of South Africa, Guidelines on Ethics for Medical Research (South Africa: Medical Research Council, 1993).

7 L.H. Glantz et al., "Research in Developing Countries: Taking 'Benefit' Seriously," Hastings Center Report 28, no. 6 (1998): 38–42; G.J. Annas and M.A. Grodin, "Human Rights and Maternal-Fetal HIV Transmission Prevention Trials in Africa," American Journal of Public Health 88 (1998): 560–63.

8 H.T. Shapiro and E.M. Meslin, "Ethical Issues in the Design and Conduct of Clinical Trials in Developing Countries," NEJM 345 (2001): 139–42. See also National Bioethics Advisory Commission, Ethical and Policy Issues in International Research: Clinical Trials in Developing Countries (Washington D.C.: U.S. Government Printing Office, 2001).

9 Annas and Grodin, "Human Rights and Maternal-Fetal HIV Transmission Prevention Trials in Africa."

10 E.J. Emanuel, D. Wendler, and C. Grady, "What Makes Clinical Research Ethical?" *JAMA* 283 (2000): 2701–711; E.J. Emanuel et al., "What Makes Clinical Research in Developing Countries Ethical? The Benchmarks of Ethical Research," *Journal of Infectious Diseases* 189 (2004): 930–37.

11 Ibid.; R.J. Levine, *Ethical and Regulatory Aspects of Clinical Research*, 2nd Edition (New Haven, Conn.: Yale University Press, 1988).

12 N. Black, "Evidence Based Policy: proceed with care," *BMJ* 323 (2001): 275–79.

13 Wilmshurst, "Scientific Imperialism," and National Consensus Conference, *Guidelines for the Conduct of Health Research Involving Human Subjects in Uganda*; L.H. Glantz et al., "Research in Developing Countries: Taking 'Benefit' Seriously"; G.J. Annas and M.A. Grodin, "Human Rights and Maternal-Fetal HIV Transmission Prevention Trials in Africa."

14 Glantz et al., "Research in Developing Countries."

15 CIOMS, *International Ethical Guidelines for Biomedical Research Involving Human Subjects*, 2nd edition (Geneva: CIOMS, 2002).

16 A. Wertheimer, *Exploitation* (Princeton, N.J.: Princeton University Press, 1999), chapter 1; N.A. Christakis, "The Ethical Design of an AIDS Vaccine Trial in Africa," *Hastings Center Report* 28 (1998): 31–37.

17 Wertheimer, *Exploitation*.

18 A.W. Wood, "Exploitation," *Social Philosophy and Policy* 12 (1995): 135–58.

19 Wertheimer, *Exploitation*.

20 Buchanan, *Ethics, Efficiency and the Market*, p.87.

21 C. Korsgaard, "The Reasons We Can Share: An Attack on the Distinction Between Agent-relative and Agent-neutral Values," in *Creating the Kingdom of Ends* (New York: Cambridge University Press, 1996).

22 Wertheimer, *Exploitation*.

23 J. Rawls, *A Theory of Justice*, 2nd edition (Cambridge, Mass.: Harvard University Press).

24 Wertheimer, *Exploitation*.

25 CIOMS, *International Ethical Guidelines for Biomedical Research Involving Human Subjects*.

26 CIOMS, *International Ethical Guidelines for Biomedical Research Involving Human Subjects*, 2nd edition.

27 I. Chalmers, "What is the Prior Probability of a Proposed New Treatment being Superior to Established Treatments?" *BMJ* 314 (1997): 74–75; B. Djulbegovic et al., "The Uncertainty Principle and Industry-Sponsored Research," *Lancet* 356 (2000): 635–38.

28 NBAC, *Ethical and Policy Issues in International Research*.

29 Participants in the 2001 Conference on Ethical Aspects of Research in Developing Countries, "Fair Benefits from Research in Developing Countries," *Science* 298 (2002): 2133–34.

30 Emanuel, Wendler, and Grady, "What Makes Clinical Research Ethical?" and Levine, *Ethical and Regulatory Aspects of Clinical Research*; J. Rawls, *The Law of Peoples* (Cambridge, Mass.: Harvard University Press, 1999).

31 T. Pogge, *World Poverty and Human Rights* (Cambridge, U.K.: Polity Press, 2002), chapters 1 and 4.

32 B.I. Innis et al., "Protection against Hepatitis A by an Inactivated Vaccine," *JAMA* 271 (1994): 1328–34.

33 C. Hoke et al., "Protection against Japanese Encephalitis by Inactivated Vaccines," *NEJM* 319 (1988): 608–614.

34 C.A. Kozik et al., "Causes of Death and Unintentional Injury among School Children in Thailand," *Southeast Asian Journal of Tropical Medicine and Public Health* 30 (1999): 129–35.

35 "Interview with Prof Natth," *Good Clinical Practice Journal* 6, no. 6 (1999): 11.

36 A. Gutmann and D. Thompson, *Democracy and Disagreement* (Cambridge, Mass.: Harvard University Press, 1996).

Chapter 54

John D. Arras

ANOTHER VOICE: FAIR BENEFITS IN INTERNATIONAL MEDICAL RESEARCH

When researchers from wealthy countries study health and disease in poverty-stricken parts of the world, what do they owe, if anything, to the people who participate in their trials?[1] This question generated a vigorous debate in the medical and bioethical literature a few years ago, and a consensus rapidly emerged that researchers and sponsors ought to assure the "reasonable availability" of any drugs proven effective in the trials. In this issue, Emanuel and colleagues challenge this regnant view. They agree that we are right to be concerned about the exploitation of host country populations, but they insist (a) that the avoidance of exploitation requires only a fair exchange of benefits, not any particular kind of benefit; and (b) that exploitation is a "micro level" feature of particular agreements that can be considered apart from background conditions of global justice.

Emanuel et al. are clearly and decisively right that the avoidance of exploitation demands "fair benefits" rather than (only) "reasonable access" to successfully developed drugs. This switch expands the menu of potential post-trial benefits to include measures to enhance the public health infrastructure and research capabilities of host countries. Even more importantly, it suggests a much more appropriate relationship between biomedical research and priority setting in both public health policy and medical services delivery. By insisting that successfully proven drugs be delivered to research subjects and local populations, the reasonable availability model subordinates local democratic decisionmaking on health policy to uncoordinated decisions bearing on the research agenda. Instead of rightly figuring as one subordinate element within a broader democratic strategy for human development, research becomes the driving engine of health policy and medical services delivery. You accept the research, you get the drug – even if some other drug or a cheaper, low-tech public health measure would have addressed the problem better. But while the fair benefits approach avoids the narrowness of the reasonable availability model, whether it too might inappropriately drive social policy decisions, albeit with a broader menu from which to choose, remains an open question.

The article's second major claim – that exploitation is a "micro level" issue separable from larger concerns about global inequity – is more debatable. While the authors are correct to note that agreements struck between rich and powerful researchers and poor and vulnerable subjects in developing countries need not be exploitative, whether larger questions of global justice can be successfully bracketed in the research context is unclear. As the authors suggest, informed consent does not solve the problem of exploitation. Democratically elected local officials representing desperately poor people can and do agree to hideously exploitative terms that are nevertheless deemed to be in their rational self interest. Anticipating this problem, the authors propose the interesting solution of a world-wide data bank that would at least let third world deliberators discover what agreements have been reached by their counterparts in other countries. But if researchers and their sponsors have structural motivations to give away as little as possible to their developing country hosts, and if local politicians and health planners in the developing world remain desperate for any kind of assistance, the proposed data bank could easily devolve into a public record of global exploitation. Accepting the horribly unjust status quo as a morally acceptable baseline for market style bargaining between rich and poor nations does not sound like a recipe for fairness.

There is, finally, the nagging question of why researchers and their sponsors should be deemed primarily responsible for development efforts in the third world. In a fully just international order that respected everyone's basic rights to security and subsistence, *researchers would do research* (hopefully on the world's most burdensome diseases), while other, more appropriate and effective bodies (governments, NGOs, the World Bank, and the like) worried about the most effective strategies for human development. But because the world is unjust, because the developed nations shamefully refuse to contribute more than a pittance toward the alleviation of world poverty, and because researchers have the knowledge and power to help alleviate human suffering and vulnerability, researchers in the world as we know it are stuck with this problem of providing fair benefits. So long as the dark cloud of global injustice looms in the backyard, however, no solution to the problem of post-trial benefits is likely to be either fully just or ultimately coherent.

Note

1 As always, my thoughts on these matters have been sharpened in conversation with Alex London.

Chapter 55

Ruth Macklin

THE DECLARATION OF HELSINKI: ANOTHER REVISION

Only eight years after a major revision of the Declaration of Helsinki (DoH), this highly regarded document providing ethical guidance for research involving human beings[1,2] has now undergone another revision.[3] This prompts the question why the World Medical Association (WMA), which issues the Declaration, decided to make changes again so soon. One can only speculate whether powerful forces exerted pressure on the WMA to change some key provisions that were viewed as unfriendly to industry and other major sponsors of multinational research. The 2008 revision strengthens the previous version in some respects and weakens it in others. The most salutary improvement is in the paragraph that stipulates when it is ethically acceptable to use placebos in a control arm of a randomised, controlled clinical trial.

A first, somewhat skeletal version of the Declaration was issued in 1964, followed by expansion and a series of amendments beginning in 1975. The amendments from 1975 through 1996 were relatively minor. A notable change occurred in 1996, when a paragraph that was to become highly controversial was amended from the previous 1989 version. The latter version contained the following clause relating to the design of a research proposal: "In any medical study, every patient − including those of a control group, if any − should be assured of the best proven diagnostic and therapeutic method." In an amendment to the Declaration in 1996, the following statement was added to that paragraph: "This does not exclude the use of placebo, or no treatment, in studies where no proven . . . method exists." At the time, this paragraph did not attract attention nor did it give rise to any debates. It was only after controversy erupted in 1997 over placebo-controlled, mother-to-child HIV transmission trials carried out in developing countries that world-wide attention focused on the Declaration of Helsinki's prohibition on the use of placebos when a proven method exists somewhere in the world. When the HIV trials were publicly criticised[4], but also strongly defended,[5] calls emerged seeking a revision of the Declaration of Helsinki.[6] In a process that took place over several years involving some inner turmoil, yet at the same time a degree of transparency, the WMA issued a significantly revised version of the Declaration in 2000.

Along with other paragraphs that gave rise to some controversy, the one that attracted most attention was the new Paragraph 29, which basically reiterated the statement in the 1996 version: "The benefits, risks, burdens, and effectiveness of a new method should be tested against those of the best current prophylactic, diagnostic, and therapeutic methods. This does not exclude the use of placebo, or no treatment, in studies where no proven prophylactic, diagnostic or therapeutic method exists." A major opponent was the US Food and Drug Administration (FDA), the regulatory agency that strongly favours the use of placebo controls in studies submitted to the agency for drug approvals. Most recently (2008) the FDA issued new regulations that abandoned its previous rule that foreign studies must comply with the provisions of the Declaration of Helsinki.[2] The FDA's new regulation replaced the DoH with the International Conference on Harmonization (ICH) Good Clinical Practice Guidance (GCP),[7] which has a much weaker provision regarding the use of placebo controls in clinical studies. In November 2001 the World Medical Association published what it called a "clarification."

Note of clarification on Paragraph 29 of The WMA Declaration of Helsinki

The WMA is concerned that paragraph 29 of the revised Declaration of Helsinki (October 2000) has led to diverse interpretations and possible confusion. It hereby reaffirms its position that extreme care must be taken in making use of a placebo-controlled trial and that in general this methodology should only be used in the absence of existing proven therapy. However, a placebo-controlled trial may be ethically acceptable, even if proven therapy is available, under the following circumstances:

- Where for compelling and scientifically sound methodological reasons its use is necessary to determine the efficacy or safety of a prophylactic, diagnostic or therapeutic method; or
- Where a prophylactic, diagnostic or therapeutic method is being investigated for a minor condition and the patients who receive placebo will not

be subject to any additional risk of serious or irreversible harm.

All other provisions of the Declaration of Helsinki must be adhered to, especially the need for appropriate ethical and scientific review.

Not only did this addition fail to "clarify" the paragraph: the first of the two conditions simply reopened the door to the very controversy that led to the revision of the Declaration in the first place. The second of the two circumstances that can permit departure from the placebo rule is relatively uncontroversial, and very few people who have objected to placebo-controlled studies are likely to reject this condition. The so-called clarification was flawed on three separate counts: first, it provided no criteria for the "compelling reasons" that could justify departure from the principle; second, the requirement for scientifically sound methodology is redundant, as it is required in every study and stated elsewhere in the Declaration of Helsinki; and third, it allowed participants in research to be subject to predictable serious or irreversible harm. In the following year, the clarification was elevated to the status of an amendment to the Declaration, where it remained until the recent revision in 2008.

With the change of one small word – replacing an "or" with an "and" – the new Paragraph 32 eliminates the worst feature of the previous Paragraph 29. The new Paragraph 32 now says:

Where for compelling and scientifically sound methodological reasons the use of placebo is necessary to determine the efficacy or safety of an intervention *and* the patients who receive placebo or no treatment will not be subject to any risk or serious or irreversible harm (italics added).

Although the condition that refers to "compelling and scientifically sound methodological reasons" remains, without further elucidation, the replacement of "or" by "and" eliminates the possibility that the design of a clinical trial with a placebo control would be ethically acceptable even if it were to subject participants to serious or irreversible harm. This change in one small word is the most salutary feature of the newly revised DoH.

Another change for the better is the addition, for the first time, of a requirement that sponsors register clinical trials: "Every clinical trial must be registered in a publicly accessible database before recruitment of the first subject" (Paragraph 19). The virtue of this requirement is that it makes transparent just which clinical trials fail to reach a successful conclusion, either because of demonstrated lack of safety or absence of efficacy. According to one commentator, "The new proposal calling for registration of clinical trials is not likely to be followed by industry."[9] The Biotechnology Industry Organization in Washington, DC, expressed the concern that "registration of all trials might jeopardize intellectual property rights and frustrate R&D efforts while providing little guidance to prescribers and patients."[9] But it is not at all clear why registration could jeopardise intellectual property rights. Proprietary information would not be included in the registration process, and that is the only element that could, realistically, jeopardise intellectual property rights. However, if companies had to disclose a phase I trial at the outset, it might affect decisions by competitors contemplating similar trials of their own products.

Another US industry trade group, The Pharmaceutical Research and Manufacturers of America (PhRMA), also objected to the registration provision in the new DoH. PhRMA said the requirement could cause delay in mounting clinical trials, and also noted that it would impose a major burden on sponsors. What is the burden that the pharmaceutical industry is worried about? If registration would, in fact, cause such a delay in initiating the trial, the company would experience a delay in realising profits from any product that proves to be successful following phase III trials. One defender of the requirement said that a benefit would be to prevent research participants from having to go through repeated testing of the same intervention[9]. It appears, then, that this pits the need to protect human participants of research against the interest of industry seeking to realise financial profits as quickly as possible.

Other changes in the DoH appear to be slight changes in wording, but they have ethical implications. One example is that of providing post-trial benefits to participants. Compare the previous version to the new one. The 2000 version says, in Paragraph 30: "At the conclusion of the study, every patient entered into the study should be assured of access to the best proven prophylactic, diagnostic and therapeutic methods identified by the study." The new version says: "At the conclusion of the study, patients entered into the study are entitled to be informed about the outcome of the study and to share any benefits that result from it, for example, access to interventions identified as beneficial in the study or to other appropriate care or benefits" (paragraph 33). Are the ethical implications of the new version stronger or weaker than those of the earlier version? It's hard to tell. The 2000 version does not mention informing participants about the outcome of the study, so that is a plus for the new version. In addition, the earlier version says that participants should be "assured of access" to methods identified by the study. But "assuring" people that they will have access is not at all the same as "ensuring access" to such methods. Nevertheless, it is possible to read the earlier version as promising more to participants than the 2008 version. The latter cites access to interventions only as "an example" of benefits resulting from the study. There may be "other appropriate care or benefits," according to the new DoH. But what might they be? And how to determine what are appropriate care or benefits? Interpretation of these paragraphs of the DoH is left to whatever bodies, be they research ethics committees or sponsors, seek to implement the provisions of the Declaration.

Another change in the revised DoH is clearly more restrictive than the earlier version. Paragraph 19 in the 2000 version said: "Medical research is only justified if there is a reasonable likelihood that the populations in which the research is carried out stand to benefit from the results of the research." In the 2008 version, Paragraph 17 says: "Medical research *involving a disadvantaged or vulnerable population or community* is only justified if the research is responsive to the health needs and priorities of this population or community and if there is a reasonable likelihood that this population or community stands to benefit from the results of the research" (italics added). Why does this justification not apply to research for all populations, but only to disadvantaged or vulnerable populations?

Other changes in the revised DoH are mostly minor modifications of the wording in the earlier version. Overall, however, the document is stylistically inelegant. It contains redundancies, as in Paragraph 11, "It is the duty of physicians who participate in medical research to protect the life, health, dignity, integrity, right to self-determination, privacy, and confidentiality of personal information of research subjects"; and Paragraph 23, "Every precaution must be taken to protect the privacy of research subjects and the confidentiality of their personal information and to minimize the impact of the study on their physical, mental and social integrity." Why is Paragraph 23 necessary? Another redundancy occurs in Paragraph 33 (discussed above) and Paragraph 14, part of which says: "The protocol should describe arrangements for post-study access by study subjects to interventions identified as beneficial in the study or access to other appropriate care or benefits." It is true that Paragraph 14 states what should be in the research protocol and Paragraph 33 addresses what the research participants are entitled to receive. Nevertheless, a better crafted document could have eliminated such redundancies and presented the items in a more logical and coherent manner.

Only time will tell whether the WMA will decide to undertake a subsequent revision of the Declaration of Helsinki, and what might prompt such a decision. For now, the Association has appointed a new working group to continue to study the matter of placebo-controlled trials.

Notes

1 Human H, Fluss SS. *The World Medical Association's Declaration of Helsinki: Historical and Contemporary Perspectives*, 2001.
2 Kimmelman J, Weijer C, Meslin EM. "The Helsinki Discords: FDA, Ethics, and International Drug Trials," *The Lancet*, in press.
3 World Medical Association Declaration of Helsinki Ethical Principles for Medical Research Involving Human Subjects. Last revised Oct 2008.
4 Lurie P, Wolfe, S. "Unethical Trials of Interventions to Reduce Perinatal Transmission of the Human Immunodeficiency Virus in Developing Countries," *N Eng J Med.* 1997 (337): Sep 18; 337: 853–6.
5 Varmus H, Satcher D. "Ethical Complexities of Conducting Research in Developing Countries,"*N Eng J Med* 1997 Oct 2; 337 (14): 1003–5.
6 Levine RJ. "The Need to Revise the Declaration of Helsinki," *N Eng J Med* Aug 1999; 341 (7): 531–4.
7 Federal Register / Vol. 73, No. 82 / Monday, April 28, 2008 / Rules and Regulations, 22800–16.
8 International Conference on Harmonization. *Guideline for Good Clinical Practice (E6).* 1996.
9 Normile D. "Clinical Trial Guidelines at Odds with US Policy," *Science* Oct 24 2008; (322): 516.

Should doctors be allowed to help patients to kill themselves?

INTRODUCTION TO PART TEN

S OME PEOPLE THINK that a patient's life should always be preserved. Two prominent reasons for thinking so are that life is sacred and that doctors have the duty to preserve their patients' lives. But many people think that this position is unrealistic because death would be a benefit to some terminally ill patients. A further issue concerns which ways of bringing about a patient's death are morally permissible. There are some important distinctions here. First, euthanasia is voluntary when a patient requests their death, non-voluntary when no relevant request has been made, and involuntary when death is against the patient's wishes. Second, active euthanasia occurs when something is done to kill a patient (for example, a lethal injection is administered), whilst passive euthanasia refers to allowing a patient to die by natural causes, for example, by withholding a life saving intervention, when their life could have been preserved. This chapter focuses on a different way of hastening a patient's death, namely physician assisted suicide (PAS; also referred to as 'assisted dying'), i.e. a patient whose prognosis is hopeless and condition intolerable asks a doctor to provide them with the medication necessary to end their own life.

The 'open letter' that starts the chapter presents a conservative response to attempts to legalise PAS in the UK. An interesting feature of the letter is that it lists pragmatic considerations as opposed to being based on religious principles, such as the doctrine of the sanctity of life, despite the fact that its authors are faith leaders. By contrast, the argument in the next selection is grounded in political philosophy. The philosophers' brief responds to two cases before the US Supreme Court involving terminally ill plaintiffs appealing for the right to choose to die rather than living lives they found intolerable. The liberal tenets of American society – principally, that each person should be allowed to devise and pursue their own conception of the good free from undue state interference – ground the right to make for oneself momentous personal decisions, such as whether to have children. When and how to end one's life is just such a decision, so it would be unconstitutional to restrict the range of end-of-life options by criminalising PAS.

Safranek queries such autonomy based arguments for PAS. According to Safranek's analysis, the basic problem is how to get from autonomy to normativity – i.e., from whether an act is autonomous to whether it ought to be permitted or performed – given that evil acts can be autonomously undertaken. A standard way of bridging the autonomy-normativity gap is the harm principle: autonomous acts should be permitted unless they harm others. But the harm principle requires us to say what constitutes harms to others, and this will vary depending on an underlying theory of the good. For example, both opponents and proponents of pornography endorse the harm principle but disagree with one another because they have 'disparate views of the human good'. Of course, one side in such a dispute could simply impose its theory of the good on the other; for example, proponents of pornography could insist that what really matters is 'the goodness of free expression [not] of women viewed as multidimensional beings'. But then the appeal to autonomy is self-refuting in that the proponent's autonomy based defence of pornography has subverted their opponent's autonomy. This analysis applies to autonomy based defences of PAS: since the dispute is based not on autonomy but on competing theories of the good, the only way an autonomy based argument could be effective is if its proponents foist their theory of the good on others, which would be self-refuting.

Dieterle suggests a structure for the PAS debate by distinguishing consequentialist considerations from deontological arguments. Dieterle's case is that, since there is no good argument of either sort, there is no reason not to legalise PAS. This could be more strongly stated. For example, he reports that the evidence, principally from Oregon and the Netherlands, mostly does not support consequentialist objections. The exception is the prediction that 'patients might be pressured by family members to seek PAS'. Dieterle's response to this exception focuses on the Cheney case. But the philosophers' brief provides two stronger responses: 'people who are dying have a right to hear and, if they wish, act on what others might wish to tell or suggest or even hint to them'; and it is ironic that the worry about being pressured by one's family (to end one's life) is used to support legislation that amounts to being pressured by the state (to continue to live).

The philosophers' brief claims that the appeal by objectors to PAS to the distinction between killing and letting die 'is based on a misunderstanding of the pertinent moral principles' and that the crucial thing is that 'the doctor acts with the same intention: to help the patient die'. In considering deontological arguments against PAS, Dieterle states: 'To say that the physicians who practice PAS intend to kill their patients is to misunderstand their intent'; he also seems to accept the premise that the doctor 'indirectly kills the patient'. These themes re-emerge in the two moral arguments against PAS which Thomson considers in detail, based on the killing/letting die distinction, and the principle of double effect, respectively. Regarding the first, Thomson argues that the killing/letting die distinction is unclear when applied to cases of withdrawing treatment; more importantly, it is irrelevant to the pertinent case of PAS because 'the doctor who supplies her patient with a lethal drug does not herself kill her patient'. Regarding the second, Thomson suggests that the principle of double effect is 'a muddle' that has been successfully rebutted (the last section of her article suggests two reasons why it nonetheless persists). She advocates separating the permissibility of an action from the intention of the agent, arguing that the latter is irrelevant to the former; given this, the argument against PAS from the principle of double effect simply falls away.

Selections

An open letter to all Members of Parliament and of the House of Lords, from leaders of British faith communities of Buddhists, Christians, Hindus, Jews, Muslims and Sikhs, expressing grave concerns at continuing and renewed efforts to legalise euthanasia.

Dworkin, R., Nagel, T., Nozick, R., Rawls, J., Scanlon, T. and Thomson, J.J. 1997. 'Assisted Suicide: The Philosophers' Brief', *New York Review of Books*, 44 (5), March 27th, 41–5.

Safranek, J.P. 1998. 'Autonomy and Assisted Suicide: the execution of freedom', *The Hastings Center Report*, 28 (4), 32–6.

Dieterle, J.M. 2007. 'Physician-assisted Suicide: A New Look at the Arguments', *Bioethics*, 21 (3), 127–39.

Thomson, J.J. 1999. 'Physician-assisted Suicide: Two Moral Arguments', *Ethics*, 109 (3), 497–518.

Chapter 56

AN OPEN LETTER TO ALL MEMBERS OF PARLIAMENT AND OF THE HOUSE OF LORDS, FROM LEADERS OF BRITISH FAITH COMMUNITIES OF BUDDHISTS, CHRISTIANS, HINDUS, JEWS, MUSLIMS AND SIKHS, EXPRESSING GRAVE CONCERN AT CONTINUING AND RENEWED EFFORTS TO LEGALISE EUTHANASIA

We, the undersigned, hold all human life to be sacred and worthy of the utmost respect and note with concern that repeated attempts are being made to persuade Parliament to change the law on intentional killing so as to allow assisted suicide and voluntary euthanasia for those who are terminally ill. As it appears likely that yet another Bill will be brought before Parliament in the near future, we consider it our duty to bring the following to the attention of Members of both Houses:

1 **Palliative care is advancing very rapidly** both in relieving the spectrum of suffering experienced by those with a terminal illness, and in supporting their families. However, such state of the art care is very unevenly distributed around the country. Providing good care does not require any change in the law but a reprioritisation of NHS resources in order to ensure that adequate training is given to doctors and nurses and that centres of specialist palliative care exist where they can be accessed by those who need them.[1] The argument that assisted suicide or euthanasia is necessary to deal with the suffering of terminal illness is false.

2 **Countries which have legalised assisted suicide or euthanasia are experiencing serious problems.** In Holland 1 in every 32 deaths arises from legal or illegal euthanasia:[2] a similar law here could lead to some 13,000 deaths a year[3] and Dutch pro-euthanasia groups are now, moreover, campaigning for further relaxations of the law – for example, to encompass people with dementia.[4] In Oregon the reluctance of many doctors to participate in legalised suicide is leading to 'doctor-shopping' with the result that many patients who receive lethal drugs, including some with psychiatric disorders, are not known to the doctors who supply them. There is also no monitoring of lethal drugs released in this way into the community.[5]

3 **The majority of doctors remain opposed to assisted dying and medical opposition has actually intensified in recent years.** The largest most recent surveys show only 22–38 per cent of doctors in favour of a change in the law.[6,7,8] This was

made very clear to the recent House of Lords Select Committee examining Lord Joffe's Assisted Dying for the Terminally Ill Bill.[9] A recent and much-publicised vote at the BMA annual conference to adopt a position of neutrality towards any future bill was unrepresentative of the Association's 134,000 members. It was carried by a very narrow majority (93 votes to 82) at a barely quorate meeting on the last day of the conference when over half of the delegates had either left or were otherwise engaged. In the debate on the matter two days before, the majority of speakers had opposed any change in the BMA's opposition to euthanasia. The Royal College of Nursing (RCN) and Royal College of General Practitioners (RCGP) are both opposed to a change in the law.

4   **Opinion polls purporting to show that a large majority of people would favour a change in the law are misleading.** They are based on answers to Yes/No or Either/ Or questions without any explanatory context and without other options – e.g. good quality palliative care – being offered. Most people have little understanding of the complexities and dangers in changing the law in this way and opinion research consists therefore to a large extent of knee-jerk answers to emotive – and often leading – questions.[10]

5   **Assisted suicide and euthanasia will radically change the social air we all breathe by severely undermining respect for life.** The previous Lords' Committee on this issue opposed assisted dying because of concern that 'vulnerable people – the elderly, lonely, sick or distressed – would feel pressure, whether real or imagined, to request early death.'[11] This concern

is just as valid today. The so-called 'right to die' would inexorably become the duty to die and potentially economic pressures and convenience would come to dominate decision-making.

We encourage all Members of both Houses to read the report[12] of the recent Select Committee on Lord Joffe's Assisted Dying for the Terminally Ill Bill. This report summarises the arguments on both sides clearly and comprehensively.

Notes

1   See HL Paper 86-I, Paragraphs 80–90.
2   Ibid, Paragraph 131.
3   Ibid, Paragraph 243.
4   See HL Paper 86-II, pp. 417–18.
5   See HL Paper 86-I, Paragraph 164.
6   "Majority of Doctors Oppose Euthanasia," *Hospital Doctor* 2003; 13 March.
7   "Doctors Oppose Assisted Suicide," *Hospital Doctor* 2003; 15 May.
8   See also frontline doctors' responses to recent BMJ editorial advocating euthanasia at http://bmj.bmjjournals.com/cgi/eletters/331/7518/0-g?ehom.
9   Ibid passim but see, for example, HL86-II pp. 96–164.
10 See HL86-I, pp. 75–80.
11 Select Committee on Medical Ethics. Report. London: HMSO, 1994. (House of Lords paper 21-I).
12 The report was published by The Stationery Office on 4 April 2005 as HL Paper 86.

Rev. Joel Edwards
General Director, Evangelical Alliance

Lama Jampa Thaya
Spiritual Director of the Dechen Community of Sakya and Kagyu centres of Buddhism in Europe

Sir Jonathan Sacks
Chief Rabbi

Peter Smith
Roman Catholic Archbishop of Cardiff

Sheikh Dr M.A. Zaki Badawi
Principal of the Muslim College and Chair, Muslim Law (Sharia) Council

His Eminence Archbishop Gregorios of Thyateira and Great Britain
(Greek Orthodox)

Bimal Krishna das
General Secretary, National Council of Hindu Temples (UK)

Tom Butler
Bishop of Southwark, Church of England

Dr Indarjit Singh
Director, Network of Sikh Organisations

Claim on why it should be illegal & change current policies.

Chapter 57

Brief of Ronald Dworkin, Thomas Nagel, Robert Nozick, John Rawls, Thomas Scanlon, and Judith Jarvis Thomson as amici curiae in support of respondents*

ASSISTED SUICIDE: THE PHILOSOPHERS' BRIEF

Interest of the amici curiae

Amici are six moral and political philosophers who differ on many issues of public morality and policy. They are united, however, in their conviction that respect for fundamental principles of liberty and justice, as well as for the American constitutional tradition, requires that the decisions of the Courts of Appeals be affirmed.[1]

Introduction and summary of argument

These cases do not invite or require the Court to make moral, ethical or religious judgments about how people should approach or confront their death or about when it is ethically appropriate to hasten one's own death or to ask others for help in doing so. On the contrary they ask the Court to recognize that individuals have a constitutionally protected interest in making those grave judgments for themselves, free from the imposition of any religious or philosophical orthodoxy by court or legislature. States have a constitutionally legitimate interest in protecting individuals from irrational, ill-informed, pressured or unstable decisions to hasten their own death. To that end, states may regulate and limit the assistance that doctors may give individuals who express a wish to die. But states may not deny people in the position of the patient-plaintiffs in these cases the opportunity to demonstrate, through whatever reasonable procedures the state might institute – even procedures that err on the side of caution – that their decision to die is indeed informed, stable, and fully free. Denying that opportunity to terminally-ill patients who are in agonizing pain or otherwise doomed to an existence they regard as intolerable could only be justified on the basis of a religious or ethical conviction about the value or meaning of life itself. Our Constitution forbids government to impose such convictions on its citizens.

Petitioners and the amici who support them offer two contradictory arguments. Some deny that the patient-plaintiffs have any constitutionally protected liberty interest in hastening their own deaths.[2] But that liberty interest flows directly from this Court's previous decisions. It flows from the right of people to make their own decisions about matters "involving the most intimate and personal choices a person may make in a lifetime, choices central to personal dignity and autonomy." *Planned Parenthood v. Casey*, 505 U.S. 833, 851 (1992).

The Solicitor General, urging reversal in support of Petitioners, recognizes that the patient-plaintiffs do have a constitutional liberty interest at stake in these cases. *See* Brief for the

United States as Amicus Curiae Supporting Petitioners at 12, *Washington v. Vacco* [hereinafter Brief for the United States] ("The term 'liberty' in the Due Process Clause . . . is broad enough to encompass an interest on the part of terminally ill, mentally competent adults in obtaining relief from the kind of suffering experienced by the plaintiffs in this case, which includes not only severe physical pain, but also the despair and distress that comes from physical deterioration and the inability to control basic bodily functions."); *see also* id. at 13 ("*Cruzan* . . . supports the conclusion that a liberty interest is at stake in this case."). The Solicitor General nevertheless argues that Washington and New York properly ignored this profound interest when they required the patient-plaintiffs to live on in circumstances they found intolerable. He argues that a state may simply declare that it is unable to devise a regulatory scheme that would adequately protect patients whose desire to die might be ill-informed or unstable or foolish or not fully free, and that a state may therefore fall back on a blanket prohibition. This Court has never accepted that patently dangerous rationale for denying protection altogether to a conceded fundamental constitutional interest. It would be a serious mistake to do so now. If that rationale were accepted, an interest acknowledged to be constitutionally protected would be rendered empty.

Argument

I. The liberty interest asserted here is protected by the Due Process Clause

The Due Process Clause of the Fourteenth Amendment protects the liberty interest asserted by the patient-plaintiffs here.

Certain decisions are momentous in their impact on the character of a person's life – decisions about religious faith, political and moral allegiance, marriage, procreation and death, for example. Such deeply personal decisions pose controversial questions about how and why human life has value. In a free society, individuals must be allowed to make those decisions for themselves, out of their own faith, conscience and convictions. This Court has insisted, in a variety of contexts and circumstances, that this great freedom is among those protected by the Due Process Clause as essential to a community of "ordered liberty." *Palko v. Connecticut*, 302 U.S. 319, 325 (1937). In its recent decision in *Planned Parenthood v. Casey*, 505 U.S. 833, 851 (1992), the Court offered a paradigmatic statement of that principle:

> matters[,] involving the most intimate and personal choices a person may make in a lifetime, choices central to a person's dignity and autonomy, are central to the liberty protected by the Fourteenth Amendment.

That declaration reflects an idea underlying many of our basic constitutional protections.[3] As the Court explained in *West Virginia State Board of Education v. Barnette*, 319 U.S. 624, 642 (1943):

> If there is any fixed star in our constitutional constellation, it is that no official . . . can prescribe what shall be orthodox in politics, nationalism, religion, or other matters of opinion or force citizens to confess by word or act their faith therein.

A person's interest in following his own convictions at the end of life is so central a part of the more general right to make "intimate and personal choices" for himself that a failure to protect that particular interest would undermine the general right altogether. Death is, for each of us, among the most significant events of life. As the Chief Justice said in *Cruzan v. Missouri*, 497 U.S. 261, 281 (1990), "[t]he choice between life and death is a deeply personal decision of obvious and overwhelming finality." Most of us see death – whatever we think will follow it – as the final act of life's drama, and we want that last act to reflect our own convictions, those we have tried to live by, not the convictions of others forced on us in our most vulnerable moment.

Different people, of different religious and ethical beliefs, embrace very different convictions about which way of dying confirms and which contradicts the value of their lives. Some fight against death with every weapon their doctors can devise. Others will do nothing to hasten death even if they pray it will come soon. Still others, including the patient-plaintiffs in these cases, want to end their lives when they think that living on, in the only way they can, would disfigure rather than enhance the lives they had created. Some people make the latter choice not just to escape pain. Even if it were possible to eliminate all pain for a dying patient – and frequently that is not possible – that would not end or even much alleviate the anguish some would feel at remaining alive, but intubated, helpless and often sedated near oblivion.

None of these dramatically different attitudes about the meaning of death can be dismissed as irrational. None should be imposed, either by the pressure of doctors or relatives or by the fiat of government, on people who reject it. Just as it would be intolerable for government to dictate that doctors never be permitted to try to keep someone alive as long as possible, when that is what the patient wishes, so it is intolerable for government to dictate that doctors may never, under any circumstances, help someone to die who believes that further life means only degradation. The Constitution insists that people must be free to make these deeply personal decisions for themselves and must not be forced to end their lives in a way that appalls them, just because that is what some majority thinks proper.

II. This court's decisions in Casey and Cruzan compel recognition of a liberty interest here

A. Casey Supports the Liberty Interest Asserted Here. In Casey, this Court, in holding that a State cannot constitutionally proscribe abortion in all cases, reiterated that the Constitution protects a sphere of autonomy in which individuals must be permitted to make certain decisions for themselves. The Court began its analysis by pointing out that "[a]t the heart of liberty is the right to define one's own concept of existence, of meaning, of the universe, and of the mystery of human life." 505 U.S. at 851. Choices flowing out of these conceptions, on matters "involving the most intimate and personal choices a person may make in a lifetime, choices central to personal dignity and autonomy, are central to the liberty protected by the Fourteenth Amendment." Id. "Beliefs about these matters," the Court continued, "could not define the attributes of personhood were they formed under compulsion of the State." Id.

In language pertinent to the liberty interest asserted here, the Court explained why decisions about abortion fall within this category of "personal and intimate" decisions. A decision whether or not to have an abortion, "originat[ing] within the zone of conscience and belief," involves conduct in which "the liberty of the woman is at stake in a sense unique to the human condition and so unique to the law." Id. at 852. As such, the decision necessarily involves the very "destiny of the woman" and is inevitably "shaped to a large extent on her own conception of her spiritual imperatives and her place in society." Id. Precisely because of these characteristics of the decision, "the State is [not] entitled to proscribe [abortion] in all instances." Id. Rather, to allow a total prohibition on abortion would be to permit a state to impose one conception of the meaning and value of human existence on all individuals. This the Constitution forbids.

The Solicitor General nevertheless argues that the right to abortion could be supported on grounds other than this autonomy principle, grounds that would not apply here. He argues, for example, that the abortion right might flow from the great burden an unwanted child imposes on its mother's life. Brief for the United States at 14–15. But whether or not abortion rights could be defended on such grounds, they were not the grounds on which this Court in fact relied. To the contrary, the Court explained at length that the right flows from the constitutional protection accorded all individuals to "define one's own concept of existence, of meaning, of the universe, and of the mystery of human life." (*Casey*, 505 U.S. at 851.)

The analysis in *Casey* compels the conclusion that the patient-plaintiffs have a liberty interest in this case that a state cannot burden with a blanket prohibition. Like a woman's decision whether to have an abortion, a decision to die involves one's very "destiny" and inevitably will be "shaped to a large extent on [one's] own conception of [one's] spiritual imperatives and [one's] place in society." *Id.* at 852. Just as a blanket prohibition on abortion would involve the improper imposition of one conception of the meaning and value of human existence on all individuals, so too would a blanket prohibition on assisted suicide. The liberty interest asserted here cannot be rejected without undermining the rationale of *Casey*. Indeed, the lower court opinions in the Washington case expressly recognized the parallel between the liberty interest in *Casey* and the interest asserted here. *See Compassion in Dying v. Washington*, 79 F.3d 790, 801 (9th Cir. 1996) (en banc) ("In deciding right-to-die cases, we are guided by the Court's approach to the abortion cases. *Casey* in particular provides a powerful precedent, for in that case the Court had the opportunity to evaluate

its past decisions and to determine whether to adhere to its original judgment."), *aff'g*, 850 F. Supp. 1454, 1459 (W.D. Wash. 1994) ("[T]he reasoning in *Casey* [is] highly instructive and almost prescriptive ..."). This Court should do the same.

B. *Cruzan Supports the Liberty Interest Asserted Here.* We agree with the Solicitor General that this Court's decision in "*Cruzan* ... supports the conclusion that a liberty interest is at stake in this case." Brief for the United States at 8. Petitioners, however, insist that the present cases can be distinguished because the right at issue in *Cruzan* was limited to a right to reject an unwanted invasion of one's body. But this Court repeatedly has held that in appropriate circumstances a state may require individuals to accept unwanted invasions of the body. *See, e.g., Schmerber v. California*, 384 U.S. 757 (1966) (extraction of blood sample from individual suspected of driving while intoxicated, notwithstanding defendant's objection, does not violate privilege against self-incrimination or other constitutional rights); *Jacobson v. Massachusetts*, 197 U.S. 11 (1905) (upholding compulsory vaccination for smallpox as reasonable regulation for protection of public health). The liberty interest at stake in *Cruzan* was a more profound one. If a competent patient has a constitutional right to refuse life-sustaining treatment, then, the Court implied, the state could not override that right. The regulations upheld in *Cruzan* were designed only to ensure that the individual's wishes were ascertained correctly. Thus, if *Cruzan* implies a right of competent patients to refuse life-sustaining treatment, that implication must be understood as resting not simply on a right to refuse bodily invasions but on the more profound right to refuse medical intervention when what is at stake is a momentous personal decision, such as the timing and manner of one's death. In her concurrence, Justice O'Connor expressly recognized that the right at issue involved a "deeply personal decision" that is "inextricably intertwined" with our notion of "self-determination." (497 U.S. at 287–89.)

Cruzan also supports the proposition that a state may not burden a terminally ill patient's liberty interest in determining the time and manner of his death by prohibiting doctors from terminating life support. Seeking to distinguish Cruzan, Petitioners insist that a state may nevertheless burden that right in a different way by forbidding doctors to assist in the suicide of patients who are not on life-support machinery. They argue that doctors who remove life support are only allowing a natural process to end in death whereas doctors who prescribe lethal drugs are intervening to cause death. So, according to this argument, a state has an independent justification for forbidding doctors to assist in suicide that it does not have for forbidding them to remove life support. In the former case though not the latter, it is said, the state forbids an act of killing that is morally much more problematic than merely letting a patient die.

This argument is based on a misunderstanding of the pertinent moral principles. It is certainly true that when a patient does not wish to die, different acts, each of which foreseeably results in his death, nevertheless have very different moral status. When several patients need organ transplants and organs are scarce, for example, it is morally permissible for a doctor to deny an organ to one patient, even though he will die without it, in order to give it to another. But it is certainly not permissible for a doctor to kill one patient in order to use his organs to save another. The morally significant difference between those two acts is not, however, that killing is a positive act and not providing an organ is a mere omission, or that killing someone is worse than merely allowing a "natural" process to result in death. It would be equally impermissible for a doctor to let an injured patient bleed to death, or to refuse antibiotics to

a patient with pneumonia – in each case the doctor would have allowed death to result from a "natural" process – in order to make his organs available for transplant to others. A doctor violates his patient's rights whether the doctor acts or refrains from acting, against the patient's wishes, in a way that is designed to cause death.

When a competent patient does want to die, the moral situation is obviously different, because then it makes no sense to appeal to the patient's right not to be killed as a reason why an act designed to cause his death is impermissible. From the patient's point of view, there is no morally pertinent difference between a doctor's terminating treatment that keeps him alive, if that is what he wishes, and a doctor's helping him to end his own life by providing lethal pills he may take himself, when ready, if that is what he wishes – except that the latter may be quicker and more humane. Nor is that a pertinent difference from the doctor's point of view. If and when it is permissible for him to act with death in view, it does not matter which of those two means he and his patient choose. If it is permissible for a doctor deliberately to withdraw medical treatment in order to allow death to result from a natural process, then it is equally permissible for him to help his patient hasten his own death more actively, if that is the patient's express wish.

It is true that some doctors asked to terminate life support are reluctant and do so only in deference to a patient's right to compel them to remove unwanted invasions of his body. But other doctors, who believe that their most fundamental professional duty is to act in the patient's interests and that, in certain circumstances, it is in their patient's best interests to die, participate willingly in such decisions: they terminate life support to cause death because they know that is what their patient wants. Cruzan implied that a state may not absolutely prohibit a doctor from deliberately causing death, at the patient's request, in that way and for that reason. If so, then a state may not prohibit doctors from deliberately using more direct and often more humane means to the same end when that is what a patient prefers. The fact that failing to provide life sustaining treatment may be regarded as "only letting nature take its course" is no more morally significant in this context, when the patient wishes to die, than in the other, when he wishes to live. Whether a doctor turns off a respirator in accordance with the patient's request or prescribes pills that a patient may take when he is ready to kill himself, the doctor acts with the same intention: to help the patient die.

The two situations do differ in one important respect. Since patients have a right not to have life support machinery attached to their bodies, they have, in principle, a right to compel its removal. But that is not true in the case of assisted suicide: patients in certain circumstances have a right that the state not forbid doctors to assist in their deaths, but they have no right to compel a doctor to assist them. The right in question, that is, is only a right to the help of a willing doctor.

III. State interests do not justify a categorical prohibition on all assisted suicide

The Solicitor General concedes that "a competent, terminally ill adult has a constitutionally cognizable liberty interest in avoiding the kind of suffering experienced by the plaintiffs in this case." Brief for the United States at 8. He agrees that this interest extends not only to avoiding pain, but to avoiding an existence the patient believes to be one of intolerable indignity or incapacity as well. *Id.* at 12. The Solicitor General argues, however, that states nevertheless have the right to "override" this liberty interest altogether, because a state could reasonably conclude that allowing doctors to assist in suicide, even

under the most stringent regulations and procedures that could be devised, would unreasonably endanger the lives of a number of patients who might ask for death in circumstances when it is plainly not in their interests to die or when their consent has been improperly obtained.

This argument is unpersuasive, however, for at least three reasons. *First*, in *Cruzan*, this Court noted that its various decisions supported the recognition of a general liberty interest in refusing medical treatment, even when such refusal could result in death. (497 U.S. at 278–79.) The various risks described by the Solicitor General apply equally to those situations. For instance, a patient kept alive only by an elaborate and disabling life support system might well become depressed, and doctors might be equally uncertain whether the depression is curable: such a patient might decide for death only because he has been advised that he will die soon anyway or that he will never live free of the burdensome apparatus, and either diagnosis might conceivably be mistaken. Relatives or doctors might subtly or crudely influence that decision, and state provision for the decision may (to the same degree in this case as if it allowed assisted suicide) be thought to encourage it.

Yet there has been no suggestion that states are incapable of addressing such dangers through regulation. In fact, quite the opposite is true. In *McKay v. Bergstedt*, 106 Nev. 808, 801 P.2d 617 (1990), for example, the Nevada Supreme Court held that "competent adult patients desiring to refuse or discontinue medical treatment" must be examined by two non-attending physicians to determine whether the patient is mentally competent, understands his prognosis and treatment options, and appears free of coercion or pressure in making his decision. *Id.* at 827–28, 801 P.2d at 630. *See also id.* (in the case of terminally-ill patients with natural life expectancy of less than six months, patient's right of self-determination shall be deemed to prevail over state interests, whereas non-terminal patient's decision to terminate life-support systems must first be weighed against relevant state interests by trial judge); In re Farrell, 108 N.J. 335, 354, 529 A.2d 404, 413 (1987) (terminally-ill patient requesting termination of life-support must be determined to be competent and properly informed about prognosis, available treatment options and risks, and to have made decision voluntarily and without coercion). Those protocols served to guard against precisely the dangers that the Solicitor General raises. The case law contains no suggestion that such protocols are inevitably insufficient to prevent deaths that should have been prevented.

Indeed, the risks of mistake are overall greater in the case of terminating life support. Cruzan implied that a state must allow individuals to make such decisions through an advance directive stipulating either that life support be terminated (or not initiated) in described circumstances when the individual was no longer competent to make such a decision himself, or that a designated proxy be allowed to make that decision. All the risks just described are present when the decision is made through or pursuant to such an advance directive, and a grave further risk is added: that the directive, though still in force, no longer represents the wishes of the patient. The patient might have changed his mind before he became incompetent, though he did not change the directive, or his proxy may make a decision that the patient would not have made himself if still competent. In Cruzan, this Court held that a state may limit these risks through reasonable regulation. It did not hold – or even suggest – that a state may avoid them through a blanket prohibition that, in effect, denies the liberty interest altogether.

Second, nothing in the record supports the conclusion that no system of rules and regulations could adequately reduce the risk of mistake. As discussed above, the experience of states in adjudicating requests to have life-sustaining treatment removed indicates the opposite.[4] The Solicitor General has provided no persuasive reason why the same sort of procedures could not be applied effectively in the case of a competent individual's request for physician-assisted suicide.

Indeed, several very detailed schemes for regulating physician-assisted suicide have been submitted to the voters of some states[5] and one has been enacted.[6] In addition, concerned groups, including a group of distinguished professors of law and other professionals, have drafted and defended such schemes. See, e.g., Charles H. Baron, et. al., A Model State Act to Authorize and Regulate Physician-Assisted Suicide, 33 Harv. J. Legis. 1 (1996). Such draft statutes propose a variety of protections and review procedures designed to insure against mistakes, and neither Washington nor New York attempted to show that such schemes would be porous or ineffective. Nor does the Solicitor General's brief: it relies instead mainly on flat and conclusory statements. It cites a New York Task Force report, written before the proposals just described were drafted, whose findings have been widely disputed and were implicitly rejected in the opinion of the Second Circuit below. See generally Quill v. Vacco, 80 F.3d 716 (2d Cir. 1996). The weakness of the Solicitor General's argument is signalled by his strong reliance on the experience in the Netherlands which, in effect, allows assisted suicide pursuant to published guidelines. Brief for the United States at 23–24. The Dutch guidelines are more permissive than the proposed and model American statutes, however. The Solicitor General deems the Dutch practice of ending the lives of people like neo-nates who cannot consent particularly noteworthy, for example, but that practice could easily and effectively be made illegal by any state regulatory scheme without violating the Constitution.

The Solicitor General's argument would perhaps have more force if the question before the Court were simply whether a state has any rational basis for an absolute prohibition; if that were the question, then it might be enough to call attention to risks a state might well deem not worth running. But, as the Solicitor General concedes, the question here is a very different one: whether a state has interests sufficiently compelling to allow it to take the extraordinary step of altogether refusing the exercise of a liberty interest of constitutional dimension. In those circumstances, the burden is plainly on the state to demonstrate that the risk of mistakes is very high, and that no alternative to complete prohibition would adequately and effectively reduce those risks. Neither of the Petitioners has made such a showing.

Nor could they. The burden of proof on any state attempting to show this would be very high. Consider, for example, the burden a state would have to meet to show that it was entitled altogether to ban public speeches in favor of unpopular causes because it could not guarantee, either by regulations short of an outright ban or by increased police protection, that such speeches would not provoke a riot that would result in serious injury or death to an innocent party. Or that it was entitled to deny those accused of crime the procedural rights that the Constitution guarantees, such as the right to a jury trial, because the security risk those rights would impose on the community would be too great. One can posit extreme circumstances in which some such argument would succeed. See, e.g., Korematsu v. United States, 323 U.S. 214 (1944) (permitting United States to detain individuals of Japanese ancestry during wartime). But these circumstances would be extreme indeed, and the Korematsu ruling has been widely and severely criticized.

Third, it is doubtful whether the risks the Solicitor General cites are even of the right character to serve as justification for an absolute prohibition on the exercise of an important liberty interest. The risks fall into two groups. The first is the risk of medical mistake, including a misdiagnosis of competence or terminal illness. To be sure, no scheme of regulation, no matter how rigorous, can altogether guarantee that medical mistakes will not be made. But the Constitution does not allow a state to deny patients a great variety of important choices, for which informed consent is properly deemed necessary, just because the information on which the consent is given may, in spite of the most strenuous efforts to avoid mistake, be wrong. Again, these identical risks are present in decisions to terminate life support, yet they do not justify an absolute prohibition on the exercise of the right.

The second group consists of risks that a patient will be unduly influenced by considerations that the state might deem it not in his best interests to be swayed by, for example, the feelings and views of close family members. Brief for the United States at 20. But what a patient regards as proper grounds for such a decision normally reflects exactly the judgments of personal ethics – of why his life is important and what affects its value – that patients have a crucial liberty interest in deciding for themselves. Even people who are dying have a right to hear and, if they wish, act on what others might wish to tell or suggest or even hint to them, and it would be dangerous to suppose that a state may prevent this on the ground that it knows better than its citizens when they should be moved by or yield to particular advice or suggestion in the exercise of their right to make fateful personal decisions for themselves. It is not a good reply that some people may not decide as they really wish – as they would decide, for example, if free from the "pressure" of others. That possibility could hardly justify the most serious pressure of all – the criminal law which tells them that they may not decide for death if they need the help of a doctor in dying, no matter how firmly they wish it.

There is a fundamental infirmity in the Solicitor General's argument. He asserts that a state may reasonably judge that the risk of "mistake" to some persons justifies a prohibition that not only risks but insures and even aims at what would undoubtedly be a vastly greater number of "mistakes" of the opposite kind – preventing many thousands of competent people who think that it disfigures their lives to continue living, in the only way left to them, from escaping that – to them – terrible injury. A state grievously and irreversibly harms such people when it prohibits that escape. The Solicitor General's argument may seem plausible to those who do not agree that individuals are harmed by being forced to live on in pain and what they regard as indignity. But many other people plainly do think that such individuals are harmed, and a state may not take one side in that essentially ethical or religious controversy as its justification for denying a crucial liberty.

Of course, a state has important interests that justify regulating physician-assisted suicide. It may be legitimate for a state to deny an opportunity for assisted suicide when it acts in what it reasonably judges to be the best interests of the potential suicide, and when its judgment on that issue does not rest on contested judgments about "matters involving the most intimate and personal choices a person may make in a lifetime, choices central to personal dignity and autonomy." *Casey*, 505 U.S. at 851. A state might assert, for example, that people who are not terminally ill, but who have formed a desire to die, are, as a group, very likely later to be grateful if they are prevented from taking their own lives. It might then claim that it is legitimate, out of concern for such people, to deny any of them a doctor's assistance. This Court need not decide now the extent to which such paternalistic interests might override an individual's liberty

interest. No one can plausibly claim, however – and it is noteworthy that neither Petitioners nor the Solicitor General does claim – that any such prohibition could serve the interests of any significant number of terminally ill patients. On the contrary, any paternalistic justification for an absolute prohibition of assistance to such patients would of necessity appeal to a widely contested religious or ethical conviction many of them, including the patient-plaintiffs, reject. Allowing *that* justification to prevail would vitiate the liberty interest.

Even in the case of terminally ill patients, a state has a right to take all reasonable measures to insure that a patient requesting such assistance has made an informed, competent, stable and uncoerced decision. It is plainly legitimate for a state to establish procedures through which professional and administrative judgments can be made about these matters, and to forbid doctors to assist in suicide when its reasonable procedures have not been satisfied. States may be permitted considerable leeway in designing such procedures. They may be permitted, within reason, to err on what they take to be the side of caution. But they may not use the bare possibility of error as justification for refusing to establish any procedures at all and relying instead on a flat prohibition.

Conclusion

Each individual has a right to make the "most intimate and personal choices central to personal dignity and autonomy." That right encompasses the right to exercise some control over the time and manner of one's death.

The patient-plaintiffs in these cases were all mentally competent individuals in the final phase of terminal illness and died within months of filing their claims. Jane Doe described how her advanced cancer made even the most basic bodily functions such as swallowing, coughing, and yawning extremely painful and that it was "not possible for [her] to reduce [her] pain to an acceptable level of comfort and to retain an alert state." Faced with such circumstances, she sought to be able to "discuss freely with [her] treating physician [her] intention of hastening [her] death through the consumption of drugs prescribed for that purpose." *Quill v. Vacco*, 80 F.2d 716, 720 (2d Cir. 1996) (quoting declaration of Jane Doe). George A. Kingsley, in advanced stages of AIDS which included, among other hardships, the attachment of a tube to an artery in his chest which made even routine functions burdensome and the development of lesions on his brain, sought advice from his doctors regarding prescriptions which could hasten his impending death. *Id.* Jane Roe, suffering from cancer since 1988, had been almost completely bed-ridden since 1993 and experienced constant pain which could not be alleviated by medication. After undergoing counseling for herself and her family, she desired to hasten her death by taking prescription drugs. *Compassion in Dying v. Washington*, 850 F. Supp. 1454, 1456 (1994). John Doe, who had experienced numerous AIDS-related ailments since 1991, was "especially cognizant of the suffering imposed by a lingering terminal illness because he was the primary caregiver for his long-term companion who died of AIDS" and sought prescription drugs from his physician to hasten his own death after entering the terminal phase of AIDS. *Id.* at 1456–57. James Poe suffered from emphysema which caused him "a constant sensation of suffocating" as well as a cardiac condition which caused severe leg pain. Connected to an oxygen tank at all times but unable to calm the panic reaction associated with his feeling of suffocation even with regular doses of morphine, Mr. Poe sought physician-assisted suicide. *Id.* at 1457.

A state may not deny the liberty claimed by the patient-plaintiffs in these cases without providing them an opportunity to demonstrate, in whatever way the state might reasonably think wise and necessary, that the conviction they expressed for an early death is competent, rational, informed, stable and uncoerced.

Affirming the decisions by the Courts of Appeals would establish nothing more than that there is such a constitutionally protected right in principle. It would establish only that some individuals, whose decisions for suicide plainly cannot be dismissed as irrational or foolish or premature, must be accorded a reasonable opportunity to show that their decision for death is informed and free. It is not necessary to decide precisely which patients are entitled to that opportunity. If, on the other hand, this Court reverses the decisions below, its decision could only be justified by the momentous proposition – a proposition flatly in conflict with the spirit and letter of the Court's past decisions – that an American citizen does not, after all, have the right, even in principle, to live and die in the light of his own religious and ethical beliefs, his own convictions about why his life is valuable and where its value lies.

Notes

* On Writ of Certiorari to the United States Courts of Appeal for the Ninth Circuit in State of Washington, *et al.*, *Petitioners*, vs. Harold Glucksberg, *et al.*, *Respondents*, and for the Second Circuit in Dennis C. Vacco, Attorney General of New York, *et al.*, *Petitioners*, vs. Timothy E. Quill, *et al.*, *Respondents*. In the Supreme Court of the United States, October Term, 1996, Nos. 95–1858, 96–110.

1 This brief is filed on the consent of all parties. *See* S. Ct. R. 37.3(a). It is not filed on behalf of any corporation. *See* S. Ct. R. 29.1.

2 *See, e.g.*, Brief for the Petitioners at 25–33, *Washington v. Glucksberg*; Brief for Petitioners Vacco and Pataki at 19–20, *Vacco v. Quill*; Brief Amici Curiae of the United States Catholic Conference, New York Catholic Conference, *et al.* in Support of Petitioners

at 7–21, *Washington v. Glucksberg*; Brief for the Institute for Public Affairs of the Union of Orthodox Jewish Congregations of America and the Rabbinical Council of America as Amici Curiae in Support of Petitioners at 5–13, *Vacco v. Quill* and *Washington v. Glucksberg*.

3 In *Cohen v. California*, 403 U.S. 15, 24 (1971), for example, this Court held that the First Amendment guarantee of free speech and expression derives from "the belief that no other approach would comport with the premise of individual dignity and choice upon which our political systems rests." Interpreting the religion clauses of the First Amendment, this Court has explained that "[t]he victory for freedom of thought recorded in our Bill of Rights recognizes that in the domain of conscience there is a moral power higher than the State." *Girouard v. United States*, 328 U.S. 61, 68 (1946). And, in a number of Due Process cases, this Court has protected this conception of autonomy by carving out a sphere of personal family life that is immune from government intrusion. See, e.g., *Cleveland Bd. of Educ. v. LeFleur*, 414 U.S. 632, 639 (1974) ("This Court has long recognized that freedom of personal choice in matters of marriage and family life is one of the liberties protected by the Due Process Clause of the Fourteenth Amendment."); *Eisenstadt v. Baird*, 405 U.S. 438, 453 (1973) (recognizing right "to be free from unwarranted governmental intrusion into matters so fundamentally affecting a person as the decision to bear and beget a child"); *Skinner v. Oklahoma*, 316 U.S. 535, 541 (1942) (holding unconstitutional a state statute requiring the sterilization of individuals convicted of three offenses, in large part because the state's actions unwarrantedly intruded on marriage and procreation, "one of the basic civil rights of man"); *Loving v. Virginia*, 388 U.S. 1, 12 (1967) (striking down the criminal prohibition of interracial marriages as an infringement of the right to marry and holding that "[t]he freedom to marry has long been recognized as one of the vital personal rights essential to the orderly pursuit of happiness by free men").

These decisions recognize as constitutionally immune from state intrusion that realm in which individuals make "intimate and personal" decisions that define the very character of their lives. See Charles Fried, *Right and Wrong* 146–47 (1978) ("What a person is, what he wants, the determination of his life plan, of his concept of the good, are the most intimate expressions of self-determination, and by asserting a person's responsibility for the results of this self-determination, we give substance to the concept of liberty.").

4 When state protocols are observed, sometimes the patient is permitted to die and sometimes not. See, e.g., *In re Tavel*, 661 A.2d 1061 (Del. 1995) (affirming finding that petitioner-daughter had proven by clear and convincing evidence that incompetent patient would want life support systems removed); *In re Martin*, 450 Mich. 204, 538 N.W.2d 399 (1995) (holding that wife's testimony and affidavit did not constitute clear and convincing evidence of incompetent patient's pre-injury decision to decline life-sustaining medical treatment in patient's present circumstances); *DiGrella v. Elston*, 858 S.W.2d 698, 710 (Ky. 1993) ("If the attending physician, the hospital or nursing home ethics committee where the patient resides, and the legal guardian or next of kin all agree and document the patient's wishes and condition, and if no one disputes their decision, no court order is required to proceed to carry out [an incompetent] patient's wishes"); *Mack v. Mack*, 329 Md. 188, 618 A.2d 744 (1993) (holding that wife failed to provide clear and convincing evidence that incompetent husband would want life support removed); *In re Doe*, 411 Mass. 512, 583 N.E.2d 1263 (applying doctrine of substituted judgment and holding that evidence supported finding that, if incompetent patient were capable of making a choice, she would remove life support).

5 For example, 46 percent of California voters supported Proposition 161, which would have legalized physician-assisted suicide, in November 1992. The measure was a proposed amendment to Cal. Penal Code § 401 (1992) which currently makes assisted suicide a felony. Those who did not vote for the measure cited mainly religious reasons or concerns that the proposed law was flawed because it lacked safeguards against abuse and needed more restrictions that might be easily added.

such as a waiting period and a psychological exami-
nation. Alison C. Hall, *To Die With Dignity: Comparing
Physician-Assisted Suicide in the United States, Japan, and the
Netherlands*, 74 Wash. U. L. Q. 803, 817 n.84 (1996).

6 In November 1994, Oregon voters approved the
Oregon Death With Dignity Act through voter initiative, legalizing physician-assisted suicide under
limited circumstances. Oregon Death With Dignity
Act, Or. Rev. Stat. §§ 127.800–.827 (1995). Under
the Oregon Act, a capable adult resident of the state,
who has been determined by the attending
physician and consulting physician to be suffering
from a terminal disease, and who has voluntarily
expressed his or her wish to die, may make a
written request for medication for the purpose of
ending his life in a humane and dignified manner
in accordance with [the provisions of the Act].

Or. Rev. Stat. § 127.805 (1995). The Act provides
specific definitions of essential terms such as "incapable" and "terminal disease." The Act also provides
numerous other regulations designed to safeguard
the integrity of the process.

Chapter 58

John P. Safranek

AUTONOMY AND ASSISTED SUICIDE: THE EXECUTION OF FREEDOM

For the last two thousand years, most western civilizations have proscribed assisted suicide and other forms of euthanasia as a violation of innocent human life. The Hippocratic Oath, professed by doctors through the centuries, explicitly condemns administering a deadly drug or even suggesting it. Yet in recent decades scholars have criticized this traditional position for violating personal autonomy[1] and precluding beneficent treatment of the suffering.[2] Their appeals to beneficence have not proven decisive, in part because the traditional position counsels beneficence: it permits physicians to alleviate suffering even when their treatment jeopardizes patients' lives. It is rather the concept of personal autonomy, recently ascendant as a principle of human action, that has transformed the social discourse of assisted suicide.

Opponents of euthanasia have not criticized assisted-suicide claims grounded on autonomy. Instead they resort to arguments based on human dignity,[3] the sacredness of life,[4] or slippery slopes[5] – none of which directly addresses the potent autonomy claim. This article challenges the argument for autonomy: it distinguishes the descriptive and ascriptive aspects of autonomy and probes whether either can ground a right to assisted suicide or logically delimit such a right. By differentiating these two aspects of autonomy, this discussion manifests the shortcomings of many autonomy-based moral and legal theories.

Autonomy is a protean concept, but most ethicists employ the term in either a descriptive or an ascriptive sense. The descriptive sense of autonomy encompasses those conditions (for example, a choice of options, competence to act)[6] and characteristics (for example, moral authenticity or integrity)[7] that scholars associate with autonomy. These characteristics and conditions are essential components of self-governed action, but they cannot justify acts of assisted suicide (or most other human acts) because they are bereft of normativity: they merely describe necessary conditions of morality, viz., voluntary agency,[8] without specifying the moral character of any particular act. If both the virtuous and the vicious can act autonomously, then the mere possession of autonomy neither specifies an agents moral character nor justifies his acts.

Although the descriptive sense of autonomy is insufficient to justify – either morally or legally – acts of assisted suicide, many scholars use the term in an ascriptive sense by grounding individual rights on autonomy.[9] Ethicists and legal scholars employ the ascriptive sense of autonomy synonymously with liberty in grounding moral or legal claims. For example, the eminent liberal philosopher Joseph Raz claims that personal autonomy "is essentially about the freedom of persons to choose their own lives,"[10] while John Stuart Mill states that the principle of liberty "requires liberty of tastes and pursuits; of framing the plan of our life to suit our own character."[11]

Jurists also identify autonomy with liberty: the Supreme Court asserts that "choices central to personal dignity and autonomy are central to the liberty protected by the Fourteenth Amendment. At the heart of liberty is the right to define one's own concept of existence . . ."[12] And writing in defense of a right to euthanasia, the respected legal scholar Ronald Dworkin claims that individuals' right to autonomy is "a right to make important decisions defining their own lives for themselves."[13] Thus in both ethics and law, autonomy and liberty are similarly concerned with an individual's freedom to make important choices for himself, unfettered by social proscriptions. This ascriptive sense of autonomy grounds the claim for a "right" to assisted suicide by guaranteeing an individual's freedom to enlist assistance in his suicide.

Although this purported right has achieved some measure of public support, its autonomy-based justification faces several formidable challenges. First, proponents of assisted suicide assert that autonomy is a fundamental good that must be protected, yet they advocate an act that extinguishes the basis of autonomy. The same conundrum prompted John Stuart Mill, a stalwart champion of individual liberty, to favor legal proscription of voluntary slavery. Mill claimed that an individual cannot freely renounce his freedom without violating that good.[14] Similarly, autonomous acts of assisted suicide annihilate the basis of autonomy and thereby undermine the very ground of their justification.

A second criticism is that proponents of ascriptive autonomy have not articulated criteria that distinguish opprobrious from acceptable autonomous acts. As noted above, society does not morally or legally sanction acts merely because they are performed autonomously. In fact, immoral and depraved acts are more objectionable if performed autonomously than if coerced.[15] Thus even if assisted-suicide proponents are granted the claim that society must recognize individuals' autonomy to regulate their own lives, they must formulate a method or principle for distinguishing those autonomous acts that are permissible, since moral agents can use their autonomy perniciously.

Most proponents of assisted suicide have not addressed the first criticism, that an autonomous act of assisted suicide destroys the basis of its justification, but many respond to the second criticism by invoking Mill's harm principle to instill normativity into autonomy. The harm principle states that a person's autonomy (or equivalently, his liberty) to perform an act can be circumscribed only if the act harms another.[16] A corollary of the harm principle is that an individual is not subject to others' moral principles if he is not harming them. Thus scholars claim that the harm principle prevents individuals from imposing their view of the good on a person who seeks assisted suicide since he is not harming others.[17]

Yet these scholars themselves ineluctably subject others to a peculiar view of the good when they ascribe autonomy-based rights to acts considered "self-regarding," but not to those perceived as "harmful." To specify an act as harmful is to judge it injurious, but what constitutes injury will depend on one's theory of the good: both the hedonist and the Aristotelian virtuous man would endorse the harm principle, yet they would offer divergent examples of harmful acts. The harm principle is wholly formal insofar as it does not articulate the criteria of a harmful act, or correlatively, which acts should be proscribed. Therefore any scholar who wields the harm principle must turn elsewhere to generate practical moral or legal precepts, namely, to a moral theory. As Joseph Raz notes,

> Since "causing harm" entails by its very meaning that the action is prima facie wrong, it is a normative concept acquiring its specific concrete content from the moral theory within which it is embedded. Without such a connection to a moral theory the harm principle is a formal principle lacking specific meaning and leading to no policy conclusions.[18]

Thus the autonomy proponent can impart normativity to the harm principle – and to autonomy – only by advancing a theory of the good. But in imposing his peculiar theory of the good on adherents of a discrepant theory, the autonomy proponent thereby undermines their autonomy.

This critique of autonomy is more radical than those that criticize autonomy proponents for proposing a view of human nature that is atomistic[19] or self-regarding,[20] for even if ascriptive autonomy does not require an individualistic view of human beings, it entails a profound dilemma. The principle of autonomy or liberty requires a "harm" principle to justify prohibiting certain types of autonomous acts, but whether an act is specified as harmful or harmless will depend on the preferred theory of good. Therefore the normative use of the principle of autonomy is performatively self-refuting: when scholars proscribe certain autonomous acts in the name of harm, or defend other autonomous acts judged harmless, they impose an axiology and subvert autonomy.[21]

Autonomy is inextricably linked to a theory of good, even if the esteemed theory is cast in "thin" terms.[22] Thin theories of the good, which protect only a few basic values such as life or property, must still specify when these goods are inviolable, such as in defense of one's country, life, material goods, etc. (A thin theory that refuses such specification is – like the harm principle – devoid of practical import and therefore unfeasible as a moral or political theory.) Thus even a thin theory of the good must proscribe certain human acts, such as stealing another's property or killing a thief to retain some trivial material object. But any individual who would rather violate these thin proscriptions capitulates his autonomy if he instead acquiesces to the heteronomous legal proscription.

To be sure, autonomy must be circumscribed to maintain the good of society or protect human life, but this appeal to goods violates the autonomy of those who retain a divergent axiology. A society must criminalize certain acts that threaten its survival, but it thereby circumscribes the autonomy of individuals who prefer to perform these threatening acts.

Moreover, this appeal to the good of society or human life underscores the instrumental character of autonomy: it is respected when invoked to protect acts that most people judge as good, for example, the preservation of society or human life, and is usually proscribed when employed in an invidious manner, as in acts of murder or robbery. Hence autonomy is necessary for the existence of a moral act but is insufficient to justify one. The justification of the act will hinge on the end to which autonomy is employed: if for a noble end, then it is upheld; if depraved, then it is proscribed. It is not autonomy per se that vindicates an autonomy claim but the good that autonomy is instrumental in achieving. Therefore an individual cannot invoke autonomy to justify an ethical or legal claim to acts such as assisted suicide; rather he must vindicate the underlying value that the autonomous act endeavors to attain.

The imposition of a theory of the good does not always undermine autonomy, and indeed can procure it. A child or mentally ill individual, incapable as each may be of rational thought, might perhaps subsequently engage in autonomous action if currently prevented from engaging in immediate life-threatening or mind-altering behavior.[23] But since these individuals did not retain sufficient rationality when incurring the imposition, their autonomy could not be violated. In contrast, a rational adult who is denied such choices sacrifices his autonomy at the moment another restricts his choice. Even if circumscription of a rational individual's autonomy is conducive to future autonomous choices, his extant autonomy is nevertheless subverted when he is prevented from enacting a criminal or mind-altering practice that was rationally chosen. Thus an imposed theory of the good violates the autonomy of those who fulfill the descriptive requisites of autonomy; if it did not, it would not be termed an imposition.

This subordination of autonomy to a view of the good generates many prevalent social controversies. Consider the conflict surrounding pornography: free speech advocates favor legalization of pornography because of the harm incurred by curtailing free expression,[24] while many feminists oppose pornography because it demeans and sexually exploits women.[25] Though both sides endorse the harm principle, their divergent moral and legal principles arise from disparate views of the human good, the former upholding the goodness of free expression, the latter the goodness of women viewed as multidimensional beings. Each group attempts to legislate its view of the good, and regardless of the outcome, one group's autonomy to attain its view of the good will be violated. Legislators or jurists must ultimately choose between the alternative views of the good, but they will inevitably harm (at least) one group by constraining its autonomy to realize its theory of the good. Certainly individuals can disapprove of certain acts but refuse to proscribe them legally, but even the refusal is usually grounded on a view of the good, such as the socially deleterious effects of abridging free speech.

Similarly, in the assisted-suicide debate, the disputants seek to legislate conflicting moral views of human life for society. Assisted-suicide proponents endorse a subjective valuation of life's worth[26] or extol an analgesic death.[27] Hence they claim that the individual should be free to terminate his devalued or painful existence. Their arguments are equivalent to the moral claims, "A competent person ought to be assisted in suicide upon request if he does not value his life," or "An individual ought to be free to terminate a painful existence."

Opponents of assisted suicide view innocent human life as intrinsically good[28] or fear a social slide to more dubious types of killing.[29] Their claims reduce to, "It is wrong to intentionally kill innocent human beings, even if they seek to die," or "Society ought to avoid traversing a slippery slope." Both sides of the debate propound moral principles embedded in theories of the good, the former extolling a subjective valuation of life's worth or the avoidance of pain, the latter innocent human life or extant social conditions.

Thus the debate over assisted suicide is a conflict between competing theories of the good, and not a dispute between proponents of autonomy and the sanctity or dignity of life. Because each side seeks the autonomy to attain its respective theory of the good, neither can invoke autonomy to vindicate its attempt to achieve its good. Therefore a proponent of assisted suicide cannot claim that restrictions violate autonomy by dictating a theory of the good.[30] For to justify assisted suicide by autonomy or to prohibit depraved acts by the harm principle, the proponent himself abrogates the autonomy of those who subscribe to a discrepant set of goods: he *a fortiori* undermines his own principle of autonomy.

These serious shortcomings of autonomy-based justifications of assisted suicide are nowhere more apparent than in the restrictions placed on the act by proponents of autonomy. Most support the act conditionally, but their stipulations underscore the tension that persists between the liberating character of autonomy and the social necessity of limiting certain autonomous acts.

This tension persists because autonomy justifies assisted suicide for nearly any reason.[31] If a person retains the autonomy to be killed, then it is irrelevant whether he is suffering from a terminal disease or a temporary illness – his autonomy is violated when he is denied death in either condition.[32] Moreover, society cannot coherently limit assisted suicide to those suffering from physical disease; to deny a person liberation from psychological suffering would likewise infringe his autonomy. Hence the Netherlands consistently extended the right to assisted suicide to those facing bleak futures.[33]

Furthermore, the principle of autonomy prevents any distinction on the basis of age: a competent twenty-year-old individual denied assisted suicide – for physical or mental suffering – is denied his autonomy as much as an eighty-year-old. The former's killing seems even more beneficent because he may face a more extended trial of suffering. And if an individual can be assisted in suicide when he autonomously deems his life valueless, then he should also be granted the autonomy to sell his organs for their fair market value or to dispose of his life in any other manner.

Thus proponents of autonomy can restrict the right to assisted suicide only by implicitly or explicitly articulating normative claims, for example, that the young or the physically healthy should not be assisted in suicide because their lives are worthwhile or might later improve. They mandate these normative views of the good for other autonomy supporters who subscribe to a divergent axiology, for example, that the young or healthy can be assisted in suicide if they autonomously so choose. If

proponents of assisted suicide – either of limited or unconditional scope – are in fact legislating a particular view of the good, then they must defend their proscription of alternative views of the good, including those that extol the goodness of innocent human life or extant social conditions.

Therefore proponents of assisted suicide face a dilemma in appealing to autonomy, whether of the descriptive or ascriptive stripe. Descriptive autonomy, devoid of normativity, defies restrictions because society thwarts the autonomy of an individual when it denies his "authentic" or "self-creative" choice to die, or any other choice. Ascriptive autonomy retains normativity, but only by capitulating its moral neutrality and violating the autonomy of adherents of conflicting moral theories. Certainly society must limit the harm that autonomous acts could yield, but any limitation implicates a view of the good and undermines opponents' autonomy.

Thus autonomy-based arguments for assisted suicide are self-refuting in two regards: first, acts of assisted suicide committed in the name of autonomy annihilate the very basis of individual autonomy; second, arguments grounded on autonomy ultimately depend on a view of the good that, if socially prescribed, would subvert individuals' autonomy to attain alternative views of the good.

Proponents of assisted suicide justify the act in moral and legal contexts by appealing to the concept of autonomy. However, they have overlooked the problematic nature of this claim, particularly its dependence on a theory of the good. The debate involving assisted suicide, like so many other social disputes, hinges on discrepant views of the good, rather than on autonomy or beneficence. Only by focusing on the conflicting views of the goods at stake – and abandoning or re-formulating the argument for autonomy – will ethicists and legal scholars resolve this controversial social issue.[34]

Acknowledgements

I would like to thank Louis Safranek, M.D., Ph.D., for his substantive and technical critiques.

Notes

1 Dan Brock, "Voluntary Active Euthanasia," *Hastings Center Report* 22, no. 2 (1992): 10–22; Baruch Brody, "Voluntary Euthanasia and the Law," in *Beneficent Euthanasia*, ed. M. Kohl (Buffalo, N.Y.: Prometheus Books, 1975); "Physician-Assisted Suicide and the Right to Die with Assistance", *Harvard Law Review* 105 (1991): 2021–40.

2 Richard L. Risley, *A Humane and Dignified Death: A New Law Permitting Physician Aid-in-Dying* (Glendale, Calif.: Americans Against Human Suffering, 1987); Sidney Wanzer, "Maintaining Control in Terminal Illness: Assisted Suicide and Euthanasia," *Humane Medicine* 6, no. 3 (1990): 186–88.

3 Leon Kass, "Death with Dignity and the Sanctity of Life," *Commentary* (March 1990): 33–43; John Mahony, *Bioethics and Belief* (London: Sheed and Ward, 1984).

4 Richard Gula, *Euthanasia: Moral and Pastoral Perspectives* (New York: Paulist Press, 1994), pp. 24–28; Joseph Boyle, "Sanctity of Life and Suicide: Tensions and Developments Within Common Morality," in *Suicide and Euthanasia*, ed. Baruch Brody (Boston: Kluwer Academic Publishers, 1989), pp. 221–50; Richard Roach, "Medicine and Killing: The Catholic View," *The Journal of Medicine and Philosophy* 4, no. 4 (1979): 383–97.

5 Richard Fenigsen, "A Case against Dutch Euthanasia," *Ethics and Medicine* 6, no. 1 (1990): 11–18; Kathleen Foley, "Competent Care for the Dying Instead of Physician-Assisted Suicide," *NEJM* 336, no. 1 (1997): 54–58.

6 Richard H. Fallon, Jr., "Two Senses of Autonomy," *Stanford University Law Review* 46 (1994): 875–99, at 875.

7 Joel Feinberg, *Harm to Self* (Oxford: Oxford University Press, 1986), ch. 18.

8 Aristotle, *Nicomachean Ethics*, trans. Richard McKeon (New York: Random House, 1941), III, I.

9 Jos Welie, "The Medical Exception: Physicians, Euthanasia and the Dutch Criminal Law," *The Journal of Medicine and Philosophy* 17, no. 4 (1992): 419–37, at 419.

10 Joseph Raz, *The Morality of Freedom* (Oxford: Oxford University Press, 1985), p. 370.

11 John Stuart Mill, On Liberty in *Utilitarianism, On Liberty, Considerations on Representative Government*, ed. H. B. Acton (London: J. M. Dent & Sons, 1972), p. 81.

12 Planned Parenthood v. Casey, 112 S.Ct. 2791, 2807 (1992).

13 Ronald Dworkin, *Life's Dominion* (New York: Alfred Knopf, 1993), p. 222.

14 Mill, *On Liberty*, p. 172.

15 William Galston, *Justice and the Common Good* (Chicago: University of Chicago Press, 1980), p. 127.

16 Mill, *On Liberty*, p. 78.

17 Max Charlesworth, *Bioethics in a Liberal Society* (Cambridge: Cambridge University Press, 1993), p. 4.

18 Raz, *The Morality of Freedom*, p. 414.

19 Alasdair Macintyre, *After Virtue* (Notre Dame, Ind.: University of Notre Dame Press, 1981).

20 Charles Taylor, *Sources of the Self: The Making of the Modern Identity* (Cambridge, Mass.: Harvard University Press, 1989).

21 John Safranek and Stephen Safranek, "Can the Right to Autonomy Be Resuscitated after Glucksberg," *University of Colorado Law Review* 69 (1998): 744–49.

22 H.L.A. Hart, *Essays in Jurisprudence and Philosophy* (Oxford: Clarendon Press, 1983), pp. 80–82.

23 Richard Arneson, "Autonomy and Preference Formation," in *In Harm's Way: Essays in Honor of Joel Feinberg*, ed. Jules Coleman and James Buchanan (Cambridge: Cambridge University Press, 1994), pp. 43–75, at 47.

24 Carrie Benson Fischer, "Employee Rights in Sex Work: The Struggle for Dancer's Rights as Employees," *Law and Inequality* 14 (1996): 521–35.

25 Catherine Mackinnon, *Feminism Unmodified: Discourses on Life and Law* (Cambridge, Mass.: Harvard University Press, 1987), ch. 11–16; Rosemarie Tong, *Women, Sex, and the Law* (Totowa, N.J.: Rowman and Allenheld Press, 1984), ch. 1.

26 Peter Singer, *Practical Ethics* (Cambridge: Cambridge University Press, 1993), p. 195; Jonathan Glover, *Causing Death, Saving Lives* (New York: Penguin Books, 1977), p. 192.

27 Timothy E. Quill, "Death and Dignity: A Case of Individualized Decision Making," *NEJM* 329 (1990): 1881–83; James Rachels, *The End of Life* (New York: Oxford University Press, 1986), pp. 152–54; Marcia Angell, "The Supreme Court and Physician-Assisted Suicide – The Ultimate Right," *NEJM* 336 (1997): 50–53, at 51.

28 Byron L. Sherwin, "Jewish Views of Euthanasia," in *Beneficent Euthanasia*, pp. 3–11.

29 Leon Kass, "Neither for Love Nor Money: Why Doctors Must Not Kill," *Public Interest* 94 (Winter 1989): 25–46; Martin Gunderson and David J. Mayo, "Altruism and Physician Assisted Death," *The Journal of Medicine and Philosophy* 18, no. 3 (1993): 281–95 at 289–90; Edmund Pellegrino, "Doctors Must Not Kill," *The Journal of Clinical Ethics* 3, no. 2 (1992): 95–102, at 99–101.

30 H. Tristam Engelhardt, Jr., "Death by Free Choice," in *Suicide and Euthanasia*, ed. Baruch Brody (Boston: Kluwer Academic Publishers, 1989), pp. 251–80.

31 Compassion in Dying v. State of Washington, 49 F.3d 586 (9th Cir. 1995).

32 Yale Kamisar, "Against Assisted Suicide – Even a Very Limited Form," *University of Detroit-Mercy Law Review* 72, no. 4 (1995): 735–69, at 740.

33 Guus de Haas, "Euthanasia and the Legal Situation in the Netherlands," *Nursing Times* 91, no. 20 (1995): 30–31.

34 The arguments for assisted suicide grounded on dignity and beneficence incur some of the same criticisms as autonomy: what constitutes beneficent or dignified action varies according to each individual's view of the good. Assisted suicide is not considered beneficent or dignified for those who esteem the good of innocent human life or fear a precipitous social decline. Therefore proponents of dignity or beneficence must justify their violation of their opponents' view of dignity and beneficence, a view, that is, of the good. To appeal to such formal concepts as beneficence and dignity in justifying assisted suicide is to beg the question of why this peculiar view of the good should be legislated, particularly if others' dignity or sense of beneficence is subverted in the process. Ultimately, the debate must be engaged in terms of goods rather than formal concepts such as dignity, beneficence, or autonomy.

Chapter 59

J.M. Dieterle

PHYSICIAN-ASSISTED SUICIDE: A NEW LOOK AT THE ARGUMENTS

There has long been a debate over the morality of physician-assisted suicide (PAS), both in philosophical literature and in the media. Arguments against PAS come in a variety of forms. One of the most common forms of argument cites the possible negative consequences of the practice as a reason not to legalize it. These arguments rely on empirical claims about the future and thus their strength depends on how likely it is that the predictions will be realized.

Physicians in the Netherlands have actively practiced PAS since 1977 and have done so legally since 1992. The data show that most of the aforementioned empirical predictions have not come to pass. However, it is not clear that the Netherlands can serve as a good test case for the empirical claims, for two reasons. First of all, the Netherlands has a socialized medical system. All citizens have adequate health care coverage, including rehabilitative and nursing home care. Some of the more dire predictions about PAS are less likely to occur against this backdrop than they would be in a free-market health care system such as that in the United States. Secondly, the Netherlands legalized active euthanasia at the same time it legalized PAS. Active euthanasia and PAS differ in an important respect: in cases of PAS, the patient him or herself is the instigator of death, whereas in cases of active

euthanasia, the physician is the instigator of death. Given this, certain dire predicted consequences of the legalization of PAS are more likely to occur in the Netherlands than they would be in a situation where only PAS is legal. So, although the Netherlands can help us with the some of the empirical predictions, it is not an ideal test case for PAS laws.

Fortunately, we now have access to data from a source which is a good test case: Oregon. Oregon's Death with Dignity Act has been in place since 1997. If the empirical predictions have any merit, the data from Oregon should provide evidence to support them.

Other arguments against PAS are based on deontological considerations, such as the intent of the physician or the inherent wrongness of killing. No amount of empirical data will either support or refute these arguments.

This paper serves as a limited defense of PAS. In Part I, I explain Oregon's Death with Dignity Act. I then turn to the arguments against PAS. In Part II, I discuss the consequentialist arguments and their empirical predictions. In Part III, I examine a specific consequentialist argument against PAS – Susan M. Wolf's feminist critique of the practice. In both Parts II and III, my primary source of evidence is Oregon. I also note the evidence from the Netherlands, although this evidence must be considered in context, given the aforementioned difficulties with using the Netherlands as a test case. In part IV, I discuss the two most prominent deontological arguments against PAS. Ultimately, I conclude that no anti-PAS argument has merit. Although I do not provide positive arguments for PAS, if none of the arguments against it are strong, we have no reason *not* to legalize it.

I. Oregon's Death with Dignity Act

Any resident of Oregon who is at least 18 years old, is capable of making and communicating health care decisions, and has been diagnosed with a terminal illness with less than 6 months to live can request a prescription for lethal medication. The request must be made to a licensed Oregon physician.

The following provisions are written into the Oregon law:

- The patient must make two oral requests to his or her physician, separated by at least 15 days.
- The patient must provide a written request to his or her physician, signed in the presence of two witnesses.
- The prescribing physician and a consulting physician must confirm the diagnosis and prognosis.
- The prescribing physician and a consulting physician must determine whether the patient is capable.
- If either physician believes the patient's judgment is impaired by a psychiatric or psychological disorder, the patient must be referred for a psychological examination.
- The prescribing physician must inform the patient of feasible alternatives to assisted suicide including comfort care, hospice care, and pain control.

- The prescribing physician must request, but may not require, the patient to notify his or her next-of-kin of the prescription request.[1]

If all steps are followed and a lethal prescription is written, the prescribing physician must file a report with the Oregon Department of Human Services. Physicians are not required to participate in PAS, nor are pharmacists or health care systems.[2]

II. Projected consequences of PAS

All of the following projected consequences have appeared in literature opposing the legalization of PAS:

1   If we legalize PAS, we will start down the slippery slope to nonvoluntary euthanasia and eventually we will end up with a social policy endorsing involuntary euthanasia.[3]
2   Abuses of the law are likely:
    a   patients might be pressured by family members or insurance companies to seek PAS.[4]
    b   vulnerable groups – the elderly, minorities and the poor – will be more likely to take advantage of PAS, due to discrimination.[5]
    c   people without insurance will request PAS because they don't see that they have other options.[6]
3   The legalization of PAS will corrupt medicine and its practitioners.[7]
4   Acceptance of PAS will weaken the prohibition on killing.[8]
5   Patients will give up too easily – they will abandon hope and kill themselves.[9]
6   Improvements in palliative and terminal care will cease.[10]
7   Citizens will begin to fear hospitals and medical personnel.[11]

I will address each of these proposed consequences in turn.

1. The slippery slope to involuntary euthanasia

Gay-Williams, a proponent of this slippery slope
argument, says:

> A person apparently hopelessly ill may be
> allowed to take his own life. Then he may be
> permitted to deputize others to do it for him
> should he no longer be able to act. . . . It is
> only a short step, then, from voluntary eutha-
> nasia (self-inflicted or authorized), to
> directed euthanasia administered to a patient
> who has given no authorization, to involun-
> tary euthanasia conducted as part of a social
> policy.[12]

We have seen nothing like this happen in
Oregon. The Death with Dignity Act is very
specific:

> In order to receive a prescription for medica-
> tion to end his or her life in a humane and
> dignified manner, a qualified patient shall
> have made an oral request and a written
> request, and reiterate the oral request to his
> or her attending physician no less than fifteen
> (15) days after making the initial oral request.
> At the time the qualified patient makes his or
> her second oral request, the attending physi-
> cian shall offer the patient an opportunity to
> rescind the request.[13]

The patient must request the lethal medication
him or herself. The request must be made three
times – twice orally and once in writing. There
are no provisions whatsoever for substituted
judgment. The law does not and will not allow
nonvoluntary or involuntary euthanasia.

Proponents of this slippery slope argument
might reply that we just have not gotten there

yet; that, eventually, residents of the state of
Oregon will call for the law to be expanded so
that nonvoluntary euthanasia is legally permissible. Eventually involuntary euthanasia will be
legal as well.

Slippery slope arguments are notoriously
fallacious unless one can show that there is a
good reason to think that taking the first step
will actually lead to all of the alleged remaining
steps. The residents of Oregon have taken the
first step. The law has been in place since 1997
and none of the other steps have followed. Is
there evidence at this point to warrant the
conclusion that any of the other steps will
follow?

Critics of PAS often point to the Netherlands
for such evidence. Cases of nonvoluntary euthanasia in the Netherlands have been widely
reported. Each year, roughly 1000 patients die
due to the result of an end-of-life decision made
without their explicit consent.[14] However, this
number can be misleading. The overwhelming
majority of these cases involve non-treatment
decisions (passive euthanasia) and the alleviation of symptoms with possible life-shortening
effects – for example, giving a patient a high
dose of morphine for pain, knowing that it
could hasten death. Both of these practices are
legal and widely practiced outside of the
Netherlands.[15] While there have been cases of
active euthanasia without the patient's explicit
request, a study showed that:

> [D]ecisions to end life without the patient's
> request covered a wide range of situations,
> with a large group of patients having only a
> few hours or days to live, whereas a small
> number had a longer life expectancy but
> were obviously suffering greatly, with verbal
> contact no longer possible. The characteristics
> [of these patients] suggest that most of the
> cases in which life was ended without the
> patient's explicit request were more similar to

cases involving the use of large doses of opioids than to cases of euthanasia.[16]

These are not the sort of horrific examples that PAS opponents often envision. Nevertheless, perhaps we do have reason to be wary. The crucial question here is whether there is evidence that legalization of PAS will lead down the slippery slope to cases of nonvoluntary and perhaps involuntary euthanasia. Since we have documented cases of nonvoluntary active euthanasia in the Netherlands, does this show that the slippery slope argument is cogent?

No. A study of end-of-life decisions in six European countries was published in 2003. Cases involving a physician giving the patient high doses of pain medication to alleviate symptoms, taking into account that this might hasten death, and cases of non-treatment were not grouped with cases of ending life without the patient's explicit request. The findings are significant. 0.6 percent of all deaths in the Netherlands were cases of ending life without the patient's explicit request. However, so were 1.5 percent of deaths in Belgium, 0.67 percent of deaths in Denmark, and 0.42 percent of deaths in Switzerland.[17] A similar study in Australia showed that 3.5 percent of all deaths in that country occurred as a result of a physician ending a patient's life without that patient's explicit request.[18] PAS was not legal in any of these countries except the Netherlands at the time of the studies. What these studies seem to show, then, is that non-voluntary euthanasia is widespread. It is not a consequence of PAS laws.

Furthermore, and perhaps more importantly, with legalization in the Netherlands we actually see a *decrease* in the willingness of physicians to perform euthanasia without the explicit request of a patient. PAS and active euthanasia were legalized in 1991; the law went into effect in 1992:

The proportion of physicians who were ever engaged in the ending of life without a patient's explicit request decreased from 27% in 1990 to 23% in 1995, and further to 13% in 2001. Furthermore, physicians unwillingness to ever do so increased, especially after 1995, from 45% in 1995 to 71% in 2001.[19]

With legalization, we also get the following results: 'In 2001/2002, physicians reported more often than in previous years that they had become more restrictive about euthanasia, and less often that they had become more permissive,'[20] and 'The requirements of due care are being met more extensively than previously, and public control has increased further'.[21]

So, if the Netherlands offers us any evidence at all about this slippery slope, it seems to be that legalization allows one to get a foothold and climb back up. And this makes sense. If the practice is legal and out in the open, it is easier to put safeguards in place and make sure they are followed.

2. Abuses of the law are likely

(a) Patients might be pressured by family members or insurance companies to seek PAS[22]

There have been no allegations that insurance companies have pressured anyone into PAS. There is no decisive evidence that anyone in Oregon has gone through with PAS because of pressure from family members. However, PAS opponents have used a case involving an 85 year old woman dying of terminal cancer to argue that the system has been abused. In 1999, Kate Cheney requested a prescription for lethal medication. Her physician arranged for her to undergo a psychiatric evaluation, to make sure that she was competent to make the decision to end her life. 'The psychiatrist noted that although assisted suicide seemed consistent with Kate's values throughout her life, "she does not seem to be explicitly pushing for this" '.[23] The psychiatrist's evaluation concluded that Ms. Cheney did not have the level of competency needed to make the decision. Ms. Cheney's daughter, Erika, was actively involved in the process and became angry over the competency decision.

Ms. Cheney chose to have a second competency evaluation, this time by a clinical psychologist. The psychologist concluded that 'Kate's "choices may be influenced by her family's wishes and her daughter, Erika, may be somewhat coercive". But she wrote that Kate "demonstrated the capacity to weigh the differences and articulate her own values" '.[24] The psychologist deemed Ms. Cheney competent.

Opponents of Oregon's Death with Dignity act have used this case as evidence of abuse. For example, Wesley J. Smith writes:

Earnest euthanasia advocates – generally abetted by a compliant media – spun the myth that assisted suicide would invariably be a rational 'choice,' strictly regulated by the state, a last resort of dying patients when nothing else could be done to alleviate their suffering. But the more we learn about how doctor-facilitated death is actually being practiced in Oregon, the clearer it becomes that these assurances were false.[25]

Smith goes on to talk about the Cheney case and concludes that Kate Cheney was coerced by her daughter into taking the lethal medication. But Smith leaves much out of his account. Kate Cheney agreed to tell her story to the Oregonian so that others could understand how the law works. Ms. Cheney is quoted as saying, '[Erika] makes more noise than I do. But that doesn't make me any less serious'.[26] After the second competency evaluation but before the prescription was written, Ms. Cheney's physician consulted with

other family members. A grand daughter, Pat, was called in. She stressed that the decision had been Ms. Cheney's – not anyone else's. Later, after being given the prescription, Kate says, 'It's not that I have any plans to use it right away. I may never use it. But it's just to have the decision in my hands, rather than anyone else's'.[27] Furthermore, Ms. Cheney's physician was convinced both that she was competent to make the decision and that she was acting on her own beliefs and desires. He says, 'I had no reason to believe that this was anyone's agenda but her own'.[28] We are not entitled to conclude, as Smith suggests, that this is a case of coercion. While I admit that there is a possibility that this case turned on pressure from a family member, it looks more like a case where a daughter was playing the role of advocate for her mother.

Between 1998 and 2004, 208 people died after ingesting a lethal dose of medication prescribed under Oregon's Death with Dignity Act. Of those 208 people, 74 (36 percent) cited 'burden on family, friends/caregivers' as one of their reasons for seeking PAS.[29] But this statistic can be misconstrued. It does not show that the family actually put pressure on the patient; it merely shows that the patient did not want to be a burden to the family. Furthermore:

Physicians indicated that patient requests for lethal medications stemmed from multiple concerns with nine in 10 patients having at least three concerns. The most frequently mentioned end-of-life concerns during 2004 were: a decreasing ability to participate in activities that made life enjoyable, loss of autonomy, and a loss of dignity.[30]

A survey of Oregon physicians noted that, 'A request for assistance with suicide was less likely to be honored if the patient perceived himself or herself to be a burden to others'.[31] The survey shows that the physicians 'were reluctant to accede to requests for assistance under these circumstances'.[32]

It is worth noting that in all 50 states in the US, patients have the right to refuse treatment and be allowed to die. Any competent adult can refuse life-saving intervention. There are no controls on voluntary passive euthanasia. Furthermore, all 50 states have procedures in place for allowing substituted judgment for the refusal of treatment. Nonvoluntary passive euthanasia is practiced every day. While there are controls in place for this practice, the controls vary from state to state – some are far more stringent than others. There is thus much more potential for abuse of passive euthanasia laws than for PAS, given the controls Oregon has in place. A family member could, for example, put pressure on a patient to refuse treatment. A family member could refuse treatment on behalf of an incompetent patient, even if that wouldn't be what the patient would have wanted.

Some might object that my comparison is unfair. Allowing a patient to die is respecting a negative right (a right of non-interference), it might be objected, whereas the 'right' to PAS would be a positive right – a right to be provided with lethal medication. Positive rights can justifiably be limited because they involve a claim on the resources of others. Negative rights involve no such claim and are thus not limited in the same way. The abuse argument is strong enough to override the positive right but not the negative right.

This objection misconstrues PAS policies. PAS laws do not grant patients positive rights against physicians to be given lethal prescriptions. No resident of Oregon has a valid claim against any physician with regard to such a prescription. Instead, PAS laws allow willing physicians to prescribe lethal medication.[33] Thus, PAS laws can be seen as granting physicians negative rights of non-interference, but no positive right is invoked whatsoever.

Furthermore, the potential abuse argument is a consequentialist argument. Opponents of PAS cite the possibility of abuse as the reason or one of the reasons for not legalizing the practice. But passive euthanasia is equivalent to PAS with regard to the possibility of abuse. On purely consequentialist grounds, we cannot distinguish between the two.

I do not mean to discount the very real possibility of abuse of PAS laws. However, every social policy has the potential for abuse. When other important rights are at stake, we cannot let the mere potential for abuse of the law keep us from enacting said law. We have to try to foresee those possible abuses and put controls in place so that the abuses rarely, if ever, occur. And it seems that this is just what Oregon has done. Ann Jackson, executive director of the Oregon Hospice Association, was initially opposed to the law. In an interview in March 2005, she said, 'The fears we had about not enough safeguards are unfounded'.[34]

(b) Vulnerable groups – the elderly, minorities and the poor – will be more likely to take advantage of PAS, due to discrimination

This prediction has turned out to be off the mark. In Oregon, 'Terminally ill younger persons were significantly more likely to use PAS than their older counterparts'.[35] The age range of the 208 people who died after ingesting a lethal medication prescribed under Oregon's Death with Dignity Act between 1998 and 2004 was 25–94, with only 16 (8 percent) over the age of 85. The median age was 69.[36] 203 of the 208 patients were white and 5 were Asian.[37] These patients were not poor or uneducated. 61 percent of them had at least some college.[38] The statistics on those seeking PAS parallel the statistics on those who actually died after ingesting the lethal medication. 97 percent were white and the median age was 68.[39]

Data from the Netherlands is further evidence that this prediction is off the mark. The 2001 report on PAS and euthanasia states that 'There are no signs indicating an increase in life-terminating treatment among vulnerable patient groups'.[40] In fact, euthanasia was most common among those with the highest socio-economic status.[41]

(c) People without insurance will request PAS because they don't see that they have other options[42]

Again, the statistics do not bear out this prediction. 129 of the 208 in Oregon had private health insurance (63 percent) and 74 were covered by Medicare or Medicaid (36 percent). Only two (1 percent) of the 208 had no health insurance at all.[43] Of those seeking PAS, only 2 percent lacked insurance.[44]

3. The legalization of PAS will corrupt medicine and its practitioners

Gay-Williams says that euthanasia:

> could have a corrupting influence so that in any case that is severe doctors and nurses might not try hard enough to save the patient. They might decide that the patient would simply be 'better off dead' and take the steps necessary to make that come about. This attitude could then carry over to their dealings with patients less seriously ill. The result would be an overall decline in the quality of medical care.[45]

This is, of course, another slippery slope argument. We have no reason to believe that taking the first step of legalizing PAS will lead to any of the other steps. Without evidence of the inevitable slide down the slope, we have no reason to believe we can't stand at the top. And, in fact,

there is evidence that terminal care has actually improved in both Oregon and the Netherlands. After passage of the Oregon law, 88 percent of those physicians responding to a survey said that they had 'sought to improve their knowledge of the use of pain medications in the terminally ill "somewhat" or "a great deal".'[46] 86 percent said that 'their confidence in the use of pain medications in the terminally ill had improved "somewhat" or "a great deal".'[47]

The Dutch government launched several initiatives in the 1990s to improve terminal care. A study published in 2003 cites rapid improvement over the last decade, with further governmental measures in place to spur additional improvements.[48]

4. Acceptance of PAS will weaken the prohibition on killing

If this were the case, then we would expect to see an increase in the homicide rate in Oregon since passage of the law. And, in fact, we have not seen such an increase. Instead, the homicide rate has gone down. In 1985, the homicide rate in Oregon was 4.7 per 100,000 residents. In 1995, it was 4.1. The Death with Dignity Act was passed in 1997; the homicide rate for that year was 2.9. In 2000, the rate was 2.0.[49] The violent crime rate has also been declining significantly. In 1985, the violent crime rate was 551.1 per 100,000 residents. In 1995, it was 522.4. In 1997, the rate was 444.4 and in 2000 it was 350.7.[50]

We cannot, of course, isolate the causal factors involved in the decline in homicide and violent crime rates in Oregon. The crime rate dropped throughout the United States during the time period in question. A proponent of this argument might counter that the rates might have dropped more had it not been for the Death with Dignity Act. But then we would expect the corresponding crime rates in other states – or at least neighboring states – to have dropped significantly more than Oregon's. But, again, this is not the case.[51]

The homicide rate in the Netherlands is also very low. In 2000, there were 1.42 homicides per 100,000 residents. 25 countries had higher homicide rates, including Canada (1.76), Belgium (1.54) and Sweden (1.97).[52]

5. Patients will give up too easily – they will abandon hope and kill themselves

This is another prediction that has turned out to be way off the mark. In 2004, PAS accounted for about 1 in 800 deaths in Oregon.[53] In the seven years between 1998 and 2004, 208 terminally ill patients took lethal medications while 64,706 other Oregonians were dying of the same diseases.[54] Terminally ill patients are thus not going out in droves and killing themselves. The law is available for those who choose to use it.

The figures are higher in the Netherlands, but it is still not the case that hordes of people are giving up and killing themselves. Roughly 140,000 people died in the Netherlands in 2000. Approximately 55,000 of them died after a prolonged illness. 2054 were cases of PAS or active euthanasia.[55]

Furthermore, and most importantly, the people seeking PAS are not depressed and devoid of hope:

> A [. . .] surprise has been the kind of people who use the [Oregon] law. They are not so much depressed as determined, said Linda Ganzini, a professor of psychiatry at Oregon Health Sciences University. She led a recent survey of 35 doctors who had received requests for suicide drugs. The doctors described the patients as 'feisty' and 'unwavering'.[56]

And, in fact, there is anecdotal evidence that the existence of a law allowing PAS actually gives people more hope. Because Oregonians and citizens of the Netherlands know that, should life become unbearable, they have the option of PAS, they can focus on the life that they have left. Instead of worrying about a future of unbearable pain and diminished autonomy, they can focus their energy on more productive endeavors.[57]

6. Improvements in palliative and terminal care will cease

This proposed consequence is cited by Steven G. Potts:

> If euthanasia had been legal 40 years ago, it is quite possible that there would be no hospice movement today. The improvement in terminal care is a direct result of attempts made to minimise suffering. If that suffering had been extinguished by extinguishing the patients who bore it, then we may never have known the advances in the control of pain, nausea, breathlessness and other terminal symptoms that the last twenty years have seen.[58]

The assumption behind Potts' assertion is that once we introduce PAS, everyone (or most people) will make use of it. But we have seen that this is not the case. Only 208 people died after ingesting lethal medication under Oregon's Death with Dignity Act between 1998 and 2004. 64,706 Oregonians dying of the same diseases did not choose PAS.[59]

The overwhelming majority of the 208 patients – 86 percent – were enrolled in hospice programs.[60] Additionally, as noted earlier, physicians attempted to improve their knowledge on end-of-life care after passage of the law including the use of pain medications in the terminally ill.[61] Finally, Oregon's Seventh Annual Report on the Death with Dignity Act notes that the availability of PAS 'may have spurred Oregon doctors to address other end-of-life concerns more effectively'.[62]

Palliative care has also improved significantly in the Netherlands. A government sponsored report on end-of-life care in the Netherlands notes that 'During the nineteen-nineties social and political priority for the palliative care of terminally ill people increased significantly'.[63] Three separate initiatives were launched in the 90s, whose main goals were:

1   research into and innovation in palliative care
2   stimulation and guidance of palliative care
3   integration of hospice facilities in regular health care.[64]

Thus there is no reason to believe that improvements in terminal and palliative care will cease with passage of PAS laws.

7. Citizens will begin to fear hospitals and medical personnel

The assumption here is that once a physician is legally permitted to assist in dying, citizens will become fearful of going to see their physician or going to the hospital, because they will be afraid that they will be killed against their will. This is not just a restatement of the slippery slope argument discussed earlier. It is a prediction about the psychological state of the populace – whether their fears are justified is quite a different matter. However, we have seen no evidence of this kind of fear in the citizens of Oregon, nor have we seen it in the Netherlands.

I have now considered all of the proposed consequences listed at the beginning of this section. The evidence has shown that nearly every prediction was wrong. The only prediction for which there is even a shred of evidence is (2a), that patients might be pressured by family

members to seek PAS. However, even here we have only one purported case and the evidence in that case does not warrant the conclusion that the patient in question was coerced. A more likely explanation, given the testimony of the patient herself and her physician, is that the daughter was advocating on behalf of her mother.

Before we move on, however, there is an objection I have not considered, i.e. there might very well be an underground practice of PAS in Oregon.[65] We know that PAS and active euthanasia occur underground elsewhere. The important question, then, is whether such underground practices are more likely to occur in Oregon than in places where PAS is not legal. If the answer to this question is yes, then this would give us reason to be wary of legalizing PAS. On the other hand, if the answer is no, then this objection would not constitute a reason against legalizing PAS (since PAS laws would not be the cause of such practices).

As yet, there has been no reputable study of any underground PAS practices in Oregon. However, we do have evidence that PAS and active euthanasia occur underground across the US. A survey done in 1996 showed that 11 percent of physicians 'reported that under current legal constraints, there are circumstances in which they would prescribe a medication for a competent patient to use with the primary intention of ending his or her life'.[66] Note that in 1996, PAS was illegal in all 50 states. The same study showed that 3.3 percent had actually written a prescription for a lethal dose of medication and 4.7 percent had administered a lethal injection.[67]

Earlier, I cited a study of end-of-life decisions in six European countries. This study showed that PAS occurred in all of the six except Italy, and it occurred more often in Switzerland than in the Netherlands.[68] PAS is not legal – and thus, all cases are underground – in Switzerland. The same study showed that ending a patient's life without his or her explicit request was more common in Belgium and Denmark than in the Netherlands.[69] PAS is not legal in Belgium or Denmark. Furthermore, the authors note:

> We cannot exclude the possibility that non-response has to some extent affected our results, especially for Italy. Whereas under-reporting of socially undesirable behavior is a more general occurrence in sociological research than over-reporting, such bias, if present, will probably result mainly in conservative estimates of the rates of end-of-life decisions.[70]

Additionally, there was a significant decrease in the willingness of physicians to end the life of a patient without his or her explicit request with legalization in the Netherlands. Another study showed that the rates of non-voluntary euthanasia are higher in Australia than the Netherlands.[71] PAS is not legal in Australia.

This evidence shows that the underground occurrence of PAS and active euthanasia is widespread, and it does offer *some* support for a negative answer to our question of whether underground practices are more likely to occur in places where PAS is legal. However, we cannot draw any definitive conclusions from the data available. Note, though, that if PAS is legally available to those who seek it, patients who wish to end their lives on their own terms do not have to go underground to find physicians who are willing to prescribe a lethal dose of medication. Furthermore, if the practice is out in the open, it is easier to put safeguards in place and make sure they are followed.

III. Susan Wolf's feminist critique of PAS

Susan Wolf has argued that, due to cultural ideology, socialization and sexism, the practice of PAS may be gendered. Women may be harmed by PAS in ways that men would not be harmed. She warns that PAS should not be legalized, at least not under the current cultural conditions.[72] Wolf says:

> Indeed, it would be surprising if gender had no influence. Women in America still live in a society marred by sexism, a society that particularly disvalues women with illness, disability, or merely advanced age.[73]

She predicts a number of gender effects of PAS. The first effect has to do with who would seek PAS. Wolf's article was written in 1996, before Oregon's law was in place. Her prediction was that women would seek PAS more often than men.[74] This has turned out to be false. 52 percent of patients who died after ingesting lethal medication under Oregon's Death with Dignity Act were male, 48 percent were female.[75] The statistics for those seeking PAS are identical to those who actually died of PAS: 52 percent male, 48 percent female.[76] Data from the Netherlands is similar – 55 percent of those who requested PAS or euthanasia in the Netherlands between 1977 and 2001 were male, 45 percent female.[77]

The second gender effect predicted by Wolf is that women may have different reasons for requesting PAS than men. Because women, historically, are less likely to receive adequate pain relief,[78] have a higher rate of depression,[79] are more likely than men to be poor,[80] and are more likely than men to lack adequate health insurance,[81] these issues may figure prominently in their decision to request PAS. However, as we saw earlier, both those who sought PAS and those who actually ingested the lethal medication were not, for the most part, poor or without health insurance.[82] Oregon's Seventh Annual Report contains a list of end-of-life concerns given by the 208 patients who ingested lethal medication. Unfortunately, the responses are no broken down by gender. We cannot tell from this list which concern was had by whom, nor can we tell where on the list of concerns any particular item fell for any particular patient (just one patient in 2004 had only one end-of-life concern – the remaining 36 had at least two and up to six concerns[83]). However, we can see that only 45 patients (22 percent) between 1998 and 2004 listed 'inadequate pain relief or concern about it' and only 6 patients (3 percent) listed 'financial implications of treatment'.[84] The three most common concerns were 'Losing autonomy,' 'Less able to engage in activities making life enjoyable,' and 'Loss of dignity'.[85]

The issue of depression is more difficult to assess. According to a survey done of Oregon physicians, 20 percent of those seeking PAS had symptoms of depression.[86] However, this statistic may be misleading because physicians might not recognize the symptoms. Perhaps the best way to address this concern is to ensure that physicians who participate in PAS are educated on the symptoms of depression and are aware that depression affects women more often than men. And, in fact, a survey of Oregon physicians reveals that 76 percent had 'sought to improve their ability to recognize psychiatric illnesses such as depression in the terminally ill "somewhat" or "a great deal" ' after passage of the law.[87]

Wolf cites another reason women might seek PAS more often than men: women have been socialized to be self-sacrificing. Because of this, Wolf argues, women's consent might not be truly autonomous. A woman might request PAS when it isn't really what she wants. Or, perhaps it wouldn't be what she would want had she not been influenced by cultural gender ideology.

Most assuredly, gender socialization and sexism play an enormous role in influencing the choices women make in our culture. Furthermore, the choice to end one's life is final and irrevocable. We want to make certain that autonomous consent is present. However, this

granted, a wholesale ban on PAS does not seem warranted. Instead, we ought to be trying to ensure that physicians participating in PAS are aware of gender issues and consider them when determining candidacy for PAS. After all, if we were to ban PAS because of these considerations, what we would, in effect, be paternalistically saying to women is, 'We won't let you make the choice to end your life as you see fit because we think that cultural conceptions of gender have unduly influenced your decision making process. Were it not for these cultural conceptions, you would not want to die'.[88]

The third gender effect cited by Wolf is that gender will play a role in physicians' decisions regarding PAS. When a female patient requests PAS, Wolf predicts, physicians will be more likely to grant her request than they will a similar request made by a male. The reason is twofold. First of all, in granting women's requests for PAS, physicians will be, according to Wolf, 'affirming women's negative self-judgments'.[89] Secondly, the *psychological dynamics* between male-physician and female-patient may make the physician more likely to grant the request. These dynamics 'may be a complex combination of rescue fantasies and the desire to annihilate'.[90]

Were Wolf correct about these predictions, we would expect the proportion of women dying after ingesting lethal medication to women requesting the prescription to be much higher than that of men. In other words, if physicians are more likely to grant a woman's request for PAS, then we would expect more women who requested PAS to be given the prescription than men who requested PAS. But we do not see such a trend. The statistics for those seeking PAS are identical to those who actually died of PAS: 52 percent male, 48 percent female.[91]

Ultimately, Wolf's critique of PAS gives us reason to be cautious about the influence of gender on PAS decisions. It does not, however, give us reason to avoid legalizing it.

IV. Deontological arguments

In this section, I consider the two most prominent deontological arguments against PAS.

1. The physician's intent

> Doctors should not kill; this is prohibited by
> the Hippocratic Oath. The physician is bound
> to save life, not take it.[92]

It is important to note that if PAS is wrong, its
wrongness cannot be *constituted* by its conflict
with the Hippocratic Oath. After all, the
Hippocratic Oath itself is just a bunch of words.
Without moral reasons to back them up, those
words cannot dictate medical ethics or physicians' duties.[93] Put another way, if PAS is, in fact,
wrong, it cannot be just because the Hippocratic
Oath forbids it. Instead, the Hippocratic Oath
must forbid PAS because it is wrong. Thus, to
show that a physician should not participate
in PAS, one must show more than just that
the practice conflicts with the Oath.

The idea behind the argument is that the
intent of a physician should never be to kill; the
intent should always be to heal or cure. In cases
where PAS would even be an option, healing or
curing is no longer possible. What should the
intent be in such cases? The intent should be to
ease the patient's suffering, to do what is in the
best interests of the patient, or to respect the
patient's autonomy. But that is exactly what PAS
does. To say that the physicians who practice PAS
intend to kill their patients is to misunderstand
their intent.

Passive euthanasia occurs when we allow
someone to die. A case counts as euthanasia only
if it were possible to prevent death – if some
intervention could have been done to save the
patient that was not done – and the patient is
allowed to die. In such cases, we do not say
that the intent of the physician is to kill the
patient, even though the patient died because
of an omission on the physician's part. We
deem the omission permissible because we
think that the patient is better off (or no worse
off) dead.[94] In such cases, we recognize that the
physician's intent is to respect the patient's
wishes, to ease suffering, or to do what is in the
best interests of the patient. We should recognize
that this is the physician's intent in cases of PAS
as well.

2. The inherent wrongness of killing

The argument: killing an innocent person is
inherently wrong. The physician indirectly kills
the patient in cases of PAS by providing the
lethal medication. So, the physician is doing
something inherently wrong.[95]

Certainly most cases of killing innocents are
wrong, but most cases of killing innocents
are cases in which the context is vastly different
from cases of PAS. Death is not desired or imma-
nent in most cases of killing innocents. PAS differs
greatly from such wrongful killings, because (at
least in Oregon) the patient has requested the
medication at least three times, the patient is
terminally ill with less than 6 months to live, and
the patient thinks that death would be better than
the life he or she is living. In short, death is not a
harm to those who seek and are granted PAS.[96] To
claim that it is nonetheless wrong begs for further
argumentation. Why is it wrong?

One argument that is fairly often heard is
religious: killing is against God's will.[97] However,
we live in a pluralistic society. No one should be
bound by the dictates of another's religion and
thus religious arguments have no place in public
policy debates.

A second argument for the wrongness of
killing is found in Gay-Williams' piece. He says:

> Every human being has a natural inclination
> to continue living. Our reflexes and responses

fit us to fight attackers, flee wild animals, and dodge out of the way of trucks. In our daily lives we exercise the caution and care necessary to protect ourselves. Our bodies are similarly structured for survival right down to the molecular level.[98]

Euthanasia or PAS, Gay-Williams continues, 'does violence to this natural goal of survival'.[99] From this he concludes that it is wrong.

There are two things to note here. First of all, to draw a normative conclusion from a description of the workings of nature is fallacious. One cannot validly conclude that something ought to be the case, morally, just because it works that way in nature. Secondly, it is difficult to see how these considerations apply to those who seek and would be candidates for PAS. They are terminally ill and they have less than six months to live. Survival just is not possible. They want to end their lives in the way that they see fit, not lingering on, in tremendous pain as 'nature' would have it.

In the absence of a viable secular argument for the claim that killing innocents is always wrong, I see no reason to accept it. Certainly killing innocents is wrong in most cases. But in those cases where the person is terminally ill and wants to die – that person has autonomously chosen to die – it does them no harm to assist them in dying. Death is not a harm to these persons. If it is not a harm, then it is difficult to see how it could be wrong.

V. Conclusion

This paper examined the arguments against PAS. We considered the consequentialist arguments against the backdrop of both Oregon and the Netherlands. We saw that the overwhelming majority of predicted consequences of PAS laws have not come to pass. Only one predicted consequence – that family members might put pressure on patients to end their lives – had even a shred of evidence in its favor, and that evidence is far from conclusive. The one case cited could just as easily be seen as a case of a family member advocating for the patient. We also considered Susan Wolf's feminist analysis of PAS and saw that her predicted consequences with regard to gender have not come to pass. Wolf's analysis gives us reason to be cautious of gender effects of PAS, but it does not warrant a wholesale ban on the practice. Finally, we considered the two most prominent deontological arguments against PAS. Neither of those arguments is convincing.

I haven't advanced any positive arguments for PAS, but this paper can nonetheless be seen as a partial defense of the practice.[100] If none of the arguments against PAS are strong, then we have no reason not to legalize it.[101]

Notes

1 Bulleted points are quoted directly from the Seventh Annual Report on Oregon's Death with Dignity Act. Oregon Department of Human Services, Office of Disease Prevention and Epidemiology. March 10, 2005, p. 8.

2 Seventh Annual Report, op. cit. note 1, p. 8.

3 This consequence is cited by: D. Brock. 1992. "Voluntary Active Euthanasia." Reprinted in W.H. Shaw, ed. 2005. Social and Personal Ethics. Belmont, CA: Wadsworth: pp. 79–89. J. Gay-Williams. 1992. "The Wrongfulness of Euthanasia." Reprinted in J. Olen et al. eds. 2003. Applying Ethics, 8th Edition. Belmont, CA: Wadsworth: pp. 180–182. Y. Kamisar. 1958. "Against Legalizing Euthanasia." Reprinted in L. Pojman, ed. 2000. Life and Death, 2nd Edition. Belmont, CA: Wadsworth: pp. 176–183.

4 J. Arras. 1997. "Physician Assisted Suicide: A Tragic View." Reprinted in B. Steinbock et al., eds. 2003. Ethical Issues in Modern Medicine, 6th Edition. Boston: McGraw Hill: pp. 394–400. D. Callahan. 1992. "When Self-Determination Runs Amok." Reprinted in R. Munson, ed. 1996. Intervention and

Reflection: Basic Issues in Medical Ethics, 5th Edition. Belmont, CA: Wadsworth: pp. 175–180. See also S. Potts. 1988. "Objections to the Institutionalization of Euthanasia." Reprinted in J.E. White, ed. 2003. *Contemporary Moral Problems*. Belmont, CA: Wadsworth: 218–221.

5   Arras, *op. cit.* note 4, p. 397.

6   Potts, *op. cit.* note 4, p. 220.

7   Gay-Williams, *op. cit.* note 3, p. 182.

8   Brock, *op. cit.* note 3, p. 86. Potts, *op. cit.* note 4, p. 219.

9   Potts *op. cit.* note 4, p. 218.

10   Ibid.

11   Ibid.

12   Gay-Williams, *op. cit.* note 3, p. 182.

13   Text of the Oregon Death With Dignity Act, available at http://egov.oregon.gov/DHS/ph/pas/ors.shtml [Accessed 12 November 2006].

14   B. Onwuteaka-Philipsen et al. "Euthanasia and Other End-of-Life Decisions in the Netherlands in 1990, 1995, and 2000," *Lancet*. 2003. Available at http://image.thelancet.com/extras/03art3297web.pdf [Accessed 12 November 2006].

15   See A. van der Heide et al. "End-of-Life Decision Making in Six European Countries: Descriptive Study for data on Switzerland, Denmark, Sweden, Belgium, Italy and the Netherlands," *Lancet*. 2003. Available at http://image.thelancet.com/extras/03art3298web.pdf [Accessed 12 November 2006].

16   P. van der Maas et al. "Euthanasia, Physician Assisted Suicide, and Other Medical Practices Involving the End of Life in the Netherlands, 1990–1995," *N Engl J Med* 1996; 335: 1699–1706.

17   van der Heide, *op. cit.* note 15.

18   H. Kuhse et al. "End-of-Life Decisions in Australian Medical Practice," *Med J Aust* 1997; 166; 191–196.

19   Onwuteaka-Philipsen, *op. cit.* note 14.

20   From the Third 'Remmelink' Report – the English summary of the 2001 report on euthanasia in the Netherlands. See www.worldrtd.net/news/world/?id=587 [Accessed 12 November 2006].

21   Ibid.

22   Data from the Netherlands is not relevant here, since nursing home care is free and open to all. Neither insurance companies nor families are financially liable for the patient's care.

23   E. Barnett. "A Family Struggle: Is Mom Capable of Choosing to Die?" *Oregonian*, Oct. 17, 1999.

24   Ibid.

25   W.J. Smith. *Oregon Assisted Suicide Abuses*. Available at www.lifeissues.net/writers/smit/smit_18oregonassistedsui.html [Accessed 12 November 2006].

26   Barnett, *op. cit.* note 23.

27   Ibid.

28   Ibid.

29   Seventh Annual Report, *op. cit.* note 1, p. 24.

30   Seventh Annual Report, *op. cit.* note 1, p. 5.

31   L. Ganzini et al. "Physicians Experiences with the Oregon Death with Dignity Act," *N Engl J Med* 2000; 342: 557–564.

32   Ibid.

33   D. Raymond makes a similar point in " 'Fatal Practices': A Feminist Analysis of Physician Assisted Suicide and Euthanasia," *Hypatia* 1999; 14(2): 1–25.

34   M. Vitez. "Oregon is the Laboratory in Assisted-Suicide Debate," *Philadelphia Inquirer*, March 13, 2005.

35   Seventh Annual Report, *op. cit.* note 1, p. 13.

36   Ibid: 22.

37   Ibid. I reserve discussion of women until the next section.

38   Ibid.

39   Ganzini, *op. cit.* note 31.

40   Third 'Remmelink' Report, *op. cit.* note 20.

41   Ibid.

42   Data from the Netherlands is not relevant here, since they have a system of socialized medicine.

43   Seventh Annual Report, *op. cit.* note 1, p. 24. The numbers do not add up to 208 because there is no data available for three of the patients.

44   Ganzini, *op. cit.* note 31.

45   Gay-Williams, *op. cit.* note 3, p. 182.

46   Ganzin, *op. cit.* note 31.

47   Ibid: 2–3.

48   A.L. Francke. 2003. *Palliative Care for Terminally Ill Patients*. International Publication Series, Ministry of Health, Welfare and Sport, No. 16.

49   www.disastercenter.com/crime/orcrime.htm [Accessed 12 November 2006].

50   Ibid.

51 See www.disastercenter.com/crime/ [Accessed 12 November 2006] for crime rates for the United States and for each individual state.

52 www.angelfire.com/rnb/y/homicide.htm#murder [Accessed 12 November 2006].

53 *Seventh Annual Report, op. cit.* note 1, p. 16.

54 Ibid: 5.

55 Francke, *op. cit.* note 48.

56 J. Schwartz & J. Estrin. "In Oregon, Choosing Death Over Suffering," *The New York Times,* June 1, 2004.

57 See www.deathwithdignity.org/fss/stories.asp [Accessed 20 June 2005]. For anecdotal evidence regarding the Netherlands law, see the PBS Frontline documentary *An Appointment with Death* (2002).

58 Potts, *op. cit.* note 4, p. 218.

59 *Seventh Annual Report, op. cit.* note 1, p. 5.

60 Ibid: 24.

61 See above, Section II. 3.

62 *Seventh Annual Report, op. cit.* note 1, p. 17.

63 Francke, *op. cit.* note 48, p. 14.

64 Ibid: 15.

65 I thank Georg Bosshard for this point.

66 D.E. Meier et al. "A National Survey of Physician-Assisted Suicide and Euthanasia in the United States," *N Engl J Med* 1998; 338: 1193–1202.

67 Ibid.

68 van der Heide, *op. cit.* note 15, p. 3.

69 Ibid: 2.

70 Ibid: 5.

71 Kuhse, *op. cit.* note 18.

72 S.M. Wolf. 1996. "A Feminist Critique of Physician Assisted Suicide." Reprinted in J.E. White, ed. 2003. *Contemporary Moral Problems.* Belmont, CA: Wadsworth: 222–234. Wolf cites four possible consequences of PAS, of which I discuss three. The fourth has to do with society's conception of PAS, given that most of the media attention has been on female patients.

73 Ibid: 222.

74 Ibid: 223.

75 *Seventh Annual Report, op. cit.* note 1, p. 20.

76 Ganzini, *op. cit.* note 31.

77 R.L. Marquet et al. "Twenty Five Years of Requests for Euthanasia and Physician Assisted Suicide in Dutch General Practice: Trend Analysis," *Br Med J* 2003; 327: 201.

78 Wolf cites P. Steinfels, "Help for Helping Hands in Death," *The New York Times,* February 14, 1993, sect. 4, pp. 1, 6.

79 "Women are twice as likely as men to suffer from depression." New Report on Women and Depression: Latest Research Findings and Recommendations. Available at www.apa.org/releases/depressionreport.html [Accessed 12 November 2006].

80 See www.census.gov/hhes/www/poverty.html [Accessed 12 November 2006].

81 See www.census.gov/hhes/wvvw/poverty03.html [Accessed 30 June 2005].

82 See above, section II 2b.

83 *Seventh Annual Report, op. cit.* note 1, p. 15.

84 Ibid: 24.

85 Ibid.

86 Ganzini, *op. cit.* note 31.

87 Ibid.

88 See Raymond, *op. cit.* note 33, for a feminist response to Wolf's argument. Raymond offers an account of female autonomy on which autonomous consent to PAS is possible.

89 Wolf, *op. cit.* note 72, p. 223.

90 Ibid: 228.

91 Ganzini, *op. cit.* note 31.

92 M. Pabst Battin. 1995. "Physician-Assisted Suicide." Reprinted in M.C. Brannigan & J. Boss, eds. 2001. *Healthcare Ethics in a Diverse Society.* Mountain View, CA: Mayfield: 524–532. Pabst Battin is not, herself, endorsing this argument. Quote taken from p. 525. See also Potts, *op. cit.* note 4, p. 219.

93 If we took the original Hippocratic Oath to be constitutive of a physician's ethical duties, then surgery would be unethical, as Pabst Battin notes (*op. cit.* note 79). The Oath says, 'I will not use the knife, not even on sufferers from stone, but will withdraw in favor of such men as are engaged in this work'. For the full text, see www.pbs.org/wgbh/nova/doctors/oath_classical.html [Accessed 12 November 2006].

94 J. Rachels. 1986. "Active and Passive Euthanasia," in *The Elements of Moral Philosophy.* Boston: McGraw Hill: 90–103.

95 See Pabst Battin, *op. cit.* note 92, p. 524. She does not endorse this argument.

96 See P. Singer. 1993. "Justifying Voluntary Euthanasia," in *Practical Ethics*. Cambridge: Cambridge Univ. Press: 176–200.

97 The only occurrence of this argument that I could find in the philosophical literature occurs in Gay-Williams, *op. cit.* note 3, p. 181. It often appears in popular arguments against PAS. Examples: www.bfl.org/cbb/assisted_suicide. htm [Accessed 22 June 2005], www.epm.org/ articles/physuici.html [Accessed 12 November 2006], www.leaderu.com/orgs/probe/docs/ euthanas.html [Accessed 12 November 2006].

98 Gay-Williams, *op. cit.* note 3, p. 181.

99 Ibid.

100 For an excellent defense of PAS, see Dworkin et al. 1997. "Assisted Suicide: The Philosophers' Brief," *The New York Review of Books*, Vol XLIV, No. 5, March 27, 1997.

101 I thank the students in my Spring 2005 Medical Ethics course, especially Randy Keller and Linda Verhun, for helpful discussions. I also thank Lori Watson for discussion on issues in this paper, especially Susan Wolf's argument. I thank Georg Bosshard and Willem Landman for helpful comments on an earlier draft. Finally, I thank Eric Buckhave for his support.

Chapter 60

Judith Jarvis Thomson

PHYSICIAN-ASSISTED SUICIDE: TWO MORAL ARGUMENTS*

What I will discuss are two of the moral arguments that have been put forward as reasons for objecting to the legalization of physician-assisted suicide. They have been taken seriously by a great many people and have had a powerful impact on the state of American law in this area.[1] I will argue that they are bad arguments.

I should say at the outset, however, that even if these are bad arguments, there may be others that are better. Many people oppose the legalizing of physician-assisted suicide on the ground that (as they think) there is no way of constraining the practice so as to provide adequate protections for the poor and the weak. They may be right, and if they are, then all bets are off. Alternatively, they may be wrong. I will simply bypass this issue.

I

It pays to begin by spelling out what practice is in question here, because the term 'physician-assisted suicide' is not transparent, and because some of what those who support legalizing would have us permit is not really describable as suicide at all.

What I in fact begin with, however, are two kinds of case in which action is not now illegal.

In the first kind, the patient requests that life-saving treatment currently in progress be discontinued. The cases I have in mind here are those in which the doctor who accedes does not merely cease to supply further drugs or other things that the patient needs: rather she positively intervenes – she shuts off, or removes the patient from, the equipment that is keeping him alive.[2] I will call these disconnecting cases.

In the second kind of case, the patient requests that life-saving treatment not be undertaken – as, for example, where he requests that he not be placed on life-saving equipment or that he not be subjected to aggressive efforts at resuscitation in case of cardiac arrest. I will call these nonconnecting cases.

Acceding to a patient's request in cases of those two kinds is legally permitted in the United States, and just about everyone, I think, regards that fact as morally satisfactory. The life-support procedures involved in those cases are intrusive and invasive, and just about everyone agrees that a patient has the right to refuse permission to continue such treatment if it is already under way, or to refuse permission to undertake it if it is not yet under way – and that the patient not merely has this right but that it must be respected. Certainly the doctor, or the hospital administration, must make sure that the patient really does want to exercise this right and that he has not been seduced or coerced into exercising it by those to whom he has become a nuisance in one or another way. But given that this condition has been met, that settles the matter: the patient's wishes must be granted.

What many people who would have physician-assisted suicide legalized are concerned with is primarily a third kind of case, one in which acceding to the patient's request is not legally permitted in most states, but in which — on their view — it should be. In this third kind of case, the patient requests a lethal drug which he himself can take at a time of his own choosing; I will call these drug-providing cases. In these cases, the doctor who accedes provides the patient with the means of committing suicide; and if the patient does commit suicide, what has happened can plainly and properly be called physician-assisted suicide.

A fourth kind of case also calls for attention. In this fourth kind, the patient is not capable of taking the lethal drug himself, and what he requests is that the doctor inject it. I will call these drug-injecting cases. In these cases, the doctor who accedes does not provide the patient with the means of committing suicide; the patient does not kill himself, it is instead the doctor who kills the patient. So these are not really cases of suicide and, a fortiori, they are not really cases of physician-assisted suicide.

Is acceding to the patient's request in this fourth kind of case illegal? Many people think it is, but whether they are right is not so clear, and I will return to that fact later. At all events, many of those who would have physician-assisted suicide legalized would have acceding in this fourth kind of case legalized too. In other words, they would have drug-injecting as well as drug-providing made legal. They speak misleadingly when they describe themselves as in favor of legalizing physician-assisted suicide; but on their view there is no significant moral difference between drug-providing and drug-injecting

and, other things being equal, what goes for the one should equally go for the other.

So we have four kinds of case before us. Acceding to a patient's request in a disconnecting or nonconnecting case is legally permissible, and just about everyone agrees that that situation is morally satisfactory. But acceding to a patient's request in a drug-providing case is on any view illegal in most states, and doing so in a drug-injecting case is (or, anyway, is widely thought to be) so too; and many people think that that situation is also morally satisfactory. That is, they have moral reasons for thinking that drug-providing and drug-injecting should be illegal. I will discuss two of their arguments.

Before turning to them, however, we should take note of two points; they may be obvious, but they are worth explicit mention even if they are.

(1) There is a familiar reason for legalizing disconnecting and nonconnecting that cannot be brought to bear in support of legalizing drug-providing and drug-injecting. Why do people think it morally satisfactory for disconnecting and nonconnecting to be legally permitted? I am sure they think this for the reason I mentioned, namely, that the life-saving treatment such patients wish to be free of is intrusive, and traditional ideas about autonomy prohibit imposing such treatment without the patient's consent. Put another way: refusing to accede to the patient's wishes in disconnecting and nonconnecting cases is a battery. By contrast — as has often been pointed out — refusing to provide or inject a lethal drug is not a battery. So that reason for legalizing disconnecting and nonconnecting cannot be brought to bear in support of legalizing drug-providing and drug-injecting.

It should be clear, however, that the fact that that reason for legalizing disconnecting and nonconnecting cannot be brought to bear in support of legalizing drug-providing and drug-injecting is no reason at all for keeping drug-providing and drug-injecting illegal. The existence of this difference is entirely compatible with there being good reason for legalizing drug-providing and drug-injecting.

Is there good reason for legalizing drug-providing and drug-injecting? That is not my topic in what follows: I will not examine arguments for legalizing drug-providing and drug-injecting. We are to be concerned throughout, in all four kinds of case, with people who are terminally ill and who wish to be helped; I take it that placing restrictions on what can be done for them is a serious infringement of liberty and, for my own part, therefore, I think that the restrictor has the burden of proof. But I will neither pick up the burden of proof nor argue that the restrictor must do so. Many would-be restrictors have themselves picked up the burden of proof, and I will discuss only whether they have succeeded in carrying it.

(2) The second point that should be taken note of is that an argument for illegalizing drug-providing and drug-injecting is not a good argument for that conclusion if it is either underbroad or overbroad – that is, if it either fails to count against both drug-providing and drug-injecting or counts against disconnecting or nonconnecting or against other acts we think should not be illegalized. To be a good argument, the argument must, on the one hand, hit the whole of its target and, on the other hand, not hit what should not be hit. The importance of this second point will emerge shortly.

II

The first of the two arguments I will discuss relies on a distinction that has come in for a considerable amount of discussion in philosophy in recent years, outside medical ethics as well as inside it, namely, the distinction between killing and letting die. In the medical context,

the idea has two parts. First, a doctor's killing her patient is different from a doctor's letting her patient die. Second, that difference makes a moral difference, indeed, it makes the following major moral difference: a doctor's killing her patient is always morally impermissible, whereas a doctor's letting her patient die is, in suitable circumstances, morally permissible.

This idea has a long tradition behind it. A doctor killing her patient has traditionally been called active euthanasia and has been regarded as morally impermissible precisely because a doctor's killing a patient is morally impermissible. A doctor's letting her patient die has traditionally been called passive euthanasia and has been regarded as morally permissible, in suitable circumstances, precisely because it is not killing but is merely letting die.

And then here is how the idea is brought to bear on our four kinds of case. If the circumstances are suitable – thus, in particular, if a doctor's patient really does wish to be disconnected from or nonconnected to intrusive life-saving treatment – then it is morally permissible for the doctor to accede to his request, because the doctor who accedes in those kinds of case does not kill but merely lets her patient die. Drug-providing and drug-injecting, by contrast, involve killing the patient and are therefore always morally impermissible.

It is then concluded that this moral difference justifies differential legal treatment, namely, legalizing action in the first two kinds of case and illegalizing action in the third and fourth.

Let us begin with the first two. It seems clear, on any ordinary understanding of the phrases 'kill' and 'let die', that the doctor who nonconnects does not kill her patient but instead merely lets him die. Is it clear that this is also true of the doctor who disconnects? As I stressed at the outset, the doctor who disconnects does not stand by, doing nothing, she positively intervenes – she shuts off, or removes the patient from, the equipment that is keeping him alive. Why is that merely letting the patient die?

Many friends of this argument reply: when a doctor nonconnects her patient, she merely "lets nature take its course," she merely lets her patient die of the underlying medical condition that is currently threatening his life. Similarly, when a doctor disconnects her patient, the doctor merely "lets nature take its course," she merely lets her patient die of the underlying medical condition that threatened his life, because of which threat he was placed on life-saving equipment. In both of those kinds of case, it is the patient's disease that causes his death. And of course they add that in drug-providing and drug-injecting, by contrast, it is not the patient's disease that causes his death, but rather the drug provided or injected.[3]

Will this do? That is, is disconnecting like nonconnecting in this respect? No doubt the patient who is disconnected dies of the disease because of which he needed life-saving equipment, but does the doctor who disconnects him merely *let* this happen? Does she merely "let nature take its course"? If the patient is currently being kept alive by (as it might be) a respirator, then nature's taking its course is currently being prevented by the respirator. The doctor who disconnects him from the respirator removes what is preventing nature from taking its course. She intervenes – and seems to be most plausibly seen as not merely letting nature take its course but rather causing it to.[4]

If I knock out the main beam that is currently preventing the fall of a roof, I do not merely let gravity take its course and the roof therefore fall on those locked in the house. I intervene – I cause gravity to take its course.

Plainly, if the doctor disconnects the patient, the patient dies sooner than he otherwise would, just as if I knock out the main beam, those locked in the house die sooner than they otherwise would.

So there is a difficulty here. While it is clear that the doctor who nonconnects merely lets her patient die, it is not immediately obvious that the doctor who disconnects merely lets her patient die. According to friends of the first argument, it is the fact that disconnecting and non-connecting are both merely letting die that immunizes them against the first argument for illegalizing drug-providing and drug-injecting; how is it to be secured that there is such a fact?

What we have so far, however, is merely appetizer. I said that the first argument relies on the difference between killing and letting die and that that difference has come in for a considerable amount of discussion in recent years. Philosophers have disputed about what exactly the difference is and whether it (whatever exactly it is) has the moral significance it is thought to have – indeed, whether it has any moral significance at all.[5] I will discuss, briefly, a move that some philosophers make at this point.

They say it is not necessary for the doctor to let her patient die that she merely "let nature take its course," if that is understood in the most natural way, thus as requiring that she do nothing at all.

They say it is necessary that (i) the patient dies of the underlying medical condition that threatens his life, for the doctor who injects a lethal drug plainly does not merely let her patient die. But they say that while meeting condition (i) is necessary, it is not sufficient. If a patient is currently on a respirator and a rival of his breaks in at night and disconnects him, the patient dies of the underlying medical condition that threatens his life, but the rival does not let him die, the rival kills him.[6] They say that the reason why the rival kills him, whereas the doctor who disconnected him would merely let him die, lies in this: if the patient is disconnected, he loses a certain stretch of life, a stretch of life that he would have with the doctor's aid (To be more precise, we should say: with the aid of the life-saving equipment supplied by the hospital whose agent she is. For brevity, I ignore the need for this precisification.) So if the doctor disconnects him, he loses what he would have with her aid. By contrast, if the rival disconnects him, he does not lose what he would have with the rival's aid: the rival has no aid to give, and the patient loses what he would have with the doctor's aid.

This points to a second necessary condition. We are to say that a doctor lets her patient die only if (i) the patient dies of the underlying medical condition that threatens his life and (ii) the patient loses what he would have with her aid. That meeting this second condition is necessary seems to me to be right: you can't let a person die unless you have, by way of alternative, a means of helping him to remain alive. The doctor who nonconnects, as well as the doctor who disconnects, meets both of these two conditions.

So it does seem right to think that meeting these two conditions is necessary. Is meeting them sufficient? No. For suppose a patient is getting along comfortably on his respirator and has declared his desire to remain on it. Suppose, now, that his doctor disconnects him. (Why? Perhaps she is overtired, or her records are in a state of confusion.) He then (i) dies of the condition that threatens his life and (ii) loses what he would have with her aid. But I think it plain that she kills him.[7] So something more is required than meeting conditions (i) and (ii).

You might think that the further necessary condition is obvious enough: it must be that the patient requests the disconnecting. That is an intuitively plausible idea. It needs generalizing, however, since the patient might be incapable of making such a request, and it be made by his guardian instead. To generalize, we need to see how the patient's, or his guardian's, request matters.

Let us go back to (ii). What I think attracts people about the idea that meeting (ii) is necessary is the thought that if I am supplying aid to a person, then I have a liberty-right to stop doing so. That is very often true. If I am supplying aid to a person — my efforts or my equipment or both — having made no commitment to continue doing so, then I do have a liberty-right to stop. (Though stopping may be morally wrong, as where continuing is easy and I have no good reason to stop.) But that is not always true. If I have committed myself to continuing, then, other things being equal, I do not have a liberty-right to stop. Consider again the doctor who disconnects her patient when he wishes not to be disconnected. She does what she has no liberty-right to do. Hospitals do not say to patients they place on life-saving equipment, "It's our respirator, we're paying for the electricity that keeps it going, and we reserve the liberty-right to disconnect you at will." No doubt the hospital owns the respirator and pays for the electricity, but a patient who is placed on the respirator has a claim-right to continued use of it, as you have a claim-right to continued use of my house if I have given you an indefinite lease on it.

So that is how the patient's request matters: in requesting that his doctor disconnect him, the patient gives her a liberty-right to do so. The generalization is now forthcoming: the third condition that has to be met is (iii) the doctor has a liberty-right to engage in the behavior (action or inaction) that issues in the patient's death.

Is meeting (iii) really a necessary condition for a doctor's letting her patient die? Is meeting an appropriate generalization of (iii) necessary for anyone's letting anyone die? These ideas seem to me intuitively very plausible.[8]

If meeting (iii) — as well as (i) and (ii) — is necessary, then something of interest follows. The difference between killing and letting die has been widely thought to be itself a nonmoral difference, which nevertheless makes a moral difference. If an agent's having a liberty-right to act as she does is required for letting die, then the difference between killing and letting die is not wholly nonmoral. That seems to me no objection to the idea that a liberty-right is required for letting die. In fact, the difference's being partially moral would tidily help to explain why and how it makes a moral difference. We should remember, moreover, that an agent's having a liberty-right to do a thing does not by itself fix that it is morally permissible for him to do it. As I said in a parenthetical remark above, it may be morally wrong to exercise a liberty-right: a liberty-right supplies only a rebuttable presumption, or defeasible ground, for moral permissibility.

But I merely make suggestions here, for all of this is under dispute. What emerges is really only that if a doctor's disconnecting her patient is her letting him die, as friends of the first argument (among others) would have us agree, then it is all the same not clear what makes that so, because it is not clear what the difference between killing and letting die consists in. Friends of the first argument would have us agree also that the difference between killing and letting die makes a moral difference, and not only a moral difference, but a major moral difference. To the extent to which we are unclear what the difference between killing and letting die consists in, we should at a minimum be suspicious of the proposal that it does make that major moral difference.

I cannot resist adding, however, that I believe that people who are primarily concerned with the morality of medical decisions have been overfascinated by the question whether the doctor who disconnects her patient kills him or lets him die. We only need to have that she lets him die if we think that if she kills him, she acts morally impermissibly. That a doctor's killing her patient is morally impermissible is, of course, exactly what the first argument says. I am in the process of arguing that we should reject

this first argument, but perhaps it pays to draw attention along the way to the many things that can be said in support of the doctor's acting morally permissibly when she disconnects her patient, even if we should find ourselves having to agree that she kills him by doing so. After all, if she kills him, it is only in that she disconnects him that she kills him. And there are all the facts about the intrusiveness of the equipment and the patient's condition and wishes, and there is the fact that the doctor is an agent of the hospital and that the relevant appropriately formed hospital committee has explicitly permitted the disconnecting on such and such grounds. Surely we need not say that these facts make it the case that the doctor who disconnects does not kill and *therefore* acts morally permissibly; we can pass directly from these facts to the conclusion that the doctor who disconnects acts morally permissibly, leaving it to those with a taste for philosophical analysis to mull over whether she kills, and if not, why not.

Or so I think. Indeed, I think it would conduce to moral clarity to proceed in that way.

In the following sections I ignore the difficulties I have been drawing attention to in this section. I will be concerned only with cases in which it is entirely clear that a doctor does or does not kill her patient. I will argue that the first argument fails to hit all of its target and also that it hits what it should not hit.

III

For let us turn now to drug-providing and drug-injecting.

First, drug-providing. In describing (in Sec. II) how the first argument is brought to bear on our four kinds of case, I said: the doctor who disconnects or nonconnects merely lets her patient die, but drug-providing and drug-injecting, by contrast, involve killing the patient and are therefore always morally impermissible

I had to use that vague word "involve", since one thing that is surely plain is that the doctor who supplies her patient with a lethal drug does not herself kill her patient. On the one hand, the patient may not take the drug at all. (Many a patient wants to be supplied with the drug only to have the comfort of knowing that if his condition becomes unbearable, so that he wishes to end his life, he will be able to do so; such a patient may in the end not take the drug.) On the other hand, even if the patient does take the drug, it is the patient (and not the doctor) who kills the patient.

Friends of the first argument might reply that while the doctor does not herself kill the patient in a drug-providing case, such a case anyway involves a killing if the patient takes the drug, for it involves the patient's killing himself. The doctor anyway assists in a killing if the patient kills himself, and whether or not he does, she enables him to kill himself. And, they might go on, it is not merely a doctor's killing her patient that is always morally impermissible, so also is suicide always morally impermissible – and where it is morally impermissible for a person to do a thing, so also is it morally impermissible to assist him in doing it or to enable him to do it.[9]

We should be clear, however, that the reply I envisage supplies a different argument against drug-providing. It does not rely on the premise that a doctor's killing her patient is always morally impermissible; it relies on the premise that a patient's killing himself is always morally impermissible. I will therefore be brief about it.

Many people do believe that suicide is always morally impermissible. I think that most who believe this believe it on religious grounds. But some believe it on nonreligiously-based moral grounds; Kant believed it on the ground of the intrinsic worth of rational beings. Many others disbelieve it. (I am myself among the disbelievers.) But I will not argue the matter. What we should notice is a question this idea raises.

Suppose you believe that suicide is always morally impermissible and, therefore, that assisting in a suicide is always morally impermissible. Can you reasonably argue for illegalizing assisting in a suicide on the ground of its moral impermissibility without thinking yourself called on to argue for illegalizing suicide? After all, you only think that assisting in a suicide is morally impermissible because you think suicide is.

Suicide once was a crime.[10] It is so no longer, and I assume that no one wishes it re-criminalized, even those who think it morally impermissible. So if you think that assisting in a suicide should remain a crime, that had better be on stronger ground than that it too is morally impermissible.

Considerably stronger grounds are certainly available for criminalizing the conduct of just anyone who helps just anyone commit suicide. In the case of physician-assisted suicide in particular, if it should really turn out to be the case that there is no way of constraining the practice so as to protect the poor and the weak, then that would be very strong ground for refusing to legalize it. But as I said at the outset, I am bypassing that ground for opposing the legalization of physician-assisted suicide.

Let us anyway return to the argument that we were looking at, which relies on the premise that a doctor's killing her patient is always morally impermissible. That argument plainly fails to hit the drug-providing part of its target.

Well, doesn't it at any rate hit the other part of its target, the drug-injecting part? Let us turn to drug-injecting.

The doctor who injects a lethal drug, on any view, kills her patient, so if a doctor's killing her patient is morally impermissible, then drug-injecting is morally impermissible. So the argument does hit this part of its target.

Indeed, drug-injecting cases should be cases in respect of which opponents of physician-assisted suicide who rely on the moral impermissibility of a doctor's killing her patient are most firm: the opponents might have been expected to say, unambiguously, that since the doctor who injects the drug kills her patient and since, also, a doctor's killing her patient is always morally impermissible, she must not inject the drug. Period.

In fact, however, most opponents of physician-assisted suicide do not say this. They are content to allow the doctor to give what is in fact a lethal dose of morphine when nothing less than that dose will relieve the patient's pain or even make it bearable. They do not regard a doctor's doing this as morally impermissible. Nor do they think that a doctor's doing this should be illegal.

Is it in fact illegal? I mentioned earlier that it is not clear that acceding to the patient's request in this fourth kind of case is illegal. I gather that if the doctor acts in accord with a certain proviso, it is not in fact illegal. During oral argument on the two physician-assisted suicide cases before the U.S. Supreme Court, the then acting solicitor general, Walter Dellinger, who was arguing the case against physician-assisted suicide, said: "We agree that state law may . . . not only allow withdrawal of medical treatment but also allow physicians to prescribe medication in sufficient doses to relieve pain even when the necessary dose will hasten death." He added a proviso: physicians may do this, he said, "so long as the physician's intent is to relieve pain and not cause death."[11]

I will invite you to have a closer look at that proviso in Section IV below. For the moment, however, we should be clear about the following. Suppose that a doctor could not relieve her patient's pain by anything less than a lethal dose of morphine and that the patient knew this and nevertheless requested the morphine. Suppose then that the doctor injected the drug, intending only to relieve the pain and not to cause death. Suppose finally that the drug-injection did cause

the patient's death. Then the doctor caused the patient's death. Indeed, the doctor killed the patient. The fact that the doctor did not intend to cause death, or to kill, is no reason at all for thinking she did not do these things. If Dellinger is right about existing state law, however, her doing so is not illegal.

This means that while the first argument does hit the drug-injecting part of its target, opponents of physician-assisted suicide do not themselves really think that this part of its target should be hit – or anyway, they do not think that all of this part of its target should be hit.

In sum, opponents of physician-assisted suicide had better forgo the first argument against it. The argument fails to hit the drug-providing part of its target, and while it does hit the drug-injecting part, this part (or anyway part of this part) should not be hit, on their own view of the matter.

IV

The remark I quoted from Dellinger's oral testimony points to the second of the two arguments against legalizing physician-assisted suicide that I will discuss. This argument (like the first) has two parts. First, a doctor's doing something intending to cause her patient's death is different from a doctor's doing something foreseeing that his death will ensue. Second, that difference makes a moral difference. Indeed, it makes the following major moral difference: a doctor's doing something intending to cause her patient's death is always morally impermissible, whereas a doctor's doing something foreseeing that his death will ensue is, in suitable circumstances, morally permissible.

And how is that idea to be brought to bear on our four kinds of case? Consider disconnecting and nonconnecting. No doubt a doctor may know that if she disconnects or nonconnects her patient his death will ensue. But if she accedes, she does not intend to cause her patient's death; she intends only to relieve her patient of an unwanted bodily intrusion, merely foreseeing that his death will ensue. That fixes that acceding in disconnect and nonconnect cases is morally permissible.

Not so (the explanation goes on) in a drug-providing case. There the doctor who accedes intends to cause her patient's death. That fixes that acceding in a drug-providing case is morally impermissible.

By contrast (finally), the doctor who accedes in a drug-injecting case may or may not intend to cause her patient's death. If she does, then that fixes that she acts morally impermissibly. If she does not — if she intends only to relieve her patient's pain, merely foreseeing that the patient's death will ensue — then that fixes that she acts morally permissibly.

It is then concluded that this moral difference justifies differential legal treatment, namely, legalizing action in disconnect and nonconnect cases, illegalizing action in drug-providing cases, and legalizing action in some drug-injecting cases, namely, those in which the doctor does not intend but merely foresees that the death will ensue.

What undergirds this argument is a quite general principle that is familiar to everyone who has looked into the contemporary litera-ture on medical ethics and on what counts as just and unjust military action in wartime: the Principle of Double Effect (PDE). Those who accept this principle draw attention to the fact that all acts have a great many effects and that it may be that a given act would have both a good effect and a bad one. Suppose that is true of a given act. PDE tells us that if the good effect of the act is proportionately good enough, then an agent may morally permissibly perform the act if, while foreseeing the bad effect, he intends only the good effect and does not intend the bad one, either as an end (i.e., for its own sake) or as a means to the good effect.

In particular, the good effect of a patient's being relieved of a bodily intrusion, or of pain, may well be proportionately good enough, despite the bad effect of the patient's death, to make it morally permissible for the doctor to do what will have both effects — though only, of course, if she intends only the good effect and not the bad one.

This principle is plainly at work in people who think that the moral permissibility or impermissibility of acceding in our cases turns on the doctors' intentions. Dellinger implies that it is also enshrined in the law governing these matters.[12]

Many philosophers think that the principle is a muddle; I join them in thinking so and will say a bit about why. For my part, what is interesting about it is only that it is still taken so seriously, despite the many rebuttals it has been subjected to.[13] I will suggest a hypothesis about why that is so in Section V.

But let us first look at five sources of concern about the way in which the putative moral difference between intending and foreseeing is brought to bear on the kinds of case at hand.

(1) You might well wonder about those assumptions about the doctors' intentions in our cases. What do we really know about doctors' intentions in acceding in cases of any of those kinds?

Moreover, consider drug-providing cases. (I presume that, licit or not, there are doctors who accede in such cases.) I think it very plausible that some doctors who provide drugs, or who would do so if it were legal to do so, do or would do so intending only to provide the patient with the comfort of knowing that if his condition becomes unbearable, so that he wishes to end his life, he will be able to do so. So the assumption about their intentions is not only unjustified, but very implausible.[14]

Existing law prohibits drug-providing; would opponents of physician-assisted suicide have us emend it so as to permit drug-providings in which the doctor does not intend to cause the patient's death?

It will not do to reply that it would be too hard to find out what the doctor's intentions are in drug-providing cases. It might well be hard to find this out. But how, then, can it consistently be held that the doctor's intentions are crucial in drug-injecting cases? Can anyone plausibly think it would be harder to find out what the doctor's intentions are in drug-providing cases than in drug-injecting cases?

(2) Bringing PDE to bear on our cases requires us to suppose that the doctor's act has a good effect and a bad effect: in our cases, the bad effect of the doctor's act is the patient's death.

But we should ask, why is that effect of the doctor's act bad? Isn't it likely that in the case of at least some of those patients, their condition is such as to make it better for them to die? Indeed, such as to make it good for them to die? Suppose a patient's remaining life will be engulfed in extreme and otherwise unrelievable pain. It is hard to see how anyone could plausibly insist that however terrible a patient's remaining life will otherwise be, it would still be, not good for him, but instead bad for him to die. After all, what is good or bad for a person depends on his condition. (Taking a lot of aspirin is good for a person with arthritis and bad for a person with ulcers.) And it is surely plausible to think that this person's condition is such as to make it good for him to die.

A patient might know how terrible his remaining life will otherwise be and nevertheless think that it would not be good for him to die. (If we suppose that there is an element of subjectivity in what is good for a person, we may well think, therefore, that it would not be good for this patient to die.) Or he might think that it would be good for him to die but not want to die. In either of those two cases, he will

presumably not request a lethal drug. But we are concerned here only with patients who do request that the doctor act. Even those who accept PDE would not have the doctor act where the patient (or his guardian if the patient is incompetent) does not request it.

So we are supposing a patient who does request action; and let us suppose that he makes this request because he thinks that it would be good for him to die, and he wants to die. Why, then, should we think that his death would be a bad effect of the doctor's act? If it would not be a bad effect of the doctor's act, PDE has no bearing on the case.

There is room for a reply. On some views, the question whether an event would be good for a person does not settle that it would be a good event. I do not point here to the possibility that an event might be good for one person and bad for another; that is certainly true. What I point to is, rather, the idea that an event might be good for a person and bad for no one, but all the same not be a good event. The idea here is that in addition to the property of being good for a person, which an event may have or lack, there is a further property of being pure, unadulterated good and that that property may be lacked by an event which is good for one and bad for no one. (This property is standardly called intrinsic goodness. As I said, whether an event would be good for a person turns on his condition; whether an event would be intrinsically good does not.) Similarly, there is a property of being pure, unadulterated bad (intrinsic badness) which an event may have even if it is good for one and bad for no one. Those who like this idea may then say that even if the patient's dying would be good for him (and bad for no one), it is nevertheless a bad event. They may add that the patient's being relieved of pain is a good event. So the doctor's acceding to his wishes would have a bad effect as well as a good one, and PDE does, after all, have a bearing on the case.

I mention this idea only because it is on offer in philosophy and not because I think there is any plausibility in it. My own view is that a thing is good only if it is good in this or that way: it is a good book, or a good chess player, or is good to look at, or is good for Jones or England or the tree in my backyard. The idea that there is also pure goodness, which is above and beyond goodness in this or that way, seems to me to be a serious mistake.[15] At best it is obscure, and if the application of PDE to our cases (or to other cases) requires reliance on it, then its applicability to those cases is at best unclear.

(3) Let us bypass the difficulty I pointed to in subsection (2) and focus on the fact that the application of PDE to our cases requires supposing that the doctor must not intend the patient's death. What exactly does that rule out? PDE tells us that two things are ruled out: a doctor's intending the death as an end and a doctor's intending the death as a means to something else, say the patient's being relieved of pain. These are different.

It is standard to take it that for a person to X, intending an event E, is for him to X because he thinks his doing so will cause E, and he wants E for one or another reason. Thus a boy might hit a ball hard (X) intending the crowd's admiring him (E): that consists in his hitting the ball hard because he thinks his doing so will cause the crowd's admiring him, and he wants the crowd's admiring him.

I take it that for a person to X, intending E as a means, is for him to X, intending an event E, where he wants E because he thinks it will cause or include an event F, which he wants. Our boy might hit the ball hard intending the crowd's admiring him as a means. That is so if he wants the crowd's admiring him (E) because he thinks the crowd's admiring him will cause donations to the team to rise (F), which he wants. So also if he wants the crowd's admiring him (E) because he thinks the crowd's admiring him

will include the school's secret society members' admiring him (F), which he wants. (They are scattered somewhere in the crowd.[16])

I postpone intending E as an end for subsection (4) below, for we need to look first at doctors who intend their patients' deaths as a means.

Suppose that injecting drug D will straightaway cause a patient's death and a fortiori his being relieved of pain. That just tells us about a drug and not about anyone's intentions in injecting it. Suppose Alice knows about drug D and injects it. I should think it can be true that she is intending her patient's death as a means to his being relieved of pain: for it can be true that she injects D because she thinks her injecting D will cause the patient's death, and she wants her patient's death because she thinks it will cause his being relieved of pain – or because she thinks it will include his being relieved of pain – and she wants his being relieved of pain. (Is death an event that causes relief from pain? Or is death a complex event that includes relief from pain among other events? I leave this open. More to the point, I leave it open for Alice to think either.)

If PDE is correct, Alice acts morally impermissibly. Can that be right? We need to notice two consequences of accepting that it is.

First, Alice acts morally impermissibly even if D is the only drug she has on hand that will relieve her patient's pain, and however terrible the patient's pain may be. I doubt that the lawyers who have been attracted by PDE have thought through the fact that basing law on it would call for illegalizing her injecting D in such circumstances.

Second, suppose that Alice has an alternative to drug D on hand, a drug C, that will not straightaway cause the patient's death: injecting C will instead cause the patient's falling into a coma, followed by his death some time later. If she injects C, her patient's death will follow his relief from pain and thus will plainly neither cause nor include it; so if she chooses C, she will presumably not intend the death as a means but will merely foresee it. If PDE is correct, then Alice must choose C. But do the patient's wishes not matter? By hypothesis, if the patient is injected with C he will live longer than if he is injected with D. By hypothesis also, however, that stretch of additional life will be unconscious life, and the patient might prefer not to live it. Does morality, and should law, require him to? This does not seem to me at all clear.

(4) Let us now turn to a doctor's intending death as an end. I take it that for a person to X, intending E as an end, is for him to X, intending E, where he wants E because of some feature he thinks it will itself have. Our boy might hit the ball hard intending the crowd's admiring him as an end. That is so if he wants the crowd's admiring him (E) because he thinks the crowd's admiring him will itself be pleasant.

Similarly for a doctor's intending her patient's death as an end. Consider Barbara, whose patient has been in a coma for some years, kept alive by life-saving equipment. We can suppose she thinks his continuing to live is a senseless indignity and that his dying would be good for him. Discovering in the hospital's files a hitherto lost record of his requesting that he be disconnected if he fell into a coma, and obtaining permission from the patient's guardian and the appropriate hospital official, she disconnects him. We can suppose that Barbara intends her patient's death as an end, for we can suppose that she wants his death because she thinks his death would be good for him.

What this draws attention to is that you can intend a death as an end and not do so out of malice. Carol, in a situation similar to Barbara's, might intend her patient's death as an end, but not because she thinks his death will be good for him, rather because she thinks his death will be appropriate revenge for wrongs she thinks he did her. Carol acts out of malice; Barbara does not.

All the same, if PDE is correct, Barbara acts morally impermissibly. That cannot be right.

(5) Do a doctor's intentions matter to the moral permissibility of what she does in our four kinds of case? And should they matter to the legal permissibility of what she does?

Let us begin with law. Disconnecting is legally permissible. Should the law be changed? For suppose that if a patient's doctor disconnects him, she will do so intending his death as a means or end. Would we have the hospital official in charge of supervising such decisions say to the patient (or to his guardian, if he is in a coma): "It would be illegal for her to do the disconnecting. You'll just have to wait until we can find a doctor who won't disconnect with that intention."

Consider drug-injecting and imagine the following scenario. You are a doctor, and your patient is near death, in terrible pain that cannot be alleviated by any less than a lethal dose of morphine, and for that reason, he asks for the injection. Suppose that, pursuant to the hospital's rules, you ask me – I am the appropriate hospital official – whether it is legally permissible for you to inject the lethal dose. I reply: "Well, I don't know. I can't tell unless you tell me what your intention would be in injecting the drug. If you would be injecting to cause death, either as means or end, then no. But if you would be injecting only to cause relief from pain, then yes." This is an absurdity. How on earth could it be thought proper for the legal permissibility of acceding to turn on what the doctor intends to bring about by acceding? Surely it should turn on the patient's condition and wishes. As it should in the case of disconnecting.

It is certainly true that the law takes an agent's intentions seriously.[17] In some cases, that is because a difference in the agent's intention may mark him as having committed a more or less grave crime: compare the difference between murder and manslaughter. (The relative gravity of the crime matters to us because of our desire that the agent be appropriately, and no more than appropriately, punished.) Are there cases in which the law takes an agent's intention seriously because what he does is a crime only if he has this or that intention? An agent's intention may be of interest because of what it shows about what he knew at the time of acting. Thus, if you walk off with my watch, and it turns out that you intended only to be walking off with your own watch, then that is of interest because it shows that you did not know the watch was mine and thus did not know that you were walking off with my watch; and if you did not know this and, moreover, had no reason to know it, then what you did was not theft. What of cases in which an agent is ignorant of nothing relevant about what he does or about what he will thereby cause? Consider cases in which an agent knows that he is X-ing, and knows that his X-ing will cause outcomes O_1 and O_2 among others. Are there any such cases in which he commits a crime if he Xs intending O_1 and does not commit a crime if he Xs intending O_2 but not O_1? If Dellinger is right about the law, then there is at least one such kind of case, namely, drug-injectings − for the doctor in a drug-injecting case knows perfectly well that she will cause both death and relief from pain, and he says that her injecting to cause death is a crime but her injecting only to relieve pain is not. I know of no others.

Similarly for moral permissibility and impermissibility. According to PDE, the question whether it is morally permissible for the doctor to inject a lethal drug turns on whether the doctor would be doing so intending death or only intending relief from pain. That is just as absurd an idea. If the only available doctor would inject to cause the patient's death, or is incapable of becoming clear enough about her own intentions to conclude that what she intends is only to relieve the patient's pain, then − according to PDE − the doctor may not proceed, and the patient must therefore continue to suffer. That cannot be right.[18]

If a doctor will inject her patient intending his death as an end and, moreover, wants his death only because his death will constitute revenge, then that does matter morally. But we have to be careful about how it matters. I suggest that it has no bearing on whether it is morally permissible for her to act. Whatever her intention may be, the patient, we are supposing, desperately wants her to inject the drug.

If we love him, we too want her to inject the drug. We can consistently believe it would be morally impermissible for her to act, while nevertheless wanting her to. But morality calls for us to feel ashamed of ourselves if we do. What is morally impermissible is, after all, exactly that: morality requires that the agent not do the thing, the agent must not do it. So if we really do believe it morally impermissible for the doctor to inject the drug, then it is a bad business in us to want her to, wanting this for the sake of the benefit to be got by the patient if she does. (Compare wanting Smith to murder your uncle, wanting this because of the estate you will inherit at your uncle's death.) I am certain, however, that no shame or guilt is called for in us if we want the doctor to inject the drug in this case. No one will be harmed, and the patient will benefit.

That she will inject for that reason matters morally, not by way of fixing that it is morally impermissible for her to proceed, rather by showing something morally bad about her.[19]

V

This brings us to the first of what I take to be the two main sources of the persistence of PDE.

What I refer to is a failure to take seriously enough the fact – I think it is plainly a fact – that the question whether it is morally permissible for a person to do a thing just is not the same as the question whether the person who does it is thereby shown to be a bad person. The doctor who injects a lethal drug to get revenge or out of hatred is a bad person. We can add that she acts badly if she acts for that reason. That is compatible with its being morally permissible for her to inject the drug.

It is not clear to me why people fail to take this seriously enough. It may be due to a desire for simplicity, that is, a desire that all moral concepts should be simply reducible to one. Alternatively, and more interesting, it may itself be due to what I take to be the second of the two sources of the persistence of PDE.

What I have in mind is that many people think that if we agree that the agent's intentions in acting are irrelevant to the question whether he may act, then we must become consequentialists. Many people think that if intentions are irrelevant to moral permissibility in action, then all that remains for moral permissibility in action to rest on is whether the consequences of the agent's act would be, on balance, good.[20]

But I think that this merely shows lack of imagination. For example, acting in a certain way might constitute an infringement of a right, which, as Ronald Dworkin put it some years ago, trumps utilities; and what fixes whether an act would infringe a right is not the agent's intentions but rather his circumstances, thus what, in virtue of his circumstances, he would in fact be doing to, or not doing for, his victim.

Rights trump utilities in the sense that they make maximizing utility impermissible – or at all events, stringent rights make maximizing utility impermissible. Other kinds of consideration may make refraining from maximizing utility permissible even if they do not make maximizing utility impermissible. It is certainly

permissible to devote one's life to relieving human needs; it is not morally required of us that we do so. No doubt a man's refusing to aid others in order that they suffer marks him as a bad person; the refusal itself, however, may be permissible – he is not required to give aid just on the ground that his intention in refusing would be a bad one.

The questions what rights we have and how stringent they are are of course hard ones. So also is the question how much morality requires us to do for others. These are among the deepest questions of moral theory. The important point here, however, is just that they are real questions: we are not forced to become consequentialists by virtue of a commitment to distinguishing between the moral worth of a person who does a thing and the moral permissibility or impermissibility of his doing it.

In short, the persistence of PDE is due to an over-simple conception of the resources of morality.

I suggest, then, that this second argument against physician-assisted suicide also fails. Just as our conclusions about these four kinds of case should not be driven by a concern about whether the doctor kills or merely lets die, so too they should not be rested on what the doctor would or would not be intending in acting. It is therefore unfortunate that these two arguments have had so powerful an impact on the state of American law in this area.

Notes

* An early version of parts of this article was presented in April 1997 as the Fifth Annual McGill Lecture in Jurisprudence and Public Policy, McGill University Law School. A later version was presented in April 1998 as a lecture in the Kennedy School's Program in Ethics and the Professions lecture series. I am grateful to the participants on both occasions for their comments and criticism.

am also grateful to Elizabeth Prevett for providing me with the relevant legal materials and for helpful discussion. It might pay to add that this article is not intended as an expansion on the moral-theoretical views expressed in sec. II.B of Ronald Dworkin et al., "Assisted Suicide: The Philosophers' Brief," *New York Review of Books* 44 (March 27, 1997): 41–47, which I joined in the preparation of; I do not know how much of what follows would be accepted by my colleagues in that enterprise.

1 See, e.g., the Supreme Court's opinion, by Chief Justice Rehnquist, in Vacco v. Quill, 117 S.Ct. 2293 (1997), which endorses the reasonableness of both arguments. All quotations below that are attributed to Rehnquist are from that opinion.

2 I assume throughout that the doctor is female and the patient is male.

3 Here is Rehnquist: "When a patient refuses life-sustaining medical treatment, he dies from an underlying fatal disease or pathology; but if a patient ingests lethal medication prescribed by a physician, he is killed by that medication."

4 In an interesting recent article, Patrick D. Hopkins asks why disabling a patient's artificial pulmonary system should count as merely letting him die, whereas disabling a person's natural, flesh-and-blood pulmonary system would presumably count as killing him. He asks: what's so special about the natural? See his "Why Does Removing Machines Count as 'Passive' Euthanasia?" *Hastings Center Report* 27 (1997): 29–37.

5 The philosophical literature on the difference between killing and letting die – and the still more general difference between doing and allowing (as it is commonly referred to) – is by now enormous. For a wealth of examples and argument, see Frances M. Kamm, *Morality, Mortality*, vol. 2 (Oxford: Oxford University Press, 1996), pt. 1. Kamm there focuses primarily on the moral significance of the difference between killing and letting die. For interesting recent discussions of the nature of the difference between killing and letting die, and the more general doing and allowing, see Jeff McMahan's "Killing, Letting Die, and Withdrawing Aid," *Ethics* 103 (1993): 250–79, and "A Challenge to Common Sense Morality," *Ethics* 108 (1998):

394–418. (The latter is a review essay on Jonathan Bennett, *The Act Itself* [Oxford: Oxford University Press, 1995]. Bennett had argued in that book, as in earlier work, that there is no morally significant difference between doing and allowing.) That condition (ii) must be met if a doctor is to let her patient die – see the text below – is argued for by both Kamm and McMahan.

I think it also pays to mention that it is very likely that there is no such thing as *the* difference between doing and allowing. In my own review essay on Bennett's book (in *Noûs* 30 [1996]: 545–57), I drew attention to a difference between cases in which a person causes an outcome O by doing something (engaging in an enterprise that causes O) and cases in which a person causes O but not by doing something. That seems to me to mark a difference between *a* pair of notions, "doing" and "allowing," though if so, they are leaner notions than Kamm and McMahan are concerned with. (A doctor's disconnecting her patient falls under "doing" rather than "allowing" in my scheme, but under "allowing" in theirs.) Perhaps the difference I point to should not be thought of as *the* difference between act and omission. In any case, if there is no such thing as *the* difference between doing and allowing, then each difference has to earn its own moral significance.

6 The case of the rival is due to Shelly Kagan; see his *The Limits of Morality* (Oxford: Oxford University Press, 1989), p. 101. Anthony Woozley had drawn attention to disconnectings by unauthorized agents in "A Duty to Rescue: Some Thoughts on Criminal Liability," *Virginia Law Review* 69 (1983): 1297. Kagan and Woozley both say, surely rightly, that in such cases, the agent kills.

7 Woozley drew attention to the fact that the doctor kills her patient in such a case in "A Duty to Rescue," p. 1297.

8 All of the clear cases of letting die that I have come across in the literature are cases in which the agent has a liberty-right to behave as he does and, moreover, it is the agent's having that liberty-right that seems to me to be what makes them be clear cases.

Not all cases are clear cases, however, and in light of some of them, it is arguable that my suggestion above should be weakened. What I have in mind is that action and inaction may work differently in this respect: perhaps we should say that meeting (iii) is required for letting die where action is in question but not where inaction is. Thus, it seems to me to be plain that the doctor I described in the text who disconnects her patient, having no liberty-right to disconnect him, does not let her patient die but instead kills him. What of a doctor who nonconnects her patient, having no liberty-right to nonconnect him? (As, e.g., where the patient wants connection, and there is no good reason for her to nonconnect him: among other things, no other patient has a prior claim on the equipment.) Does the non-connector kill her patient? Or does she instead wrongfully let him die? I lean toward the former, but would not be surprised if others leaned toward the latter. If they are right, then meeting condition (iii) is not required for letting die where inaction is in question.

Here is another pair of cases. In one, the town's electrical supply is down, and a baby's parents have been keeping it wrapped against the cold. They now change their minds: they deliberately remove the wraps, and the baby therefore dies of the cold. As responsible for it, they have no liberty-right to remove the wraps. It seems to me plain that they kill the baby. In a second case, a baby's parents deliberately stop feeding it, thereby starving it to death. As responsible for it, they have no liberty-right to stop feeding it. Do they kill it? Or do they instead wrongfully let it die? Once again, I lean toward the former, but I do not find it surprising that some writers have leaned toward the latter – see, for example, the articles by McMahan cited in n. 5 above. If they are right, then (once again) meeting condition (iii) is not required for letting die where inaction is in question.

My impression, however, is that any theory of these matters is going to have a cost in that it will draw lines at some places where intuitions differ. A good theory would be worth the cost; the best would explain why intuitions differ where they do.

9 The possible reply I describe in the text is only one of two that suggest themselves. The other is hinted at by the passage I quoted from Rehnquist in n. 3 above. That is, it might be said that drug-providing involves a killing (if the patient takes the drug) in this way: the drug kills the patient. By contrast, when a doctor nonconnects or disconnects, the patient merely "dies from an underlying fatal disease or pathology" that was threatening his life, and thus nothing actually kills him. I suggest that we bypass this idea, since the (putative) fact that a fatal disease does not kill a person who dies of it is too soft to support the moral weight it would be required to bear.

10 Suicide? Nobody was available to be prosecuted for it! But the deceased's estate was available for attachment.

Since suicide was a crime, so also was attempted suicide (cf. murder and attempted murder), and there the attempter was himself available to be prosecuted. Like suicide, attempted suicide is no longer a crime.

Moreover, since suicide was a crime, so also was assisting in a suicide. (Abetting a crime is standardly itself a crime.) What is in question here is only what ground is to be given for saying that assisting in a suicide should remain a crime while neither suicide nor attempted suicide does.

11 In her concurring opinion in Vacco v. Quill and its companion case, Washington v. Glucksberg, Justice O'Connor writes: "The parties and amici agree that in these States [New York and Washington] a patient who is suffering from a terminal illness and who is experiencing great pain has no legal barriers to obtaining medication, from qualified physicians, to alleviate that suffering, even to the point of causing unconsciousness and hastening death." She does not explicitly add the proviso Dellinger added though it may well be thought to lurk in the expression "to alleviate that suffering" (emphasis added) which implies that alleviating the suffering is the intent of supplying the medication. Rehnquist' opinion, by contrast, explicitly endorses th reasonableness of the proviso; see n. 12 below.

12 In n. 11 of his opinion, Rehnquist quotes a brie that itself contained the following quotation

"Although proponents of physician-assisted suicide and euthanasia contend that terminal sedation is covert physician-assisted suicide or euthanasia, the concept of sedating pharmacotherapy is based on informed consent and the principle of double effect." Rehnquist adds (*in propria persona*): "Just as a State may prohibit assisting suicide while permitting patients to refuse unwanted lifesaving treatment, it may permit palliative care related to that refusal, which may have the foreseen but unintended 'double effect' of hastening the patient's death."

13 For a recent example, see Bennett, chap. 11.

14 In n. 12 of his opinion, Rehnquist says that the opinion does not say that the doctor who drug-provides will *always* have a significantly different intention from the doctor who disconnects or drug-injects; he says the opinion says only that the doctor who disconnects or drug-injects does, *or may*, intend only to respect the patient's wishes or to relieve the patient's pain. The footnote then concludes: "In the absence of omniscience, however, the State is entitled to act on the reasonableness of the distinction." I am not sure what this means. Does it mean that in the absence of omniscience, the State is entitled to act on the reasonableness of those assumptions about the doctors' intentions? What would entitle the State to do this?

15 I have argued in a number of places that it is a mistake, most recently in "The Right and the Good," *Journal of Philosophy* 94 (1997): 273–98.

16 The latter example is adapted from an example of Thomas Nagel's, in which a military officer drops a bomb on a village, intending the event that consists in the deaths of everyone in the village, wanting that because he thinks the deaths of everyone will (not cause but) include the deaths of some guerrillas who are hiding in the village. See Thomas Nagel, "War and Massacre," reprinted in his *Mortal Questions* (Cambridge: Cambridge University Press, 1979), p. 61.

17 Here is Rehnquist: "The law has long used actors' intent or purpose to distinguish between two acts that have the same result." He goes on to quote approvingly from an earlier case: "The ... common law of homicide often distinguishes ...

between a person who knows that another person will be killed as a result of his conduct and a person who acts with the specific purpose of taking another's life." And he adds (*in propria persona*): "The law distinguishes between actions taken 'because of' a given end from actions taken 'in spite of' their unintended but foreseen consequences." What I ask in the text is: in which cases does that difference *by itself* mark crime off from non-crime?

A difference in intention may make a difference as to what will in fact happen, and in that way may bear on permissibility in action. For an example, see n. 19 below. Here, however, we are concerned with a difference that is only in the agent's intention.

18 I argue here only that what an agent intends (as opposed to merely foreseeing) in acting is irrelevant to the moral permissibility or impermissibility of his action. (James Rachels argued for this conclusion in his *The End of Life* [Oxford: Oxford University Press, 1986], chap. 6.) I believe we should also accept the stronger conclusion that the agent's beliefs are also irrelevant, but I do not argue for it here.

19 Rehnquist quotes the following example in which an action is taken, not "because of" but "in spite of" its unintended but foreseen consequences (see n. 17 above): "When General Eisenhower ordered American soldiers onto the beaches of Normandy, he knew that he was sending many American soldiers to certain death.... His purpose, though, was to ... liberate Europe from the Nazis."

Suppose we are there at the time of Eisenhower's ordering the invasion to begin and he whispers to us, "If truth be told, all I'm really intending in issuing the order is to cause the deaths of a lot of American soldiers." And suppose we believe him. Presumably we should telephone Roosevelt and say, "Cancel it!" For if that's all Eisenhower is intending in issuing the order, then there is real ground for worry about his planning of the invasion, and thus about what will go on in it: it is likely that there will be more deaths than are needed, not more than are needed for his purpose, of course, but more than are needed for the liberation of Europe. But if, perhaps *per impossibie*, these concerns could be proved groundless, if, i.e., we could become convinced that all will go exactly as it would go if that were not his intention, then there would be no reason to cancel the order. If which intention he acts with will make no difference to what happens, then his intention bears, not on whether he may act, but only on him.

20 It was G. E. M. Anscombe who first introduced PDE into secular moral theory in her "Modern Moral Philosophy," *Philosophy*, vol. 33 (1958), which was reprinted in her *Collected Philosophical Papers*, vol. 3 (Oxford: Blackwell, 1981). The idea that we are committed to consequentialism if we fail to notice the distinction between intending and foreseeing, or fail to recognize its bearing on moral permissibility, emerges clearly in that essay.

Index